LONDON MATHEMATICAL SOCIETY LECTURE NOTE SERIES

Managing Editor: Professor J.W.S. Cassels, Department of Pure Mathematics and Mathematical Statistics, University of Cambridge, 16 Mill Lane, Cambridge CB2 1SB, England

The titles below are available from booksellers, or, in case of difficulty, from Cambridge University Press.

34 Representation theory of Lie groups, M.F. ATIYAH *et al*
36 Homological group theory, C.T.C. WALL (ed)
39 Affine sets and affine groups, D.G. NORTHCOTT
46 p-adic analysis: a short course on recent work, N. KOBLITZ
50 Commutator calculus and groups of homotopy classes, H.J. BAUES
59 Applicable differential geometry, M. CRAMPIN & F.A.E. PIRANI
66 Several complex variables and complex manifolds II, M.J. FIELD
69 Representation theory, I.M. GELFAND *et al*
74 Symmetric designs: an algebraic approach, E.S. LANDER
76 Spectral theory of linear differential operators and comparison algebras, H.O. CORDES
77 Isolated singular points on complete intersections, E.J.N. LOOIJENGA
79 Probability, statistics and analysis, J.F.C. KINGMAN & G.E.H. REUTER (eds)
83 Homogeneous structures on Riemannian manifolds, F. TRICERRI & L. VANHECKE
86 Topological topics, I.M. JAMES (ed)
87 Surveys in set theory, A.R.D. MATHIAS (ed)
88 FPF ring theory, C. FAITH & S. PAGE
89 An F-space sampler, N.J. KALTON, N.T. PECK & J.W. ROBERTS
90 Polytopes and symmetry, S.A. ROBERTSON
92 Representation of rings over skew fields, A.H. SCHOFIELD
93 Aspects of topology, I.M. JAMES & E.H. KRONHEIMER (eds)
94 Representations of general linear groups, G.D. JAMES
95 Low-dimensional topology 1982, R.A. FENN (ed)
96 Diophantine equations over function fields, R.C. MASON
97 Varieties of constructive mathematics, D.S. BRIDGES & F. RICHMAN
98 Localization in Noetherian rings, A.V. JATEGAONKAR
99 Methods of differential geometry in algebraic topology, M. KAROUBI & C. LERUSTE
100 Stopping time techniques for analysts and probabilists, L. EGGHE
101 Groups and geometry, ROGER C. LYNDON
103 Surveys in combinatorics 1985, I. ANDERSON (ed)
104 Elliptic structures on 3-manifolds, C.B. THOMAS
105 A local spectral theory for closed operators, I. ERDELYI & WANG SHENGWANG
106 Syzygies, E.G. EVANS & P. GRIFFITH
107 Compactification of Siegel moduli schemes, C-L. CHAI
108 Some topics in graph theory, H.P. YAP
109 Diophantine analysis, J. LOXTON & A. VAN DER POORTEN (eds)
110 An introduction to surreal numbers, H. GONSHOR
113 Lectures on the asymptotic theory of ideals, D. REES
114 Lectures on Bochner-Riesz means, K.M. DAVIS & Y-C. CHANG
115 An introduction to independence for analysts, H.G. DALES & W.H. WOODIN
116 Representations of algebras, P.J. WEBB (ed)
117 Homotopy theory, E. REES & J.D.S. JONES (eds)
118 Skew linear groups, M. SHIRVANI & B. WEHRFRITZ
119 Triangulated categories in the representation theory of finite-dimensional algebras, D. HAPPEL
121 Proceedings of *Groups - St Andrews 1985*, E. ROBERTSON & C. CAMPBELL (eds)
122 Non-classical continuum mechanics, R.J. KNOPS & A.A. LACEY (eds)
125 Commutator theory for congruence modular varieties, R. FREESE & R. MCKENZIE
126 Van der Corput's method of exponential sums, S.W. GRAHAM & G. KOLESNIK
127 New directions in dynamical systems, T.J. BEDFORD & J.W. SWIFT (eds)
128 Descriptive set theory and the structure of sets of uniqueness, A.S. KECHRIS & A. LOUVEAU
129 The subgroup structure of the finite classical groups, P.B. KLEIDMAN & M.W.LIEBECK
130 Model theory and modules, M. PREST
131 Algebraic, extremal & metric combinatorics, M-M. DEZA, P. FRANKL & I.G. ROSENBERG (eds)
132 Whitehead groups of finite groups, ROBERT OLIVER
133 Linear algebraic monoids, MOHAN S. PUTCHA
134 Number theory and dynamical systems, M. DODSON & J. VICKERS (eds)
135 Operator algebras and applications, 1, D. EVANS & M. TAKESAKI (eds)
136 Operator algebras and applications, 2, D. EVANS & M. TAKESAKI (eds)
137 Analysis at Urbana, I, E. BERKSON, T. PECK, & J. UHL (eds)
138 Analysis at Urbana, II, E. BERKSON, T. PECK, & J. UHL (eds)
139 Advances in homotopy theory, S. SALAMON, B. STEER & W. SUTHERLAND (eds)

140 Geometric aspects of Banach spaces, E.M. PEINADOR & A. RODES (eds)
141 Surveys in combinatorics 1989, J. SIEMONS (ed)
142 The geometry of jet bundles, D.J. SAUNDERS
143 The ergodic theory of discrete groups, PETER J. NICHOLLS
144 Introduction to uniform spaces, I.M. JAMES
145 Homological questions in local algebra, JAN R. STROOKER
146 Cohen-Macaulay modules over Cohen-Macaulay rings, Y. YOSHINO
147 Continuous and discrete modules, S.H. MOHAMED & B.J. MÜLLER
148 Helices and vector bundles, A.N. RUDAKOV *et al*
149 Solitons, nonlinear evolution equations and inverse scattering, M. ABLOWITZ & P. CLARKSON
150 Geometry of low-dimensional manifolds 1, S. DONALDSON & C.B. THOMAS (eds)
151 Geometry of low-dimensional manifolds 2, S. DONALDSON & C.B. THOMAS (eds)
152 Oligomorphic permutation groups, P. CAMERON
153 L-functions and arithmetic, J. COATES & M.J. TAYLOR (eds)
154 Number theory and cryptography, J. LOXTON (ed)
155 Classification theories of polarized varieties, TAKAO FUJITA
156 Twistors in mathematics and physics, T.N. BAILEY & R.J. BASTON (eds)
157 Analytic pro-*p* groups, J.D. DIXON, M.P.F. DU SAUTOY, A. MANN & D. SEGAL
158 Geometry of Banach spaces, P.F.X. MÜLLER & W. SCHACHERMAYER (eds)
159 Groups St Andrews 1989 volume 1, C.M. CAMPBELL & E.F. ROBERTSON (eds)
160 Groups St Andrews 1989 volume 2, C.M. CAMPBELL & E.F. ROBERTSON (eds)
161 Lectures on block theory, BURKHARD KÜLSHAMMER
162 Harmonic analysis and representation theory, A. FIGA-TALAMANCA & C. NEBBIA
163 Topics in varieties of group representations, S.M. VOVSI
164 Quasi-symmetric designs, M.S. SHRIKANDE & S.S. SANE
165 Groups, combinatorics & geometry, M.W. LIEBECK & J. SAXL (eds)
166 Surveys in combinatorics, 1991, A.D. KEEDWELL (ed)
167 Stochastic analysis, M.T. BARLOW & N.H. BINGHAM (eds)
168 Representations of algebras, H. TACHIKAWA & S. BRENNER (eds)
169 Boolean function complexity, M.S. PATERSON (ed)
170 Manifolds with singularities and the Adams-Novikov spectral sequence, B. BOTVINNIK
171 Squares, A.R. RAJWADE
172 Algebraic varieties, GEORGE R. KEMPF
173 Discrete groups and geometry, W.J. HARVEY & C. MACLACHLAN (eds)
174 Lectures on mechanics, J.E. MARSDEN
175 Adams memorial symposium on algebraic topology 1, N. RAY & G. WALKER (eds)
176 Adams memorial symposium on algebraic topology 2, N. RAY & G. WALKER (eds)
177 Applications of categories in computer science, M. FOURMAN, P. JOHNSTONE, & A. PITTS (eds)
178 Lower K- and L-theory, A. RANICKI
179 Complex projective geometry, G. ELLINGSRUD *et al*
180 Lectures on ergodic theory and Pesin theory on compact manifolds, M. POLLICOTT
181 Geometric group theory I, G.A. NIBLO & M.A. ROLLER (eds)
182 Geometric group theory II, G.A. NIBLO & M.A. ROLLER (eds)
183 Shintani zeta functions, A. YUKIE
184 Arithmetical functions, W. SCHWARZ & J. SPILKER
185 Representations of solvable groups, O. MANZ & T.R. WOLF
186 Complexity: knots, colourings and counting, D.J.A. WELSH
187 Surveys in combinatorics, 1993, K. WALKER (ed)
188 Local analysis for the odd order theorem, H. BENDER & G. GLAUBERMAN
189 Locally presentable and accessible categories, J. ADAMEK & J. ROSICKY
190 Polynomial invariants of finite groups, D.J. BENSON
191 Finite geometry and combinatorics, F. DE CLERCK *et al*
192 Symplectic geometry, D. SALAMON (ed)
193 Computer algebra and differential equations, E. TOURNIER (ed)
194 Independent random variables and rearrangement invariant spaces, M. BRAVERMAN
195 Arithmetic of blowup algebras, WOLMER VASCONCELOS
196 Microlocal analysis for differential operators, A. GRIGIS & J. SJÖSTRAND
197 Two-dimensional homotopy and combinatorial group theory, C. HOG-ANGELONI, W. METZLER & A.J. SIERADSKI (eds)
198 The algebraic characterization of geometric 4-manifolds, J.A. HILLMAN
199 Invariant potential theory in the unit ball of C^n, MANFRED STOLL
200 The Grothendieck theory of dessins d'enfant, L. SCHNEPS (ed)
201 Singularities, JEAN-PAUL BRASSELET (ed)
202 The technique of pseudodifferential operators, H.O. CORDES
203 Hochschild cohomology, A. SINCLAIR & R. SMITH
204 Combinatorial and geometric group theory, A.J. DUNCAN, N.D. GILBERT, J. HOWIE (eds)
207 Groups of Lie type and their geometries, W.M. KANTOR & L. DI MARTINO (eds)

London Mathematical Society Lecture Note Series. 202

The Technique of Pseudodifferential Operators

H.O. Cordes
Emeritus, University of California, Berkeley

CAMBRIDGE
UNIVERSITY PRESS

CAMBRIDGE UNIVERSITY PRESS
Cambridge, New York, Melbourne, Madrid, Cape Town, Singapore, São Paulo

Cambridge University Press
The Edinburgh Building, Cambridge CB2 8RU, UK

Published in the United States of America by Cambridge University Press, New York

www.cambridge.org
Information on this title: www.cambridge.org/9780521378642

First published 1995

A catalogue record for this publication is available from the British Library

ISBN 978-0-521-37864-2 paperback

Transferred to digital printing 2008

To my 6 children, Stefan and Susan

Sabine and Art, Eva and Sam

TABLE OF CONTENTS

PREFACE

It is generally well known that the Fourier-Laplace transform converts a linear constant coefficient PDE P(D)u=f on $\mathbb{R}^n$ to an equation $P(\xi)u^{\sim}(\xi)=f^{\sim}(\xi)$, for the transforms $u^{\sim}$, $f^{\sim}$ of u and f, so that solving P(D)u=f just amounts to division by the polynomial $P(\xi)$. The practical application was suspect, and ill understood, however, until theory of distributions provided a basis for a logically consistent theory. Thereafter it became the Fourier-Laplace method for solving initial-boundary problems for standard PDE. We recall these facts in some detail in sec's 1-4 of ch.0.

The technique of pseudodifferential operator extends the Fourier-Laplace method to cover PDE with variable coefficients, and to apply to more general compact and noncompact domains or manifolds with boundary. Concepts remain simple, but, as a rule, integrals are divergent and infinite sums do not converge, forcing lengthy, often endlessly repetitive, discussions of 'finite parts' (a type of divergent oscillatory integral existing as distribution integral) and asymptotic sums (modulo order $-\infty$).

Of course, pseudodifferential operators (abbreviated ψdo's) are (generate) abstract linear operators between Hilbert or Banach spaces, and our results amount to 'well-posedness' (or normal solvability) of certain such abstract linear operators. Accordingly both, the Fourier-Laplace method and theory of ψdo's, must be seen in the context of modern operator theory.

To this author it always was most fascinating that the same type of results (as offered by elliptic theory of ψdo;'s) may be obtained by studying certain examples of Banach algebras of linear operators. The symbol of a ψdo has its abstract meaning as Gelfand function of the coset modulo compact operators of the abstract operator in the algebra.

On the other hand, hyperbolic theory, generally dealing with a group exp(Kt) (or an evolution operator U(t)) also has its manifestation with respect to such operator algebras: conjugation with exp(Kt) amounts to an automorphism of the operator algebra, and of the quotient algebra. It generates a flow in the symbol space essentially the characteristic flow of singularities. In [C_1], [C_2] we were going into details discussing this abstract approach.

We believe to have demonstrated that ψdo's are not necessary to understand these fact. But the technique of ψdo's, in spite of its endless formalisms (as a rule integrals are always 'distribution integrals', and infinite series are asymptotically convergent, not convergent), still provides a strongly simplifying principle, once the technique is mastered. Thus our present discussion of this technique may be justified.

On the other hand, our hyperbolic discussions focus on invariance of ψdo-algebras under conjugation with evolution operators, and do not touch the type of oscillatory integral and further discussions needed to reveal the structure of such evolution operators as Fourier integral operators. In terms of Quantum mechanics we prefer the Heisenberg representation, not the Schroedinger representation.

In particular this leads us into a discussion of the Dirac equation and its invariant algebra, in chapter X. We propose it as algebra of observables.

The basis for this volume is (i) a set of notes of lectures given at Berkeley in 1974-80 (chapters I-IV) published as preprint at U. of Bonn, and (ii) a set of notes on a seminar given in 1984 also at Berkeley (chapters VI-IX). The first covers elliptic (and parabolic) theory, the second hyperbolic theory. One might say that we have tried an old fashiened PDE lecture in modern style.

In our experience a newcomer will have to reinvent the theory before he can feel at home with it. Accordingly, we did not try to push generality to its limits. Rather, we tend to focus on the simplest nontrivial case, leaving generalizations to the reader. In that respect, perhaps we should mention the problems (partly of research level) in chapters I-IV, pointing to manifolds with conical tips or cylindrical ends, where the 'Fredholm-significant symbol' becomes operator-valued.

The material has been with the author for a long time, and was subject of many discussions with students and collaborators. Especially we are indebted to R. McOwen, A.Erkip, H. Sohrab, E. Schrohe, in chronclogical order. We are grateful to Cambridge University Press for its patience, waiting for the manuscript.

Berkeley, November 1993 Heinz O. Cordes

Chapter 0. INTRODUCTORY DISCUSSIONS.

In the present introductory chapter we give comprehensive discussions of a variety of nonrelated topics. All of these bear on the concept of pseudo-differential operator, at least in the author's mind. Some are only there to make studying ψdo's appear a natural thing, reflecting the author's inhibitions to think along these lines.

In sec.1 we discuss the elementary facts of the Fourier transform, in sec.'s 2 and 3 we develop Fourier-Laplace transforms of temperate and nontemperate distributions. In sec.4 we discuss the Fourier-Laplace method of solving initial-value problems and free space problems of constant coefficient partial differential equations. Sec.5 discusses another problem in PDE, showing how the solving of an abstract operator equation together with results on hypo-ellipticity and "boundary-hypo-ellipticity" can lead to existence proofs for classical solutions of initial-boundary problems. Sec.6 is concerned with the operator e^{Lt} , for a first order differential expression L . Sec.'s 7 and 8 deal with the concept of characteristics of a linear differential expression and learning how to solve a first order PDE. Sec.9 gives a mini-introduction to Lie groups, focusing on the mutual relationship between Lie groups and Lie algebras. (Note the relation to ψdo's discussed in ch.8).

We should expect the reader to glance over ch.0 and use it to have certain prerequisites handy, or to get oriented in the serious reading of later chapters.

0. Some special notations.

The following notations, abbreviations, and conventions will be used throughout this book.

(a) $\kappa_n = (2\pi)^{-n/2}$, $đx = \kappa_n dx_1 dx_2 \ldots dx_n = \kappa_n dx$.

(b) $\langle x\rangle = (1+|x|^2)^{1/2}$, $\langle\xi\rangle = (1+|\xi|^2)^{1/2}$, etc.

(c) Derivatives are written in various ways, at convenience: For $u=u(x)=u(x_1,\dots,x_n)$ we write $u^{(\alpha)}=\partial_x^\alpha u=\partial_{x_1}^{\alpha_1}\partial_{x_2}^{\alpha_2}\dots u = =\partial^{\alpha_1}/\partial x^{\alpha_1}\dots\partial^{\alpha_n}/\partial x^{\alpha_n}u$. Or, $u_{|x_j}=\partial_{x_j}u$, $u_{|x}$ to denote the n-vector with components $u_{|x_j}$, $\nabla_x^k u$ for the k-dimensional array with components $u_{|x_{i_1}x_{i_2}\dots}$. For a function of $(x,\xi)=(x_1,\dots,x_n,\xi_1,\dots,\xi_n)$ it is often convenient to write $u_{(\beta)}^{(\alpha)}=\partial_\xi^\alpha\partial_x^\beta u$.

(d) A multi-index is an n-tuple of integers $\alpha=(\alpha_1,\dots,\alpha_n)$. We write $|\alpha|=|\alpha_1|+\dots+|\alpha_n|$, $\alpha!=\alpha_1!\dots\alpha_n!$, $\binom{\alpha}{\beta}=\binom{\alpha}{\beta}\dots\binom{\alpha}{\beta}$, $x^\alpha=x_1^{\alpha_1}\dots x_n^{\alpha_n}$, etc., $\mathbb{N}^n=\{\text{all multi-indices}\}$.

(e) Some standard spaces: $\mathbb{R}^n$ = n-dimensional Euclidean space $\mathbb{B}^n$=directional compactification of $\mathbb{R}^n$ (one infinite point ∞x added in every direction (of a unit vector x).

(f) Spaces of continuous or differentiable complex-valued functions over a domain or differentiable manifold X (or sometimes only $X=\mathbb{R}^n$): C(X) = continuous functions on X ; CB(X)= bounded continuous functions on X; CO(X)= continuous functions on X vanishing at ∞ ; CS(X) = continuous functions with directional limits; $C_0(X)$ = continuous functions with compact support; $C^k(X)$= functions with derivatives in C, to order k, (incl. $k=\infty$). $CB^\infty(X)$="all derivatives exist and are bounded". The Laurent-Schwartz notations $\mathcal{D}(X)=C_0^\infty(X)$, $\mathcal{E}(X)=C^\infty(X)$ are used. Also $\mathcal{S}=\mathcal{S}(\mathbb{R}^n)$= "rapidly decreasing functions" (All derivatives decay stronger as any power of x). Also, distribution spaces $\mathcal{D}'$, $\mathcal{E}'$, $\mathcal{S}'$.

(g) L^p-spaces: For a measure space X with measure $d\mu$ we write $L^p(X)=L^p(X,d\mu)=\{$measurable functions u(x) with $|u|^p$ integrable$\}$ for $1\le p<\infty$; $L^\infty(X)=\{$essentially bounded functions$\}$.

(h) Maps between general spaces: C(X,Y) denotes the continuous maps $X\to Y$. Similar for the other symbols under (f), i.e., CB(X,Y) ,.... .

(i) Classes of linear operators (X= Banach space) : $L(X)$ ($K(X)$)= continuous (compact) operators; $GL(X)$ ($U(H)$) = invertible (unitary) operators of $L(X)$ (of $L(H)$, H=Hilbert space); $U_n=U(\mathbb{C}_n)$. For operators $X\to Y$, again, $L(X,Y)$, etc.

(j) The convolution product: For $u,v\in L^1(\mathbb{R}^n)$ we write $w(x) =(u*v)(x)=\kappa_n\int dy\,u(x-y)v(y)$ (Note the factor $\kappa_n=(2\pi)^{-n/2}$).

(k) Special notation: " $X\subset\subset Y$ " means that X is contained in a compact subset of Y .

(l) For technical reason we may write $\lim_{\varepsilon\to 0} a(\varepsilon) = a|_{\varepsilon\to 0}$.

(m) Abbreviations used: ODE (PDE) = ordinary (partial) differential equation (or "expression"). FOLPDE (or folpde)= first order linear partial differential equation (or "expression"); ψdo= pseudodifferential operator.

(n) Integrals need not be existing (proper or improper) Riemann or Lebesgue integrals, unless explicitly stated, but may be distribution integrals By this term we mean that either (i) the integral may be interpreted as value of a distribution at a testing function-the integrand may be a distribution, or (ii) the limit of Riemann sums exists in the sense of weak convergence of a sequence of (temperate) distributions, or (iii) the limit defining an improper Riemann integral exists in the sense of weak convergence, as above, or (iv) the integral may be a 'finite part' (cf. I,4).

(o) Adjoints: For a linear operator A we use 'distribution adjoint' $A^{\sim}$ and 'Hilbert space adjoint' A^*, corresponding to transpose A^T and adjoint $\bar{A}^T = A^*$, in case of a matrix $A=((a_{jk}))$, respectively. For a symbols $a(x,\xi)$, a^* (or a^+) may denote the symbol of the adjoint ψdo of $a(x,D)$, as specified in each section.

(p) supp u (sing supp u (or s.s.u)) denotes the (singular) support of the distribution u.

1. The Fourier transform; elementary facts.

Let $u \in L^1(\mathbb{R}^n)$ be a complex-valued integrable function. Then we define the Fourier transform $u^\wedge = Fu$ of u by the integral

$$u^\wedge(x) = \int_{\mathbb{R}^n} d\!\!\!/\xi\, u(\xi) e^{-ix\xi} \quad , \ x \in \mathbb{R}^n \ , \tag{1.1}$$

with $x\xi = x.\xi = \sum_{j=1}^n x_j \xi_j$, an existing Lebesgue integral. Clearly,

$$|u^\wedge(x)| \le \|u\|_{L_1} = \int_{\mathbb{R}^n} d\!\!\!/x\, |u(x)| \ . \tag{1.2}$$

Note that $u^\wedge$ is uniformly continuous over $\mathbb{R}^n$: We get

$$\begin{aligned} |u^\wedge(x) - u^\wedge(y)| &\le 2\int d\!\!\!/\xi\, |\sin(x-y)\xi/2|\,|u(\xi)| \\ &\le N|x-y|\,\|u\|_{L^1} + 2\int_{|\xi|\ge N} d\!\!\!/\xi\, |u(\xi)| \ , \end{aligned} \tag{1.3}$$

where the right hand side is $<\varepsilon$ if N is chosen for $\int_{|\xi|\ge N} < \varepsilon/4$, and then $|x-y| < \varepsilon/(2N\|u\|_{L^1})$. Moreover, we get $u^\wedge \in CO(\mathbb{R}^n)$, i.e., $\lim_{|x|\to\infty} u^\wedge(x)=0$, a fact, known as the <u>Riemann-Lebesgue lemma</u>.

To prove the latter, we reduce it to the case of $u \in C_0^\infty(\mathbb{R}^n)$: The space C_0^∞ is known to be dense in L^1. By (1.1) we get

$$(1.4) \qquad |u^\wedge(x)-v^\wedge(x)| \le \|u-v\|_{L^1} < \varepsilon/2 \text{ , as } v \in C_0^\infty \text{ , } \|u-v\|_{L^1}<\varepsilon/2.$$

Hence $\lim_{|x|\to\infty} v^\wedge(x)=0$ implies $|u^\wedge| \le |u^\wedge-v^\wedge| + |v^\wedge| < \varepsilon$ whenever x is chosen according to $|v^\wedge| < \varepsilon/2$.

But for $v \in C_0^\infty$ the Fourier integral extends over a ball $|\xi| \le N$ only, since $v=0$ outside. We may integrate by parts for

$$(1.5) \quad |x|^2 u^\wedge(x) = -\int \Delta_\xi(e^{-ix\xi})v(\xi)đ\xi = -\int đ\xi e^{-ix\xi}(\Delta v)(\xi) = -(\Delta v)^\wedge(x) \text{ ,}$$

with the Laplace differential operator $\Delta_\xi = \sum_{j=1}^n \partial_{\xi_j}^{\ 2}$. Clearly we have $\Delta v \in C_0^\infty \subset L^1$ as well, whence (1.1) applies to Δv, for

$$(1.6) \qquad |v^\wedge(x)| \le \|\Delta v\|_{L^1}/|x|^2 \to 0 \text{ , as } |x| \to \infty \text{ ,}$$

completing the proof.

The above partial integration describes a general method to be applied frequently in the sequel. (1.6) may be derived under the weaker assumptions that $v \in C^2$, and that all derivatives $v^{(\alpha)}$, $|\alpha| \le 2$, are in L^1 (cf. pbm.5). On the other hand, there are simple examples of $u \notin L^1$ such that $u^\wedge$ does not decay as rapidly as (1.6) indicates. In particular, $u \in L^1$ exists with $u^\wedge \notin L^1$ (cf.pbm.4).

This matter becomes important if we think of inverting the linear operator $F: L^1 \to CO$ defined by (1.1), because formally an inverse seems to be given by almost the same integral. Indeed, define the (complex) <u>conjugate Fourier transform</u> $\bar{F}: L^1 \to CO$ by $\bar{F}u = \overline{(Fu)}$, or, $u^\vee = \bar{F}u$, where

$$(1.7) \qquad u^\vee(x) = \int đ\xi e^{ix\xi} u(\xi) \quad \text{, } u \in L^1(\mathbb{R}^n) \text{ .}$$

Then, in essence, it will be seen that $\bar{F}$ is the inverse of the operator F. More precisely we will have to restrict F to a (dense) subspace of L^1, for this result. Or else, the definition of the operator $\bar{F}$ must be extended to certain non-integrable functions, for which existence of the Lebesgue integral (1.7) cannot be expected. Both things will be done, eventually.

It turns out that F induces a unitary operator of the Hilbert space $L^2(\mathbb{R}^n)$: We have <u>Parseval's relation</u>:

$$(1.8)\qquad \int_{\mathbb{R}^n} đx|u^\wedge(x)|^2 = \int_{\mathbb{R}^n} đx|u(x)|^2 \ , \text{ for all } u \in L^1(\mathbb{R}^n)\cap L^2(\mathbb{R}^n) \ .$$

Formula (1.8) is easier to prove as the <u>Fourier inversion formula</u>, asserting $u^{\wedge\vee}=u^{\vee\wedge}=u$ for certain u: We may write

$$(1.9)\qquad \int_{Q_N} đx\bar{u}^\wedge(x)v^\wedge(x) = \int đ\xi đ\eta \bar{u}(\xi)v(\eta)\Pi_{j=1}^n \int_{-N}^N e^{ix_j(\xi_j-\eta_j)} đx_j \ ,$$

assuming that $u,v\in L^1(\mathbb{R}^n)$, with the 'cube' $Q_N=\{|x_j|\le N, j=1,..,N\}$, some integer N>0. Indeed, the interchange of integrals leading to (1.9) is legal, since the integrand is $L^1(Q_N\times\mathbb{R}^n\times\mathbb{R}^n)$.

Note that $\int_{-N}^N e^{ist}dt = 2\frac{\sin sN}{s}$, $s\ne 0$, $= 2N$, $s=0$, allowing evaluation of the inner integrals at right of (1.9). With $\int đ\xi đ\eta = \int đ\xi \int đ\eta$, and $\eta = \xi - \zeta/N$, $đ\eta = N^{-n}đ\zeta$, (1.9) assumes the form

$$(1.10)\qquad \int_{Q_N} đx\bar{u}^\wedge(x)v^\wedge(x) = \int đ\xi\bar{u}(\xi)\int đ\zeta v(\xi-\zeta/N)\Pi_{j=1}^n\varphi(\zeta_j) \ ,$$

where $\varphi(t) = (2/\pi)^{1/2}\frac{\sin t}{t}$, $t\ne 0$, continuously extended into t=0.

For $v\in C(\mathbb{R}^n)$, as $N\to\infty$, the function $v(\xi-\zeta/N)$ will converge to $v(\xi)$, independent of ζ . Thus one expects the inner integral at right of (1.10) to converge to $v(\xi)\int\Pi_{j=1}^n\varphi(\zeta_j)đ\zeta_j = v(\xi)$, since

$$(1.11)\qquad \int_0^\infty \sin t\ dt/t = \pi/2 \ .$$

Legalization of this argument will confirm Parseval's relation, since the right hand converges to the right hand side of (1.8), as $N\to\infty$. With $u\in L^1$ and $v\in C_0^\infty$ (setting $\varphi_n(\zeta)=\Pi\varphi(\zeta_j)$) write

$$(1.12)\qquad \int đ\xi\bar{u}(\xi)\int đ\zeta\varphi_n(\zeta)(v(\xi-\zeta/N)-v(\xi)) = \int_{Q_N} đx\bar{u}^\wedge v^\wedge - \int_{R^n} đx\bar{u}v \ .$$

To show that the inner integral at left goes to 0 as $N\to\infty$ it is more skilful to use the integration variable $\theta=\zeta/N$, $d\zeta=N^n d\theta$. For $n=1$, $\int \sin N\theta\ (v(\xi-\theta)-v(\xi))d\theta/\theta = \int_{|\theta|\le\delta} + \int_{|\theta|\ge\delta} = I_0 + I_\infty$.

Here we get (with $w(\theta) = (v(\xi-\theta)-v(\xi))/\theta$)

$$|I_0| \le\delta\|v'\|_{L^\infty},\ I_\infty=\frac{1}{N}((w(\theta)\cos(N\theta)|_{\theta=-\delta}^{\theta=\delta} + \int_{|\theta|\ge\delta}\cos(N\theta)w_{|\theta}(\theta)d\theta).$$

The latter gives $I_\infty \le \frac{c}{N\delta}(\|v\|_{L^\infty}+ \|v'\|_{L^\infty})$, with a constant c, only depending on the volume of supp v, i.e., it is fixed after fixing v . The estimates imply the inner integral to go to 0, uniformly as $x \in \mathbb{R}^n$. For $u \in L^1$ the Lebesgue theorem then implies the left hand side of (1.12) to tend to 0, as $N \to \infty$, for each fixed $v \in C_0^\infty$.

For general n the proof is a bit less transparent, but remains the same: Split the inner integral into a sum of integrals over a small neighbourhood of 0 and its complement. In the first term use differentiability of v; in the second an integration by parts.

We now have a 'polarized' Parseval relation, in the form

$$(1.13) \qquad \int_{R^n} d\!\!\!/x \bar{u}^\wedge v^\wedge = \int_{R^n} d\!\!\!/x \bar{u} v \ , \text{ for } u \in L^1 \ , \ v \in C_0^\infty .$$

For $u \in L^1 \cap L^2$ pick a sequence $u_j \in C_0^\infty$ with $\|u-u_j\|_{L^1} \to 0$, $\|u-u_j\|_{L^2} \to 0$, as is possible. Then, since $u_j - u_l \in C_0^\infty \subset L^2$, (1.13) with $u=v=u_j-v_j$ implies $\|u_j^\wedge - u_l^\wedge\|_{L^2} = \|u_j - u_l\|_{L^2} \to 0$, $j,l \to \infty$. In other words, u_j and $u_j^\wedge$ both converge in L^2. Clearly, $u_j^\wedge \to u^\wedge$. Indeed, initially we showed uniform convergence over $\mathbb{R}^n$, while the L^2-limit $z = \lim u_j^\wedge$ satisfies $(u^\wedge, \varphi) = \int \bar{z} \varphi dx$ for all $\varphi \in C_0^\infty$. This yields $\int (u^\wedge - z)\varphi dx = 0$ for all such φ, hence $u^\wedge = z$ (almost everywhere), since C_0^∞ is dense in L^2. Substituting $u=v=u_j$ in (1.13), letting $j \to \infty$, it follows that (1.8) is valid for all $u \in L^1 \cap L^2$, confirming Parseval's relation.

Clearly (1.13) also holds for all $u, v \in L^1 \cap L^2$. We use it to prove the Fourier inversion Let n=1. For $v \in L^1 \cap L^2$, $u = \chi_{[0,x_0]}$, some $x_0 > 0$ apply (1.13). Confirm by calculation of the integral that

$$(1.14) \qquad (2\pi)^{1/2} u^\wedge(x) = (e^{-ixx_0} - 1)/(-ix) = h_{x_0}(x) \ , \quad x \neq 0 \ ,$$

hence

$$(1.15) \qquad \int_0^{x_0} v(x)dx = \int d\!\!\!/x v^\wedge(x) \bar{h}_{x_0}(x)dx \quad .$$

The Fourier inversion formula is a matter of differentiating (1.15) for x_0 under the integral sign, assuming that this is legal Consider the difference quotient:

$$(1.16) \qquad (2\delta)^{-1} \int_{x_0-\delta}^{x_0+\delta} v(x)dx = \int d\!\!\!/x v^\wedge(x) e^{ixx_0} \frac{\sin \delta x}{\delta x} .$$

Assuming only that v, $v^\wedge$ both are in L^1, it follows indeed that

$$(1.17) \quad \lim_{\delta \to 0} (2\delta)^{-n} \int_{Q_{x_0,\delta}} v(x)dx = \int d\!\!\!/x v^\wedge(x) e^{ixx_0} = (v^\wedge)^\vee(x_0), \ x_0 \in \mathbb{R}^n.$$

(Actually, our proof works for n=1, $x_0 > 0$ only, but can easily be extended to all x_0, and general n. One must replace the derivative d/dx_0 by a mixed derivative $\partial^n/(\partial x_{0,1} \ldots \partial x_{0,n})$.) Indeed, letting $\delta\to 0$ in (1.17) we obtain (1.15), using that $\sin(\delta x)/(\delta x) \to 1$ uniformly on compact sets, and boundedly on $\mathbb{R}$, as $\delta\to 0$.

If v is continuous at x_0 then clearly the left hand side of (1.17) equals $v(x_0)$, giving the Fourier inversion formula, as it is well known. For n=1, if v has a jump at x_0 then the left hand side of (1.17) equals the mean value $(v(x_0+0)+v(x_0-0))/2$.

Again for n=1 a limit of (1.16), as $\delta\to 0$ exists, if only

$$(1.18) \qquad \lim_{\alpha\to\infty} \int_{-\alpha}^{+\alpha} v^{\wedge}(x)đx ,$$

the principal value, exists (cf. pbm.6), without requiring $v^{\wedge} \in L^1$.

We summarize our results thus far:

Proposition 1.1. The Fourier transform $u^{\wedge}$ of (1.1) and its complex conjugate $u^{\vee} = (\bar{u}^{\wedge})^{-}$ are defined for all $u \in L^1(\mathbb{R}^n)$, and we have $u^{\wedge}, u^{\vee} \in CO(\mathbb{R}^n)$. For $u \in L^1(\mathbb{R}^n) \cap L^2(\mathbb{R}^n)$ we have Parseval's relation (1.8). If both $u \in L^1(\mathbb{R}^n)$ and $u^{\wedge} \in L^1(\mathbb{R}^n)$ hold, then we have $u^{\wedge\vee}(x) = u^{\vee\wedge}(x) = u(x)$ for almost all $x \in \mathbb{R}^n$.

It is known that the Banach space $L^1(\mathbb{R}^n)$ is a commutative Banach algebra under the convolution product

$$(1.19) \qquad u*v = w , \ w(x) = \int đy\, u(x-y)w(y) = \int đy\, v(x-y)u(y) .$$

Indeed,

$$(1.20) \qquad \|w\|_{L^1} = \int |w(x)|dx \le \kappa_n \int dx \int dy |u(x-y)| |v(y)| = \kappa_n \|u\|_{L^1} \|v\|_{L^1} .$$

Prop.1.2, below, clarifies the role of the Fourier transform F for this Banach-algebra: F provides the Gelfand homomorphism.

Proposition 1.2. For $u,v \in L^1$ let $w = u*v$. Then we have

$$(1.21) \qquad w^{\wedge}(\xi) = u^{\wedge}(\xi)v^{\wedge}(\xi) , \ \xi \in \mathbb{R}^n .$$

Proof. We have

$$w^{\wedge}(\xi) = \int đx e^{-ix\xi} \int đy\, u(x-y)v(y) = \int đy e^{-iy\xi} \int dx\, u(x-y)e^{-i(x-y)\xi} .$$

The substitution x-y=z , dy=dz thus confirms (1.21), q.e.d.

The importance of the Fourier transform for PDE's hinges on

Proposition 1.3. If $u^{(\beta)} \in L^1$ for all $\beta \le \alpha$ then

$$(1.22) \qquad u^{(\alpha)\wedge}(\xi) = i^{|\alpha|}\xi^{\alpha} u^{\wedge}(\xi) , \ \xi \in \mathbb{R}^n .$$

Proof. Partial integration gives $\int dx e^{-ix\xi} u^{(\alpha)}(\xi) = (-1)^{|\alpha|} \int dx \partial_x^{\alpha}(e^{ix\xi})$ (with vanishing boundary integrals), implying (1.21), q.e.d.

Given a linear partial differential equation

$$(1.23) \qquad P(D)u = f \quad , \; P(D) = \sum_{|\alpha| \le N} a_\alpha D_x^\alpha \; ,$$

where $f \in L^1(\mathbb{R}^n)$, $D_{x_j} = -i\partial_{x_j}$, one might attempt to find solutions by taking the Fourier transform. With proper assumptions (1.21) gives

$$(1.24) \qquad P(\xi)u^\wedge(\xi) = f^\wedge(\xi) \; .$$

Assuming that $e = (\frac{1}{P(x)})^\vee$ exists, (1.24) will assume the form

$$(1.25) \qquad u^\wedge(\xi) = e^\wedge(\xi) f^\wedge(\xi) \; ,$$

which by prop.1.2 (and Fourier inversion) is equivalent to

$$(1.26) \qquad u(x) = \int d\!\!\!/ y\, e(x-y) f(y) \; .$$

Presently, (1.26) can only have a formal meaning, since normally $(1/P) \notin L^1$, or $f \notin L^1$, or $u \notin L^1$, in practical applications.

However, as to be discussed in the sections below, the Fourier transform may be extended to more general classes of functions and to generalized functions. Then (1.26) yields a powerful tool for solving problems in constant coefficient PDE's (cf. sec.4).

Problems. 1) For n=1 obtain the Fourier transforms of the functions a) $(a^2+x^2)^{-1}$, $a>0$; b) $(\sin^2 ax)/x^2$, $a>0$; c) $1/\cosh x$.

2) For general n obtain the Fourier transform of $e^{-\alpha x^2}$, $\alpha>0$.
3) Obtain the Fourier transform of $f(x) = (1+|x|^2)^{-\nu}$, where $\nu>n/2$ (Hint: A knowledge of Bessel functions is required for this problem). 4) Construct a function $f(x) \in L^1(\mathbb{R}^n)$ such that $f^\wedge \notin L^1$.
5) The Riemann-Lebesgue lemma states that $f^\wedge \in C0$ whenever $f \in L^1$. Is it true that even $f^\wedge(x) = O(\langle x \rangle^{-\varepsilon})$ for each $f \in L^1$ with some $\varepsilon>0$? 6) Combining some facts, derived above, show that, for n=1, every piecewise smooth function $f(x) \in L^1(\mathbb{R})$ has a Fourier transform satisfying $f(x) = O(1/x)$, as $|x|$ is large, and satisfying

$$(1.27) \qquad (f(x+0)+f(x-0))/2 = \lim_{\alpha\to\infty} \int_{-\alpha}^{\alpha} d\!\!\!/ y\, e^{ixy} f^\wedge(y) \; , \; x \in \mathbb{R} \; .$$

Here 'piecewise smooth' means, that $\mathbb{R}$ may be divided into finitely many closed subintervals in each of which f is C^1 , possibly after changing its value at boundary points.

2. Fourier analysis for temperate distributions on $\mathbb{R}^n$.

We assume that the reader is familiar with the concept of distribution, as a continuous linear functional on the space $\mathcal{D}(\mathbb{R}^n) = C_0^\infty(\mathbb{R}^n)$. A linear functional $f:\mathcal{D} \to \mathbb{C}$ is said to be continuous if $\langle f,\varphi_j\rangle \to 0$ whenever $\varphi_j \to 0$ in $\mathcal{D}$. The latter means that (i) $f_j \in \mathcal{D}$, $j=1,2,\ldots$, (ii) $\operatorname{supp} \varphi_j \in K \subset\subset \mathbb{R}^n$, K independent of j, (iii) $\sup\{|\varphi^{(\alpha)}(x)| : x \in \mathbb{R}^n\} \to 0$, as $j \to \infty$, for every α. The space of distributions on $\mathbb{R}^n$ is called $\mathcal{D}'=\mathcal{D}'(\mathbb{R}^n)$. The space $L^1_{loc}(\mathbb{R}^n)$ of locally integrable functions is naturally imbedded in $\mathcal{D}'$ by defining

$$(2.1) \qquad \langle f,\varphi\rangle = \int f(x)\varphi(x)dx \ , \text{ for } f \in L^1_{loc} .$$

The derivatives $f^{(\alpha)} = \partial_x^\alpha f$ of a distribution $f \in \mathcal{D}'$ are defined by

$$(2.2) \qquad \langle f^{(\alpha)},\varphi\rangle = (-1)^{|\alpha|}\langle f,\varphi^{(\alpha)}\rangle \ , \ \varphi \in \mathcal{D} ,$$

the product of a distribution $f \in \mathcal{D}'$ and a $C^\infty(\mathbb{R}^n)$ function g by

$$(2.3) \qquad \langle gf,\varphi\rangle = \langle f,g\varphi\rangle \ , \ \varphi \in \mathcal{D} .$$

Thus Lf is defined for any distribution $f \in \mathcal{D}'(R^n)$ and linear differential operator $L=\sum a_\alpha \partial_x^\alpha$ with coefficients $a_\alpha(x) \in C^\infty(\mathbb{R}^n)$.

While the value f(x) of a distribution at a point x is a meaningless concept, one may talk about the restriction $f|\Omega$ of $f \in \mathcal{D}'(\mathbb{R}^n)$ to an open subset Ω , and its properties: First of all, the space $\mathcal{D}'(\Omega)$ of distributions over Ω consists of the continuous linear functionals on $\mathcal{D}(\Omega)=C_0^\infty(\Omega)$, with continuity defined as for $\mathbb{R}^n$. For $f \in \mathcal{D}'(\mathbb{R}^n)$, the restriction $f|\mathcal{D}(\Omega)$ defines a distribution of $\mathcal{D}'(\Omega)$, denoted by $f|\Omega$. Thus, for example, it is meaningful to say that $f \in \mathcal{D}'(\mathbb{R}^n)$ is a function (a $C^k(\Omega)$-function, etc.) in an open set $\Omega \subset \mathbb{R}^n$ - it means that $f|\Omega$ has this property. For a distribution $f \in \mathcal{D}'(\Omega)$ on an open set the derivatives and product with $g \in C^\infty(\Omega)$ is defined as in (2.2) , (2.3) . The support supp f (singular support sing supp f) of $f \in \mathcal{D}'$ is defined as the smallest closed set E (intersection of all closed sets E) such that f=0 (such that f is C^∞) in the complement of E .

The concept of Fourier transform can be generalized to distributions on $\mathbb{R}^n$, with multiple benefit: Some non-L^1-functions will get distributions as Fourier transforms. Certain distributions will get functions as Fourier transforms. The Fourier inversion formula and many assumptions (limit interchanges) will simplify.

We used the Fourier integral of (1.1) only for $u \in L^1(\mathbb{R}^n)$. It is practical to introduce a growth restriction for $u \in \mathcal{D}'(\mathbb{R}^n)$ if we want $u^\wedge$ to be a distribution again. Later on (sec.3) we also define $u^\wedge$ for general $u \in \mathcal{D}'(\mathbb{R}^n)$, but it no longer will be a distribution in $\mathcal{D}'(\mathbb{R}^n)$. We follow [$\text{Schw}_1$] here, but [GS] in sec.3.

The growth restriction is imposed by requesting that $u \in \mathcal{D}'$ allows an extension to a larger space of testing functions called $\mathcal{S}$. Here $\mathcal{S}$ - the space of rapidly decreasing functions- consists of all $\varphi \in C^\infty(\mathbb{R}^n)$ such that for all multi-indices α and $k=1,2,\ldots$,

$$\varphi^{(\alpha)}(x) = O(\langle x\rangle^{-k}) . \tag{2.4}$$

- the derivatives of φ decay faster than every power $\langle x\rangle^{-k}$.

Note that, equivalently, we could have prescribed that for every α one (and the same) of the following conditions be satisfied:

$$\langle x\rangle^k u^{(\alpha)}(x) \text{ (for every } k=1,2,..\text{), or } x^\beta u^{(\alpha)}(x) \text{ (for every } \beta\text{)},$$

$$\text{or } (x^\beta u(x))^{(\alpha)} \text{ (for every } \beta\text{), is } O(1) \text{ , or is } o(1) \text{ , or is} \tag{2.5}$$

$$CB \text{ , or } CO \text{ , or } L^2 \text{ , or } L^p \text{ (for some } 1\le p\le\infty\text{)} .$$

Indeed, for a given α one of these conditions may be weaker or stronger than the other. However for <u>all</u> α simultaneously all conditions are equally strong. One must use Leibniz' formula to handle interchanges of ∂_x^α and multiplications (cf. lemma 2.8).

The above at once gives the following:

<u>Theorem 2.1</u>. We have $\mathcal{S} \subset L^1(\mathbb{R}^n)$, so that $u^\wedge$ of (1.1) (and $u^\vee$) are defined on $\mathcal{S}$. Moreover, for $u \in \mathcal{S}$, we have $u^\wedge, u^\vee \in \mathcal{S}$, and

$$(u^\wedge)^\vee(x) = (u^\vee)^\wedge(x) = u(x) \text{ , } x \in \mathbb{R}^n . \tag{2.6}$$

The Fourier transform and its conjugate therefore define bijective linear maps $\mathcal{S} \leftrightarrow \mathcal{S}$, inverting each other.

<u>Proof</u>. Using repeated partial integration and $x^\alpha e^{-ix\xi} = i^{|\alpha|}\partial_\xi^\alpha e^{-ix\xi}$, $\partial_x^\beta e^{-ix\xi} = i^{-|\beta|}\xi^\beta e^{-ix\xi}$ get $\int dx e^{-ix\xi} x^\alpha u^{(\beta)}(x) = i^{|\alpha|}\partial_\xi^\alpha \int dx e^{-ix\xi} u^{(\beta)}(x)$

$= i^{|\alpha|+|\beta|}\partial_\xi^\alpha \xi^\beta \int dx e^{-ix\xi} u(x)$, hence

$$(x^\beta u^{(\alpha)})^\wedge(\xi) = i^{|\alpha|+|\beta|}(\xi^\alpha u^\wedge(\xi))^{(\beta)} . \tag{2.7}$$

In fact, we get $x^\beta u^{(\alpha)} \in L^1$, for every α,β, by the equivalence

(2.5) , for $u \in S$. Therefore the right hand side is in CO , for every α,β , so that $u^\wedge \in S$, again by the equivalence (2.5). Thus we get $u^\wedge \in S$ for all $u \in S$. Similarly for "$^\vee$" . Also, the Fourier inversion formula holds for $u \in S$, and the left hand side of (1.17) equals v(x). This implies (2.5), also by taking complex conjugates. The bijectivity then follows at once, q.e.d.

Following Schwartz we introduce distributions with controlled growth at infinity - so called <u>temperate</u> <u>distributions</u> - over $\Omega=\mathbb{R}^n$ as <u>continuous</u> <u>linear</u> <u>functionals</u> <u>over</u> S. The space of all temperate distributions is denoted by S'. Clearly, $S \supset D$, so that a functional u over S induces a functional over D - its restriction $u|D$.

<u>Definition 2.2</u>. A sequence of functions $\varphi_j \in S$ is said to converge to 0 (in S) if for every multi-index α and k = 0,1,2,... the sequence $\langle x\rangle^k\varphi_j^{(\alpha)}(x)$ converges to zero uniformly for all $x \in \mathbb{R}^n$.

<u>Definition 2.3</u>. A linear functional u over S is said to be continuous if $\varphi_j \in S$, $\varphi_j \to 0$ in S implies $\langle u,\varphi_j\rangle \to 0$.

<u>Temperate</u> <u>distributions</u> <u>are</u> <u>distributions</u>. More precisely speaking: For $u\in S'$ the restriction $u|D$ determines u uniquely, and $u|D\in D'(\mathbb{R}^n)$. To confirm this we must prove:

<u>Lemma 2.4</u>. a) If $\varphi_j\in D$, $\varphi_j\to 0$ in D, then we also have $\varphi_j\to 0$ in S.
b) For $\varphi\in S$ there exists a sequence $\varphi_j\in D$ such that $\varphi-\varphi_j\to 0$ in S .

From lemma 2.4 it follows that for $u\in S'$ the restriction $v=u|D$ is continuous over D : If $\varphi_j\to 0$ in D , then $\varphi_j\to 0$ in S (by (a)), hence $\langle v,\varphi_j\rangle=\langle u,\varphi_j\rangle \to 0$. Hence $v\in D'$. Furthermore, if u, $w\in S'$ have $u|D=w|D=v\in D'$, then for $\varphi\in S$ let φ_j be a sequence of (b) above. Get $u-w\in S'$, $\langle u-w,\varphi-\varphi_j\rangle \to 0$. Hence $0=\langle u-w,\varphi_j\rangle=\langle v-v,\varphi_j\rangle\to\langle u-w,\varphi\rangle$, implying that $\langle u,\varphi\rangle=\langle v,\varphi\rangle$ for all $\varphi\in S$, or u=v, so that indeed $u\in S'$ is uniquely determined by its restriction $v=u|D\in D'$.

<u>Proof of lemma 2.4</u>. (a): If $\varphi_j\in D$, $\varphi_j\to 0$ in D then supp $\varphi_j^{(\alpha)} \subset K\subset\subset \mathbb{R}^n$, while the functions $\langle x\rangle^k$ are bounded in K. Thus the uniform convergence $\langle x\rangle^k\varphi_j^{(\alpha)}(x)\to 0$ in $\mathbb{R}^n$ follows from the uniform convergence $\varphi_j^{(\alpha)}(x)\to 0$ in $\mathbb{R}^n$, and we have $\varphi_j\to 0$ in S, proving (a).

To prove (b), let $\chi(x)\in C_0^\infty(\mathbb{R}^n)$ satisfy $\chi(x)=1$ near 0. For a $\varphi\in S$ define $\varphi_j(x)=\varphi(x)\chi(x/j)$, j=1,2,... , so that $\varphi_j\in D$. Setting $\omega_j(x)=1-\chi(x/j)$, get $\psi_j=\varphi-\varphi_j=\varphi\omega_j=0$ in $|x|\le 1$ for large j. Note, $\langle x\rangle^k\psi_j^{(\alpha)}$ is a linear combination of $\theta_{\beta\gamma,j}=\langle x\rangle^k\varphi^{(\beta)}\omega_j^{(\gamma)}$, $\beta+\gamma=\alpha$,

where $\sup\{|\theta_{\beta\gamma}{}^{,j}(x)|:x \in \mathbb{R}^n\} \le \sup\{|\omega_j^{(\gamma)}|\}\sup\{\langle x\rangle^k\varphi^{(\beta)}:|x|\ge 1\}$. Since $\varphi\in S$ the second sup at right goes to 0 as $1\to\infty$ (i.e.,as $j\to\infty$). Also, $\sup\{|\omega_j^{(\gamma)}|\}=j^{-|\gamma|}\sup\{\omega(x):x\in \mathbb{R}^n\}\le c$. Thus $\psi_j\to 0$ in S, q.e.d.

Note that polynomials, and delta functions $\delta^{(\alpha)}(x-a)$ are examples of temperate distribution. However, $e^x \notin \mathcal{S}(\mathbb{R})$ (pbms.2,3).

To generalize F we still require the following.

Corollary 2.5. The transforms F and $\bar{F}$ both have the property that

$\varphi_j \in \mathcal{S}$, $\varphi_j \to 0$ in $\mathcal{S}$ implies $F\varphi_j \to 0$ $\bar{F}\varphi_j \to 0$ in $\mathcal{S}$.

It is sufficient to prove this for F. Again we need an equivalence like (2.2), now for the property '$\varphi_j \to 0$ in $\mathcal{S}'$' :

Proposition 2.6. Let $\varphi_j \in \mathcal{S}$, $j=1,\ldots$. Then '$\varphi_j \to 0$ in $\mathcal{S}$ ' is equivalent to each of the following conditions:

$$\langle x\rangle^k \varphi_j^{(\alpha)} \to 0 \text{ , or } x^\beta \varphi_j^{(\alpha)} \to 0 \text{ , or } (x^\beta \varphi_j)^{(\alpha)} \to 0$$

(2.8) for all multi-indices α , β , or $k=0,1,2,\ldots$, in one (and the same) of the norms of $CB(\mathbb{R}^n)$ or $L^p(\mathbb{R}^n)$, $1 \le p \le \infty$.

For the proof cf. lemma 2.8.

Using prop.2.6, lemma 2.5 is a matter of (1.2), and (2.7). Indeed, if $\varphi_j \to 0$ in $\mathcal{S}$, we have $\|x^\beta \varphi_j^{(\alpha)}\|_{L^1} \to 0$, $j \to \infty$, hence $\|(x^\alpha \varphi_j^\wedge)^{(\beta)}\|_{CB} \to 0$, implying $\varphi_j^\wedge \to 0$, q.e.d.

For a given $u \in \mathcal{S}'$, observe that $u^\wedge$, defined by

$$(2.9) \qquad \langle u^\wedge, \varphi\rangle = \langle u, \varphi^\wedge\rangle \text{ , } \varphi \in \mathcal{S} \text{ ,}$$

defines a functional in $\mathcal{S}'$, since $\varphi_j \to 0$ in $\mathcal{S}$ implies $\varphi_j^\wedge \to 0$ in $\mathcal{S}$ (by cor.2.5) , hence $\langle u, \varphi^\wedge\rangle \to 0$. If $u \in L^1(\mathbb{R}^n)$ then it follows that $u \in \mathcal{S}'$ (cf. pbm.3). In that case we have

$$(2.10) \qquad \langle u, \varphi^\wedge\rangle = \int đx\, u(x) \int đ\xi\, \varphi(\xi) e^{-ix\xi} = \langle \int đ\xi\, e^{-ix\xi} u(\xi), \varphi\rangle \text{ , } \varphi \in \mathcal{S} \text{ ,}$$

by Fubini's theorem, since the integrand is $L^1(\mathbb{R}^{2n})$. Thus, for $u \in L^1$, (2.10) implies that the functional (2.9) coincides with that of the Fourier transform $u^\wedge$ of (1.1). Accordingly, <u>for a general</u> $u \in \mathcal{S}'$ <u>we define the Fourier transform</u> $u^\wedge$ as the functional of (2.9) and the conjugate Fourier transform $u^\vee$ by

$$(2.11) \qquad \langle u^\vee, \varphi\rangle = \langle u, \varphi^\vee\rangle \text{ , } \varphi \in \mathcal{S} \text{ .}$$

It is clear at once that we have

Theorem 2.7. The (conjugate) Fourier transform coincides with the (conjugate) Fourier transform previously defined for L^1-functions (cf.(1.1), and (1.7)). We have the Fourier inversion formula

$$(u^\wedge)^\vee = (u^\vee)^\wedge = u \ , \text{ for all } u \in \mathcal{S}' . \tag{2.12}$$

Also, for $u \in \mathcal{S}'$ we have $x^\alpha u^{(\beta)} \in \mathcal{S}'$, and (2.7) holds as well.

Prop.2.6 and (2.2) follow from the (evident) lemma, below.

Lemma 2.8. a) We have (using Leibniz' formula and its adjoint)

$$(x^\alpha u)^{(\beta)} = \sum c_{\alpha\beta\gamma} x^{\alpha-\gamma} u^{(\beta-\gamma)} \ , \ x^\alpha u^{(\beta)} = \sum d_{\alpha\beta\gamma} (x^{\alpha-\gamma} u)^{(\beta-\gamma)} \ , \tag{2.13}$$

with finite sums and constants $c_{\alpha\beta\gamma}$, $d_{\alpha\beta\gamma}$.

b) We have

$$|x^\alpha| \le \langle x \rangle^{|\alpha|} \ , \text{ and } \langle x \rangle^k \le c_k \sum_{|\alpha| \le k} |x^\alpha| \text{ with a constant } c_k . \tag{2.14}$$

c) We have

$$\|u\|_{L^p} \le \|\langle x \rangle^{-k}\|_{L^p} \|\langle x \rangle^k u\|_{L^\infty} \ , \ 1 \le p < \infty \ , \ k > n/p . \tag{2.15}$$

d) We have

$$\|u\|_{L^\infty} \le c\|u^\wedge\|_{L^1} \le c\|(1+|x|)^{n+1} u^\wedge\|_{L^\infty} \le c \sum_{|\alpha| \le n+1} \|x^\alpha u^\wedge\|_{L^\infty} \tag{2.16}$$

$$= c \sum_{|\alpha| \le n+1} \|(u^{(\alpha)})^\wedge\|_{L^\infty} \le c \sum_{|\alpha| \le n+1} \|u^{(\alpha)}\|_{L^1} .$$

Problems: 1) Show that the following functionals define distributions in $\mathcal{D}'(\mathbb{R}^n)$: a) $\langle f, \varphi \rangle = \varphi^{(\alpha)}(x^0)$, for given multiindex α and $x^0 \in \mathbb{R}^n$; b) $\langle f, \varphi \rangle = \int_{|x|=1} \varphi(x) dS$, dS=surface measure ; c) $\langle \text{p.v.}\frac{1}{x}, \varphi \rangle = \lim_{\eta \to 0} \int_{|x| \ge \eta} \varphi(x) \frac{dx}{x}$ (for n=1). 2) Obtain the first partials of the distributions of pbm.1. 3) Show that distributions $f_\pm \in \mathcal{D}'(\mathbb{R})$ are defined by $\langle f_\pm, \varphi \rangle = \lim_{\varepsilon \to 0, \varepsilon > 0} \int_{-\infty}^{\infty} \varphi(x) \frac{dx}{x \pm i\varepsilon}$. Relate $f_\pm$ with p.v.$\frac{1}{x}$ of pbm.1. 4) The distribution derivative satisfies Leibniz' formula and its adjoint (cf. [C₁],I,(1.23)). 5) Show that a distribution $f \in \mathcal{D}'(\Omega)$ with $f^{(\alpha)} \in C(\Omega)$, $|\alpha| \le k$ is a function in $C^k(\Omega)$. 6) Let L^1_{pol} be the class of all $u \in L^1_{loc}(\mathbb{R}^n)$ with $\langle x \rangle^{-k} u \in L^1(\mathbb{R}^n)$ for some $k = k(u)$. Show that $L^1_{pol} \subset \mathcal{S}'$. 7) Show that $p(x) = \sum_{|\alpha| \le m} a_\alpha x^\alpha \in L^1_{pol} \subset \mathcal{S}'$. Also that $CB(\mathbb{R}^n) \subset L^1_{pol}$, and $L^p(\mathbb{R}^n) \subset L^1_{pol}$, $1 \le p \le \infty$. 8) Show that e^{ax}

$\in \mathcal{D}'(\mathbb{R})$, but $e^{ax} \notin \mathcal{S}'$, as Re a $\neq 0$. 9) Let $\mathcal{T}_{pol}$ be the class of all $a \in C^\infty(\mathbb{R}^n)$ with $a^{(\alpha)}(x) = O(\langle x \rangle^{k_\alpha})$, for some $k_\alpha \in \mathbb{Z}$, for every α. Show that differentiation and multiplication by $a \in \mathcal{T}_{pol}$ leaves $\mathcal{S}'$ invariant. That is, for $u \in \mathcal{S}'$, $a \in \mathcal{T}_{pol}$, $\alpha \in \mathbb{Z}_+^n$ we have $au \in S'$, $u^{(\alpha)} \in \mathcal{S}'$. 10) Obtain the Fourier transform of the following distributions (If necessary, show, they are in $\mathcal{S}'$): a) x^α, $\alpha \in \mathbb{Z}_+^n$; b) $\delta_{x_0}^{(\beta)}$, $\beta \in \mathbb{Z}_+^n$; c) e^{iax} , $a \in \mathbb{R}^n$. 11) Obtain $(\text{p.v.}\frac{1}{x})^\wedge$, for the distribution of pbm.1. 12) Define a distribution $\text{p.v.}\frac{1}{\sinh x} \in \mathcal{S}'$, using the same kind of 'principal-value integral' as in pbm 1. Calculate $(\text{p.v.}\frac{1}{\sinh x})^\wedge$. 13) Obtain the Fourier transform of a 2π-periodic $C^\infty(\mathbb{R})$- function a(x). Hint: Use that a(x) has a uniformly convergent <u>Fourier series</u> $a(x) = \sum_{-\infty}^{\infty} a_m e^{imx}$, $a_m = \frac{1}{2\pi}\int_0^{2\pi} a e^{-imx} dx$

14) Let $f(x) = |\sin x|$. Show that $f \in \mathcal{S}'$ and evaluate $f^\wedge$.

<u>3. The Paley-Wiener theorem; Fourier transform of a general</u> $u \in \mathcal{D}'$.

The support of a distribution $u \in D'$ was defined as smallest closed set Q with u=0 in $\Omega \setminus Q$. We now consider u with supp $u \subset\subset \Omega$.

A simple but important remark is that a compactly supported distribution $u \in D'(\Omega)$, as linear functional over $\mathcal{D}(\Omega)$, admits a natural extension to the larger space $\mathcal{E} = C^\infty(\Omega)$. (The notation was introduced by Schwartz again.) Indeed, for a given $\chi(x) \in C_0^\infty(\Omega)$ with $\chi(x) = 1$ near supp u, define the extension of $\langle u,. \rangle$ to $\mathcal{E}$ by

$$\langle u,\varphi \rangle = \langle u,\chi\varphi \rangle \ , \text{ for all } \varphi \in \mathcal{E}(\Omega) = C^\infty(\Omega) \ . \tag{3.1}$$

This defines an extension: if $\varphi \in \mathcal{D}(\Omega)$, then $(1-\chi)\varphi \in \mathcal{D}(\Omega)$, and supp $(1-\chi)\varphi \subset$ supp $(1-\chi)$ is disjoint from supp u, hence $\langle u,(1-\chi)\varphi \rangle =$ 0, or, $\langle u,\varphi \rangle = \langle u,\chi\varphi \rangle$. The extension is independent of the choice of χ. If $\theta \in \mathcal{D}(\Omega)$ has the property of χ then $\tau - \chi = 0$ near supp u, $\Rightarrow$

$$\langle u,\theta\varphi \rangle = \langle u,\chi\varphi \rangle \ , \text{ for all } \varphi \in \mathcal{E}(\Omega) \ . \tag{3.2}$$

The class of all distributions $u \in \mathcal{D}'(\Omega)$ with compact support is commonly denoted by $\mathcal{E}'(\Omega)$. We have seen that $\mathcal{E}'(\Omega)$ is naturally identified with a class of linear functionals on the space $\mathcal{E}(\Omega)$.

<u>Proposition 3.1</u>. The set $\mathcal{E}'(\Omega)$ of all (above extensions of) compactly supported $u \in \mathcal{D}'(\Omega)$ coincides with the set of continuous lin-

ear functionals over $\mathcal{E}(\Omega)$ (i.e., the functionals u over $\mathcal{E}(\Omega)$ such that $\varphi_j \in \mathcal{E}$, $\varphi_j \to 0$ in $\mathcal{E}$ implies $\langle u,\varphi_j\rangle \to 0$). Here $\varphi_j \to 0$ in $\mathcal{E}$ means that $\varphi_j^{(\alpha)}(x) \to 0$ uniformly on compact sets of Ω, for all α .

Clearly the extension (3.1) to $\mathcal{E}$ of $u \in \mathcal{D}'$ with supp $u \subset\subset \Omega$ is a continuous linear functional over $\mathcal{E}$, in the above sense: If $\varphi_j \in \mathcal{E}$, $\varphi_j \to 0$ in $\mathcal{E}$, then $\chi\varphi_j \to 0$ in $\mathcal{D}$, as a consequence of Leibniz' formula. Vice versa, for a continuous linear functional u over $\mathcal{E}$ the restriction $v=u|\mathcal{D}$ is a distribution in $\mathcal{D}'$, since $\varphi_j \in \mathcal{D}$, $\varphi_j \to 0$ in $\mathcal{D}$ trivially implies $\varphi_j \to 0$ in $\mathcal{E}$. Prop.3.1 follows if we can show that supp $v \subset\subset \Omega$. Suppose not, then a sequence of balls B_j may be constructed such that $u \neq 0$ in B_j, while every set $K \subset\subset \Omega$ is disjoint from all but finitely many of the B^j. Construct $\varphi_j \in \mathcal{D}$, supp $\varphi_j \subset B_j$ with $\langle u,\varphi_j\rangle = 1$. Observe that $\varphi_j \to 0$ in $\mathcal{E}$ while $\langle u,\varphi_j\rangle = 1$ does not tend to zero, a contradiction. Q.E.D.

For a compactly supported distribution on $\mathbb{R}^n$ we always have a Fourier transform in the sense of sec.2, i.e., we get $\mathcal{E}'(\mathbb{R}^n) \subset \mathcal{S}'$:

<u>Theorem 3.2</u>. All compactly supported distributions over $\mathbb{R}^n$ are temperate. Moreover, for $u \in \mathcal{E}' \subset \mathcal{S}'$, $u^\wedge$ is a C^∞-function given by

$$(3.3) \qquad u^\wedge(x) = \int d\!\!{}^-\xi e^{-ix\xi} u(\xi) = \langle u, e_x\rangle \ , \ e_x(\xi) = e^{-ix\xi} \ ,$$

with a distribution integral, given by the third expression (3.3).

In fact, the function $u^\wedge(x)$ is entire analytic, in the n complex variables x_j, in the sense that $v(z) = \langle u, e_z\rangle$, $e_z(x) = e^{-izx}$, is meaningful for all $z \in \mathbb{C}^n$, (not only $\mathbb{R}^n$), and defines an extension of $u^\wedge$ of (3.3) to $\mathbb{C}^n$ having continuous partial derivatives in the complex sense with respect to each of the variables $z_1,\dots,z_n$.

Note that formula (3.3) is meaningful only by virtue of our extension (3.1) of $u \in \mathcal{E}'$ to all of $\mathcal{E}$.

<u>Proof</u>. For $u \in \mathcal{D}'(\mathbb{R}^n)$, supp $u \subset\subset \mathbb{R}^n$, the natural extension to $\mathcal{E}$ may be restricted to $\mathcal{S}$ again to provide a continuous linear functional on $\mathcal{S}$, since "$\varphi_j \to 0$ in $\mathcal{S}$ " implies "$\varphi_j \to 0$ in $\mathcal{E}$". Hence $u \in \mathcal{S}'$. The function $v(z)$ indeed is meaningful for all $z \in \mathbb{C}^n$. Existence of $\partial v/\partial z_j$ is a matter of the continuity of the functional u over $\mathcal{E}$: For a fixed z , $h \in \mathbb{C}^n$, form the difference quotient

$$(3.4) \qquad w_\varepsilon = (v(z+\varepsilon h)-v(z))/\varepsilon = \langle u,(e_{z+\varepsilon h}-e_z)/\varepsilon\rangle \ , \ \varepsilon > 0 \ .$$

For the directional derivative $\nabla_h e_z$ of e_z at z , we get

$$(3.5) \qquad \psi_\varepsilon = (e_{z+\varepsilon h}- e_z)/\varepsilon - \nabla_h e_z \to 0 \text{ in } \mathcal{E} \ ,$$

Indeed, this only means that $\partial_x^\alpha \psi_\varepsilon \to 0$ uniformly on $K \subset\subset \mathbb{R}^n$, as readily verified for $e_z(\xi)$. Continuity of u then implies

$$\lim_{\varepsilon\to 0,\varepsilon\neq 0} w_\varepsilon = \langle u,\nabla_h e_z\rangle \ , \tag{3.6}$$

confirming that v(z) is analytic for all z. Formally we then get

$$\langle u^\wedge,\varphi\rangle = \langle u,\int d\!\!{}^-\xi e_\xi \varphi(\xi)\rangle = \int d\!\!{}^-\xi\langle u,e_\xi\rangle\varphi(\xi) = \int d\!\!{}^-\xi v(\xi)\varphi(\xi) \ , \tag{3.7}$$

with v(x) as defined, where the interchange of limit leading to the second equality remains to be confirmed. Clearly (3.7) implies $u^\wedge=v$, i.e., (3.3) and thm.3.2 follows. For the interchange of limit show existence of the improper Riemann integral $\int d\!\!{}^-\xi e_\xi\varphi(\xi)$ in the sense of convergence in $\mathcal{E}$: For $K\subset\subset \mathbb{R}^n$ we must show that $\int_K d\!\!{}^-\xi e_\xi\varphi(\xi)-S_k\to 0$ in $\mathcal{E}$, as $k\to\infty$. Here S_k is any sequence of Riemann sums, with maximum partition diameter tending to 0 as $k\to\infty$. Also, that $\int_{\mathbb{R}^n\setminus K} d\!\!{}^-\xi e_\xi\varphi(\xi)\to 0$, as K runs through a sequence K_j with $\cup K_j=\mathbb{R}^n$, again, with convergence in $\mathcal{E}$. Again, convergence in $\mathcal{E}$ just means local uniform convergence with all derivatives. One confirms easily the local uniform convergence in the parameter x , since the function $e_\xi(x) = e^{-ix\xi}$ is continuous. Similarly for the x-derivatives, again continuous in x and ξ . This, and the fact that the x-derivatives of the Riemann sums are Riemann sums again, indeed allows to confirm the desired convergences. Q.E.D.

As examples for Fourier transforms of compactly supported distributions we mention those of the delta-function and its derivatives. As seen in 2,pbm.5 we get $\delta_0^{(\alpha)\wedge} = i^{|\alpha|}\kappa_n x^\alpha$. In fact, this is an immediate consequence of (3.3), above.

We observe that the entire analytic function $u^\wedge(z)$ of (3.3), as a function of complex arguments z , has a growth property which characterizes the Fourier transforms of compactly supported distributions. The result is called the <u>Paley-Wiener theorem</u>.

<u>Theorem 3.3</u>. An entire analytic function v(z) over $\mathbb{C}^n$ is the Fourier transform of a compactly supported distribution $u\in \mathcal{D}'(\mathbb{R}^n)$ if and only if there exists an integer k > 0 and a real $\eta>0$ such that

$$v(z) = O(\langle z\rangle^k e^{\eta|\operatorname{Im} z|}) \text{ for all } z\in\mathbb{C}^n \ , \ \langle z\rangle=(1+\textstyle\sum|z_j|^2)^{1/2}. \tag{3.8}$$

Moreover, the constant η may be chosen as the radius of the smallest ball $|x|\le r$ containing supp u . Furthermore, $u\in \mathcal{D}(\mathbb{R}^n)$ if and only if (3.8) holds for <u>all</u> k with $\eta=\max\{|x|: x\in \text{supp } u\}$.

<u>Proof</u>. For $u \in \mathcal{E}'$ we must have

(3.9) $$|\langle u,\varphi\rangle| \le c \sup\{|\varphi^{(\alpha)}(x)| : x \in K , |\alpha|\le k \} .$$

for some c, k, and some compact $K \supset \text{supp}\, u$ and all $\varphi \in \mathcal{E}$. Otherwise for every c=k=j and $|x|\le j$ there exists $\varphi=\varphi_j \in \mathcal{E}$ with $\langle u,\varphi_j\rangle=1$, and ">" holds in (3.9). Or, $|\varphi_j^{(\alpha)}(x)| \le \frac{1}{j}$ for all $|\alpha|\le j$, $|x|\le j$, j=1, 2,..., implying uniform convergence $\varphi_j^{(\alpha)}(x) \to 0$, $j\to\infty$, a contradiction, since $1=\langle u,\varphi_j\rangle$ does not tend to 0.

We get $u^\wedge(z)=\langle u,\chi_z e_z\rangle$, $\chi_z=\chi(|z|(|x|-\eta))$ where $\chi \in C^\infty(\mathbb{R})$, $\chi(t)=1$, $t<\frac{1}{2}$, $=0$, $t>1$, χ decreasing. It follows that $\text{supp}\,\chi_z \subset \{|x|\le\eta+\frac{1}{|z|}\}$ so that $(\chi_z e_z)^{(\alpha)}(x)= O(e^{\eta|\text{Im} z|+1}\langle z\rangle^k)$. Combining this with (3.9) we get (3.8) with the proper constant η.

Next assume $u\in C_0^\infty(\Omega)$. We trivially get (3.9) with k=0 and $K=\text{supp}\, u \subset\subset \Omega$, since $u \in L^1$. Similarly for $u^{(\alpha)}$. Accordingly, for all α we get $|z^\alpha u^\wedge(z)|=|\langle u,e_z^{(\alpha)}\rangle|=O(e^{\eta|\text{Im}\, z|})$, hence (3.8) for all k.

Vice versa, (3.8) for all k implies $x^\alpha v|\mathbb{R}^n \in L^1 \subset \mathcal{S}'$. Then $v^\vee$ is given by the conjugate Fourier integral. We have $u=(v|\mathbb{R}^n)^\vee \in CO$, and even $u^{(\alpha)} \in CO$, i.e., $u \in C^\infty(\mathbb{R}^n)$. To show that $\text{supp}\, u \subset\subset \mathbb{R}^n$ write

(3.10) $$u(x) = \int d\!\!\!{}^-\xi e^{ix\xi} v(\xi) .$$

If $\theta \in \mathbb{R}^n$ is given arbitrary then we also may write (3.10) as

(3.11) $$u(x) = \int d\!\!\!{}^-\xi e^{ix(\xi+i\theta)} v(\xi+i\theta) .$$

Indeed, this is a matter of Cauchy's integral theorem, applied for a rectangle in the complex ζ_j-plane with sides $\text{Re}\,\zeta_j=\pm A$, $\text{Im}\,\zeta_j=0$ or θ_j. In such a rectangle the integrand $e^{ix\zeta}v(\zeta)$ is holomorphic as a function of ζ_j for constant other variables, so that the complex integral over the boundary vanishes. For $A\to\infty$ the integrals over $\text{Re}\,\zeta_j=\pm A$, $0<\text{Im}\,\zeta_j<\theta_j$, tend to zero, in view of (3.8) for k=-2 for example. The integration pathes have length θ_j and the integrand is $O(\langle A\rangle^{-2} e^{(\eta-x)|\theta|})$. The integral (3.10) may be written as n-fold iterated integral over $\mathbb{R}$. The above proceedure allows the transfer of the integration from $\mathbb{R}$ to the line $\{\xi_j+i\theta_j : x_j \in \mathbb{R}\}$.

Next let us estimate (3.11):

(3.12) $$u(x) = O(\int d\!\!\!{}^-\xi e^{\eta|\theta|-x\theta} \langle \xi+i\theta\rangle^{-k}) = O(e^{\eta|\theta|-x\theta}) ,$$

setting k=n+1 (it holds for every k), and using that $\langle \xi+i\theta\rangle \ge \langle\xi\rangle$. The 'O(.)- constant' is independent of θ. Hence we can set $\theta=tx$, $t>0$, for $u(x)=O(e^{t|x|(\eta-|x|)})$. The exponent is <0 as $|x|>\eta$, and u(x)=0 follows. Thus supp u$\subset\{|x|\le\eta\}$, u$\in$ **D**, if (3.8) for all k.

Finally, if (3.8) holds for <u>some</u> k, let $\chi(x)\in$ **D**, supp $\chi\subset\{|x|\le 1\}$, $\chi(x)\ge 0$, $\int\chi(x)đx=1$. For $\varepsilon>0$ let $\chi_\varepsilon(x)=\varepsilon^{-n}\chi(\frac{x}{\varepsilon})$. Note that $\chi_\varepsilon{}^\wedge(\xi)=\chi^\wedge(\varepsilon\xi)\to\chi^\wedge(0)=1$, as $\varepsilon\to 0$. Moreover, for any $\varphi\in$ **S** get $\varphi\chi_\varepsilon{}^\wedge-\varphi\to 0$ in **S**. Since supp $\chi_\varepsilon\subset\{|x|\le\varepsilon\}$ we have (3.8) for $\chi_\varepsilon{}^\wedge(\zeta)$ with η replaced by ε for all k. Hence the product $v(z)\chi_\varepsilon{}^\wedge(z)$ satisfies (3.8) with η replaced by $\eta+\varepsilon$ again for all k. It follows that $v\chi_\varepsilon{}^\wedge=u_\varepsilon{}^\wedge$ with $u_\varepsilon\in$ **D**, supp $u_\varepsilon\subset\{|x|\le\eta+\varepsilon\}$. Also $\langle u_\varepsilon,\varphi\rangle=\langle u_\varepsilon{}^\wedge,\varphi^\vee\rangle=\langle v,\chi_\varepsilon{}^\wedge\varphi^\vee\rangle\to\langle v,\varphi^\vee\rangle=\langle u,\varphi\rangle$, for all $\varphi\in$ **S**. The latter implies that supp u$\subset\{|x|\le\eta+\varepsilon\}$, all $\varepsilon>0$. It follows that u$\in$ **E**, supp u$\subset\{|x|\le\eta\}$, q.e.d.

Let **Z** denote the space of all entire analytic functions v(z) $=v(z_1,\dots,z_n)$ in n complex variables such that for k=0,1,2,..., and some $\eta\ge 0$ we have (3.8) satisfied. We shall say that a sequence $v_j\in$ **Z** tends to 0 in **Z** if (i) estimates (3.8) hold with constants independent of j, and (ii) $m_j=\mathrm{Max}\{|v_j(x)|: x\in\mathbb{R}^n\}\to 0$, as $j\to\infty$.

<u>Corollary 3.4</u>. The Fourier transform F: u $\to$ u$^\wedge$ establishes a linear bijection **D** $\leftrightarrow$ **Z** which is continuous in either direction, in the sense that $u_j\to 0$ in **D** holds if and only if $u_j{}^\wedge\to 0$ in **Z** .

<u>Proof</u>. After thm.3.3 we focus on continuity only. If $u_j\to 0$ in **D** then supp $u_j\subset\{|x|\le a\}$ for a independent of j . This yields (3.8) with η=a independent of j, by thm.3.3. Inspecting the first part of thm.3.3's proof we also find the O(.) constant independent of j.

Vice versa, if $v_j\to 0$ in **Z**, then (3.8) with η independent of j implies supp $u_j{}^\vee\subset\{|x|\le\eta\}$. But (3.8), for real z=x, implies $v_j=O(\langle x\rangle^{-k})$, uniformly in x and j, for every k. Thus conclude from cdn.(ii) that $\|x^\alpha v_j\|_{L^1}\to 0$, as $j\to\infty$. For the inverse Fourier transform $u_j=v_j{}^\wedge$ we get $\|u_j{}^{(\alpha)}\|_{L^\infty}\to 0$, so that indeed $u_j\to 0$ in **D**. Q.E.D.

Following [GS] we now define a <u>Fourier transform of a general distribution</u> f $\in$ **D**'($\mathbb{R}^n$) regardless of growth at infinity, as a continuous linear functional $f^\wedge$: **Z**$\to\mathbb{C}$. Here of course "$f^\wedge$ continuous" means that "$\langle f^\wedge,\varphi_j\rangle\to 0$, whenever $\varphi_j\to 0$ in **Z**". We define $f^\wedge$ by

(3.13) $$\langle f^\wedge,\varphi\rangle=\langle f,\varphi^\wedge\rangle\ ,\ \text{for all } \varphi\in \mathbf{Z}\ ,$$

taking into account that $\varphi^\wedge=(\bar{\varphi}^\vee)^-\in$ **D** for $\varphi\in$ **Z** .

This definition is compatible with the earlier ones. Indeed, we have **Z** $\subset$ **S** , in the sense that for u$\in$ **Z** the restriction u$|\mathbb{R}^n$ determines u uniquely and is contained in **S** . Moreover, **Z** is dense in **S**, since **Z**=**D**$^\wedge$, and **D** is dense in **S** while F and F^- are continuous maps $\mathcal{S}\to\mathcal{S}$. Also $\varphi_j\to 0$ in $\mathcal{Z}$ implies $\varphi_j\to 0$ in $\mathcal{S}$. For $u\in\mathcal{S}'$ the restriction $u|\mathcal{Z}$ determines u and we have $u|\mathcal{Z}\in\mathcal{Z}'$. Hence get a natural imbedding $\mathcal{S}'\to\mathcal{Z}'$. For $u\in\mathcal{S}'\subset\mathcal{Z}'$ we earlier defined $\langle u^\wedge,\varphi\rangle=\langle u,\varphi^\wedge\rangle$ for $\varphi\in\mathcal{S}$. The restriction $u^\wedge|\mathcal{Z}$ gives our present functional,q.e.d.

Notice that $u^\wedge$, for $u\in\mathcal{D}'$ in general is not a distribution, as defined in sec.2. It is a linear functional on $\mathcal{Z}$, not on $\mathcal{D}$.

Recall that for a function $f\in L^1_{loc}(\mathbb{R})$ with f=0 in x<0 and $f=O(e^{cx})$, some c, one commonly defines the Laplace transform by

$$(3.14)\qquad f^\sim(\zeta)=\int_0^\infty d\!\!/x\, e^{-ix\zeta}f(x)\ ,\ \operatorname{Im}\zeta<-c\ ,$$

where the integral exists and defines a holomorphic function in the complex half-plane $\operatorname{Im}\zeta<-c$ (we have modified the standard definition, by a factor i). The inverse transform then is given by

$$(3.15)\qquad f(x)=\int_{-\infty+i\gamma}^{+\infty+i\gamma} d\!\!/\zeta\, e^{ix\zeta}f^\sim(\zeta)\ ,$$

with a complex curve integral along the parallel $\operatorname{Im}\zeta=\gamma<-c$.

We now will identify $f^\sim$ with the Fourier transform $f^\wedge\in\mathcal{Z}'$ of the distribution $f\in\mathcal{D}'$. For $\varphi\in\mathcal{D}$, $\operatorname{supp}\varphi\subset\{|x|\le\eta\}$, we know that $\varphi^\vee$ is entire analytic, satisfying (3.8). For $\gamma<-c$ we have

$$(3.16)\qquad \int_{\operatorname{Im}\zeta=\gamma} f^\sim(\zeta)\varphi^\vee(\zeta)d\zeta=\int_0^\infty d\!\!/x f(x)\int_{\operatorname{Im}\zeta=\gamma} e^{-ix\zeta}\varphi^\vee(\zeta)d\zeta=\langle f,\varphi\rangle\ .$$

The integral $dx\,d|\zeta|$ exists absolutely, hence the interchange, by Fubini's theorem. Also, we get $\int_{\operatorname{Im}\zeta=\gamma}=\int_{\mathbb{R}}$, at right, by the properties of the (analytic) integrand. Then (3.16) follows from Fourier inversion for functions in $\mathcal{D}$. Or, $f^\wedge\in\mathcal{Z}'$ may be written as

$$(3.17)\qquad \langle f^\wedge,\varphi\rangle=\int_{\operatorname{Im}\zeta=\gamma} f^\sim(\zeta)\varphi(z)d\zeta\ ,\ \varphi\in\mathcal{Z}\ ,$$

where we must choose $\gamma<-c$ with c of (3.14) (for f).

Thus for a function $f\in L^1_{loc}(\mathbb{R})$ of exponential growth and =0 in x<0 the Fourier transform $f^\wedge$ is given as the complex integral (3.17) involving the Laplace transform $f^\sim$ of f.

Problems. 1) Obtain the Laplace transforms of the following functions (Each is extended zero for x<0). a) x^k k=0,1,..., b) e^{ax}; c) cos bx ; d) $e^{ax}\sin bx$; e) $\frac{\sin x}{x}$. In each case, discuss the Fourier transform - i.e., the linear functional on $\mathcal{Z}$. 2) Obtain the inverse Laplace transform of a) $\frac{1}{\sqrt{z-a}}$; b) $\log(1+\frac{1}{z^2})$. (In each case specify a branch of the (multi-valued) function well defined in a half-plane $\operatorname{Im} z<\gamma$.) 3) For $u\in\mathcal{D}'(\mathbb{R}^n)$ with $\operatorname{supp} u\subset\{x_1\ge0\}=$

$\mathbb{R}^n_+$ and $e^{cx_1}u \in \mathcal{S}'$, for some c, show that u^ may be defined by a complex integral like (3.17), with u˜ replaced by u_ℓ˜ , "ℓ"="^" with respect to $(x_2,\dots,x_n)$. 4) The convolution product w=u*v, so far defined for $u,v \in L^1(\mathbb{R}^n)$, by (1.19), may be defined for general distributions $u,v \in \mathcal{D}'(\mathbb{R}^n)$ under a support restriction -for example (i) if supp $u \subset\subset \mathbb{R}^n$, supp v general, or (ii) if supp $v \subset \{x_1 \geq 0\}$, supp $v \subset \{|x| \leq cx_1\}$. One then defines $\langle w,\varphi\rangle = \iint dxdy\, u(x)v(y)\varphi(x+y)$, with a distribution integral (for precise definition cf.[Schw_1], or, [C_1],I,(8.1)). Show that (1.21) is valid for this convolution product as well, assuming in case (ii) the cdns. of pbm.3 for u,v. 5) Let T_A be the space of all entire functions $\chi(z)$ satisfying (3.8) for some k. Show that $\chi\varphi \in \mathcal{Z}$, for $\varphi \in \mathcal{Z}$, $\chi \in \mathcal{T}_A$. Moreover, show that $f \in \mathcal{Z}'$ allows definition of a product $\chi f \in \mathcal{Z}'$, setting $\langle \chi f,\varphi\rangle = \langle f,\chi\varphi\rangle$, $\varphi \in \mathcal{Z}$. All polynomials p(x) belong to $\mathcal{T}_A$. 6) Show that (1.22) is valid for general distributions $u \in \mathcal{D}'(\mathbb{R}^n)$.

4. The Fourier-Laplace method; examples.

We now will discuss the 'Fourier-Laplace method' for 'free space'-problems of the following constant coefficient operators:

(4.1) $\Delta = \sum_{j=1}^n \partial_{x_j}^{\ 2}$ (the Laplace operator) ,

(4.2) $\Delta + k^2$ (the Helmholtz operator) ,

(4.3) $H = \partial_{x_0} - \Delta = \partial_t - \Delta$ (the heat operator) ,

(4.4) $\Box = \partial_{x_0}^{\ 2} - \Delta = \partial_t^2 - \Delta$ (the wave operator) ,

(4.5) $\Box + m^2$ (the Klein-Gordon operator) .

The last 3 operators act on the n+1 variables $x_0 = t$, $(x_1,\dots,x_n) = x$. The first two act on x only, distinguishing x_0 from the others.

The discussion around (1.23)-(1.26) was a formal attempt to solve constant coefficient PDE in free space (in all $\mathbb{R}^n$). We found $e = (\frac{1}{P(x)})^\vee$, for a P(D), of special interest. Now we are prepared to implicate this technique, called the Fourier-Laplace method.

Certain initial-boundary problems may be converted into free space problems: (a) An initial-value problem for (4.3),(4.4), or

(4.5) seeks to find solutions u of $P(D)u=f$ in some half-space, say, $t=x_0>0$, where f is given in $t\geq 0$, together with initial data of u at $t=0$. Such problem may be written as a free space problem by extending $u=0$ and $f=0$ into $t<0$, letting v and g be the extended functions. We will not have $P(D)v=g$ then, but, rather, $P(D)v=g+h$, with a distribution h, supp $h\subset\{t=0\}$, since normally v will jump at $t=0$. The initial conditions on u often are well posed if they allow to determine h, making the initial-value problem equivalent to the free space problem $P(D)v=g+h$, where $g+h$ is given.

(b) Another example: If $\Delta u=f$ (Δ of (4.1)) is to be solved in a half-space under Dirichlet bondary conditions - say, $\Delta u=f$ in $x_1>0$, $u=0$ as $x_1=0$, then consider the odd extensions of u and f to $\mathbb{R}^n$: $v(x_1,\dots,x_n)=u(x)$, $x_1>0$, $=-u(-x_1,x_2,\dots,x_n)$, $x_1<0$, similarly g extending f. It follows that $\Delta v=g$ in $\mathbb{R}^n$, again converting the half-space Dirichlet problem of Δ to a free space problem over $\mathbb{R}^n$. Similarly for the Neumann problem, using even extensions.

Technique (a) works as well for a more general initial surface $t=\Theta(x)$, $x\in\mathbb{R}^n$. Both techniqes may be combined to reduce certain initial-boundary problems to free space problems.

The above will emphasize the power of the Fourier Laplace method. (4.1)-(4.5) are a crossection of popular PDE's. We control parabolic and hyperbolic initial value problems, elliptic boundary problems and initial-boundary problems in half spaces, etc., with Green's (Riemann) functions, using results on special functions.

From now on interprete the equation $P(D)v=g$, $x\in\mathbb{R}^n$, as a PDE involving distributions $v,g\in\mathcal{D}'(\mathbb{R}^n)$. The Fourier transform exists without restrictions: Using 3,pbm.6, we get $P(x)v^\wedge=g^\wedge$, where $v^\wedge, g^\wedge\in\mathcal{Z}'$. If $e\in\mathcal{D}'(\mathbb{R}^n)$ solves $P(D)e=(2\pi)^{n/2}\delta$ we get $P(x)e^\wedge=1$.

In the cases corresponding to (4.1)-(4.5) we get, respectively,

$$(4.6)\quad P(x)=-|x|^2\ ,\ =k^2-|x|^2\ ,\ =it+|x|^2\ ,\ =|x|^2-t^2\ ,\ =m^2+|x|^2-t^2,$$

where $t=x_0$ again. Generally, $\frac{1}{P(x)}\notin L^1_{loc}$, except for (4.3) and (4.1), $n\geq 3$, due to zeros of P. Some pbm's of sec's 2,3 (and, more generally, [C₁],II) discuss distributions p.v.a associated to $a\notin L^1_{loc}$. $P(x)z=1$ may have many solutions $z\in\mathcal{D}'$ (or $\in\mathcal{Z}'$). For (4.3)-(4.5) we will be interested in $z=e^\wedge$, $P(x)z=1$, with supp $e\subset\{t\geq 0\}$, because then supp $e*g\subset\{t\geq 0\}$ whenever supp $g\subset\{t\geq 0\}$, so that $u=(e*(g+h))|\{t\geq 0\}$ will solve the initial-value problem for $P(D)u=f$, $t\geq 0$, we started with in (a) above. Indeed, a proper z exists: In pbm's 1 and 3 of sec.2 we defined p.v.$\frac{1}{x}$, and $f_\pm$, all 3 distinct, $xf=1$. Only f_- has its inverse Fourier transform $=0$ for $x<0$.

For (4.3)-(4.5) we will construct such $e\in \mathcal{D}'(\mathbb{R}^{n+1})$ solving $P(D)e=\sqrt{2\pi}^{n+1}\delta$, supp $e \subset \{t\geq 0\}$, using the setup of sec.3, pbm.3: Such e, if $e^{cx}e\in \mathcal{S}'$, will have a Fourier transform in $x=(x_1,..)$ and a Laplace transform in $x_0=t$. Accordingly we must seek an inverse Laplace transform of an inverse Fourier transform of a suitable solution z of Pz=1, or vice versa, in appropriate variables. Our proofs will be sketchy, in part, due to overflow of details.

The lemma below is convenient, due to spherical symmetry of P.

Lemma 4.1. Given a spherically symmetric function $f(x)=\omega(|x|)$, where $\omega(r) \in L^1(\mathbb{R}_+,r^{n-1}dr)$. Then the transforms $f^\wedge$ and $f^\vee$ are spherically symmetric as well: $f^\wedge(x)=f^\vee(x)=\chi(|x|)$, where $\chi(r)$ and $\omega(r)$ are related by the Hankel transform $\mathcal{H}_\nu$, $\nu=\frac{n}{2}-1$. In detail we have

$$\text{(4.7)}\quad \begin{aligned} r^{(n-1)/2}\chi(r) &= \mathcal{H}_{n/2-1}(r^{(n-1)/2}\omega(r)) ,\\ r^{(n-1)/2}\omega(r) &= \mathcal{H}_{n/2-1}(r^{(n-1)/2}\chi(r)) ,\end{aligned}$$

where

$$\text{(4.8)}\quad \mathcal{H}_\nu(\lambda(r))(\rho) = \int_0^\infty \sqrt{\rho r}\ J_\nu(r\rho)\ \lambda(r)\ dr \quad , \ \mathrm{Re}\ \nu > -\tfrac{1}{2} ,$$

with the Bessel function $J_\nu(z)$. The second formula (4.7) is valid if in addition $\chi \in L^1(\mathbb{R}_+,r^{n-1}dr)$.

Proof. For an orthogonal nxn-matrix O get $f^\wedge(O\xi) = \int d\!\!/ x e^{-ix^TO\xi}f(\xi)=$ $\int d\!\!/ x e^{-i(O^Tx)^T\xi}\omega(|O^Tx|) = \int d\!\!/ y e^{-iy\xi}f(y)=f^\wedge(\xi)$. Thus $f^\wedge$ has the same symmetry: $f^\wedge(\xi)=\chi(|\xi|)$, with some $\chi(\rho)$. We may set

$$\text{(4.9)}\quad f^\wedge(\xi)=f^\wedge(|\xi|,0,\dots,0)=\int d\!\!/ x e^{-i\rho x_1}\omega(r)=\kappa_n\int_0^\infty r^{n-1}\omega(r)\int e^{-ir\rho z_1}dS$$

where the inner integral I is over the unit sphere $|z|=1$. Evaluate this inner integral by converting it to an integral on the n-1-dimensional ball $|\lambda|^2 \leq 1$, setting $z=(z_1,\lambda)$. We know that $dS=\frac{d\lambda}{\sqrt{1-\lambda^2}}$. With a contribution from the upper and lower hemisphere where $z_1 = \pm\sqrt{1-\lambda^2}$, writing $d\lambda=\sigma^{n-2}d\sigma d\Sigma$, $\sigma=|\lambda|$, etc., we get

$$I=2\int\sigma^{n-2}\frac{d\sigma}{\sqrt{1-\sigma^2}}d\Sigma\cos(r\rho\sqrt{1-\sigma^2})=2a_{n-1}\int_0^1\sigma^{n-2}\frac{d\sigma}{\sqrt{1-\sigma^2}}\cos(r\rho\sqrt{1-\sigma^2}) .$$

A substitution $\sigma=\sin\theta$ of integration variable yields

$$\text{(4.10)}\quad I = 2a_{n-1}\int_0^{\pi/2}d\theta\ \sin^{n-2}\theta\ \cos(r\rho\cos\theta) ,$$

with $a_{n-1}=\frac{2}{\Gamma((n-1)/2)}\pi^{(n-1)/2}$, the area of the n-1-dimensional unit sphere. Using Poisson's formula ([MOS],p.79) we get

$$\int_0^{\pi/2}\cos(r\rho\cos\theta)\sin^{n-2}\theta\, d\theta = 2^{n/2-2}\sqrt{\pi}\,\Gamma(\tfrac{n-1}{2})J_{n/2-1}(r\rho)\ . \tag{4.11}$$

Substituting into (4.10) and I into (4.9) confirm (4.7). No change if $e^{-ix\xi}$ in (4.9) is replaced by $e^{ix\xi}$. Thus lemma 4.1 follows.

Recall also (1.26), now under the aspect of 3,pbm.4. In details, the convolution product $v*w$ of two distributions $v,w\in \mathcal{D}'(\mathbb{R}^n)$ may be defined by setting (with a distribution integral)

$$\langle v*w,\varphi\rangle = \int v(x)w(y)\varphi(x+y)dxdy = \langle v\otimes w,\psi\rangle\ ,\ \psi(x,y)=\varphi(x+y)\ , \tag{4.12}$$

for all $\varphi\in \mathcal{D}(\mathbb{R}^n)=C_0^\infty(\mathbb{R}^n)$, if v and w satisfy the support condition

$$K_a = (\operatorname{supp} v \times \operatorname{supp} w)\cap\{|x+y|\le a\} \text{ is compact for every } a>0\ . \tag{4.13}$$

Here supp v and supp w are regarded as subsets of $\mathbb{R}^n$ of the variables x and y, respectively (cf. [Schw_1] or [C_1], I,8).

The distribution $v=e*g$ is defined for e as constructed above whenever $g\in \mathcal{E}'$ for (4.1)-(4.3) , and $g\in \mathcal{D}'(\mathbb{R}^{n+1})$, $g=0$ as $t<0$, for (4.4) and (4.5) , since condition (4.13) holds under these assumptions. Moreover, $P(D)v=g$ follows, leading to a solution of the free space problem, and the related initial-boundary problems.

Now let us attempt a detailed construction of the proper e.

Ia) Consider the operator Δ of (4.1), i.e., the potential equation $\Delta u=f$. For $n\ge3$ the function $-\frac{1}{|x|^2}$ is L^1_{pol}, hence a distribution in $\mathcal{S}'$. This is a homogeneous distribution of degree -2 . Hence $e=-(\frac{1}{|x|^2})^\vee$ is homogeneous of degree 2-n . It is also spherically symmetric. Conclusion: $e(x)=c_n|x|^{2-n}$, with a constant c_n. Clearly $e\in L^1_{pol}$. The constant c_n may be evaluated by looking at

$$(2\pi)^{n/2}\varphi(0)=\langle e,\Delta\varphi\rangle = c_n\int\Delta\varphi(x)|x|^{2-n}dx = c_n\lim_{\varepsilon\to0}\int_{r=\varepsilon} dS_1\varphi\frac{d}{dr}r^{2-n}$$

$=c_n a_n(2-n)\varphi(0)$. It follows that

$$e_n(x) = \frac{-1}{(n-2)\omega_n}|x|^{2-n}\ ,\qquad \omega_n=\frac{2}{\Gamma(n/2)}\sqrt{\pi}^n\ . \tag{4.14}$$

For n=1 we first define a distribution

$$e^\wedge = -\text{p.f.}\frac{1}{x^2} = \frac{d}{dx}(\text{p.v.}\frac{1}{x})\ , \tag{4.15}$$

involving the distribution derivative and p.v.$\frac{1}{x}$ of pbm.1,sec.2. We confirm that $e^\wedge$ solves $-x^2e^\wedge=1$. Using $e=e^{\wedge\vee}$ one finds that

$$e(x) = \sqrt{\pi/2}\,|x|\ . \tag{4.16}$$

For n=2 we can define (cf.[C_1],II,(2.11), for l=2)

$$(4.17)\quad e^{\wedge}(x)=-\text{p.f.}\frac{1}{|x|^2}=-\sum_{j=1}^{2}\left(\frac{1}{|x|^2}x_j\log|x|\right)_{|x_j}=-\frac{1}{2}\Delta((\log|x|)^2)\ .$$

again with distribution derivatives. However, it is easier to confirm directly that

$$(4.18)\qquad e(x) = \kappa_1\log|x|$$

is a spherically symmetric L^1_{loc}-function satisfying $\Delta e=\sqrt{2\pi}\delta$ (just evaluate the integral $\langle\Delta e,\varphi\rangle=\langle e,\Delta\varphi\rangle$, using partial integration).

Ib) Consider (4.2), i.e, the Helmholtz equation $(D+\lambda)u=f$, also known as the time-independent wave equation if $\lambda=k^2>0$, and as the resolvent equation of the Laplace operator Δ if $\lambda\in\mathbb{C}$, $\lambda\neq k^2$. In the latter case get $e^{\wedge}(x)=(\lambda-|x|^2)^{-1}\in C^\infty\cap L^1_{pol}$. An evaluation of e(x) is possible, using lemma 4.1, as long as $n\le 3$. We get

$$(4.19)\qquad e(x) = |x|^{1-n/2}\int_0^\infty \rho^{n/2}\frac{d\rho}{\lambda-\rho^2}\,J_{n/2-1}(\rho|x|)\ .$$

For larger n this integral ceases to exist. However, it still will exist as improper integral in the sense of distributions - that is, as $\lim_{A\to\infty}\int_0^A$, where the limit exists in weak convergence of $\mathcal{D}'$ (i.e., $\langle\lim.,\varphi\rangle = \lim\langle\,.,\varphi\rangle$). For odd n the Bessel function $J_{n/2-1}$ may be expressed by trigononmetric functions. For example, in case n=3 we get $J_{1/2}(z) = \sqrt{2}\,\frac{\sin z}{\sqrt{\pi z}}$. Or,

$$(4.20)\qquad e(x) = \sqrt{\frac{2}{\pi}}\frac{1}{|x|}\int_0^\infty\frac{\rho d\rho}{\lambda-\rho^2}\sin\rho|x|\ .$$

We may write $\lambda = \kappa^2$, picking the root κ with Im $\kappa>0$. Then $\int_0^\infty\frac{\rho d\rho}{\lambda-\rho^2}\sin\rho r = \frac{1}{2}\int_{-\infty}^\infty\frac{d\rho}{\kappa-\rho}\sin\rho r = \frac{1}{4i}\int_{-\infty}^\infty e^{ir\rho}\frac{d\rho}{\kappa-\rho} = -\frac{\pi}{2}e^{i\kappa|x|}$. Hence

$$(4.21)\qquad e(x) =-\sqrt{\pi/2}\ e^{i\kappa|x|}/|x|\ .$$

(4.21) may be confirmed, noting that $e=e^{i\kappa r}/r$ solves $(\Delta+\lambda)e=2\pi\delta$.

For $\kappa=k$ real $P(x)=k^2-|x|^2$ vanishes at the set $|x|=k$, and $\frac{1}{P}$ is not L^1_{loc} . Then look at p.f.$(\frac{1}{P(x)})$. Or else, observe that

$$(4.22)\qquad \lim_{\varepsilon\to 0,\varepsilon>0} e_{k+i\varepsilon}(x) = e(x)\ ,\ e_\kappa(x)\text{ as }e(x)\text{ in (4.21)}\ ,$$

in the sense of distributions. This implies that

$$(4.23)\qquad e_\pm(x) = -\sqrt{\pi/2}e^{\pm ik|x|}/|x|$$

both will solve $(\Delta+k^2)e=(2\pi)^{3/2}\delta$. The proper sign may be chosen by imposing a 'radiation condition' at ∞ .

For general $n\geq 2$ we still may evaluate the integral (4.19). Using a formula by Sonine and Gegenbauer (cf. [MOS], p.105) we get

$$(4.24) \qquad e(x)=-\left(\frac{\kappa}{|x|}\right)^{n/2-1}K_{n/2-1}(\kappa|x|) \ , \ \kappa^2=-\lambda \ , \ \operatorname{Re}\kappa > 0 \ ,$$

with the modified Hankel function $K_\nu(z)$. Again get (4.24) more directly, observing that $e(x)=\gamma(|x|)$ solves $(\Delta+\lambda)e=0$, hence $\gamma(r)$ solves the ODE $\gamma''+\frac{n-1}{r}\gamma'-\kappa^2\gamma=0$. Substituting $\gamma=r^{-\nu}\delta$, $\nu=\frac{n}{2}-1$, we obtain the modified Bessel equation $\delta''+\frac{1}{r}\delta'-(1+\frac{\kappa^2}{r^2})\delta=0$, showing that the only spherically symmetric solutions of $(\Delta+\lambda)u=0$ in $\mathcal{S}'$ are the multiples of (4.24). A partial integration shows that $\int e(\Delta+\lambda)\varphi dx= \varphi(x)$ for all $\varphi\in \mathcal{D}$, fixing the remaining multiplicative constant.

II) In the case (4.3) of the heat equation $Hu=u_t-\Delta u=f$ we may use (4.24): Applying F_x^{-1} to $\frac{1}{p}$ for the second and third polynomial (4.6) gives the same result, if we set $\lambda=k^2=-\kappa^2=it$. That is, $\kappa= e^{-i\pi/4}\sqrt{t}$, $\operatorname{Re}\kappa >0$, in (4.24) will define $F_x^{-1}(\frac{1}{p})$, and we then must obtain the inverse Laplace transform.

It is more practical, however, to first obtain $F_t^{-1}(\frac{1}{p})$.

Note $\theta(\tau)=(i\tau+a)^{-1}$ has inverse Laplace transform $\theta^\Delta(t)=\sqrt{2\pi}e^{-at}$, $t\geq 0$, $=0$, $t<0$, calculating $\int_0^\infty e^{-at}e^{-i\tau t}dt$. For $e^\backprime=F^{-1}(\frac{1}{p})$ get

$$(4.25) \qquad e^\backprime(t) = (2\pi)^{1/2}e^{-t|x|^2} \ , \text{ as } t\geq 0 \ , \ e^\backprime(t)=0 \ , \text{ as } t<0 \ .$$

Recall that $(e^{-|x|^2/2})^\vee=e^{-|x|^2/2}$ (in n dimensions), by a complex integration. Also for the function $g_\sigma(x) = g(\sigma x)$ we get

$$(4.26) \qquad g_\sigma^\vee(x) = \sigma^{-n}g^\vee(x/\sigma) \ , \quad \sigma\in \mathbb{R}_+ \ ,$$

as shown by an integral substitution. Choosing $\sigma=\sqrt{2t}$ we thus get

$$(4.27) \qquad e(t,x) = \frac{\sqrt{2\pi}}{(\sqrt{2t})^n} e^{-|x|^2/4t} \ , \ t\geq 0 \ , \ =0 \ , \ t<0 \ .$$

This is the well known <u>fundamental solution of the heat operator</u>. It is not $L^1(\mathbb{R}^{n+1})$. Use it to solve the <u>initial value problem</u>

$$(4.28) \qquad Hu = \partial_{x_0}u-\Delta u = f \ , \ x_0\geq 0 \quad , \ u(x_0,x) = \varphi(x) \ ,$$

where f,φ are given C_0^∞-functions. Setting u and f zero in $t<0$ to obtain functions v and g we get

$$(4.29) \qquad Hv = g + \delta(t)\otimes\varphi(x) = h \; .$$

Thus $v=e*h$, or,

$$(4.30) \quad u(t,x)=\kappa_{n+1}\int_{\tau\le t} d\!\!\!{}^-t\, d\!\!\!{}^-y\, e(t-\tau,x-y)f(y) + \kappa_{n+1}\int d\!\!\!{}^-y\, e(t,x-y)\varphi(y).$$

III) Now we look at (4.4), or, the wave equation

$$(4.31) \qquad u = (\partial_t^2 - \Delta)u = f \; .$$

We apply the Fourier-Laplace method as for (II): The function

$$(4.32) \qquad \frac{1}{P} = \frac{1}{|x|^2-t^2} = \frac{1}{2|x|}\left\{\frac{1}{t+|x|} - \frac{1}{t-|x|}\right\}$$

has inverse Laplace transform (in t) given by

$$(4.33) \qquad \frac{\sqrt{2\pi}}{2i\,|x|}\left(e^{-it|x|}-e^{it|x|}\right) = \sqrt{2\pi}\;\frac{\sin(|x|t)}{|x|} \;, \; t\ge 0 \;,$$

(and zero for t<0). Looking for F_x^{-1} of the function (4.33) we cannot apply a Fourier integral, since the function (4.33) is not L^1. First set n=2. Writing $F_x^{-1}w=w^{\Delta}$ for a moment, we have

$$(4.34) \quad \langle e,\varphi\rangle = \sqrt{2\pi}\,\langle \frac{\cos rt}{r^2}, \varphi_{|t}{}^{\Delta}\rangle = \langle f,\varphi_{|t}{}^{\Delta}\rangle \;,\; \varphi\in \mathcal{D}(\mathbb{R}^n) \;,\; r=|x| \;,$$

using that $(\cos |x|t)_{|t} = -|x| \sin |x|t$. Now the inverse Fourier integral of $f=\frac{\cos rt}{r^2}$ may be calculated as improper Riemann integral $\lim_{A\to\infty}\int_{|x|\le A}$, although still that function is not L^1 : Using (4.7), (4.8) - where $\nu=\frac{1}{2}$, $J_\nu(z)=\sqrt{2}\,\frac{\sin z}{\sqrt{\pi z}}$, for n=3 - we get

$$f^{\Delta}=\omega(|x|) \;,\; \omega(r)=(\sqrt{2\pi})\left(\frac{\sqrt{2}}{\sqrt{\pi}}\right)\int_0^\infty \sqrt{r\rho}\left(\frac{\rho}{r}\right)\frac{\sin r\rho}{\sqrt{r\rho}}\,\frac{\cos \rho t}{\rho^2}\,d\rho \;.\; \text{Or,}$$

$$\omega(r) = \frac{2}{r}\int_0^\infty \sin r\rho \cos \rho t\,\frac{d\rho}{\rho} = \frac{1}{r}\int_0^\infty (\sin \rho(t-r) - \sin \rho(t+r))\,\frac{d\rho}{\rho}$$

$$= \frac{1}{r}\left\{\frac{\pi}{2}\,\text{sgn}(t-r) - \frac{\pi}{2}\right\}$$. Conclusion:

$$(4.35) \quad e(t,x) = \frac{\pi}{|x|}\,\partial_t H(t-|x|) \;,\; \text{as } t>0 \;,\; =0 \;,\; \text{as } t<0 \;,\; n=3 \;,$$

with the distribution derivative ∂_t , and the Heaviside function $H(t)=1$, $t\ge 0$, $H(t)=0$, $t<0$. We are tempted to write $\partial_t H(t-|x|)$ as $\delta(t-|x|)$, but then must remember the proper interpretation.

Converting the Cauchy problem for the wave equation,

(4.36) $$u_{|tt} - \Delta u = \varphi \ , \ t>0 \ , \ u=\varphi \ , \ u_{|t} = \psi \text{ at } t=0 \ ,$$

into a full space problem we get

(4.37) $$v_{|tt} - \Delta v = \delta'(t)\otimes\varphi(x) + \delta(t)\otimes\psi(x) + g = h \ ,$$

where v and g are u and f , extended zero for t<0 . Evaluating e*h we get the solution of (4.36) in the form

(4.38) $$u(t,x) = \frac{1}{4\pi}\int_{|x-y|\le t} \frac{dy}{|x-y|} f(t-|x-y|,y) + \frac{1}{4\pi}\,\partial_t\Big(t\int_{|z|=1}\varphi(x+tz)dS_z\Big) + \frac{1}{4\pi}\int_{|z|=1}\psi(x+tz)dS_z \ .$$

(4.38) is known as the <u>Kirchhoff formula</u>.

For general $n\ge 2$ let $j=[\frac{n-1}{2}]$. As in (4.35) write $\langle e,\varphi\rangle = \sqrt{2\pi}\langle f_j, ((\partial_t)^j\varphi)^\wedge\rangle$, where $f_j = \pm\frac{\sin \rho t}{\rho^{1+j}}$ or $f_j = \pm\frac{\cos \rho t}{\rho^{1+j}}$ (with $\rho=|x|$) for even (odd) j. The inverse Fourier integral of f_j exist at least as $\lim_{A\to\infty}\int_{|x|\le A}$; we apply lemma 4.1 as above. For odd n we get

(4.39) $$f_j^\wedge(t,x) = \omega_j(t,|x|) \ , \ \omega_j(t,r) = \pm r^{-\nu}\int_0^\infty J_\nu(r\rho)\begin{Bmatrix}\sin \rho t\\ \cos \rho t\end{Bmatrix}\frac{d\rho}{\sqrt{\rho}} \ ,$$

for even n this will be of the form

(4.40) $$f_j^\wedge(t,x) = \omega_j(t,|x|) \ , \ \omega_j(t,r) = \pm r^{-\nu}\int_0^\infty J_\nu(r\rho)\begin{Bmatrix}\sin \rho t\\ \cos \rho t\end{Bmatrix} d\rho \ ,$$

with a convergent improper Riemann integral. More precisely, in (4.39) and (4.40) we have "$(-1)^l\cos \rho t$ " as j is odd, and "$(-1)^l\sin \rho t$ " as j is even, with $l=[\frac{j}{2}]$. Our distribution e then is given by (4.41) below, with a distribution derivative ∂_t^j :

(4.41) $$e(t,x) = (-1)^j\sqrt{2\pi}\,\partial_t^j\omega_j(t,|x|) \ ,$$

Now we must consult different formulas on Bessel function integrals where as a general reference we quote [MOS] (or [MO]). Case (a): n even, j odd. Then $j=[\frac{n-1}{2}]=\frac{n}{2}-1=\nu$, $l=\frac{j-1}{2}$, $n\equiv 0 \pmod 4$). We use a Weber-Schafheitlin integral ([MOS] p.99, last formula). Since $\nu=j$ is odd, we have for t>r the formula

(4.42) $$I = \int_0^\infty J_\nu(r\rho)\cos \rho t \, d\rho = (r^2-t^2)^{-1/2}\cos(\nu\sin^{-1}(\tfrac{t}{r})) = \tfrac{1}{r}P_{j-1}(\tfrac{t}{r}),$$

with a certain polynomial $P_{j-1}(t)$ of degree j-1. Note that the integral enters (4.41) under the derivative ∂_t^j , so that e=0 for

$t>|x|$. For $r<t$ we get $I=\frac{(-1)^l}{\sqrt{t^2-r^2}}r^{-j}(t+\sqrt{t^2-r^2})^{-j}$, by the same formula. Summarizing: For $n\equiv 0 \pmod 4$ we get

$$(4.43)\qquad e(t,x)=-\sqrt{2\pi}\,\partial_t^{2l+1}(Z_l(t,r)) , \quad \text{with}$$

$$Z_l(t,r)=r^{-2l}Q_{2l}(\tfrac{t}{r}),\ t>r\ ,\ =\frac{1}{\sqrt{t^2-r^2}\,(t+\sqrt{t^2-r^2})^{2l+1}}\ ,\ t<r\ ,$$

where $\quad Q_{2l}(\tau)=\sum_{m=0}^{l}(-1)^{l-m}\binom{2l+1}{2m}\tau^{2m}(1-\tau^2)^{l-m}$.

Case (b): n even, j even. We still have $j=\nu=\frac{n}{2}-1$, $l=\frac{j}{2}$, $n\equiv 2 \pmod 4$. Use formula 3 on p. 51 of [MO]. The result is as in (4.43) with

$$Z_l(t,r)=r^{-n/2}Q_{2l-1}(\tfrac{t}{r})\ ,\ t>r\ ,\ =\frac{1}{\sqrt{t^2-r^2}\,(t+\sqrt{t^2-r^2})^{2l}}\ ,\ t<r\ ,$$

$$(4.43)_b\qquad Q_{2l-1}(\tau)=\sum_{m=0}^{l-1}(-1)^{l-m}\binom{2l}{2m+1}\tau^{2m+1}(1-\tau^2)^{l-m-1}\ ,\ l=\frac{n-2}{4}>0,$$

while for $n=2$ (i.e., $l=0$) we set $Q_{2l-1}\equiv 0$.

Case (c): n, j odd, $[\frac{n}{2}-1]=\frac{n-1}{2}=j$, $n=2j+1$, $l=\frac{j-1}{2}$, $j=2l+1$, $n\equiv 3 \pmod 4$.

Let $I=\int_0^\infty J_\nu(r\rho)\cos\rho t\frac{d\rho}{\sqrt{\rho}}=\sqrt{\frac{t\pi}{2}}\int_0^\infty J_{j-1/2}(r\rho)J_{-1/2}(t\rho)d\rho$. Use formula 2, p.50 of [MO] with $(a,b)=(r,t)$, $(\nu,n)=(l+\frac{1}{2},l)$. Get $\nu-n=2$, $I=0$, $t>r$,

$$(4.44)\qquad I=\sqrt{\frac{\pi}{2r}}\,\frac{1\cdot 3\cdot 5\cdots(2l-1)}{2\cdot 4\cdot 6\cdots 2l}\,{}_2F_1(l+\tfrac{1}{2},-l;\tfrac{1}{2};(\tfrac{t}{r})^2)=\sqrt{\frac{\pi}{2r}}\,Q_{2l}(\tfrac{t}{r}),\ t<r$$

where the hypergeometric function (hence $Q_{2l}(z)$)

$$(4.45)\qquad {}_2F_1(l+\tfrac{1}{2},-l;\tfrac{1}{2};z)=1-l(2l+1)z+\frac{l(l-1)}{1\cdot 3}(2l+1)(2l+3)\frac{z^2}{2!}+\ldots$$

is a polynomial in z of degree $l=\frac{n-3}{4}$. Combining (4.39) and (4.44),

$$(4.46)\qquad e(t,x)=(-1)^{l+1}\pi r^{-2l-1}\partial_t^{2l+1}\{Q_{2l}(\tfrac{t}{r})H(r-t)\}\ ,\ l=\frac{n-3}{4}\ ,\ r=|x|.$$

Case (d): n odd, j even, $j=\frac{n-1}{2}$, $n=2j+1$, $l=\frac{j}{2}$, $n\equiv 1 \pmod 4$. Now we need $I=\int_0^\infty J_\nu(r\rho)\sin\rho t\frac{dr}{\sqrt{\rho}}=\sqrt{\frac{\pi t}{2}}\int_0^\infty J_{j-1/2}(r\rho)J_{1/2}(t\rho)d\rho$. Again use the formula of case (c), now with $(a,b)=(r,t)$, $(n,\nu)=(l-1,l+\frac{1}{2})$. Get

$$(4.47)\qquad I=\sqrt{\frac{\pi}{2r}}(\tfrac{t}{r})\frac{3\cdot 5\cdots(2l+1)}{2\cdot 4\cdots(2l-2)}\,{}_2F_1(l+\tfrac{1}{2},l-1;\tfrac{3}{2};\tfrac{t^2}{r^2})=\sqrt{\frac{\pi}{2r}}Q_{2l-1}(\tfrac{t}{r}),\ t<r,$$

and $I=0$ for $t>r$, with a polynomial $Q_{2l-1}(z)$ of degree $2l-1$. Hence

$$(4.48)\qquad e(t,x)=(-1)^l\pi r^{-2l}\partial_t^{2l}\{Q_{2l-1}(\tfrac{t}{r})H(r-t)\}\ ,\ l=\frac{n-1}{4}\ ,\ r=|x|\ .$$

IV) Finally, look at the Klein-Gordon equation (4.5), n=3 only

(4.49) $$(\Box + m^2)u = u_{tt} - \Delta u + m^2 u = f \quad .$$

Equation (4.32) now assumes the form

(4.50) $$\frac{1}{P} = \frac{1}{\lambda^2 - t^2} = \frac{1}{2\lambda}\left\{\frac{1}{t+\lambda} - \frac{1}{t-\lambda}\right\} \; , \; \lambda(x) = \sqrt{m^2 + |x|^2} \quad .$$

Accordingly,

(4.51) $$F_x e = \sqrt{2\pi}\, \frac{\sin t\sqrt{x^2+m^2}}{\sqrt{x^2+m^2}} \quad .$$

For $n=3$, $\nu=\frac{1}{2}$, $j=\frac{n-1}{2}=1$ use (4.7),(4.8) on $\chi(|x|)$ of (4.51), for

(4.52) $$\sqrt{r}\,\omega(r) = \pi\sqrt{t}\int_0^\infty J_{1/2}(r\rho)J_{1/2}(t\sqrt{m^2+\rho^2})\rho^{3/2}/\sqrt{m^2+\rho^2}\; d\rho \quad .$$

This integral diverges, just as for m=0. Applying a formula of Sonine (1880) and Gegenbauer (1884) (first on p.104 of [MOS]), for $(a,b)=(t,r)$, $(\mu,\nu)=(\frac{1}{2},\frac{1}{2}+\varepsilon)$, $m=z$, $\varepsilon>0$, we get

(4.53) $$e(t,x) = \pi\sqrt{\tfrac{t}{r}}\lim_{\varepsilon\to 0}\int_0^\infty J_{1/2}(r\rho)J_{1/2+\varepsilon}(t\sqrt{m^2+\rho^2})\frac{\rho^{1/2}\,d\rho}{\sqrt{\rho^2+m^2}^{\,1/2}}$$
$$= \lim_{\varepsilon\to 0,\varepsilon>0} H(r-t)\sqrt{\tfrac{r}{t}}\left(\frac{m}{\sqrt{t^2-r^2}}\right)^{1-\varepsilon} J_{\varepsilon-1}(m\sqrt{t^2-r^2}) \; ,$$

with existing weak limit in $\mathcal{D}'(\mathbb{R}^3)$. Or,

(4.54) $$e(t,x) = \pi m \lim_{\varepsilon\to 0,\varepsilon>0}\{H(t-|x|)J_{\varepsilon-1}(m\sqrt{t^2-r^2})/\sqrt{t^2-r^2}^{\,\varepsilon-1}\} .$$

Recall, for $\nu=\varepsilon-1$ we have

(4.55) $$J_\nu(z) = \frac{(z/2)^{\varepsilon-1}}{\Gamma(\varepsilon)} - \sum_{m=0}^\infty \frac{(z/2)^{1+2m}}{\Gamma(1+m+\varepsilon)(m+1)!} \; , \; \varepsilon>0 \quad .$$

The second term, at right, goes to $-J_1(z)$, as $\varepsilon\to 0$, uniformly on $K \subset\subset \mathbb{R}^n$, and weakly in $\mathcal{D}'$. For a $\varphi\in\mathcal{D}$ write $\langle e,\varphi\rangle = T_1 + T_2$, by the decomposition (4.55). Then $T_2 \to -\pi m\langle J_1(m\sqrt{t^2-r^2})/\sqrt{t^2-r^2},\varphi\rangle$, $T_1 \to$ $2\pi \lim_{\varepsilon\to 0}\int_{|x|\le t}\frac{dxdt}{\Gamma(\varepsilon)}\frac{\varphi(t,x)}{(t^2-r^2)^{1-\varepsilon}}$, where $\Gamma(\varepsilon)=\frac{1}{\varepsilon}(1+c_1\varepsilon+\ldots)$. Hence

(4.56) $$T_1 = 2\pi \lim_{\varepsilon\to 0}\left\{\int_{|x|\le t} dxdt\frac{\varphi(t,x)}{t+r}\frac{\varepsilon}{(t-r)^{1-\varepsilon}} = 2\pi\lim\int\frac{dxdt}{t+r}\{(t-r)^\varepsilon\}_t\right.$$
$$= -2\pi\lim_{\varepsilon\to 0}\int_{|x|\le t} dxdt\,\partial_t\left\{\frac{\varphi(t,x)}{r+t}\right\}(t-r)^\varepsilon = -2\pi\int H(t-r)\left\{\frac{\varphi}{t-r}\right\}_t dtdx \quad .$$

In other words, $T_1 = \pi\langle\frac{1}{r}\partial_t H(t-r),\varphi\rangle$. Result:

(4.57) $$e(t,x) = \frac{\pi}{r}\partial_t H(t-|x|) - m\pi\frac{H(t-|x|)}{\sqrt{t^2-|x|^2}}\,J_1(m\sqrt{t^2-x^2}) \quad .$$

5. Abstract solutions and hypo-ellipticity.

In sections 1,2,3,4 we have deployed the Fourier-Laplace method of solving certain problems involving constant coefficient PDE's under sufficiently simple initial boundary conditions. The focus was on the Fourier transform. In a distribution setting it reduces the problem of solving $P(D)v=g$ to the division problem $P(x)v^{\wedge}=g^{\wedge}$, providing an 'inverse' for $P(D)$ in the form $(\frac{1}{P})^{\vee}*$, speaking roughly. In later chapters, this type of inverse construction will be extended to variable coefficients - and more general boundaries, but the central role of the Fourier transform is maintained, and the inverses obtained have similar features.

On the other hand, the theory of constructing an inverse of an abstract operator between two linear spaces is well studied, in present times. It will be ever present in the background of our discussions. The role of certain Hilbert spaces- L^2-Sobolev spaces - in theory of ψdo's will have to be studied. Theory of (bounded and unbounded) Fredholm operators in Hilbert spaces will be crucial for elliptic ψdo's, as well as compactness properties.

Often existence of a generalized solution to a PDE (a boundary problem) can be derived by purely abstract arguments. But then theory of ψdo's may have to be used to derive properties of such solution -even differentiability, and that we have a solution in the classical sense. Let us discuss an example.

For an open domain $\Omega \subset \mathbb{R}^n$, consider the Hilbert space $\mathbf{H}=L^2(\Omega)$ with norm and inner product

$$(5.1) \qquad (u,v) = \int_{\Omega}\bar{u}v\,dx \ , \ \|u\|^2 = (u,u) \ , \ u,v \in \mathbf{H} \ .$$

The Laplace operator $\Delta=\sum_1^n \partial_{x_j}^{\ 2}$ defines a linear map $C_0^{\infty}(\Omega)\to C_0^{\infty}(\Omega)\subset \mathbf{H}$ which may be interpreted as an <u>unbounded operator</u> H_0 of $\mathbf{H}$ with domain dom $H_0=C_0^{\infty}(\Omega)$, using that $C_0^{\infty}(\Omega)$ is a dense subspace of $\mathbf{H}$.

In fact, H_0 is hermitian and negative, i.e., we have

$$(5.2) \qquad (u,H_0u) \text{ real and } \le 0 \ , \text{ for every } u \in \text{dom } H_0 \ .$$

This implies that H_0 has self-adjoint extensions. A distinguished such extension H , called the Friedrichs extension, may be constructed using a rather simple form-closing principle (cf. $[Ka_1]$,[RN],$[Wm_1]$,$[Yo_1]$, or $[C_2]$,I, thm.2.7).

To express the above in different terms: H is also a linear

map dom H → $\mathcal{H}$ with $\mathcal{H} \supset$ dom H $\supset$ dom $H_0 = C_0^\infty(\Omega)$, and $H_0 = H|(\text{dom } H_0)$. We have (5.2) for H instead of H_0, but H is maximal with respect to this property: No proper extension of H still satisfies (5.2).

This maximality has important consequences: Any operator H satisfying (5.2) must be hermitian. A maximal such operator must be self-adjoint- it possesses a spectral measure, as we will not discuss here. Moreover, it follows that the unbounded operator K= 1-H is invertible in the following sense: There exists a continuous operator $R \in L(\mathcal{H})$, called $K^{-1} = (1-H)^{-1}$, such that $R:\mathcal{H} \to$ dom H $\subset \mathcal{H}$ is a bi-jection inverting the linear map K:dom H → $\mathcal{H}$.

Accordingly, due to existence of the inverse R, it follows that the linear equation

$$(5.3) \qquad (1-H)u = f \quad , \; f \in \mathcal{H} \; ,$$

admits one and only one solution $u \in$ dom H $\subset \mathcal{H}$, for every $f \in \mathcal{H}$.

After this abstract discussion we now ask: Does u of (5.3), a function of $\mathcal{H}$ whose existence is shown by a chain of abstract arguments, solve the differential equation $u-\Delta u=f$? Also, does u satisfy any kind of boundary condition?

The corresponding question is meaningful not only for Δ , but for any smooth differential expression a(x,D) defined in Ω such that the corresponding operator $L_0: C_0^\infty \to C_0^\infty \subset \mathcal{H}$ satisfies (5.2).

For general a(x,D), with the Friedrichs extension L of L_0 , the unique $u \in$ dom $L \subset \mathcal{H}$ solving $(1-L)u=f$, for given $f \in \mathcal{H} \subset L^1_{\text{loc}}(\Omega) \subset D'(\Omega)$ is a distribution solution of $u-a(x,D)u = f$, since

$\langle u,\varphi-a(x,D)^{-}\varphi\rangle = (\bar{u},\varphi)-((Lu)^{-},\varphi)=((u-Lu)^{-},\varphi)=\langle f,\varphi\rangle$, $\varphi \in D(\Omega)$, using that $(Lu,\varphi)=(u,a(x,D)\varphi)$, $\varphi \in D(\Omega)=C_0^\infty(\Omega)$.

Question: (a) If, in addition, we have $f \in C^\infty(\Omega)$, will u be smooth - and a classical solution of $u-a(x,D)u=f$? (b) If, in addition, Ω has a smooth boundary Γ, and $f \in C^\infty(\Omega \cup \Gamma)$, will also u be $C^\infty(\Omega \cup \Gamma)$, and will it satisfy boundary conditions?

For $a(x,D)=\Delta$ both questions have a positive answer. In particular, u must satisfy the Dirichlet condition u=0 on Γ. The same remains true if a(x,D) is elliptic and of second order (cf. ch.5). Generally, if a(x,D) is <u>hypoelliptic</u> the answer to (a) will be positive. An answer to (b) is given in V,4. We speak of a concept called boundary hypoellipticity, in that respect.

<u>6. Exponentiating a first order partial differential operator</u>.

In this section we discuss the formal linear operator e^G, with a first order differential expression (a 'folpde')

(6.1) $$G = \sum_{j=1}^{n} b^j \partial_j + p \ , \ \partial_j = \partial/\partial_{x_j} \ ,$$

with real C^∞-coefficients $b_j(x)$, $j=1,\dots,n$, and complex-valued $p(x) \in C^\infty$, all defined in some domain $\Omega \subset \mathbb{R}^n$. By definition e^G will be the solution operator of the initial-value problem

(6.2) $$\partial_t u = \partial u/\partial t = Gu \ , \ x \in \Omega \ , \ t \geq 0 \ , \ u = u_0 \ , \ t = 0 \ ,$$

i.e., $u(x,t)=e^{Gt}u_0(x)$, $u_0 \in C^\infty(\Omega)$, is the unique solution of (6.2).

The single first order PDE $\partial_t u=Gu$ becomes an ODE along certain curves, called characteristic curves, defined as solutions of a system of ODE's, the system of characteristic equations,

(6.3) $$t^{\cdot} = 1 \ , \ x_j^{\cdot} = -b^j(x) \ , \ j = 1,\dots,n \ , \ "^{\cdot}" = d/ds \ ,$$

for $(t(s),x(s))=(t(s),x_1(s),\dots,x_n(s))$ with real-valued functions t,x_j of a parameter s. The first equation (6.3) gives t-s=c=const. Thus set t=s, reducing (6.3) to the autonomous system (for $x_j(t)$)

(6.4) $$x^{j\cdot} = -b^j(x^1,\dots,x^n) \ , \ j = 1,\dots,n \ , \ "^{\cdot}" = d/dt \ .$$

Since the functions $b_j(x)$ are real-valued and C^∞ it is clear that the orbits (i.e. the curves in Ω, given by the parametric representation $x=x(t)$, for $x(t)$ solving (6.4)), provide a family of non-intersecting smooth curves covering the entire domain Ω . A solution $x(t)$ with $x(0)=x^0$ will exist in a maximal interval $0 \leq t < t_0$, where either $t_0=\infty$ or $x(t) \in \Omega \backslash K$ for $t \in [t_K,t_0)$ for every compact set $K \subset \Omega$ with some $t_K<t_0$. In particular $t_0=\infty$ whenever for every $t_1>0$ a compact set $K_{t_1} \subset \Omega$ can be found such that the 'apriori estimate' $x(t) \in K_{t_1}$, as $0 \leq t \leq t_1$, can be verified from the fact that $x(t)$ solves (6.4) and $x(0)=x^0$. An orbit may degenete to a point, or, may be closed -i.e., $x(t)$ may be periodic.

The relation between (6.2) and (6.3) is this: A function $u \in C^1$, defined for (t,x) near (t^0,x^0), $x^0 \in \Omega$, solves the PDE (6.2) if and only if along every solution curve $(t,x(t))$ of (6.3) the composite function $u(t,x(t))=\varphi(t)$ solves the ODE

(6.5) $$\frac{d\varphi}{dt} = \gamma(t)\varphi \ , \quad \gamma(t)=p(x(t)) \ .$$

Indeed, $\frac{dx}{dt}j(t) = -b_j(x(t))$ implies that

(6.6) $$\frac{d}{dt}(u(t,x(t)))=\{u_{|t}+\Sigma x_j^{\cdot} u_{|x_j})\}(t,x(t)=(u_{|t}-\Sigma b^j u_{|x_j})(t,x(t)) \ .$$

Hence if u solves (6.2), we will get (6.5) along every curve x(t)

for which the composition $u(t,x(t))$ is defined. Vice versa, if $u(t,x)$ is a function defined and C^1 near (t^0,x^0) such that $\varphi(t)$ satisfies (6.5) for every curve $x(t)$ solving (6.4) in some neighbourhood of x^0. Then $u(t,x)$ solves (6.2) near (t^0,x^0).

Let $u(0,x)=u_0(x)$ be given for all $x\in\Omega$. Then a unique solution $u(t,x)$ of (6.2) may be constructed as follows: At each $x^0\in\Omega$ construct the solution $x(t)$ of (6.4) through x^0, defined for a maximal t-interval (t_-,t_+), $t_\pm=t_\pm(x^0)$. Define $\gamma(t)=p(x(t))$, and $\varphi(t)= u_0(x^0)\exp\{\int_0^t \gamma(\tau)d\tau\}$, $t\in(t_-,t_+)$. Then set $u(t,x(t))=\varphi(t)$ along the curve $x=x(t)$, $t\in(t_-,t_+)$, and do this for all such curves.

Assume that we have $t_\pm=\pm\infty$ for every $x^0\in\Omega$. Then the map $x^0\to x(t^0,x^0)$ defined by following the orbit through x^0 from $t=0$ to $t=t^0$ defines a diffeomorphism $\nu_{t^0}:\Omega\to\Omega$. Indeed, this map $\Omega\to\Omega$ is 1-1 by construction, and it may be inverted by solving the reverse initial-value problem (follow $x(t)$ through x^0 for a t-interval of length t^0 in negative t-direction). The map ν_t is C^∞ by standard results of dependence of solutions of an ODE on initial values.

We thus obtain a 1-parameter family $\mathcal{F}=\{\nu_t : t\in\mathbb{R}\}$ of diffeomorphisms $\nu_t : \Omega\to\Omega$, having the group property

$$\nu_t\circ n_\tau = n_{t+\tau}\ ,\ t,\ \tau\in\mathbb{R}\ , \tag{6.7}$$

since following the curves for $t+\tau$ units has the same effect than following first for t then for τ units. Such a group of diffeomorphisms is commonly called a <u>flow</u>. The flow defined by (6.4) is called the <u>characteristic</u> <u>flow</u> of the PDE (6.2) .

In this terminology we have proven:

<u>Theorem 6.1</u>. Assume that all solutions of the characteristic system (6.4) extend for $t\in(-\infty,\infty)$. Then the problem (6.2) has a unique solution $u(t,x)\in C^\infty(\mathbb{R}\times\Omega)$, for $u_0\in C^\infty(\Omega)$, where u is given by

$$u(t,x) = (u_0\circ\nu_{-t})(x)\cdot\exp\{\int_0^\tau (p\circ\nu_{-\tau})(x)d\tau\}\quad . \tag{6.8}$$

Indeed our above description of $u(t,x)$ translates to (6.8). In particular the diffeomorphism ν_t is inverted by ν_{-t}, hence u is defined for all $x\in\Omega$ and all t , and (6.8) follows.

Note that (6.8) establishes an abelian group $\{e^{tG} : t\in\mathbb{R}\}$ of linear maps $e^{tG}:C^\infty(\Omega)\to C^\infty(\Omega)$, defined by

$$e^{tG}u_0(x) = u(t,x)\ ,\ u_0\in C^\infty(\Omega)\ ,\ t\in\mathbb{R}\ . \tag{6.9}$$

We also have

$$(6.10)\qquad e^{tG} : C_0^\infty(\Omega) \to C_0^\infty(\Omega) \ , \ t \in \mathbb{R} .$$

The relation between equations (6.2) and (6.4) remains invariant under a coordinate transform of Ω, i.e., the characteristic equations (6.4) go into those of the transformed equation (6.2). Accordingly theorem 6.1 also holds for a folpde G defined on a differentiable manifold Ω with local representation (6.2) and characteristic equations locally given by (6.4) , assuming that our condition remains satisfied - that all solutions of $x^\cdot=-b(x)$, regardless of the choice of initial value $x^0 \in \Omega$, extend into an infinite time interval, passing of $x(t)$ between charts of different coordinates being permitted.

Now we want to address the problem of finding a more concrete condition for the assumption of thm.6.1 that all solutions of the characteristic system (6.3) extend indefinitely. Assume that Ω carries a Riemmannian metric $ds^2=\Sigma h^{jk}dx_j dx_k$, under which it is complete. Let $d(x,x^0)$ denote the distance from x to x^0 on Ω .

Proposition 6.2. Assume that the principal part tensor (b^j) of the expression G of (6.1) satisfies the estimate

$$(6.11)\qquad |b(x)| = \{(\Sigma h_{jk}\bar{b}^j b^k)(x)\}^{1/2} = O(1+d(x,x_0)) \ , \ x \in \Omega ,$$

with some fixed $x_0 \in \Omega$. Then all $x(t)$ of (6.4) extend to $-\infty<t<\infty$.

Proof. By well known results on continuation of solutions of ODE's it suffices to get an apriori estimate. Let $x(t)$ solving (6.4) be defined for $t \in [t_0,t_1]$, $x(t_0)=x^0$. Let $\gamma(t)=1+d(x^0,x(t))$. We know $\gamma(t)$ is Lipschitz continuous in t ,and that

$$(6.12)\qquad |d\gamma/dt| \le \{\Sigma h_{jk}x_j^\cdot x_k^\cdot\}^{1/2} = |b(x(t))| = O(\gamma(t)) \ ,$$

noting that (6.11) trivially holds for each fixed $x_0 \in \Omega$,if it is valid for only one x_0. Now (6.12) implies $\log \gamma(t) = O(t-t_0)$, or

$$(6.13)\qquad x(t) \in \{d(x,x_0) \le e^{c(t-t_0)} -1 \} .$$

Since Ω is complete, the sphere (6.13) is compact for $t=t_1$, and we have shown that $x(t)$ stays in it. This proves the proposition.

Now we look at G and e^G in a Hilbert space $H=L^2(\Omega,d\mu)$, with

$$(6.14)\qquad (u,v) = \int_\Omega \bar{u}v d\mu \quad , \ \|u\|=\sqrt{(u,u)}$$

as inner product and norm, $d\mu$ denoting a positive C^∞-measure on Ω, locally of the form $d\mu=\kappa dx$, $0<\kappa \in C^\infty$. We still assume that the characteristic flow $\mathcal{F}$ is defined: All solutions of (6.4) extend indefinitely. Clearly the restriction $G_0=G|C_0^\infty(\Omega)$ may be interpreted as unbounded operator G_0 of $\mathcal{H}$ with domain dom $G_0 = C_0^\infty(\Omega)$.

Assume G skew-selfadjoint: with real-valued b_j, p we have

$$(6.15) \qquad G = \Sigma b^j\partial_j + 1/2\ \kappa^{-1}(\kappa b^j)_{|x^j} + ip \quad ,$$

then G_0 with domain $C_0^\infty(\Omega)$ clearly is skew-hermitian: We have

$$(6.16) \qquad (u,G_0u) + (G_0u,u) = 0 \quad , u \in \text{dom } G_0 = C_0^\infty(\Omega) .$$

If u(t,x) solves (6.2) , for some $u_0 \in C_0^\infty(\Omega)$, (6.10) implies that $u(t,.) \in \mathcal{D}(\Omega)$, for each fixed t. The $\gamma(t)=\|u(t,.)\|^2 \in C^\infty(\mathbb{R})$, and

$$(6.17) \quad d\gamma/dt = (u,u_{|t}) + (u_{|t},u) = (u,G_0u) + (G_0u,u) = 0 \ , \ t\in \mathbb{R} .$$

Accordingly,

$$(6.18) \qquad \|e^{tG}u_0\| = \|u(t,.)\| = \|u_0\| = \text{const., for all } u_0 \in C_0^\infty(\Omega).$$

This shows that the operator e^{tG}: $C_0^\infty(\Omega) \to C_0^\infty(\Omega)$ defines an isometry in the norm of $\mathcal{H}$. Clearly this is an invertible isometry, using the group property of e^{tG}. Let us first assume that p of (6.15) vanishes. Then $v(x)=e^{tG}u_0$, for some fixed t, is of the form

$$(6.19) \qquad v(x) = (u_0\circ\chi)(x)\omega(x) \quad , \ \chi(x) = \nu_{-t}(x) \ ,$$

with a real-valued positive C^∞-function $\omega(x)$ independent of u_0, by (6.8), with $p=(2\kappa)^{-1}(\kappa b^j)_{|x^j}$. From (6.18) and (6.19) we conclude

$$(6.20) \quad \|v\|^2 = \int_\Omega |(u_0\omega)(\chi(x))|^2 d\mu = \int_\Omega |u_0(x)|^2 d\mu = \|u_0\|^2, u_0\in C_0^\infty.$$

With new integration variables $y=\chi(x)$ in the first integral we get

$$(6.21) \quad \int_\Omega |u_0(x)|^2(\omega^2(x)(d\mu\circ\nu_t)/d\mu - 1) = 0 \ , \text{ for all } u_0\in C_0^\infty(\Omega).$$

But (6.21) implies that

$$(6.22) \qquad \omega(\chi(x)) = \{(d\mu\circ\nu_{-t})/d\mu\}^{1/2} \quad .$$

For p general, $e^{tG}u_0(x)$ will carry an additional exponential factor. All other functions remain the same. We have proven

<u>Proposition 6.3</u>. For a skew-selfadjoint G (6.15) assumes the form

$$(6.23) \quad e^{tG}u_0(x) = ((u_0\circ\nu_{-t})\{\partial\mu\circ\nu_{-t}/\partial\mu\}^{1/2})(x)\exp\{i\int_0^t p\circ\nu_{-\tau}(x)d\tau\},$$

defining an invertible isometry $C_0^\infty \to C_0^\infty$ the sense of the norm of $\mathcal{H}$.

Then $e^{Gt}:C_0^\infty \to C_0^\infty$ admits a continuous extension to H, a unitary operator. We get a group $\{U(t)=(e^{G_0 t})^{closure} : t\in \mathbb{R}\}$ of unitary operators, strongly continuous in t. Its infinitesimal generator

$$(6.24)\quad G_1=dU/dt(0)\ ,\ \mathrm{dom}\ G_1=\{u\in H:\ \lim_{\varepsilon\to 0}\frac{U(\varepsilon)u-u}{\varepsilon}=G_1 u \text{ exists in } H\}$$

is a skew-selfadjoint realization of the folpde G , extending G_0. This realization is unique, since dom $e^{G_0 t}=C_0^\infty$ is dense in H. G_1 is the closure of G_0, and iG_0 is essentially self-adjoint.

Finally consider a dissipative expression G (i.e. $G_0+G_0^*\le 0$). G must have the form (6.15) with real-valued b_j but complex p satisfying Im $p(x)\ge 0$, $x\in\Omega$. Hence (6.16) now assumes the form

$$(6.25)\qquad (u,G_0u)+(G_0u,u) = -\int_\Omega d\mu(\mathrm{Im}\ p)|u|^2 \le 0\ ,\ u\in C_0^\infty(\Omega)\ .$$

Under this assumption the map e^{tG_0} , still of the form (6.23), is a contraction under the norm of H, for $t\ge 0$, i.e., $\|e^{Gt}\|\le 1$, $t\ge 0$. This follows by repeating the above argument.

7. Solving a nonlinear first order partial differential equation.

In this section we are going to consider the Cauchy problem for a single first order partial differential equation

$$(7.1)\qquad F(x,u,p) = 0\ ,\ p = u_{|x} = (\partial_{x_1}u,\ \dots\ ,\partial_{x_n}u)\ ,$$

where F is a given real-valued C^∞-function of the 2n+1 real variables $x=(x_1,\dots,x_n)$, u , $p=(p_1,\dots,p_n)$. For simplicity assume F defined in $\Omega\times\mathbb{R}\times\mathbb{R}^n$, with some domain $\Omega\subset\mathbb{R}^n$. Our main application will be the characteristic equation (8.3), in sec.8, below. There $F=a_N(x,p)$ is independent of u and a homogeneous polynomial in p.

The Cauchy problem seeks a solution of (7.1) satisfying an initial condition at some n-1-dimensional submanifold $\Gamma\subset\Omega$. Let Γ be given by $x=x(s)$, $s=(s_1,\dots,s_{n-1})\in\Sigma\subset\mathbb{R}^{n-1}$, with $\partial x/\partial s=((\partial x_j/\partial s_1))$ of rank n-1. Require as <u>initial condition</u> that

$$(7.1')\qquad u(x(s)) = \omega(s) \text{ is given for } s\in\Sigma\ .$$

Then differentiation of (7.1') for s_j determines all tangential derivatives of u at Γ, while (7.1) gives an implicit relation for normal derivative of u. It is natural to assume that the n equations resulting for $p_j=u_{|x_j}$ locally admit precisely one solution, so that $u_{|x}$ is fully determined along the hypersurface Σ.

Thus it is convenient to assume not only $u=\omega(s)$ but also $p=u_{|x}$ given for $s\in\Sigma$. Of course certain conditions on $\omega(s),p(s)$ must express the fact that $p=u_{|x}$. In other words, we assume given

$$(7.2)\qquad x = x(s)\ ,\ p = p(s)\ ,\ u = \omega(s)\quad ,\ s\ \in\Sigma\ ,$$

with $\partial x/\partial s$ of maximal rank n-1, while x(s), ω(s), p(s) satisfy

$$(7.3)\qquad \textstyle\sum_{j=1}^{n} p_j \partial_{s_l} x_j = \partial_{s_l}\omega\quad ,\ s\in\Sigma\quad ,\ l = 1,\ldots,n-1\quad ,$$

(called the <u>strip condition</u>) and

$$(7.4)\qquad F(x(s),\omega(s),p(s)) = 0\quad ,\quad s\in\Sigma\quad .$$

A (2n+1)-tuple of functions (7.2) satisfying (7.3) is called an ((n-1)-dimensional) strip: The plane in (x,u)-space through the point (x(s),u(s)) with normal vector p(s) is tangential to the n-1-surface (x(s),u(s)) by condition (7.3). If in addition (7.4) holds, then we talk about an <u>integral strip</u> of (7.1).

Instead of asking for a function u(x) satisfying (7.1) and (7.1') it is more convenient to ask for a solution of (7.1) extending a given n-1-dimensional integral strip (7.2). That is,

$$(7.5)\qquad u(x(s)) = \omega(s)\quad ,\ u_{|x}(x(s)) = p(s)\quad ,\quad s\in\Sigma\ .$$

Thus the <u>Cauchy problem</u> for (7.1) seeks to find u(x) satisfying (7.1) and (7.5) for a given '<u>initial integral strip</u>' (7.2). We shall see that this reformulation already holds the key for a solution of the problem, while in fact only ω(s) may be freely chosen in many cases, perhaps up to a finite choice of p(s).

In sec.6 we already solved this problem for a linear equation (6.2). In the present more general setting the system of characteristic equations (6.3) or(6.4) must be replaced by the following:

$$(7.6)\quad \dot{x} = F_{|p}(x,u,p),\ \dot{u} = pF_{|p}(x,u,p),\ \dot{p} = -F_{|x}(x,u,p)-pF_{|u}(x,u,p),$$

with $'\frac{d}{dt}'='\dot{}\,'$. This is a system of 2n+1 first order ODE's in 2n+1 unknowns x(t), u(t), p(t). The coefficients of (7.6) are C^∞ in x, u, p, independent of t, hence there exists a unique local solution defined for $|t|<\delta$ for sufficiently small δ, satisfying

$$(7.7)\qquad x(0) = x^0\ ,\ u(0) = u^0\ ,\ p(0) = p^0\quad ,$$

for arbitrary $x^0\in\Omega$, $u^0\in\mathbb{R}$, $p^0\in\mathbb{R}^n$. The solutions may be extended as long as they stay inside $\Omega\times\mathbb{R}\times\mathbb{R}^n$. Their orbits fill $\Omega\times\mathbb{R}\times\mathbb{R}^n$ as a nowhere intersecting family of curves. Generally we assume that

(7.8) $$F_{|p}(x,u,p) \neq 0 \quad , (x,u,p) \in \Omega \times \mathbb{R} \times \mathbb{R}^n .$$

Then the orbits are nondegenerate C^∞-curves. Moreover, $x=x(t)$ defines a nondegenerate C^∞-curve in x-space $\mathbb{R}^n$.

We claim that $(x(t),u(t),p(t))$ also may be interpreted as a <u>1-dimensional strip</u> in (x,u)-space, along which F is constant. That is, $x=x(t)$, $u=u(t)$ defines a C^∞-curve Ξ in (x,u)-space; at $(x(t),u(t))\in \Xi$, the plane through that point with normal $p(t)$, i.e.

(7.9) $$u - u(t) = \sum_{j=1}^n p_j(t)(x_j-x_j(t)) = p(t)\cdot(x-x(t))$$

defines a plane tangent to Ξ at $(x(t),u(t))$. Indeed the vector $(x^{\cdot}(t),u^{\cdot}(t))=(F_{|p},p\cdot F_{|p})$ clearly lies in the plane (7.9). We get

(7.10) $$\begin{aligned} d/dt(F(x(t),u(t),p(t)) &= F_{|x}x^{\cdot} + F_{|u}u^{\cdot} + F_{|p}p^{\cdot} \\ &= F_{|x}F_{|p} + F_{|u}(pF_{|p}) - F_{|p}(F_{|x} + pF_{|u}) = 0 \quad , \end{aligned}$$

confirming that F is constant along the strips discussed.

A <u>characteristic integral strip</u> is defined to be a 1-dimensional strip as described above,with the additional property that

(7.11) $$F(x(t),u(t),p(t)) = 0 \quad , \text{ along } \Xi .$$

For the construction of a solution of the Cauchy problem (7.1) and (7.5) we start from a given n-1-dimensional integral strip (7.2). For each $(x(s),\omega(s),p(s))$, $s\in \Sigma$, we obtain the unique characteristic integral strip through (x,ω,p). That is we construct

(7.12) $$x(s,t) = x(s_1,\dots,s_{n-1},t) \ , \ \omega(s,t) \ , \ p(s,t) \ , \ s \in \Sigma ,$$

solving (7.6) and the initial conditions

(7.13) $$x(s,0) = x(s) \ , \ \omega(s,0) = \omega(s) \ , \ p(s,0) = p(s) \ , \ s \in \Sigma .$$

The functions (7.12) exist only for $(s,t) \in \Omega_0$, with

(7.14) $$\Omega_0 = \{(s,t) : s \in \Sigma \ , \ |t| < t_1(s)\} \quad ,$$

with a suitable function $t_1(s) > 0$, defined over Σ .

Suppose the function $x(s,t)$ can be inverted near $t=0$.Since we assume $\partial x/\partial s$ of maximal rank this means that $\partial x/\partial t=x^{\cdot}$, at Γ, is linearly independent of the $\partial x/\partial s_j$. In other words, the projection of the characteristic strip through $(x(s),\omega(s),p(s))$ onto x-space must never be tangent to Γ. Under this condition we call Γ an admissible initial integral strip. The implicit function theorem then implies that x(s,t) is invertible whenever the initial strip is admissible, and $t_1(s)$ is choosen sufficiently small. Let $\lambda(x)=(s(x),t(x))$ denote the inverse. We claim that a solution of the Cauchy problem (7.1), (7.5) is defined by setting

$$u = \omega\circ\lambda \quad , \text{ i.e., } u(x) = \omega(s(x),t(x)) \ . \tag{7.15}$$

Indeed for $x \in \Gamma$ we have $t = 0$, so that $u(x)=u(x(s))=\omega(s)$. From (7.15) we get $u(x(s,t)) = \omega(s,t)$. Differentiating this for s_j and t , using (7.6) and the strip conditions (7.3) , we get

$$\sum_j u_{|x_j}\partial_{s_l}x_j = \partial_{s_l} = \sum_j p_j\partial_{s_l}x_j \tag{7.16}$$

$$\sum_j u_{|x_j}\partial_t x_j = \partial_t\omega = \sum_j p_j\partial_t x_j \quad , (s,t) \in \Sigma \ .$$

Since the Jacobian $\partial x/\partial(s,t)$ is invertible, (7.16) implies that

$$p(s,t) = u_{|x}(x(s,t)) \quad , \quad (s,t) \in \Sigma \ . \tag{7.17}$$

In particular we again get t=0 , as $x\in \Gamma$,hence $u_{|x}(x(s)) =p(s)$. Hence we have (7.5) .Also we know that $F(x(s,t),\omega(s,t),p(s,t))= 0$. Thus (7.15) and (7.17) imply that the PDE (7.1) is satisfied by u of (7.15) .Thus indeed we solved the Cauchy problem.

Next let us assume that u(x) satisfies (7.1) and (7.5) , for an admissible strip (7.2) . We then may consider the system

$$x^{\cdot} = F_{|p}(x,u(x),u_{|x}(x)) \quad , \ "\cdot" = d/dt \tag{7.18}$$

of n first order ODE's in n unknowns x(t). Let y(s,t) solve (7.18) and the initial cdn's $y(s,0)=x(s)$. Let $\kappa(s,t)= u(y(s,t))$, $q(s,t)= u_{|x}(y(s,t))$. We will show that y,κ,q solves (7.6) with the same initial conditions as our constructed $x(s,t)$, $\omega(s,t)$, $p(s,t)$, so that we must have $x=y$, $\kappa=\omega$, $p=q$, for all $s\in \Sigma$, $|t|<t_1(s)$. Since $\kappa = u\circ y = u\circ x = \omega$, we then find that $u = \omega\circ\lambda$ coincides with our previously constructed solution (7.15) of (7.1) and (7.5) , so that we have uniqueness of the solution of this Cauchy problem.

Indeed, from (7.18) get $y^{\cdot} =F_{|p}(y,\kappa,q)$, assuming $u \in C^2$. Also, $F(x,u(x),u_{|x}(x))=0$, since u solves (7.1). Differentiating we get

$$F_{|x}(x,u,u_{|x}) + F_{|u}(x,u,u_{|x}) + F_{|p}(x,u,u_{|x})u_{|xx} = 0 \ . \tag{7.19}$$

In (7.18) let $x = y(s,t)$. Then $q^{\cdot} = u_{|xx}(y)F_{|p}(y,\kappa,q) =-F_{|x}(y,\kappa,q) - F_{|u}(y,\kappa,\theta)q$ so that indeed the system (7.6) for y,κ,q follows .

We have proven the result below.

Theorem 7.1. Let $\Gamma = \{x(s) : s \in \Sigma\}$ be an n-1-dimensional C^∞-surface in $\mathbb{R}^n$ with local parametric representation, rank $\partial s/\partial x = n-1$. Then for every admissible integral strip $(x(s),\omega(s),p(s))$ over Γ the Cauchy problem (7.1),(7.5) admits a solution $u \in C^\infty(\Omega_0)$, with some neighbourhood Ω_0 of Γ. The solution is unique within $C^2(\Omega_0)$.

Remark 7.2. Suppose an n-1-dimensional surface $\Gamma=\{x\in\Omega:\psi(x)=\psi(x_0)\}$ is given, where $\psi_{|x}\neq 0$, and $F(x,\psi(x),\psi_{|x}(x))=0$ on Γ. Then, if $x=x(s)$ represents (some part of) Γ, we find that $\omega(s)=\psi(x(s))$, $p(s)=\psi_{|x}(x(s))$ defines an n-1-dimensional integral strip of F. Suppose this strip is admissible. Then thm.7.1 yields a solution $\varphi(x)$, defined near Γ, such that $\varphi(x)=\psi(x)$, and $\varphi_{|x}(x)=\psi_{|x}(x)$ on Γ. It follows that Γ also is given in the form $\varphi = c$, at least locally.

This construction will be useful in sec.8, where we ask whether a characteristic surface always may be given as a surface of constancy of a solution φ of the characteristic equation (8.3). However, a closer inspection shows that, for an equation (8.3) the above integral strip is never admissible: We get ψ=const on Γ, hence the surface normal is a multiple of $\psi_{|x}(x(s))=p(s)$ for all s On the other hand, the projection of the characteristic integral strip at x(s) has direction $x^\bullet=F_{|p}(x(s),p(s))$. Also, since $F(x,p)=a_N(x,p)$ of (8.3) is homogeneous of degree N in p, we get $p.x^\bullet = p.F_{|p}(x(s),p(s))= NF(x(s),p(s))= 0$ along a characteristic integral strip. Accordingly, $x^\bullet$ must be in the tangent space of x(s), since it is perpendicular to the surface normal.

8. Characteristics and bicharacteristics of a linear PDE.

The conventional approach to the concept of a characteristic (hyper-)surface for a differential expression

$$L = a(x,D) = \sum_{|\alpha|\le N} a_\alpha(x)D^\alpha \tag{8.1}$$

is the Cauchy-Kowalewska theorem, discussing existence of a unique analytic solution for a PDE with analytic coefficients satisfying analytic initial data. The result directs the attention to certain surfaces along which data cannot be prescribed freely.

The concepts also make sense for real-valued solutions of a PDE with real coefficients. However, one has to work with two different sets of assumptions: Either assume real C^∞-coefficients or complex analytic coefficients. The latter case is mostly analogous to the real C^∞-case, and will not be discussed in detail.

Accordingly we assume that L of (8.1) has complex C^∞-coefficients, but that the coefficients of the principal part polynomial

(8.2) $$a_N(x,\xi) = \sum_{|\alpha|=N} a_\alpha(x)\xi^\alpha \quad \neq \quad 0$$

are real-valued $C^\infty(\Omega)$, for a domain $\Omega\subset \mathbb{R}^n$, up to a common factor $0\neq\gamma\in C^\infty(\Omega)$ (here assumed 1). The characteristic equation of L is

(8.3) $$a_N(x,\varphi_{|x}) = 0 \quad ,$$

with the gradient $\varphi_{|x}$ of φ. Clearly (8.3) is a single first order PDE for the unknown (real-valued) function $\varphi(x)\in C^\infty(\Omega)$.

In sec.7 we discussed solving such PDE. Since $F(x,p,u)=a_N(x,p)$ is independent of u, the characteristic system reduces to

(8.4) $$x^{\cdot} = a_{N|\xi}(x,\xi) \quad , \quad \xi^{\cdot} = - a_{N|x}(x,\xi) \quad ,$$

a system of 2n first order ODE for 2n unknowns $x(s)$, $\xi(s)$, where we wrote $\xi(s)$ instead of $p(s)$. The equation for $u^{\cdot}$ of (7.6) is $\varphi^{\cdot}= Na_N(x,\varphi_{|x})=0$, due to (8.3), using Eulers relation for the homogeneous function of degree N in $\varphi_{|x}$). (7.6) splits into this equation and (8.4) to be solved separately. The integral strips are given as solutions $(x(s),\xi(s))$ of (8.4) together with $\varphi(s)$=const. A system of the form (8.4) is commonly called a Hamiltonian system.

The graph of a solution φ of (8.3) is fibered by strips $x(s)$ $\xi(s)$ solving (8.4), φ=const.: Through any point $(x^0,\varphi(x^0))$ on the graph there is a unique characteristic integral strip

$$(x(s),\varphi(x^0),\xi(s)) \text{ with } x(0)=x^0 \quad , \quad \xi(0)=\varphi_{|x}(x^0) \quad ,$$

where $(x(s),\xi(s))$ solves (8.4). In sec.7 we have seen how solutions of (8.3) may be composed of such strips. Moreover, since φ is constant along such strips, it is clear that they stay on the same level surface of φ. Thus also the level surfaces of solutions are fibered by characteristic integral strips of (8.3).

A characteristic (hyper-)surface Γ of the differential expression (8.1) is defined as a surface of constancy of a solution φ of the characteristic equation (8.3), with $\varphi_{|x}\neq 0$. To be precise we require that for every $x^0\in\Gamma$ there exists a solution $\varphi = \varphi^{x^0}$ of (8.3), defined in $\mathcal{B}=\{|x-x^0|<\varepsilon\}$ such that $\Gamma\cap\mathcal{B} = \{x\in\mathcal{B}: \varphi(x)=c\}$, for some constants c, $\varepsilon>0$.

Actually one may define somewhat more generally by dropping the assumption that $\varphi(x)$ solves (8.3) near the surface Γ, and requiring (8.3) on the surface Γ only.

It is clear from the above that a characteristic surface Γ is fibered by characteristic integral strips: There is precisely one strip $(x(s),c,\xi(s))$ through each point $x \in \Gamma$. These strips are called <u>bicharacteristic strips</u> of the expression L .

We speak of a <u>simply characteristic surface</u> Γ of the expression L of (8.1) if Γ is given by $\varphi(x)=0$ with $\varphi(x)=0$, $\varphi_{|x}(x)\neq 0$, $a_N(x,\varphi_{|x}(x))=0$ on Γ , as above, and in addition $a_{N|\xi}(x,\varphi_{|x}(x))\neq 0$ on Γ . <u>For a simple characteristic (surface) it is no loss of generality to assume that $\varphi(x)$ solves (8.3) near Γ</u> .

Indeed, we may assume that $a_{N|\xi_1}(x^0,\varphi_{|x}(x^0))\neq 0$, and then solve $a_N(x,\xi)=0$ for ξ_1 . Get a function $p_1(x,\pi^{\sim})$ with $p_1(x^0,\varphi_{|x}(x^0)^{\sim})=\varphi_{|x_1}(x^0)$, where $p^{\sim}=(p_2,\dots,p_n)$. The plane perpendicular to the integral strip through x^0 is noncharacteristic near x^0 . In that plane Σ_0 set $\omega(x)=\varphi(x)$, $p^{\sim}(x)=\varphi_{|x}(x)^{\sim}$, with our given $\varphi(x)$ solving (8.3) on Γ near x^0 . Also set $p_1=p_1(x,p^{\sim})$. We get an admissible integral strip, and a solution $\omega(x)$ near x^0 . We have $\omega(x)=\varphi(x^0)$ on the strip through x^0 . Also, $\omega(x)=\varphi(x)=\varphi(x^0)$ on $\Gamma\cap\Sigma_0$. Hence also $\omega(x)=\varphi(x^0)=c$ on the intersection with Γ of a small ball around x^0 . Thus $\omega(x)$ indeed solves (3.3) and is constant on Γ near x^0 .

A characteristic surface Γ along which $F_{|\xi}=a_{N|\xi}(x,\varphi_{|x})=0$ will be called a <u>multiple characteristic</u>. An example of an expression with multiple characteristics is the heat operator (example b), below). For hyperbolic equations one tends to avoid multiple characteristics, introducing the concept of strictly hyperbolic expressions (cf. VII,2). An expression L having no multiple real characteritics is said to be of <u>principal type</u> (VII,2).

As mentioned initially, characteristic surfaces are important for the <u>initial-value problem</u> of (8.1): For a given (hyper-) surface $\Gamma\subset\mathbb{R}^n$ and given 'data' ψ_α defined on Γ and f defined near Γ we ask for existence of u (defined near Γ such that

$$(8.5)\qquad Lu = f \text{ near } \Gamma,\ u^{(\alpha)}=\partial_x^\alpha u=\psi_\alpha,\ x\in\Gamma\ ,\ \text{for all } |\alpha|\le N-1\ .$$

The bicharacteristic strips, on the other hand, will be recognized as carriers of singularities of solutions of Lu = f, (cf. ch.6, ch.9). The key result is our version of Egorov's theorem (VI,thm.5.1), accessible only after a study of ψdo's.

Solving (8.5), it is clear that not all functions ψ_α may be prescribed arbitrarily. Assuming Γ and ψ_α smooth, let $\nu(x)$ be a smooth transversal vector, defined (locally) near Γ, and never tangential to Γ, let us start prescribing arbitrarily the functions $u=\psi_0$, $D_\nu u=\psi_1$, ... , $D_\nu^N u=\psi_{N-1}$, on Γ , where

$$(8.6)\qquad D_\nu=\textstyle\sum_{j=1}^n \nu_j(x)\partial_{x_j}\ .$$

This will determine all $u^{(\alpha)}$: $|\alpha|\le N-1$, on Γ, as tangential derivatives of the ψ_j: For some local parametric representation $x=x(\lambda)$ $\lambda=(\lambda_1,\ldots,\lambda_{n-1})$ of Γ the derivatives $u(\alpha)(x)$, $x\in\Gamma$, become linear expressions in the derivatives $\partial_\lambda^\alpha\psi_j$. Accordingly, $Lu=f$, along Γ, translates into a linear equation of the form

$$(8.7)\qquad pD_\nu^N u + \sum p_\alpha\partial_\lambda^\alpha\psi_j = f \quad .$$

Now if $\varphi\in C^1$ with $\varphi_{|x}\neq 0$ is constant on Γ we find that

$$(8.8)\qquad \partial_{x_j} = (\varphi_{|x_j}/D_\nu\varphi)D_\nu + T_j \ , \ j=1 \ , \ \ldots \ , n \ ,$$

with vector fields T_j satisfying $T_j\varphi = 0$. Substituting (8.8) into (8.1) one finds that p in (8.7) is given by

$$(8.9)\qquad p = (D_\nu\varphi)^{-N} \sum_{|\alpha|=N} a_\alpha(x)\varphi_{|x}^{\alpha}(x) \ .$$

If $p(x)\neq 0$ on Γ then (8.7) may be solved for $D_\nu^N u$ in some neighbourhood of Γ. Moreover, we may apply D_ν to the resulting equation for an infinite number of new relations of the form $D_\nu^m u=$ linear expression in derivatives of lower order, allowing recursive calculation of all $D_\nu^m u$ on Γ. Clearly this allows construction of a unique solution of (8.5) by its Taylor series, assuming that an analytic solution exists. The recursion may be used for estimates proving convergence of the Taylor series and existence of an analytic solution, if the coefficients are analytic. This sketches a proof of the Cauchy-Kowalewska theorem mentioned initially. Moreover the importance of the concept of characteristic surface is clear from the result below, the proof of which is evident, after the above.

<u>Proposition 8.1.</u> The function p(x) of (8.9) vanishes identically on Γ if and only if Γ is a characteristic surface. We get $p\neq 0$ on Γ if and only if Γ is nontangential to any characteristic surface.

To provide more detailed information on characteristic surfaces and bicharacteristics we return to some examples, mainly of sec.4.

a) $L = \Delta$ (of (4.1)) : The characteristic equation

$$(8.10)\qquad {\varphi_{|x}}^2 = \sum_{j=1}^n {\varphi_{|x_j}}^2 = 0$$

has only constant real-valued solutions, since it requires $\varphi_{|x}=0$. Real characteristic surfaces do not exist. However, Δ has constant (i.e., analytic) coefficients. The Hamiltonian system (8.4) may be regarded as a set of ODE for complex-valued analytic functions. There exist solutions of (8.10) in the n complex variables $x_1,\ldots,$ x_n, defined in a domain of $\mathbb{C}^n$. Setting such φ constant will give

surfaces in $\mathbb{C}^n$, called complex characteristics. Notice that the Helmholtz operator (4.2) has the same principal part, hence the same real or complex characteristics as the operator Δ .
b) $L=\partial_t-\Delta$, in the n+1 variables $x_0=t$, $x=(x_1,\dots,x_n)$, (i.e., (4.3)) The characteristic equation is (8.10) again. But we now have one more variable $t=x_0$. Thus (8.10) requires that $\varphi(t,x)$ is independent of x, but it may depend on t. It follows that the characteristic surfaces all are hyperplanes of the form t=const. Complex characteristics of more general kind may be constructed, of course

In sec.4 we solved a modified Cauchy problem, of the form

$$(8.11) \qquad Lu = f \quad , \quad t \geq t_0 \ , \ u = \psi \text{ at } t = t_0 \ ,$$

for f, $\varphi \in C_0^\infty$. Only one condition at $t=t_0$ (instead of 2) is needed to make the solution of this characteristic Cauchy problem unique. The real characteristics of L all are multiple characteristics: For $\varphi(t,x)=t-t_0$ we have $\varphi_{|x}=(1,0)$, hence $a_2(\varphi_{|x})=a_{2|\xi}(\varphi_{|x})=0$.

c) $L = \partial_t^{\,2} - \Delta$, (i.e., ex'le (4.4)) , again discussed in n+1 independent variables (t,x) . The characteristic equation is

$$(8.12) \qquad \varphi_{|t}^{\;2} = \varphi_{|x}^{\;2} = \textstyle\sum_{j=1}^n \varphi_{|x_j}^{\;2} \ .$$

Being only interested in the surfaces φ=const., for $(\varphi_{|t},\varphi_{|x})\neq 0$, we may reduce the number of variables: (8.12) implies $\varphi_{|x}=0$ unless $f_{|t}\neq 0$. Hence $\varphi=0$ may locally be solved for t. Writing t=J(x), get

$$(8.13) \qquad \textstyle\sum_{j=1}^n J_{|x_j}^{\;2} = J_{|x}^{\;2} = 1 \ .$$

This equation is often referred to as the equation of geometrical optics, or the eiconal equation. The bicharacteristic strips (solutions of (8.4), in this case) are straight lines x=at+b with constant $a,b\in\mathbb{R}^n$, $|a|=1$, and constant ξ. The base curves of the bicharacteristic strips are the light rays of geometrical optics.

The Cauchy problem (8.5) is well posed for the wave operator along non-characteristic surfaces (cf.sec.4 and ch.7.). Again the Klein-Gordon operator (4.5) has the same principal part and characteristics as the wave operator, and a similar Cauchy problem.

d) L is an elliptic operator, with C^∞-coefficients. That is, we have $a_N(x,\xi)\neq 0$,as $x\in\Omega$, $\xi\in\mathbb{R}^n$, $\xi\neq 0$. Again the characteristic equation (8.3) implies $\varphi_{|x}=0$,assuming φ real-valued. Real characteristics do not exist. Complex characteristics again may be defined only if the a_α are analytic in $x_1,\dots,x_n$.

e) L = P(D) is a constant coefficient hyperbolic operator (cf. VII,1), with respect to some given real vector $h\neq 0$. As a consequence of VII, prop.1.5. the principal part $P_N(\xi)$ also is a hyperbolic polynomial: We have $P_N(h)\neq 0$, and the algebraic equation $P_N(\xi+\lambda h)=0$, for fixed real ξ, has N not necessarily distinct real roots λ_j, $j=1,\dots,n$. The characteristic equation is of the form $P_N(\varphi_{|x})=0$. For simplicity let $h=(0,\dots,0,1)$, as always may be achieved by a linear transformation of independent variables. Then the characteristic equation decomposes into the N equations

$$(8.15) \qquad \varphi_{|x_n} = \tau_j(\varphi_{|x_1},\dots,\varphi_{|x_{n-1}}) \quad ,$$

with $\tau_j(\xi_1,\dots,\xi_{n-1}) = \lambda_j(\xi_1,\dots,\xi_{n-1},0)$ the real solutions of $P_N(\xi+\lambda h) = 0$,for $\xi = (\xi_1,\dots,\xi_{n-1},0)$.

In the presence of multiple roots we cannot expect that the τ_j depend smoothly on $\xi_1,\dots,\xi_{n-1}$. If smoothness can be arranged then each of the N equations (8.15) will give its own family of characteristic surfaces and bicharacteristics. Thus one may expect N different types of characteristics for an N-th order expression.

The N different types of real characteristics do exist if the N roots $\lambda_j(\xi)$ are all distinct, for $\xi=(\xi_1,\dots,\xi_{n-1},0)\neq 0$. Then the τ_j will be smooth. In this case P(D) is called <u>strictly hyperbolic</u> (cf. VII,2). We shall see in ch.7 that the Cauchy problem (8.5) is always well posed, even for variable coefficients, if L is strictly hyperbolic. For constant coefficients the Cauchy problem of a general hyperbolic L is well posed (cf. [Ga₁], [Hr₁]).

<u>9. Lie groups and Lie algebras, for classical analysts</u>.

Let us discuss the relationship between a Lie group *G* and *A*, its Lie algebra, using a form suitable for analysis minded readers. Generally, *G* may be represented on some $GL(\mathbb{R}^N)$ (we will not show this, although the tools are developed). For simplicity let *G* be a Lie subgroup of $GL(\mathbb{R}^N)$, as will be true for all our applications. A path $\psi(t)$, $|t|<\eta$, in *G* then is a path of N×N-matrices together with its derivative $\psi'=d\psi/dt$. For $\psi(0)=e$=unit of *G* , and $g\in G$, the path $g\psi(t)$ starts at *g* and $(g\psi(t))'=g\psi'(t)$ with matrix product.

Let $\gamma:U\to G$, or explicitly, $\gamma=\gamma(u)$, $\gamma(0)=e$, be a charted neighbourhood of *e*, with an open set $U\subset\mathbb{R}^\nu$, $\nu=\dim G$, $0\in U$. For a $g\in G$ the function $\gamma_g=g\gamma(u)$, $u\in U$, defines a charted neighbourhood of *g*.

The 'partial derivative' $\gamma_{|u_j}$ is defined as directional derivative along the coordinate line u_k=const, $k\neq j$, $u_j=t$ (i.e., of) $\psi(t)=\gamma(u_1,\dots,u_{j-1},t,u_{j+1},\dots,u_N)$. Similarly for $\gamma_{g|u_j}$.

In these coordinates the tangent space at $e=\gamma(0)$ (i.e., the corresponding Lie algebra A) consists of all $a=(a^j)\in\mathbb{R}^\nu$, representing the matrix $A_a = \sum a^j\gamma_{|u_j}(0)$. We get 1-1-correspondences $\mathbb{R}^\nu = A \leftrightarrow \{A_a : a\in\mathbb{R}^\nu\} \leftrightarrow \{a(g): a\in\mathbb{R}^\nu\}$ with the vector field $a(g)$ defined by using the same $a\in\mathbb{R}^\nu$ in the local coordinates γ_g at g : $a(g)=\sum a^j\gamma_{g|u_j}(0)$. $a(g)$ defines a global folpde L_a on G - i.e., $a(\gamma_g(u))=L_a(\gamma_g(u))$, with L_a, in coordinates $\gamma(u)$, near e, given by

$$(9.1)\quad L_a = \sum_l\Big(\sum_j a^j\theta_{l|v_j}(u,0)\Big)\gamma(u)\partial/\partial u_l,\ \text{where } \theta(u,v)=\gamma^{-1}(\gamma(u)\gamma(v)),$$

similarly for general g with γ_g . Again $A \leftrightarrow \{L_a : a\in\mathbb{R}^\nu\}$, so that $a\in A$ appears in 4 forms, as a ν-vector (a^j), a matrix A_a , or a folpde L_a on G , or again as $a(g)=\sum a^j\gamma_{g|u_j}(0)=gA_g$.

For $a\in A$ define a smooth curve $\psi(t)$ in G as solution of

$$(9.2)\qquad \frac{d\psi}{dt} = a(\psi(t)) \ ,\ t \text{ near } 0 \ ,\ \psi(0) = e \ .$$

Locally, (9.2) constitutes a system of ν ODE's of order 1, with real C^∞-coefficients. Its Cauchy problem (9.2), even at t_0 instead 0, is uniqely solvable. In fact, (9.2) amounts to $\psi'=\psi A_a$, $\psi(0)=e$, with the above A_a, and matrix multiplication. This is trivially solved by the exponential function $\psi(t)=\exp(A_a t)=\sum_0^\infty A_a{}^j t^j/j!$. By a continuation argument get $\psi(t)=\exp(A_a t)\in G$, and (9.2) for $t\in\mathbb{R}$. In particular, $\{\psi(t)\}$ defines a subgroup of G. We summarize:

<u>Proposition 9.1</u>. For each tangent vector $a\in A=T_e(G)$ there is a 1-parameter subgroup $\psi(t)=\psi_a(t)$, $t\in\mathbb{R}$, defined as unique solution of the Cauchy problem (9.2), with $a(g)\equiv L_a$ of (9.1).

<u>Corollary 9.2</u>. The connected component $G_e\subset G$ of the unit element e is generated by all the above 1-parameter subgroups, and even by

$$\{\psi_a(t):\ a\in A\ ,\ 0\le t\le\varepsilon\}\ ,\ \text{for any fixed } \varepsilon>0\ .$$

Indeed, any $g\in G_e$ may be connected to e by a path Γ , and Γ may be replaced by a 'polygon' with sides translates of 'lines' $\psi_a(t)$, $0\le t\le\varepsilon'\le\varepsilon$, for fixed $a\in A$. This shows that, indeed, $g=\psi_{a_1}(\varepsilon_1)\psi_{a_2}(\varepsilon_2)\dots\psi_{a_M}(\varepsilon_m)$, for finite M and $\varepsilon_j\le\varepsilon$.

Corollary 9.3. We have

$$\psi_a(t) = e^{A_a t} = \sum_0^\infty A_a{}^j/j! \ , \ A_a = \psi_a{}^\cdot(0) = \sum a^j \gamma_{|u_j}(0) \ . \tag{9.3}$$

The subgroup G_e is generated by all $e^{A_a t}$, $a \in A$, $0<t\le\varepsilon>0$.

Completing the structure of A, we define a bracket operation: Consider the commutators $[A_a,A_b]= A_aA_b-A_bA_a$,and $[L_a,L_b]=L_aL_b-L_bL_a$ of the matrices and (matrix-valued) first order expressions, resp. Get $\gamma(u)\gamma(v)=\gamma(\theta(u,v))$, near $u=v=0$, with $\theta(u,v)$ of (9.1). Differentiating this conclude that $[\gamma_{|u_j}(0),\gamma_{|u_l}(0)]= \sum_r \kappa^r_{jl}\gamma_{|u_r}(0)$, with $\kappa^r_{jl} = \theta_{r|u_jv_l}(0,0)-\theta_{r|u_lv_j}(0,0)$. Accordingly,

$$[A_a,A_b]=A_c \ , \ c = (c^r)=(\sum_{j,l} \kappa^r_{jl}a^jb^l) \ , \tag{9.4}$$

again belongs to the tangent space of $G \subset GL(\mathbb{R}^N)$ at $g=e$. Next, $[L_a,L_b]$ is a first order expression: Let $\left\{\begin{smallmatrix}p_j\\q_j\end{smallmatrix}\right\}(u)=\Sigma\left\{\begin{smallmatrix}a^j\\b^j\end{smallmatrix}\right\}\theta_{j|v_k}(u,0)\gamma(u)$

The matrices $p_j(u)$, $q_l(u)$ commute for $u\in U$, hence (9.1) yields

$$[L_a,L_b]= \sum_{j,l}(p_jq_{l|u_j}-q_jp_{l|u_j})\partial u_l \ . \tag{9.5}$$

At $u=0$ get $p_j=a^je$ $q_j=b^je$, $\theta(0,v)=v$, hence $\theta_{l|v_k}(0,0)=\delta_{lk}$. Thus $\left\{\begin{smallmatrix}p_j\\q_j\end{smallmatrix}\right\}_{|u_j}(0)=\left\{\begin{smallmatrix}a^j\\b^j\end{smallmatrix}\right\}\gamma_{|u_j}(0)+\Sigma\left\{\begin{smallmatrix}a^k\\b^k\end{smallmatrix}\right\}\theta_{l|v_ku_j}(0,0)$. We get

$$\sum_j(p_jq_{l|u_j}-q_jp_{l|u_j}) = \sum(a^jb^l-b^ja^l)\gamma_{|u_j}=b^lA_a-a^lA_b \ , \tag{9.6}$$

with our above A_a, using the symmetry $\theta_{l|u_jv_k}=\theta_{l|v_ju_k}$, at $u=v=0$. Accordingly, $[L_a,L_b]\gamma(0)=\Sigma(b^lA_a-a^lA_b)\gamma_{|u_l}=A_aA_b-A_bA_a=[A_a,A_b]=L_c\gamma(0)$

The argument may be repeated at general $g\in G$, using $\gamma_g(u)$ instead $\gamma(u)$. One finds that $[L_a,L_b]\gamma_g(0)=L_c\gamma_g(0)$, for all $g\in G$, with c of (9.5) . Accordingly $A_c \leftrightarrow L_c$, under our 1-1-correspondence. This induces a bracket operation in the Lie algebra A, given by

$$[a,b] \leftrightarrow[A_a,A_b]= A_c \leftrightarrow[L_a,L_b] = L_c \ , \ c \text{ of } (9.5) \ . \tag{9.7}$$

$[.,.]$ has the usual properties, skew-symmetry, and Jacobi identity.

Next we assume given a Lie matrix algebra - i.e., a linear subspace A of some $L(\mathbb{R}^N)$ containing all of its (matrix-) commutators $[A,B]=AB-BA$, $A,B\in A$. Assume $\dim A = \nu$. Prop.9.1 and cor.9.2

then suggest a direction of approach for generation of a corresponding Lie group G - a Lie subgroup of $GL(\mathbb{R}^n)$: G should contain all matrices e^{At}, $t\in\mathbb{R}$, $A\in\mathcal{A}$, and all their finite products. It should be closed under matrix multiplication and inversion. Here we are only interested in connected Lie groups (i.e. might get only the component of 1 of a larger group). In fact we will look at the minimal such set - all finite products of e^{At}, $A\in\mathcal{A}$, $t\in\mathbb{R}$.

Observe that $V_t = e^{At}e^{Bt}$, for A, B $\in\mathcal{A}$, no longer defines a 1-parameter group, unless A and B commute. For general A, B we will prove the Baker-Campbell-Hausdorff formula: For small $|t|$,

$$(9.8)\qquad V_t = e^{C(t)} \quad , \text{ where } C(t)\in\mathcal{A}$$

is a convergent power series in t (cf. lemma 9.5, below).

As a first step in this direction:

Proposition 9.4. V_t satisfies the differential equation ("'"="$\frac{d}{dt}$")

$$(9.9)\qquad V_t' = (A+e^{At}Be^{-At})V_t \ ,$$

where the coefficient $A+e^{At}Be^{-At}\in\mathcal{A}$ is a convergent power series.

Proof. (9.9) follows trivially by differentiation. Also, the family $B_t = e^{At}Be^{-At}$ satisfies the differential equation

$$(9.10)\qquad Bt' = ad_A B_t \ , \ B_0 = B \ ,$$

with the linear operator

$$(9.11)\qquad ad_A B = [A,B] \ , \ ad_A \in L(L(\mathbb{R}^N)) \ .$$

Therefore,

$$(9.12)\quad B_t = e^{ad_A t}B = \sum_{j=0}^{\infty}\frac{1}{j!}(ad_A)^j B = B+[A,B]t+[A,[A,B]]\frac{t^2}{2}+\dots \ ,$$

showing that $B_t\in\mathcal{A}$, hence $A+B_t\in\mathcal{A}$, q.e.d.

We will prove (9.8) in the following form.

Lemma 9.5. For A , B $\in\mathcal{A}$, define $C=C(t) = \log(e^{At}e^{Bt})$ by setting

$$(9.14)\qquad C(t) = X(t)-X(t)^2/2+X(t)^3/3 \pm\dots+(-1)^k X(t)^k/k \pm\dots,$$

with $\quad X(t) = e^{At}e^{Bt}-1 = (1+At+A^2\frac{t^2}{2!}+\dots)(1+Bt+B^2\frac{t^2}{2!}+\dots) -1$.

Then we have a convergent power series expansion

$$(9.15)\quad C(t) = \log(e^{At}e^{Bt}) = \sum_{j=1}^{\infty}C_j t^j \quad , \ |t|< (\log 2)/(\|A\|+\|B\|) \ ,$$

where C(t) and all coefficients C_j belong to the Lie algebra $\mathcal{A}$.

In particular,

(9.16) $C_1 = A+B$, $C_2 = \frac{1}{2}[A,B]$, $C_3 = \frac{1}{12}[A,[A,B]] + \frac{1}{12}[B,[B,A]]$.

The coefficient C_j is a finite linear combination of terms which are applications of (together j-1) operators ad_A or ad_B to A or B.

<u>Proof</u>. We use a proof of D.Djokovic [Dj] (we owe to G. Hochschild)

Let $C(t) = \log(e^{At}e^{Bt})$, defined as composition of the logarithmic power series and $X(t) = \sum_{m=1}^{\infty} X_j t^j$. Note that the matrix coefficients c_{jl} of $C(X) = \log X = \sum_{j=1}^{\infty} (-1)^j X^j / j$ are convergent power series in the N^2 complex variables x_{jk} , where $X = ((x_{jk}))_{j,k=1,\dots,N}$, whenever $\|X(t)\| < 1$. The latter holds for $e^{\|A\|t} e^{\|B\|t} - 1 = e^{(\|A\|+\|B\|)t} - 1 < 1$, i.e., $(\|A\|+\|B\|)t < \log 2$. Thus c_{jl} are complex differentiable functions of the N^2 complex variables x_{jk} , which in turn are power series of t . This implies complex differentiability of the function C(t) for t as in (9.15). So C(t) is a convergent power series , and (9.15) follows. We are only left with showing that the coefficients C_j belong to **A** . The first coefficient (cf. (9.16)) is easily verified by direct computation. For the general case we will show that C(t) solves an ODE ((9.18), below) which leads to a recursion for the C_j.

We calculate that $(e^C)'e^{-C} = \sum_{m=1}^{\infty} \sum_{k=0}^{\infty} \frac{(-1)^k}{N!k!} (C^m)'C^k$

$= \sum_{m=1}^{\infty} \sum_{n=m}^{\infty} (-1)^{n-m} \frac{1}{(n-m)!m!} (C^m)'C^{n-m} = \sum_{m=1}^{\infty} \sum_{n \geq m} \frac{1}{n!} \binom{n}{m} (C^m)'C^{n-m}$

$= \sum_{n=1}^{\infty} \frac{1}{n!} \sum_{m=1}^{n} (-1)^{n-m} \binom{n}{m} (C^m)'C^{n-m}$. Here, $(C^m)' = \sum_{k=0}^{m-1} h^k C' C^{m-k-1}$. Thus the inner sum equals $T_n = \sum_{m=0}^{n-1} \sum_{k=0}^{m} (-1)^{n-m-1} \binom{n}{m+1} C^k C' C^{n-k-1} =$ $\sum_{k=0}^{n-1} C^k C' C^{n-k-1} \sum_{m=k}^{n-1} (-1)^{n-m-1} \binom{n}{m+1} = \sum C^k C' C^{n-k-1} Z_k$, where $Z_k =$ $\sum_{j=0}^{n-k-1} (-1)^j \binom{n}{j} = (-1)^{n-1-k} \binom{n-1}{k}$, by induction. We get $T_n =$ $\sum_{k=0}^{n-1} C^k C' C^{n-k-1} (-1)^{n-1-k} \binom{n-1}{k} = (\mathrm{ad}\, C)^{n-1} C'$ whence

(9.17) $\sum_{n=1}^{\infty} \frac{1}{n!} (\mathrm{ad}\, C)^{n-1} C' = (e^C)'e^{-C} = V_t' V_t^{-1} = A + e^C B e^{-C}$, using (9.9) and $e^C=e^{At}e^{Bt}=V_t$. With (9.12) we get $e^C B e^{-C} = e^{ad_C} B$.

The left hand side of (9.17) may be written as $f(ad_C)C'$, with the power series $f(\xi)=\sum_{n=1}^{\infty}\xi^{n-1}/n! =(e^x-1)/x$. Notice that $g(x)=\frac{1}{f(x)}$ is a power series too. Applying it to (9.17) yields the desired differential equation

$$C' = g(ad_C)(A + e^{ad_C} B) . \tag{9.18}$$

Expanding both sides of (9.18) we get $C'= \sum_{n=1}^{\infty}(n+1)C_{n+1}t^n$, while the right hand side may be written in the form

$$g(\textstyle\sum_1^{\infty} t^k ad_{C_k})A + h(\sum_1^{\infty} t^k ad_{C_k})B , \ h(x)=g(x)e^x .$$

The coefficient of the k-th power of t, at right, is a finite linear combination of a finite number of applications of ad_{C_j} to A or B, where $j \le k$. In fact, the sum of all j in each such term equals k. We know that $C_1 \in \mathbf{A}$. If we assume by induction that $C_1,\ldots,C_k \in \mathbf{A}$, it follows that $(k+1)C_{k+1} \in \mathbf{A}$, hence $C_{k+1} \in \mathbf{A}$. Also C_{k+1} must be a linear combination of k applications of ad_A and ad_B to A or B, as stated. The proof is complete.

<u>Corollary 9.6</u>. Let $A_1,\ldots,A_\nu$ be a basis of $\mathbf{A}$ (as a linear space), and let $A(u)= \sum_{j=1}^{\nu}A_j u_j$, $u_j \in \mathbb{R}$. There exist ν power series $f_k(u,v)$ $k=1,\ldots,\nu$ in 2ν variables u,v convergent for $|u|,|v|<\varepsilon_0$ with some $\varepsilon_0>0$ such that $f_k(u,v) = u_k+v_k + \ldots$, $k=1,\ldots,\nu$, and

$$e^{A(u)}e^{A(v)} = e^{A(f(u,v))} , \ |u|,|v|<\varepsilon_0 . \tag{9.19}$$

<u>Proof</u>. As in the proof of lemma 9.5 we conclude that $e^{A(u)}e^{A(v)} = e^{C(u,v)}$ where C(u,v) is a convergent power series in u and v, as $\|A(u+v)\|<\log 2$. Lemma 9.5 then implies that $C(u,v) \in \mathbf{A}$ for sufficiently small $|u|,|v|$. Hence we may write $C(u,v)=A(f(u,v))$, where again $f_k(u,v)$ are convergent power series. It then is evident that $f_k(u,v) = u_k+v_k$ + higher powers. Q.E.D.

From the given linear subspace $\mathbf{A}$ of $L(\mathbb{C}^N)$ we now define a group $\mathbf{G} = \mathbf{G}(\mathbf{A})$ as the collection of finite products of matrices e^A, with $A \in \mathbf{A}$, where the group operation is matrix multiplication. Clearly $\mathbf{G}$ contains $I = e^0$. We intend to show that G is a (ν-dimensional) Lie-group. Moreover, the tangent space of G at its unit element I equals the Lie-algebra $\mathbf{A}$. And, vice versa, if we depart from a general Lie-subgroup $\mathbf{G}$ of $GL(\mathbb{R}^N)$, then define $\mathbf{A}$ as the linear space of all directional derivatives of curves in $\mathbf{G}$ starting at I , and then, with this $\mathbf{A}$, defining the above group $\mathbf{G}(\mathbf{A})$, we get $\mathbf{G} = \mathbf{G}(\mathbf{A})$.

First let us establish a coordinate chart for a neighbourhood of the identity $I \in \mathbf{G}$. With the notations of cor.9.6 define

$$(9.20) \qquad \gamma(u) = e^{u_1A_1+u_2A_2\ldots+u_\nu A_\nu} = e^{A(u)} \quad , \ |u|<\varepsilon_0 \ ,$$

where $\varepsilon_0>0$ is kept fixed. Then $\gamma(u)$ defines an invertible map of the ball $\mathbf{B}_{\varepsilon 0}=\{|u|<\varepsilon\}$ onto a set $U_I \subset \mathbf{G}$ containing I, and $\gamma(0) = I$. For general $G\in \mathbf{G}$ similarly $\gamma_G(u) = \gamma(u)G$ defines a map of $\mathbf{B}_\varepsilon$ onto a subset $U_G \subset \mathbf{G}$ with $G=\gamma_G(0) \in U_G$. Introduce a topology on $\mathbf{G}$ by using all the "balls" $\gamma_G(\mathbf{B}_\varepsilon)$ $0<\varepsilon\le\varepsilon_0$, $G\in \mathbf{G}$, as a basis. Then each set U_G is an open neighbourhood of G . Let $U_G \cap U_{G'} \neq \emptyset$. Then $e^{A(u^0)}=e^{A(v^0)}Z$, $Z=GG'^{-1}\in \mathbf{G}$, for some u^0 , $v^0 \in \mathbf{B}_{\varepsilon_0}$. Using cor. 9.6 we get $Z = e^{A(f(u^0,-v^0))}$, assuming ε_0 properly chosen. Notice that $f(u,v) = 0$ can be solved for v near $u=v=0$, giving a function $v=v(u)$, $v(0)=0$, since the Jacobian $f_{|v}(0,0)=I$ is invertible. It follows that also $f_{|u}(u^0,-v^0)$ is invertible, so $f(u,-v)=f(u^0,-v^0)$ may be solved for v giving an analytic map $v=v(u)$, $v(u^0)=u^0$, provided that u^0, v^0 are sufficiently small,- i.e., that ε_0 is chosen sufficiently small. Conclusion:

$$Z = e^{A(f(u^0,-v^0)} = e^{A(f(u,-v(u))} = e^{A(u)}e^{-A(v(u))} \text{ for u close to } u^0.$$

Or, $\gamma_G(u) = \gamma_{G'}(v(u))$ for small $|u-u^0|$ showing that the map $\gamma_G{}'^{-1} \circ \gamma_G : \gamma_G^{-1}(U_G \cap U_G{}') \to \gamma_G{}'^{-1}(U_G \cap U_{G'})$ is analytic. Thus indeed, this imposes a manifold structure (and a topology) onto $\mathbf{G}$ making it a Lie Group. Every finite product Πe^{B^j} , $B_j\in \mathbf{A}$ may be connected to I, by the path Πe^{tB^j}, $0\le t\le 1$, hence G is connected. Evidently, the tangent space of G at I (i.e.,the space of all directional derivatives of curves in $\mathbf{G}$ starting at I) coincides with $\mathbf{A}$.

Thus, indeed, we established a <u>1-1-correspondence between the connected Lie subgroups of $GL(\mathbb{R}^N)$ and the Lie subalgebras of $L(\mathbb{R}^N)$</u>.

Chapter 1. CALCULUS OF PSEUDO-DIFFERENTIAL OPERATORS

0. Introduction.

In this chapter we deal with the details of pseudo-differential operator calculus. We follow a presentation in a lecture of 1974/75 [CP], inspired by the local approach of Hoermander [Hr_2], dealing with operators on $\mathbb{R}^n$, for didactical reasons. Replacing asymptotic expansions of [CP] by Leibniz formulas with integral remainder of 1.5 is an improvement we learned from R.Beals [B_1] who uses 'weight functions' more general than our $\langle x\rangle^{m_2}\langle\xi\rangle^{m_1}$ of (3.2). Still asymptotic expansions are needed, and will be studied in 1.6.

We will discuss 4 different representations of ψdo', referenced as $a(x,D)=a(M_l,D)$, $a(M_r,D)$, $a(M_l,M_r,D)$, and $a(M_w,D)$, the first two corresponding to the left and right multiplying of Kohn and Nirenberg [KN], and the others to a representation of Friedrichs [Fr_3], and the Weyl representation.

The reader who dislikes the infinitely repeated formal discussions has our sympathy. For other approaches to the same subject cf. ch.7 where the ψdo's of certain symbol classes are identified as operators on $H=L^2(\mathbb{R}^n)$, <u>smooth</u> under action of certain Lie subgroups of $U(H)$. Or else, cf. [C_1] and [C_2], where regular and singular elliptic boundary problems are approached with tools of C^*-algebras, avoiding entirely the ψdo-calculus.

The calculus presented generalizes formal calculus of differential operators. We get a collection of Frechet algebras containing differential operators, with formulas for product and adjoint like Leibniz formulas, containing generalized inverses (so-called Green inverses) of their elliptic and hypo-elliptic operators, as seen in ch.II. The algebras are 'graded': Each of their operators has a differentiation (m_1) and a multiplication (m_2) order.

We will get the same Fredholm theory in Sobolev spaces as provided abstractly, using C^*-algebras, in [C_1],III,IV.

1. Definition of ψdo's.

In this section (and occasionally later on) we will write a(M) for the multiplication operator $u(x)\to a(x)u(x)$. More generally, for a pseudo-differential operator generated from a 'symbol' $a(x,y,\xi) \in C^\infty(\mathbb{R}^{3n})$, with $x,y,\xi \in \mathbb{R}^n$, we will write $a(M_l,M_r,D)$ to indicate an order of operation: Multiplication corresponding to the x-(y-) variable is carried out to the left (to the right) of the differentiation D of the ξ-variable, hence "M_l" (hence "M_r") (cf. cor.2.3).

Introduce the space *ST* of all $a=a(x,y,\xi) \in C^\infty(\mathbb{R}^{3n})$ satisfying

$$(1.1)\quad (D_x^\alpha D_y^\beta D_\xi^\gamma a)(x,y,\xi) = O(\langle\xi\rangle^{\kappa(k)} \cdot \langle(x,y)\rangle^{\lambda(l)}),\ |\alpha|+|\beta| \le k,\ |\gamma| \le l,$$

with nondecreasing $\kappa(k),\lambda(l)$, $k,l=0,1,\dots$, such that

$$(1.2)\qquad \lim_{j\to\infty}(\kappa(j) - j) = \lim_{j\to\infty}(\lambda(j) - j) = -\infty \quad .$$

Here, and in all of the following, we use the abbreviation

$$(1.3)\qquad \langle z\rangle = (1 + |z|^2)^{1/2} \ ,\ z \in \mathbb{R}^m \ ,$$

(particularly $\langle(x,y)\rangle=(1 + x^2 + y^2)^{1/2}$). Another useful shorthand:

$$(1.4)\qquad đx = (2\pi)^{-n/2}dx \ ,\ đ\xi = (2\pi)^{-n/2}d\xi \ ,$$

measures over $\mathbb{R}^n$. The functions of *ST* are called symbols. For $a\in$ *ST* define a linear operator $A=a(M_l,M_r,D)$ called pseudodifferential operator (ψdo) with symbol a by setting, with integrals over $\mathbb{R}^n$,

$$(1.5)\qquad (Au)(x) = \int đ\xi \int đy\ e^{i\xi(x-y)}\ a(x,y,\xi)\ u(y) \quad .$$

The precise meaning of (1.5) is clarified in thm.1.1, below.

Theorem 1.1. The right hand side of (1.5) is well defined, for $u\in$ *S*, $x\in \mathbb{R}^n$, in the following sense: The integral $\int đy$ exists as improper Riemann integral, for $x,\xi\in \mathbb{R}^n$. It supplies a $C^\infty(\mathbb{R}^{2n})$-function I. For fixed $x\in \mathbb{R}^n$, $I(x,.) \in L^1(\mathbb{R}^n)$, and $\int đ\xi$ defines a function $v=Au\in S=S(\mathbb{R}^n)$. Moreover, the map $A:S\to S$ is continuous $S\to S$.

Proof. Clearly $I(x,\xi)= \int đy\ u(y)\ e^{i\xi(x-y)}a(x,y,\xi)$ exists as described, since the exponential is bounded, while $a(x,y,\xi)=O(\langle(x,y)\rangle^{1})$ is of polynomial growth in y, for fixed x, so that $a(x,.,\xi)u(.)\in S \subset L^1(\mathbb{R}^n)$. For a useful ξ-estimate one uses the identity

$$(1.6)\quad e^{i\xi(x-y)} = \langle\xi\rangle^{-2m}(1-\Delta_y)^m e^{i\xi(x-y)} = \langle x-y\rangle^{-2m}(1-\Delta_\xi)^m e^{i\xi(x-y)} \ ,$$

$m=0,1,2,\dots$, with the Laplace operator $\Delta_y = \sum_j \partial_{y_j}^2$. A partial integration for $\int dy$ is applied over a finite ball. The boundary terms die out, as the ball tends to $\mathbb{R}^n$, by (1.1) and $u\in \mathcal{S}$. This yields

$$(1.7)\qquad I(x,\xi) = \langle\xi\rangle^{-2m} \int đy\, e^{i\xi(x-y)}\, (1-\Delta_y)^m (a(x,y,\xi)u(y)) \quad .$$

In the following we shall say that a given expression is '<u>in the span</u>' of another given set of expressions, if it is a finite linear combination with complex coefficients of these expressions. For example, the integrand of (1.7) is in the span of

$$(1.8)\quad p_{\alpha\beta}(x,y,\xi) = e^{i\xi(x-y)}(D_y^\alpha a)(x,y,\xi)u^{(\beta)}(y) \ ,\ |\alpha| + |\beta| \le 2m.$$

Note that we get

$$(1.9)\qquad p_{\alpha\beta}(x,y,\xi) = O(\langle\xi\rangle^{\kappa(2m)}\langle (x,y)\rangle^{\lambda(0)}\langle y\rangle^{-q}) \quad ,$$

using (1.1) and the decay of $u\in \mathcal{S}$, where q may be arbitrary. Thus

$$(1.10)\quad I(x,\xi) = O(\ \langle\xi\rangle^{\kappa(2m)-2m} \int \langle (x,y)\rangle^{\lambda(0)}\langle y\rangle^{-q}\, dy\ ).$$

A substitution $y = \langle x\rangle y^{\sim}$, $dy = \langle x\rangle^n dy^{\sim}$ brings "$\int$" in (1.10) to

$$(1.11)\ \langle x\rangle^{\lambda(0)+n}\int \langle y\rangle^{\lambda(0)}(1+\langle x\rangle^2|y|^2)^{-q/2} \le c\langle x\rangle^{\lambda(0)+n},\ c=\int\langle y\rangle^{\lambda(0)-q}$$

Therefore

$$(1.12)\qquad I(x,\xi) = O(\langle\xi\rangle^{\kappa(2m)-2m}\langle x\rangle^{\lambda(0)+n}\ )\ ,$$

and, with (1.2), it now is evident that

$$(1.13)\qquad \int \xi^\alpha\ I(x,\xi)\ đ\xi$$

exists as an improper Riemann integral ,for all x.The estimate

$$(1.14)\qquad \kappa(2m) - 2m < -n-|\alpha|$$

holds for large m. It insures that $I(x,.)\in L^1$, for all fixed x. For $\alpha=0$ this proves the first assertion of the theorem.

Let us clarify the dependence of (1.12) on the function u :

<u>Lemma 1.2.</u> We have

$$(1.15)\qquad |I(x,\xi)| \le c\|u\|_{k_0}\langle\xi\rangle^{\kappa(2m)-2m}\langle x\rangle^{\lambda(0)+n} \quad ,$$

with c depending on m and $a\in ST$, but not on $u\in \mathcal{S}$, where the norms

$$(1.16)\qquad \|u\|_k = \sup_{|\alpha|\le k, x\in\mathbb{R}^n}\|\langle x\rangle^k u^{(\alpha)}\|_{L^\infty}$$

of $\mathcal{S}$ are used.

This lemma is clear, since $|u^{(\alpha)}(x)|\le\|u\|_k\langle x\rangle^{-k}$, $x\in\mathbb{R}^n$, $|\alpha|\le k$, estimates $u^{(\alpha)}$ by $|x|^{-q}$, as the only way, u enters (1.12).

Let us prepare a few other simple lemmata for thm.1.1.

<u>Lemma 1.3.</u> For $a \in ST$ we get $a_{\alpha\beta\gamma}=D^x_\alpha D^y_\beta D^\xi_\gamma a \in ST$ and $P\cdot a \in ST$, for any polynomial $P=P(x,y,\xi)$. With κ,λ of(1.2) for a, corresponding functions for $a_{\alpha\beta\gamma}$ and $P\cdot a$ are $\kappa(j+|\alpha|+|\beta|)$, $\lambda(j+|\gamma|)$, and $\kappa(j+L)$, $\lambda(j+M)$, resp., with the ξ-degree L and (x,y)-degree M of P.

The verification of Lemma 1.3 is left to the reader.

<u>Lemma 1.4.</u> The integral at right of (1.5) defines a C^∞-function of x denoted by v . Moreover, the derivatives $v^{(\alpha)}$ may be calculated under the integral signs of (1.5).

<u>Proof</u>. Notice that $D^\beta_x(e^{i\xi(x-y)}a(x,y,\xi))$ is in the span of

$$e^{i\xi(x-y)}\xi^\alpha D^\gamma_x a(x,y,\xi) \quad , \quad \alpha+\gamma=\beta\ . \tag{1.17}$$

By lemma 1.3 we have $\xi^\alpha D^\gamma_x a(x,y,\xi) \in ST$. For an induction proof one only must show existence of the first derivatives and differentiability under the integral sign for $|\beta|=1$. Let $\varphi \in L^1 \cap L^\infty(\mathbb{R})$, supp $\varphi \subset\subset \mathbb{R}$. Using (1.12) and Fubini's theorem (twice) we get

$$\begin{aligned}&\int dx_j\varphi(x_j)\int d\xi\int dy\ u(y)D^\beta_x(e^{i\xi(x-y)}a(x,y,\xi))\\ &=\int d\xi\int dy\int dx_j\varphi(x_j)u(y)D^\beta_x(e^{i\xi(x-y)}a(x,y,\xi))\ ,\ D^\beta_x=-i\partial_{x_j}\ ,\end{aligned} \tag{1.18}$$

where we set $\varphi=\chi_{[a,t]}$, the characteristic function of $[a,t]$. The inner integral at right of (1.18) may be calculated explicitly. A differentiation of (1.18) for t will give the desired result.

<u>Lemma 1.5.</u> For $a \in ST$ and every multi-index γ we have

$$\begin{aligned}&\int d\xi\int dy\ u(y)\ (x-y)^\gamma e^{i\xi(x-y)}a(x,y,\xi)\\ &=(-1)^\gamma\int d\xi\int dy\ u(y)\ e^{i\xi(x-y)}(D^\gamma_\xi a)(x,y,\xi)\ .\end{aligned} \tag{1.19}$$

<u>Proof.</u> By the second identity (1.6), (1.19) is a matter of partial integration, carrying the differentiation from the exponential to $a(x,y,\xi)$. But the interchange of $\int d\xi$ and $\int dy$ is generally impossible, so that we will require the argument, below.

1) It suffices again to assume $|\gamma| = 1$, by lemma 1.3.

2) A conclusion as in lemma 1.4 may be used to verify that

$$\int d\xi\int dy\partial_{\xi_j}\{u(y)e^{i\xi(x-y)}a(x,y,\xi)\}=\int d\xi\partial_{\xi_j}\int dyu(y)e^{i\xi(x-y)}a(x,y,\xi)\ .$$

3) The right hand side of (2) is zero, because boundary integrals vanish, by (1.12). Thus (2) yields (1.19) for $|\gamma|=1$, q.e.d.

<u>Proof of thm.1.1 (continued).</u> Let $v = Au$, $u \in \mathcal{S}$, as defined by (1.5). Lemma 1.4 implies

$$(1.21)\qquad x^{\alpha}v^{(\beta)}(x) = \int d\!\!/\xi \int d\!\!/y\; u(y)((x-y)+y)^{\alpha}\partial_x^{\beta}(e^{i\xi(x-y)}a(x,y,\xi))\ ,$$

where the right hand side is in the span of

$$(1.22)\qquad \int d\xi \int dy(u(y)y^{\delta})(x-y)^{\varepsilon}(D_x^{\sigma}e^{i\xi(x-y)})(D_x^{\tau}a)(x,y,\xi),\ \delta+\varepsilon=\alpha,\ \sigma+\tau=\beta.$$

Using $D_x^{\sigma}e^{i\xi(x-y)} = (-1)^{|\sigma|}D_y^{\sigma}e^{i\xi(x-y)}$ in (1.22) we may integrate by parts, in the inner integral, to obtain expressions in the span of

$$(1.23)\qquad \begin{aligned}&\int d\xi \int dy(u(y)y^{\delta})^{(\lambda)}(x-y)^{\varepsilon-\mu}D_x^{\tau}D_y^{\nu}a(x,y,\xi)e^{i\xi(x-y)}\\ &=(-1)^{\varepsilon-\mu}\int d\xi \int dy(u(y)y^{\delta})^{(\lambda)}D_x^{\tau}D_y^{\nu}D_{\xi}^{\varepsilon-\mu}a(x,y,\xi),\ \delta+\varepsilon=\alpha,\ \sigma+\tau=\beta,\ \lambda+\mu+\nu=\sigma\end{aligned}$$

We also applied lemma 1.5 and another partial integration $d\xi$. Note that $|\varepsilon-\mu|\le|\varepsilon|\le|\alpha|$, and $|\tau|+|\nu|\le|\tau|+|\sigma|=|\beta|$. Hence lemma 1.2 and lemma 1.3 give an estimate of (1.22) by

$$(1.24)\qquad \int d\xi\ \ c\|u\|_k\langle\xi\rangle^{\kappa(2m+|\beta|)-2m}\langle x\rangle^{n+\lambda(|\alpha|)}\ .$$

Accordingly $x^{\alpha}v^{(\beta)}(x)$ also is bounded by (1.24). Applying this for all $|\alpha|\le l$, for some integer l, we get

$$(1.25)\qquad |v^{(\beta)}(x)| \le c_0\langle x\rangle^{n+\lambda(l)-l}\|u\|_k\ ,\ c_0=c\int d\xi\langle\xi\rangle^{\kappa(2m+|\beta|)-2m}\ .$$

Here m , l are integers, but m must be chosen large enough to insure existence of the integral in (1.25), by (1.2). Also, c_0 and k depend on the choice of m,l and β, but not on u. Using (1.2) it follows that for l_0 there exists k, c_0 independent of u with

$$(1.26)\qquad \|Au\|_{l_0} \le c_0\|u\|_k\ ,\ u\in\mathcal{S}\ .$$

Clearly this amounts to continuity of $a:\mathcal{S}\to\mathcal{S}$ and thm.1.1 is proven.

<u>2. Elementary properties of ψdo's.</u>

<u>Theorem 2.1.</u> Let

$$(2.1)\qquad \langle u,v\rangle = \int_{\mathbb{R}^n} u(x)v(x)d\!\!/x\ ,\ u,v\in\mathcal{S}\ .$$

Then we have, for all $a \in \mathcal{ST}$,

$$(2.2)\qquad \langle a(M_l,M_r,D)u,v\rangle = \langle u,a(M_r,M_l,-D)v\rangle\ ,\ \text{for all } u,v\in\mathcal{S}\ ,$$

with $A^{\sim}=a(M_r,M_l,-D)$, the ψdo with symbol $a^{\sim}(x,y,\xi)=a(y,x,-\xi)$.

Proof. Clearly $a^{\sim}\in ST$ as $a\in ST$. The statement is a matter of integral interchanges. Get $J=v(x)I(x,\xi)\in L^1(\mathbb{R}^{2n})$, $v\in S$, by lemma 1.2. Hence $\langle v,Au\rangle =\int d\!\!{}^-x\int d\!\!{}^-\xi v(x)I(x,\xi)= \int d\!\!{}^-\xi\int d\!\!{}^-x\int d\!\!{}^-y e^{i\xi(x-y)}v(x)u(y)a(x,y,\xi)$

$$= \int d\!\!{}^-\xi\int d\!\!{}^-y\int d\!\!{}^-x \ \ldots = \int d\!\!{}^-y\int d\!\!{}^-\xi\int d\!\!{}^-x \ \ldots \quad = \langle u,A^{\sim}v\rangle ,$$

for $A = a(M_l,M_r,D)$, with a 3-fold application of Fubini's theorem and a substitution $(x,y,\xi) \to (y,x,-\xi)$ of variables, q.e.d.

$A^{\sim}$ will be called the distribution adjoint (or dual) of the operator A, (to be distinguished from the 'formal Hilbert space adjoint' A^*, formed with the inner product $(u,v) = \int \bar{u}v dx$). Note:

$$(2.3) \qquad A^* = \bar{a}(M_r,M_l,D) \quad .$$

Since $A=a(M_l,M_r,D)$ is strongly continuous $S\to S$, by thm.1.1, an extension $B : S' \to S'$ of $A : S \to S$ is given by

$$(2.4) \qquad \langle Bu,\varphi\rangle = \langle u,A^{\sim}\varphi\rangle \quad , \quad u \in S' , \varphi \in S ,$$

with $\langle u,\varphi\rangle$, the value of $u\in S'$ at $\varphi\in S$. In other words, we may regard the ψdo either as operator $S\to S$ or, $S'\to S'$. Thm.2.1 gives $A\subset B$.

Generally we will not distinguish in notation between the operators A and B , but will regard a ψdo as an operator $A:S'\to S'$ with $A|S$ mapping to S. In fact, ψdo's more or less will be regarded like differential expressions in $[C_2]$,II with domain to be fixed.

Theorem 2.2. ST is an algebra of complex-valued functions under pointwise addition and multiplication, with (real) involution $a\to a^{\sim}$.
Proof. Evidently ST is closed under "+", "$^{\sim}$", and scalar product. Note that $D_x^{\alpha}D_y^{\beta}D_{\xi}^{\gamma}(ab)$ is in the span of (2.5), by Leibniz' formula:

$$(2.5) \quad (D_x^{\alpha'}D_y^{\beta'}D_{\xi}^{\gamma'}a)\cdot(D_x^{\alpha''}D_y^{\beta''}D_{\xi}^{\gamma''}b) \quad , \ \alpha'+\alpha''=\alpha \ , \ \beta'+\beta''=\beta \ , \ \gamma'+\gamma''=\gamma \ .$$

In (1.1) we may assume the same κ and λ, for a and b. Using (1.2) write $\kappa(j) = j - \sigma(j)$, $\lambda(j) = j - \tau(j)$, $\sigma(j) \to \infty$, $\tau(j) \to \infty$, and estimate the products of (2.5) by

$$(2.6)\langle\xi\rangle^{|\alpha|+|\beta|-\sigma(|\alpha'|+|\beta'|)-\sigma(|\alpha''|+|\beta''|)}\langle(x,y)\rangle^{|\gamma|-\tau(|\gamma'|)-\tau(|\gamma''|)}$$

Then define $\sigma'(j)=\mathrm{Min}_{l\le j}\{\sigma(l)+\sigma(j-l)\}$, $\tau'(j)=\mathrm{Min}_{l\le j}\{\tau(l)+\tau(j-l)\}$,
We get the estimates (1.1) for c with $\kappa'=j-\sigma'$, $\lambda'=j-\tau'$ instead of

κ,λ. In particular we also get $\kappa' \to \infty$, $\lambda' \to \infty$, q.e.d.

Corollary 2.3. The symbol algebra ST contains all a of the form

$$(2.7) \qquad a(x,y,\xi) = \sum_{|\alpha|\le N} b_\alpha(x) c_\alpha(y) \xi^\alpha ,$$

where b_α, $c_\alpha \in C^\infty(\mathbb{R}^n)$ satisfy the estimates

$$(2.8) \qquad D^\beta b_\alpha(x) = O(\langle x\rangle^p) , \quad D^\beta c_\alpha(x) = O(\langle x\rangle^p) , \quad x \in \mathbb{R}^n ,$$

with a constant p, independent of x, α, β. Then we get

$$(2.9) \qquad A = a(M_l, M_r, D) = \sum_{|\alpha|\le N} b_\alpha(M) D^\alpha c_\alpha(M) ,$$

a differential operator.

Proof. The Fourier transform $u^\wedge = Fu$ and its inverse $u^\vee = F^{-1}u$,

$$(2.10) \qquad u^\wedge(\xi) = \int đx\, e^{-ix\xi} u(x) , \quad u^\vee(x) = \int đ\xi\, e^{ix\xi} u(\xi) , \quad u \in S ,$$

satisfy (c.f. 0,(1.22)).

$$(2.11) \qquad F D_j F^{-1} = M_j , \quad j = 1,\dots,n.$$

Theorem 2.4. Let a,b $\in ST$ be independent of y and x, respectively. Then the ψdo's $A = a(M_l,D)$, and $B = b(M_r,D)$ may be written as

$$(2.12) \quad (Au)(x) = \int đ\xi e^{ix\xi} a(x,\xi) u^\wedge(\xi) , \quad (Bu)^\wedge(\xi) = \int dy e^{-iy\xi} b(y,\xi) u(y).$$

for $u \in S$, with the Fourier transform (2.10).

Proof. The first formula is clear, using (2.10). Also,

$$(2.13) \qquad (Bu)(x) = \int đ\xi e^{ix\xi} \int đy\, e^{-iy\xi} b(y,\xi) u(y) ,$$

where the inner integral $I(\xi)$ is in $L^1(\mathbb{R}^n)$, by lemma 1.2 (In this case I is independent of x). Hence we may write $(Bu)(x) = I^\vee(x)$. We know that $Bu \in S$. Therefore $(Bu)^\wedge = (I^\vee)^\wedge = I$, as proposed, q.e.d.

Theorem 2.5. For a,b $\in ST$ let $c(x,y,\xi) = a(x,\xi) b(y,\xi)$. We have

$$(2.14) \qquad c \in ST , \text{ and } c(M_l,M_r,D) = a(M_l,D) b(M_r,D) .$$

Proof. From Theorem 2.2 we conclude that $c \in ST$. Using (2.12),

$$A(Bu)(x) = \int đ\xi e^{ix\xi} a(x,\xi) (Bu)^\wedge(\xi) = \int đ\xi e^{ix\xi} a(x,\xi) \int đy\, b(y,\xi) e^{-iy\xi} u(y)$$

$$= (c(M_l,M_r,D)u)(x) , \quad u \in S , \text{ q.e.d.}$$

For a symbol $a \in ST$ consider the bilinear form

$$(2.15)\quad \langle v,Au\rangle = \langle v,a(M_l,M_r,D)u\rangle = \int đx\int đ\xi\int đy\, a(x,y,\xi)e^{i\xi(x-y)}v(x)u(y),$$

for $u,v \in \mathcal{S}$ .We have seen that $\int đx\int đ\xi\int đy = \int đ\xi\int đx đy$ .Note that a temperate distribution $k \in \mathcal{S}'(\mathbb{R}^{2n})$ is defined by

$$(2.16)\quad \langle k,w\rangle = \int đ\xi\int đxđy\; e^{i\xi(x-y)}a(x,y,\xi)w(x,y)\ ,\ w \in \mathcal{S}(\mathbb{R}^{2n})\ .$$

By the techniques of sec.1 we get

$$(2.17)\quad \langle k,w\rangle = \int đ\xi\langle\xi\rangle^{-2m}\int đxđy e^{i\xi(x-y)}(1-\Delta_x)^m(aw)\quad .$$

For large m this is in the span of

$$(2.18)\quad \int dxdy\left(\int d\xi e^{i\xi(x-y)}(D_x^\gamma a)(x,y,\xi)\langle\xi\rangle^{-2m}\right)D_x^\beta w(x,y)\ ,\ |\beta|+|\gamma|\le 2m,$$

where (1.1),(1.2) give existence of the inner integral J. But J is (x,y)-continuous and $O(\langle(x,y)\rangle^{\lambda(0)})$. Thus $\langle J,D_x^\beta w\rangle$ define $D_x^\beta J\in \mathcal{S}'$.

We shall call k, defined by (2.16), the <u>distribution kernel</u> of $A=a(M_l,M_r,D)$, $a\in ST$. Note $k\in \mathcal{S}'(\mathbb{R}^{2n})$ is uniquely determined by

$$(2.19)\quad \langle v,Au\rangle = \langle k,v\otimes u\rangle\quad ,\ u,v \in \mathcal{S}\ .$$

Formally, with distribution integrals, we may write

$$(2.20)\quad (Au)(x) = \int k(x,y)u(y)đy\ ,\ (A^{\sim}u)(x) = \int k(y,x)u(y)đy\ .$$

<u>Proposition 2.6</u>. We have

$$(2.21)\quad k(x,y) = k_A(x,y) = a^{\vee}(x,y,x-y) = a^{\vee}(x,y,\sigma)|_{\sigma=x-y}\ ,$$

in the following precise sense: The inverse Fourier transform $a^{\vee}$ of $a(x,y,\xi)$ with respect to its third (its 2n+1-st...3n-th) argument ξ is a distribution in $\mathcal{S}'(\mathbb{R}^{3n})$ which may be written as a continuous family $p_\sigma=a^{\vee}(x,y,(x-y)+\sigma)$ of distributions in $\mathcal{S}'(\mathbb{R}^{2n})$,

$$(2.22)\quad \langle p_\sigma,\psi\rangle = \int dxdy p_\sigma(x,y)\psi(x,y)\ ,\ \psi\in \mathcal{S}(\mathbb{R}^{2n})\ ,$$

in the form

$$(2.23)\quad \langle a^{\vee},\varphi\rangle = \int d\sigma\langle p_\sigma,\psi_\sigma\rangle\ ,\ \psi_\sigma(x,y)=\varphi(x,y,\sigma+x-y)\ ,\ \varphi\in \mathcal{D}(\mathbb{R}^{3n})\ ,$$

with a distribution integral in (2.22) defining $p_\sigma\in C(\mathbb{R}^n,\mathcal{S}'(\mathbb{R}^{2n}))$ while (2.23) is a Riemann integral with continuous integrand, of compact support. Then the distribution kernel k coincides with the distribution p_0 of (2.22) for $\sigma=0$.

Proof. Clearly a family $p_\sigma \in C(\mathbb{R}^n, \mathcal{S}'(\mathbb{R}^{2n}))$ is defined by setting

$$(2.24)\qquad \langle p_\sigma,\psi\rangle = \int đ\xi \int dxdy e^{i\xi(x-y+\sigma)} a(x,y,\xi)\psi(x,y)\ ,\ \psi\in S(\mathbb{R}^{2n})\ .$$

Indeed, we get p_σ in the form (2.17) with a replaced by $ae^{i\sigma\xi}$. Then a look at (2.18) (with this a) shows that $\langle p_\sigma,\psi\rangle \in C^0(\mathbb{R}^n)$.

We get $\int d\sigma\langle p_\sigma,\psi_\sigma\rangle = \int đ\xi \int dxdyd\sigma e^{i\xi(x-y+\sigma)} a(x,y,\xi)\varphi(x,y,\sigma+x-y)$

$$= \int đ\xi \int dxdyd\sigma e^{i\xi\sigma} a(x,y,\xi)\varphi(x,y,\sigma) = \int d\xi dxdy a(x,y,\xi)\varphi^{\vee}(x,y,\xi) = \langle a^{\vee},\varphi\rangle\ ,$$

confirming (2.23), where we have used a linear substitution of integration variable, and some Fubini-type interchanges. Q.E.D.

The distribution kernel's singular support will be of interest in the following. In sec.3 we introduce restricted symbol classes with the property that the kernel $k=k_a$ for such a symbol a has singular support contained in the 'diagonal' $\{(x,y)\in \mathbb{R}^{2n}: x=y\}$.

For general $a \in \mathcal{ST}$ this is not true: Consider $a(\xi)=\cos |\xi|^2 \in \mathcal{ST}$. Then sing supp k_a coincides with the set $\{(x,y):|x-y|=1\}$, as follows from prop.2.6 , 0,lemma 4.1 and 0,(4.42).

In the following let $\mathcal{ST}_1=\{a\in \mathcal{ST}: \partial_y^\alpha a=0\}$ be the set of $a\in \mathcal{ST}$ independent of y. Following a general convention the class of all ψdo's $A=a(M_1,D)$ with symbol in a class $\mathcal{P}$ will be denoted by Op$\mathcal{P}$.

3. Hoermander symbols; Weyl ψdo's; distribution kernels.

For differential operators A of the form (2.9) it appears that there are many representations as ψdo's, due to the fact that Leibniz' formula may be used to convert a product $D^\alpha a(M)$ to a sum of products $a_{\alpha'}(M)D^{\alpha'}$, and vice versa. In fact, it is clear that a differential operator (2.9) always may be rewritten in the form

$$(3.1)\qquad A = b(M_1,D) = b_1(M_r,D)\ ,$$

with unique symbols $b\in \mathcal{ST}$ and $b_1\in \mathcal{ST}$,depending on x,ξ (y,ξ) only.

The same will be true for a ψdo $A=a(M_1,M_r,D)$, if a satisfies stronger inequalities. In fact, a generalization of the Leibniz formulas will be developed, either involving a Taylor expansion type reminder, or an asymptotically convergent infinite series, as discussed in sec.'s 4 and 5. Later we will use only one representation $A=a(M_1,D)$ for all ψdo's (also written as $A=a(x,D)$), following a general convention. But it will be convenient to have more general forms available. The restricted symbol class $SS\subset ST$ below will have other desirable features: The distribution kernel k_A of A, for an $a\in SS$, (cf. sec.2) will have sing supp $k_A\subset\{x=y\}$. OpSS will be an algebra. Also, as seen in II,5, a ψdo in OpSS will not increase the wave front set of a temperate distribution.

Apart from the representations $A=b(M_l,D)$, and $A=c(M_r,D)$ resulting for $A=a(M_l,M_r,D)\in$ OpSS, a third representation, of the form $A=e((M_l+M_r)/2,D)=e(M_w,D)$ is useful, with a third symbol e , and $e(M_l+M_r)/2,D)=f(M_l,M_r,D)$ with $f(x,y,\xi)=e((x+y)/2,\xi)$. $e(M_w,D)$ will be called the <u>Weyl representation of</u> a ψdo. $b(M_l,D)$ and $c(M_r,D)$ are referred to as the 'left (right) multiplying representations'.

Let SS denote the class of functions $a\in C^\infty(\mathbb{R}^{3n})$ such that

$$(3.2)\qquad D_x^\alpha D_y^\beta D_\xi^\gamma a = O\left(\langle\xi\rangle^{m_1+\delta(|\alpha|+|\beta|)-\rho_1|\gamma|}\langle x\rangle^{m_2-\rho_2|\alpha|}\langle y\rangle^{m_3-\rho_3|\beta|}\right),$$

with real constants m_j,ρ_j,τ satisfying

$$(3.3)\qquad 0<\rho_j\le 1\ ,\ j=1,2,3,\ 0\le\delta<\rho_1\ .$$

Also, by $SS^{m,\rho,\delta}$ we denote the class of all $a\in ST$ satisfying (3.2) for a given $m=(m_1,m_2,m_3)$, $\rho=(\rho_1,\rho_2,\rho_3)$, δ , where only the first condition (3.3) will be required, and no longer $\delta<\rho_3$. Thus, $SS^{m,\rho,\delta}$ in general will not be a subset of SS.

Let ψh and $\psi h_{m,\rho,\delta}$ denote the classes of $a(x,\xi)\in SS$ or $SS_{m,\rho,\delta}$, (i.e., a is independent of y). Clearly, for ψh and $\psi h_{m,\rho,\delta}$, the constants m_3, ρ_3 are redundant: Any $m_3\ge 0$ and any $\rho_3\in\mathbb{R}$ may be chosen since $\partial_y^\alpha a=0$. We then will set $m_3=0$, $\rho_3=1$, or omit m_3 , ρ_3, writing $m=(m_1,m_2)$ and $\rho=(\rho_1,\rho_2)$ as 2-component vectors only, ignoring the trivial estimates (3.2) for $\beta\ne 0$. We will mostly be concerned with ψh, since it will be found that OpSS = Opψh.

We also define

$$\psi h_{-\infty} = \cap\{\psi h_{m,\rho,\delta} : m\in\mathbb{R}^2\} = \mathcal{S}(\mathbb{R}^{2n})\quad ,$$

$$\psi h_{\infty,\rho,\delta} = \cup\{\psi h_{m,\rho,\delta} : m\in\mathbb{R}^2\}\ .$$

Hoermander [Hr_2] has introduced a class $S^m_{\rho,\delta}$ of <u>local symbols</u> consisting of $C^\infty(\Omega\times\mathbb{R}^n)$-functions, with a domain $\Omega\subset\mathbb{R}^n$ using inequalities similar to (3.2). He defines $S^m_{\rho,\delta}$ by the estimates

$$(3.4)\qquad a^{(\alpha)}_{(\beta)}(x,\xi) = O(\langle\xi\rangle^{m-\rho|\alpha|+\delta|\beta|})\ ,\ \xi\in\mathbb{R}^n\ ,\ x\in K\ ,$$

for all α, on all sets $K\subset\subset\Omega$. A ψdo $A=a(x,D)$, similar to $a(M_l,D)$, is defined for $C_0^\infty(\Omega)$-functions u by the first formula (2.12).

Accordingly we shall refer to symbols in the classes $S^{m}_{\rho,\delta}$, $\boldsymbol{SS}$, $\psi\boldsymbol{h}$, $\boldsymbol{SS}^{m,\rho,\delta}$, $\psi\boldsymbol{h}_{m,\rho,\delta}$ as <u>Hoermander type symbols</u> .

From (3.2) it follows that

$$(3.5) \qquad (D^{\alpha}_{x}D^{\beta}_{y}D^{\gamma}_{\xi}a)(x,y,\xi)=O\left(\langle\xi\rangle^{m_1+\delta(|\alpha|+|\beta|)}\langle(x,y)\rangle^{|m_1|+|m_2|}\right) ,$$

recalling that $\rho_j \geq 0$. This implies (1.1), with $\kappa(k)= m_1+\delta k$, $\lambda(k)= |m_2|+|m_3|$. Moreover, (1.2) holds whenever $\delta < 1$. Accordingly,

$$(3.6) \qquad \boldsymbol{SS} \subset \boldsymbol{ST} \ , \ \boldsymbol{SS}^{m,\rho,\delta} \subset \boldsymbol{ST} \ , \text{ as } \delta < 1 \ .$$

<u>Lemma 3.1.</u> Let $a \in \boldsymbol{SS}^{m,\rho,\delta}$, $b \in \boldsymbol{SS}^{m,\rho',\delta'}$, then we get $a + b \in \boldsymbol{SS}^{m^{\sim},\rho^{\sim},\delta^{\sim}}$, $a\cdot b \in \boldsymbol{SS}^{m^{\wedge},\rho^{\wedge},\delta^{\wedge}}$, where

$$(3.7) \quad m^{\sim}_j=\text{Max}\{m_j,m'_j\},\ \delta^{\sim}=\delta^{\wedge}=\text{Max}\{\delta,\delta'\},\ m^{\wedge}_j=m_j+m'_j,\ \rho^{\sim}_j=\rho^{\wedge}_j=\text{Min}\{\rho_j,\rho'_j\} .$$

Moreover, $\cup\{\boldsymbol{SS}^{m,\rho,\delta}: m_j \in \mathbb{R}\}=\boldsymbol{SS}^{\infty,\rho,\delta}$ is an algebra under pointwise addition and multiplication. All statements remain true if $\boldsymbol{SS}$, in all expressions, is replaced by $\psi\boldsymbol{h}$.

The proof uses Leibniz' formula. It is left to the reader.

<u>Remark:</u> Note that $\boldsymbol{SS}$ and $\psi\boldsymbol{h}$ are not algebras. In particular the condition $\delta < \rho_1$ needs not to hold for sum or product of two symbols with summands or factors of different ρ and δ.

Thm.3.2 below is central for the calculus of ψdo's. Its proof will be given after some preparations, discussed in sec.5ff.

<u>Theorem 3.2.</u> Let $a \in \boldsymbol{SS}^{m,\rho,\delta}$, with m,ρ,δ satisfying (3.3) . There exists a unique $b \in \psi\boldsymbol{h}_{m^{\sim},\rho^{\sim},\delta^{\sim}}$ and $b_1 \in \psi\boldsymbol{h}_{m^{\sim},\rho^{\sim},\delta^{\sim}}$, with

$$(3.8) \quad m^{\sim}_1 = m_1 \ , \ m^{\sim}_2 = m_2+m_3 \ , \ \rho^{\sim}_1 = \rho_1 \ , \ \delta^{\sim} = \delta \ , \ \rho^{\sim}_2 = \text{Min}\{\rho_2,\rho_3\} \ ,$$

such that (3.1) holds for $A=a(M_l,M_r,D)$. Moreover, there also exists a unique $b_2\in \psi\boldsymbol{h}_{m^{\sim},\rho^{\sim},\delta^{\sim}}$, with $m^{\sim},\rho^{\sim},\delta^{\sim}$ of (3.8) such that

$$(3.9) \qquad A = a(M_l,M_r,D) = b_2((M_l+M_r)/2,D) \ .$$

Observe that $c=b_2((x+y)/2,\xi)\in \boldsymbol{ST}$, whenever $b_2 \in \psi\boldsymbol{h}_{m,\rho,\delta}$, $\delta<1$, by an estimate similar to (3.5). Thus $c(M_l,M_r,D)=b_2((M_l+M_r)/2,D)$ is well defined, but need not be in $\boldsymbol{SS}$. We will find, later on, that, with m,ρ,δ satisfying cdn.(3.3), the classes $\{a(M_l,D)\}$, $\{a(M_r,D)\}$, $\{a(M_w,D)\}$, where $a \in \psi\boldsymbol{h}_{m,\rho,\delta}$, are identical.

Note that, for $C = a((M_l+M_r)/2,D)$, we get from (2.3) that

$$(3.10) \qquad C^* = \bar{a}((M_l+M_r)/2,D) \ .$$

Note that, for $C = a((M_l+M_r)/2,D)$, we get from (2.3) that

$$C^* = \bar{a}((M_l+M_r)/2,D) \quad . \tag{3.10}$$

This points to one of the advantages of the Weyl type ψdo: <u>If the symbol a is real then the operator C above is formally selfadjoint</u> On the other hand the 'left multiplying type' $a(M_l,D)$ has other advantages: If $a(x,\xi)$ is a polynomial in ξ, the resulting differential operator $a(M_l,D)$ is in the conventional form -coefficients at left from the differentiations. Recall, for Weyl type ψdo's,

$$a(M_w,D) = a((M_l+M_r)/2,D) = c(M_l,M_r,D),\ c(x,y,\xi)=a((x+y)/2,\xi). \tag{3.11}$$

The restricted class SS produces useful distribution kernels:

<u>Theorem 3.3.</u> Let $a \in SS$. Then the distribution kernel $k \in S'(\mathbb{R}^{2n})$ of $A=a(M_l,M_r,D)$ has its singular support contained in the diagonal $\{(x,y)\in \mathbb{R}^{2n}: x=y\} = \Theta$ of $\mathbb{R}^{2n}$. Moreover, for fixed x (fixed y), the distribution $k(x,.)$ $(k(.,y))$ equals a function in $S(\mathbb{R}^n)$ outside any neighbourhood of $x=y$, and uniformly so for $x\in K$ $(y\in K)$, K any compact set (in fact, we have $\langle y\rangle^{\alpha}|\partial_x^{\beta}\partial_y^{\gamma}k|\le c_{\alpha\beta\gamma\varepsilon}$, as $x\in K$, $|x-y|\ge\varepsilon$ ($\langle y\rangle^{\alpha}|\partial_x^{\beta}\partial_y^{\gamma}k|\le c_{\alpha\beta\gamma\varepsilon}$, as $y\in K$, $|x-y|\ge\varepsilon$)).

<u>Proof</u>. Referring to prop.2.6, and the discussion in [C1],II, note that $a(x,y,.) \in M$ hence $a^{\vee}(x,y,.)\in S^0_{ps}$ (cf.[C1],II,thm.4.3). This implies the statement. Offering details (and independence from [C1]), let $w\in C_0^{\infty}(\mathbb{R}^{2n})$, $w=0$ near Θ. Then supp w has positive distance from Θ, and $|x-y|^{-m}\in C^{\infty}(\text{supp } w)$. By (1.6), with a partial integration, one may bring (2.16) to the form

$$\langle k,w\rangle = \int d\!\!\!/\xi\int d\!\!\!/x d\!\!\!/y\ e^{i\xi(x-y)}w(x,y)|x-y|^{-2m}\Delta_{\xi}^{m}a(x,y,\xi) \quad . \tag{3.12}$$

From (3.1) we get

$$\Delta_{\xi}^{m}a(x,y,\xi) = O(\langle\xi\rangle^{m_1-2m\rho_1}\langle(x,y)\rangle^{m_2+m_3}) \ ,\ m = 0,1,2,\dots . \tag{3.13}$$

If m is large enough then the integrand will be $L^1(\mathbb{R}^{3n})$, so that the integrals may be interchanged, for

$$\langle k,w\rangle = \int kw d\!\!\!/x d\!\!\!/y \ ,\ k(x,y)= \int d\!\!\!/\xi e^{i\xi(x-y)}\Delta_{\xi}^{m}a(x,y,\xi)/|x-y|^{2m}. \tag{3.14}$$

(3.14) has convergent Lebesgue integrals, if m is large. Moreover,as m gets larger and larger, the formulas

$$\partial_x^{\alpha}\partial_y^{\beta}k(x,y) = \int d\!\!\!/\xi\ \partial_x^{\alpha}\partial_y^{\beta}\left(e^{i\xi(x-y)}\Delta_{\xi}^{m}a(x,y,\xi)/|x-y|^{2m}\right) \tag{3.15}$$

give the derivatives of k(x,y) (k is independent of m; the inte-

that sing supp $k \subset \Theta$. Very similar arguments will show that $y^{\alpha}|\partial_x^{\beta}\partial_y^{\gamma}k| \le c_{\alpha\beta\gamma\varepsilon}$, as $x \in K$, $|x-y| \ge \varepsilon$, etc. Q.E.D.

4. The composition formulas of Beals.

Proposition 4.1. For $a \in C_0^{\infty}(\mathbb{R}^{2n})$, $b \in C_0^{\infty}(\mathbb{R}^{3n})$ we have the formulas

$$(4.1) \qquad a(M_r,D) = p(M_l,D) \ , \ b(M_l,M_r,D) = q(M_l,D) \ ,$$

with $p,q \in \mathcal{S}(\mathbb{R}^{2n})$ given by

$$(4.2) \quad p(x,\xi)=\int d\!\!\!/ y d\!\!\!/ \eta e^{-iy\eta}a(x-y,\xi-\eta) \ , \ q(x,\xi)=\int d\!\!\!/ y d\!\!\!/ \eta e^{-iy\eta}b(x,x-y,\xi-\eta).$$

Proof. Focus on the first relation (4.1), typical for the second. For $u \in \mathcal{S}$ and $a \in C_0^{\infty}(\mathbb{R}^{2n})$ we get

$$(4.3) \quad a(M_r,D)u(x) = \int d\!\!\!/ \xi d\!\!\!/ y e^{i\xi(x-y)}a(y,\xi)u(y) = \int dy a^{\vee}(y,x-y)u(y)$$

with "$\vee$" indicating the inverse Fourier transform with respect to the 2-nd (set of)variable(s) of $a(x,\xi)$, by integral interchange. Write $a^{\vee}(y,x-y)=a^{\vee}(x-(x-y),x-y)=c(x,x-y)$, i.e., $c(x,z)=a^{\vee}(x-z,z)= \int d\!\!\!/ \xi e^{i\xi z}a(x-z,\xi)$. Clearly $c \in C^{\infty}(\mathbb{R}^{2n})$, $c=0$ for large $|x|$, $c(x,.) \in \mathcal{S}$, for fixed x. Hence $c \in \mathcal{S}(\mathbb{R}^{2n})$; we may define $p(x,\xi)=\int d\!\!\!/ z c(x,z)e^{-i\xi z} \in \mathcal{S}(\mathbb{R}^{2n})$, and get $c(x,z)=p^{\vee}(x,z)$, $c(x,x-y)=p^{\vee}(x,x-y)$. Accordingly,

$$(4.4) \quad a(M_r,D)u(x) = \int d\!\!\!/ y p^{\vee}(x,x-y)u(y) = \int d\!\!\!/ \xi d\!\!\!/ y p(x,\xi)e^{i\xi(x-y)}u(y) \quad .$$

Clearly the right hand side of (4.4) equals $p(M_l,D)u(x)$. Also,

$$(4.5) \qquad p(x,\xi) = \int d\!\!\!/ z d\!\!\!/ \zeta a(x-z,\zeta)e^{-iz(\xi-\zeta)} \quad ,$$

which becomes (4.2), after an integral substitution $\xi-\zeta=\eta$, q.e.d.

Proposition 4.2. Let $A=a(M_l,D)$, $B=b(M_l,D)$ with $a,b \in C_0^{\infty}(\mathbb{R}^{2n})$. Then Then $C=AB=c(M_l,D)$, $A^*=a^*(M_l,D)$ with $c(x,\xi)$, $a^*(x,\xi)$ given by

$$(4.6) \quad \begin{aligned} c(x,\xi) &= \int d\!\!\!/ y d\!\!\!/ \eta a(x,\xi-\eta)b(x-y,\xi)e^{-iy\eta} \ , \\ a^*(x,\xi) &= \int d\!\!\!/ y d\!\!\!/ \eta \bar{a}(x-y,\xi-\eta)e^{-i\eta y} \ . \end{aligned}$$

In prop.4.2 A^* denotes the formal Hilbert space adjoint of A, as in (2.3), here of the form $A^*=\bar{a}(M_r,D)$. Thus the second formula follows from (4.1) and (4.2). For the product AB write

$$(4.7) \quad (ABu)(x) = \int d\!\!\!/ \xi d\!\!\!/ y d\!\!\!/ \eta d\!\!\!/ z \ e^{i\xi(x-y)}a(x,\xi)b(y,\eta) \ e^{i\eta(y-z)} \ u(z) \ .$$

We may write a 4n-fold integral, the integrand being $L^1(\mathbb{R}^{2n})$. Use

$$(4.8)\qquad e^{-i\eta(x-z)}e^{i\xi(x-y)}e^{i\eta(y-z)} = e^{-i(\xi-\eta)(y-x)} ,$$

and continue (4.7) as follows.

$$(4.9)\qquad = \int đ\eta đz\, e^{i\eta(x-z)}\, u(z) \int đ\xi đy\, a(x,\xi)b(y,\eta)\, e^{-i(\xi-\eta)(y-x)} .$$

Denoting the inner integral by $c(x,\eta)$, (4.6) follows, q.e.d.

In the trivial case of C_0^∞-symbols formulas (4.1),(4.2),(4.6) achieve the transition $M_l \leftrightarrow M_r$, and the calculation of the symbol of operator product and adjoint. We will show that the formulas remain in effect for symbols in ST and ST_1 if the Riemann integrals in (4.2) and (4.6) are replaced by a type of distribution integral we call a <u>finite part</u> - it resembles a concept of Hadamard [Hd_1].

Let us still look at the Weyl-representation for C^∞-symbols.

Clearly (3.10) gives $a(M_w,D)^* = \bar{a}(M_w,D)$, a formula like (4.6), but much simpler - it does not involve integrals. But the formula for the symbol of a product, now involves a 4n-fold integral:

<u>Proposition 4.3.</u> Let $a,b \in C_0^\infty(\mathbb{R}^n)$, and $c \in C_0^\infty(\mathbb{R}^{3n})$. We have

$$(4.10)\qquad C = c(M_l,M_r,D) = p(M_w,D) \ ,\ Q = a(M_w,D)b(M_w,D) = q(M_w,D)$$

where p , $q \in \mathcal{S}(\mathbb{R}^{2n})$ are given by the formulas

$$(4.11)\qquad \begin{aligned} p(x,\xi) &= \int đyđ\eta\, e^{-i\eta y} c(x+y/2,x-y/2,\xi-\eta) ,\\ q(x,\xi) &= \int đyđzđ\eta đ\zeta\, a(x-y+z/4,\xi-\eta+\zeta/4)b(x-y-z/4,\xi-\eta-\zeta/4)e^{-i(\eta z-\zeta y)} . \end{aligned}$$

<u>Proof</u>. For $(4.10)_1$ depart from (4.3), writing $c^{\wedge}(x,y,x-y) = d(w,z) = c^{\wedge}(w+z/2,w-z/2,z)$, with $w=(x+y)/2$, $z=x-y$. Defining $p(w,\zeta) = d^{\wedge}(w,\zeta) = \int đz\, d(w,z)e^{-iz\zeta}$, get $Cu(x) = \int đy\, p^{\wedge}(\frac{x+y}{2},x-y)u(y) = p(M_w,D)u(x)$. Again $p \in \mathcal{S}(\mathbb{R}^{2n})$; interchanges are trivial, confirming $(4.10)_1$-$(4.11)_1$.

For the second formula one writes

$$(4.12)\qquad \begin{aligned} Qu(x) &= (ABu)(x) = \int r(x,z)u(z)đz ,\\ r(x,z) &= \int đ\xi đ\eta đy\, e^{i\xi(x-y)-i\eta(z-y)} a((x+y)/2,\xi)b((y+z)/2,\eta) . \end{aligned}$$

Again write $r(x,y) = s(w,v) = r(w+v/2,w-v/2)$, $w=(x+z)/2$, $v=x-z$, and get $Q = q(M_w,D)$, with $q(w,\zeta) = s^{\wedge}(w,\zeta) = \int đv\, s(w,v)e^{-i\zeta v}$, along lines used before. Then show that q assumes the form (4.11) after an integral substitution. No trouble with integral interchanges. $q\in S$ follows.

Now we start with the discussion of a more general singular integral called the finite part (integral) since it resembles Hadamard's finite part (cf. also [C1],II). This will be a distribution integral using partial integration, based on identities like (1.6)

$$e^{-iy\eta} = \langle\eta\rangle^{-2N}(1-\Delta_y)^N e^{-iy\eta} = \langle y\rangle^{-2M}(1-\Delta_\eta)^M e^{-iy\eta} , \tag{4.13}$$

valid for all nonnegative N,M. Using (4.13) and some (legitimate) partial integrations, the first formula (4.2) assumes the form

$$\begin{aligned} p(x,\xi) &= \int dy d\eta \langle\eta\rangle^{-2N}((1-\Delta_y)^N e^{-iy\eta})a(x-y,\xi-\eta) \\ &= \int dy d\eta e^{-iy\eta}\langle\eta\rangle^{-2N} a_N(x-y,\xi-\eta) , \quad a_N(z,\zeta)=(1-\Delta_z)^N a(z,\zeta) . \end{aligned} \tag{4.14}$$

Using the other identity (4.13) we get, similarly,

$$p(x,\xi) = \int dy d\eta e^{-iy\eta}\langle y\rangle^{-2M}(1-\Delta_\eta)^M(\langle\eta\rangle^{-2N}a_N(x-y,\xi-\eta)) . \tag{4.15}$$

Here the expression $(1-\Delta_\eta)^M(\langle\eta\rangle^{-2N}a_N(x-y,\xi-\eta))$ is in the span of

$$(\langle\eta\rangle^{-2N})^{(\alpha)}a_N^{(\beta)}(x-y,\xi-\eta) , \quad |\alpha|+|\beta| \le 2M . \tag{4.16}$$

If M and N are chosen sufficiently large then the right hand side of (4.15) is meaningful, as a Riemann or Lebesgue integral, even if we only require that $a\in ST_1$. Indeed, the expressions (4.16) are $O(\langle\eta\rangle^{-2N}\langle x-y\rangle^{\lambda(2M)}\langle\xi-\eta\rangle^{\kappa(2N)})$, with κ,λ of (1.1), so that the integrand of (4.15) is $O(\langle y\rangle^{-2M}\langle x-y\rangle^{\lambda(2M)}\langle\eta\rangle^{-2N}\langle\xi-\eta\rangle^{\kappa(2N)})$. We assume $\kappa\ge0$, $\lambda\ge0$ in (1.2): they may be replaced by $\mathrm{Max}\{0,\kappa\}$, $\mathrm{Max}\{0,\lambda\}$, also satisfying (1.1), (1.2). Thus $\langle x-y\rangle^\lambda = O(\langle x\rangle^\lambda+\langle y\rangle^\lambda)$; similarly for $\xi-\eta$. Existence of the integral (4.15) follows from (1.2).

The integrals in $(4.2)_2$, (4.6), (4.11) are treated similarly, using (1.1) with identities like (4.13), and partial integration.

Note that all these integrals are of the form

$$\int ds d\sigma e^{is\sigma}w(s,\sigma) , \tag{4.17}$$

with an integral over $\mathbb{R}^{2m}$, and $w\in C_0^\infty(\mathbb{R}^{2m})$. If $w\in C^\infty(\mathbb{R}^{2m})$ satisfies

$$\partial_s^\alpha\partial_\sigma^\beta w(s,\sigma) = O(\langle s\rangle^{\lambda(|\alpha|)}\langle\sigma\rangle^{\lambda(|\beta|)}) , \quad \lambda(k)-k \to -\infty , \ k\to\infty , \tag{4.18}$$

we define the <u>finite part(integral)</u> $\mathrm{p.f.}\int ds d\sigma e^{is\sigma}w(s,\sigma)$ as follows: Use identities like (4.13) and formal partial integrations as above until an integrand in $L^1(\mathbb{R}^{2m})$ is reached. Then define $\mathrm{p.f.}\int$ as the Lebesgue integral of that integrand. That is,

$$\text{p.f.}\int ds d\sigma e^{is\sigma} w(s,\sigma)$$

$$(4.19)\qquad = \int ds d\sigma e^{is\sigma} \langle \sigma \rangle^{-2M} (1-\Delta_s)^M \left\{ \langle s \rangle^{-2N} (1-\Delta_\sigma)^N w(s,\sigma) \right\}$$

$$= \int ds d\sigma e^{is\sigma} \langle s \rangle^{-2M} (1-\Delta_\sigma)^M \left\{ \langle \sigma \rangle^{-2N} (1-\Delta_s)^N w(s,\sigma) \right\} \quad ,$$

for sufficiently large M,N, where it yet has to be shown that this definition is independent of N,M and the choice of first or second expression,at right.The latter will follow from the lemma,below.

<u>Lemma 4.4.</u> Let $w \in ST_1(\mathbb{R}^m)$ (i.e., satisfy (4.18)), and let

$$(4.20)\qquad w_j(s,\sigma) = w(s,\sigma)\chi_j(s)\chi_j(\sigma) \ , \ j = 1,2,\dots \ ,$$

with $\chi_j(s) = \chi(s/j)$, where $\chi \in C_0^\infty(\mathbb{R}^m)$ equals 1 in $|s| \le 1$. If the two expressions at right of (4.19) are denoted by $I(w) = I_{N,M}(w)$ and $J_{N,M}(w) = J(w)$, respectively, then we have

$$(4.21)\qquad \lim_{j\to\infty} I_{N,M}(w_j) = I_{N,M}(w) \ , \ \lim_{j\to\infty} J_{N,M}(w_j) = J_{N,M}(w) \ ,$$

for all sufficiently large N and M.

<u>Proof.</u> If w_j is substituted for w in $I_{N,M}$ (or $J_{N,M}$) then Leibniz' formula may be used to obtain $I(w_j)$, for example, as sum of terms

$$\int ds d\sigma \chi_j^{(\alpha)}(s)\chi_j^{(\beta)}(\sigma) O(\langle s \rangle^{-2N} \langle \sigma \rangle^{-2M} \partial_s^\gamma \partial_\sigma^\delta w(s,\sigma), |\alpha|+|\gamma| \le 2N, |\beta|+|\delta| \le 2M.$$

If N,M are chosen such that the integral $I_{N,M}(w)$ exists,i.e.,

$$(4.22)\qquad \lambda(2N) - 2N < -n \ , \ \lambda(2M) - 2M < -n \ ,$$

then each integrand is of the form $\chi_j^{(\alpha)}(s)\chi_j^{(\beta)}(\sigma) O((\langle \sigma \rangle \langle s \rangle)^{-n})$. This implies that all integrals tend to zero, except for $\alpha=\beta=0$, using that $c_j^{(\alpha)}(s) = j^{-|\alpha|}\chi^{(\alpha)}(s/j)$. On the other hand, the integral for $\alpha = 0$ converges to $I(w)$, since $c_j(s) \to 1$, $s \in \mathbb{R}^m$, q.e.d.

The lemma below will be useful to differentiate finite parts under the integral sign.

<u>Lemma 4.5</u>. For a symbol $w \in ST_1$ the function

$$h(x,\xi) = \text{p.f.}\int đy đ\eta \ w(x-y,\xi-\eta)e^{iy\eta}$$

is $C^\infty(\mathbb{R}^{2n})$, and we have

$$h_{(\beta)}^{(\alpha)}(x,\xi) = \text{p.f.}\int đy đ\eta \ w_{(\beta)}^{(\alpha)}(x-y,\xi-\eta)e^{iy\eta} \quad .$$

The proof is a calculation: Just use (4.19), with M,N suitable for both above finite parts. Then confirm that the differentiation may be taken into the Lebesgue integrals, and also commutes with the two Laplacians there. Observe $w \in ST_1$ implies that $\omega(x,\xi) = w(x^0+\lambda x, \xi^0+\mu\xi) \in ST_1$ for arbitrary $\lambda,\mu \in \mathbb{R}$, x^0, $\xi^0 \in \mathbb{R}^n$.

Lemma 4.6. If in prop.4.1 we require only $a \in ST_1$, $b \in ST$, then the integrals in (4.2) still exist as finite parts, and define C^∞-functions p,q, with all derivatives of polynomial growth:

$$(4.23) \qquad p^{(\alpha)}_{(\beta)}(x,\xi)\ ,\ q^{(\alpha)}_{(\beta)}(x,\xi) = O(\langle (x,\xi)\rangle^{N_{\alpha,\beta}}) \quad .$$

Moreover, we have (4.1) satisfied in the sense that, for $u \in S$,

$$a(M_r,D)u(x) = \int d\xi\, p(x,\xi)e^{ix\xi}u^\wedge(\xi), \quad b(M_1,M_r,D)u(x) = \int d\xi\, q(x,\xi)e^{ix\xi}u^\wedge(\xi),$$

Proof. For $a,b \in C_0^\infty$ we may differentiate p,q of (4.2) under the integral sign, getting p,q in the span of the expressions

$$(4.24) \quad \int dy\,d\eta\, e^{-iy\eta}a^{(\alpha)}_{(\beta)}(x-y,\xi-\eta)\ ,\ \int dy\,d\eta\, e^{-iy\eta}\partial_x^\gamma\partial_y^\delta\partial_\xi^\beta b(x,y,\xi)\ ,\ \gamma+\delta=\alpha.$$

For general $a \in ST_1$, $b \in ST$ again construct sequences $a_j = a\chi_j(x)\chi_j(\xi)$ $b_j = b\chi_j(y)\chi_j(\xi)$, as in lemma 4.4. Let a_j be substituted for a, and the partial integrations of the finite part be executed in (4.24). As $j\to\infty$ get limits with integrands $=O((\langle y\rangle\langle\eta\rangle)^{-2N}(\langle x-y\rangle\langle\xi-\eta\rangle^{\lambda(2N)})$, in the first case, and $=O((\langle y\rangle\langle\eta\rangle)^{-2N}(\langle (x,x-y)\rangle\langle\xi-\eta\rangle)^{\lambda(2N+k)})$, in the second, letting $k=|\alpha|+|\beta|$. We were setting N=M, and $\kappa(k)=\lambda(k)$, as always in the following. We get, with $r = 2\lambda(2N+|\alpha|+|\beta|)$,

$$(4.25) \quad p(x,\xi) = \lim_{j\to\infty}p_j(x,\xi) = \text{p.f.}\int dy\,d\eta\, e^{-iy\eta}a(x-y,\xi-\eta) = O(\langle (x,y)\rangle^r) \ ,$$

and the same formula for q. Here p_j, q_j denote the p,q for a_j,b_j. The estimate O(...) in (4.25) also holds for p_j, uniformly, since the a_j satisfy uniform estimates (1.1). Clearly (4.1), in the form above, is true for $u \in S$ and a_j, b_j, p_j, q_j. For $j\to\infty$ get the desired relation for p and also for q, by a similar conclusion, q.e.d.

We leave it to the reader to confirm that the corresponding statements are true for prop.'s 4.2 and 4.3. Summarizing:

Theorem 4.7. If the symbols a,b of prop.4.1, symbols a,b of prop. 4.2, and a,b,c of prop.4.3 are chosen in ST_1, ST ; ST_1, ST_1 ; ST_1, ST_1, ST, respectively, instead of in C_0^∞ then the integrals in (4.2), (4.6), (4.11) still exist as finite parts. They define C^∞-functions $p(x,\xi)$, $q(x,\xi)$, $c(x,\xi)$, $a^*(x,\xi)$, $p(x,\xi)$, $q(x,\xi)$, respectively, which are of polynomial growth in (x,ξ), together with all their derivatives. Moreover, (4.1), $AB=c(M_1,D)$, $A^*=a^*(M_1,D)$

and (4.10) still hold for $u \in \mathcal{S}$,in the sense of lemma 4.6.

By silent convention we omit the finite part label at the integrals and write the composition formulas for symbols (which we shall call Beals formulas) in their forms (4.2), (4.6), (4.11), even for symbols in ST and ST_1.

Note that we are not getting the estimates (1.1) back for the symbols p,q, etc. Instead we obtained the weaker estimates (4.23) of polynomial growth. It is of interest to ask for subsets of ST_1, for example, characterized by stronger inequalities, such that the composed symbols p, c, a^*, q again satisfy the stronger estimates. As a comparatively large such class we introduce

$$(4.26)\quad \psi t = \psi t_\infty = \{a \in ST_1 : a^{(\alpha)}_{(\beta)}(x,\xi) = O(\langle\xi\rangle^{m_1}\langle x\rangle^{m_2}), \text{for all } \alpha,\beta\}\ .$$

Here the pair $m = (m_1, m_2)$ will be called the order of the symbol a , and $\psi t_m \subset \psi t$ will denote the class of all symbols of order m.

Theorem 4.8. Let ΨT_l, ΨT_r, ΨT_w denote the classes of all $a(M_l,D)$, $a(M_r,D)$, $a(M_w,D)$, respectively, with $a \in \psi t$. Then we have $\Psi T_l = \Psi T_r = \Psi T_w = \Psi T = OP\psi t$, with transition formulas between representations of $A \in \Psi T$ given by thm.4.7. Moreover, ΨT is an algebra under operator multiplication containing its Hilbert space adjoints. The order m of $A \in \Psi T$ (defined as order of its symbol) is independent of the representation used. Orders add when operators are multiplied. Thm 4.7 gives formulas for symbols of adjoints and products.

For a proof we check earlier arguments. The stronger conditions imply the algebra property. Details are left to the reader.

5. The Leibniz formulas with integral remainder.

The Leibniz formulas express a product P(D)a(M) of a differential polynomial P(D) and a multiplication a(M) by a C^∞-function a as a finite sum of terms $a_j(M)P_j(D)$, with multiplication and differentiation in the other order. Similarly a(M)P(D), $(P(D)a(M))^*$. We have a(x,D)b(x,D)=c(x,D), $(a(x,D))^* = a^*(x,D)$, for differential operators a(x,D), b(x,D), with finite sums

$$(5.1)\quad \begin{aligned} c(x,\xi) &= \sum_\theta (-i)^{|\theta|}/\theta!\, a^{(\theta)}(x,\xi) b_{(\theta)}(x,\xi)\ , \\ a^*(x,\xi) &= \sum_\theta (-i)^{|\theta|}/\theta!\ a^{(\theta)}_{(\theta)}(x,\xi)\quad . \end{aligned}$$

Other formulas express p,q of (4.2), or of (4.11), in that case.

In the following corresponding formulas will be derived for

ψdo's. We focus on symbol classes $\psi h_{m,\rho,\delta}$ of Hoermander type.

Actually the formulas follow for symbols in ST or ST_1, and the tool of derivation is the ordinary Taylor formula with integral remainder. There will be a finite sum, as in (5.1), but also a remainder term, as in Taylor's formula. Unless the symbols belong to $SS_{m,\rho,\delta}$, $\delta<\rho_1$, however, the formulas may be of little value, since the remainder does not decay. The formulas will be called Leibniz formulas with integral remainder.

We recall Taylor's formula (with integral remainder):

$$a(x,\xi-\eta) = \sum_{|\theta|\le N} a^{(\theta)}(x,\xi)(-\eta)^\theta/\theta! + \rho_N , \tag{5.2}$$

$$\rho_N=(N+1)\sum_{|\theta|=N+1}(-\eta)^\theta/\theta!\rho_{\theta,N} , \quad \rho_{\theta,N} = \int_0^1 (1-\tau)^N a^{(\theta)}(x,\xi-\tau\eta)d\tau .$$

Substituting (5.2) into (4.2), (4.6), (4.11) one obtains the following typical collection of Leibniz formulas. Other formulas, for similar symbol transitions, follow just as easily.

Theorem 5.1. Let $a,b \in ST_1$, $c \in ST$, and let the symbols $a_{rl}(x,\xi)$, $c_l(x,\xi)$, $c_w(x,\xi)$, $p_l(x,\xi)$, $p_w(x,\xi)$, $a^*(x,\xi)$ be defined by

$$a_{rl}(M_l,D) = a(M_r,D) , \quad c_l(M_l,D) = c(M_l,M_r,D) = c_w(M_w,D) ,$$

$$p_l(M_l,D) = a(M_l,D)b(M_l,D) , \quad p_w(M_w,D) = a(M_w,D)b(M_w,D) , \tag{5.3}$$

$$a(M_l,D)^* = a^*(M_l,D) .$$

We have the following Leibniz formulas (with integral remainder):

$$a_{rl}(x,\xi) = \sum_{j=0}^N(-i\partial_x\cdot\partial_\xi)^j a(x,\xi)/j! + (N+1)\rho_N , \tag{5.4}$$

$$\rho_N(x,\xi) = \int_0^1 d\tau(1-\tau)^N/N!\,\text{p.f.}\int d\!\!{}^-y\, d\!\!{}^-\eta\, e^{-iy\eta}(-i\partial_x\cdot\partial_\xi)^{N+1}a(x-y,\xi-\tau\eta);$$

$$c_l(x,\xi) = \sum_{j=1}^N 1/j!\{(-i\partial_y\cdot\partial_\xi)^{N+1}c(x,y,\xi)\}_{x=y} + (N+1)\rho_N , \tag{5.5}$$

$$\rho_N(x,\xi) = \int_0^1 d\tau(1-\tau)^N/N!\,\text{p.f.}\int d\!\!{}^-y\, d\!\!{}^-\eta\, e^{-iy\eta}(i\partial_y\cdot\partial_\xi)^{N+1}c(x,x-y,\xi-\tau\eta);$$

$$c_w(x,\xi)=\sum_{j=0}^N 1/j!\ (i\partial_\xi\cdot\partial_y)^j c(x+y/2,x-y/2,\xi)|_{y=0} + (N+1)\rho_N , \tag{5.6}$$

$$\rho_N = \int_0^1 d\tau\ (1-\tau)/N!\ \text{p.f.}\int d\!\!{}^-y\, d\!\!{}^-\eta\, e^{-iy\eta}(i\partial_y\partial_\xi)^{N+1}c(x+y/2,x-y/2,\xi-\tau\eta);$$

$$p_l(x,\xi)= \sum_{j=0}^N 1/j!\ (-i\partial_\xi\cdot\partial_y)^j(a(x,\xi)b(y,\eta)|_{x=y,\xi=\eta} +(N+1)\rho_N , \tag{5.7}$$

$$(5.7)\quad \rho_N = \int_0^1 (1-\tau)^N/N!\ \frac{d\tau}{\tau}\ \text{p.f.}\int d\!\!/ y d\!\!/ \eta e^{-iy\eta}(-i\partial_y\cdot\partial_\eta)^{N+1} a(x,\xi-\tau\eta)b(x-y,\xi);$$

$$(5.8)\quad \begin{aligned} p_w(x,\xi) &= \sum_{j=0}^N \frac{1}{j!}(\tfrac{i}{2}(\partial_\xi\cdot\partial_y - \partial_x\cdot\partial_\eta))^j (a(x,\xi)b(y,\eta)|_{x=y,\xi=\eta} + \rho_N, \\ \rho_N &= M\int_0^1 (1-\tau)^N d\tau \int d\!\!/ y d\!\!/ z d\!\!/ \eta d\!\!/ \zeta e^{2i(z\eta-y\zeta)}(\tfrac{i}{2}(\partial_s\partial_\kappa - \partial_t\partial_\sigma))^{N+1} a(s,\sigma)b(t,\kappa),\end{aligned}$$

$M=N+1$, $s=x-y$, $\sigma=\xi-\eta$, $t=x-\tau z$, $\kappa=\xi-\tau\zeta$, with a finite part integral.

$$(5.9)\quad a^*(x,\xi) = \sum_{j=1}^N (-i\partial_x\cdot\partial_\xi)^j \bar{a}(x,\xi)/j! + (N+1)\rho_N ,$$

with ρ_N as in (5.4) , with a replaced by $\bar{a}$.

Note that the term ρ_N has been used in all formulas, to denote the remainder, without implying equality in different expressions.

<u>Proof.</u> In each case the equation at once is confirmed formally, by substituting (5.2) into (4.2), (4.6), (4.11), and using that

$$(5.10)\quad \int d\!\!/ \eta(-i\eta)^\theta e^{-iy\eta} = \partial_y^\rho(1)^\wedge = (2\pi)^{n/2}\partial_y^\rho\delta(y) \quad ,$$

in the distribution sense. A partial integration must be used with

$$(5.11)\quad (\partial_x\cdot\partial_y)^l/l! = \sum_{|\theta|=l} \partial_x^\theta\partial_y^\theta/\theta! \quad .$$

For c_w we must apply (5.2) onto the product ab, in 2n variables.

The detailed derivation is technical and repeats the same steps over and over. Focus on (5.7), the others follow similarly. Using (5.2) in (4.6)*1*, (with c(x,x) called $_p1(x,x)$) we get the terms

$$(5.12)\quad \text{p.f.}\int d\!\!/ y d\!\!/ \eta\ e^{-iy\eta}(-\eta)^\theta/\theta!\ a^{(\theta)}(x,\xi)b(x-y,\xi) \ .$$

For $a,b \in C_0^\infty(\mathbb{R}^{2n})$ this is easily transformed into

$$(5.13)\quad \begin{aligned} &= (-i)^{|\theta|}/\theta!\ \int d\!\!/ y d\!\!/ \eta\ \partial_y^\theta(e^{-iy\eta})a^{(\theta)}(x,\xi)b_N(x-y,\xi)\langle\eta\rangle^{-2N} \\ &= (-i)^{|\theta|}/\theta! \int d\!\!/ y d\!\!/ \eta\ e^{-iy\eta}a^{(\theta)}(x,\xi)b_{N(\theta)}(x-y,\xi)\langle\eta\rangle^{-2N} \ ,\end{aligned}$$

with N sufficiently large, but with the η-partial integrations reversed to the status M=0. This still gives a well convergent integral, as long as a,b are C_0^∞ . Recall that $b_N = (1-\Delta_x)^N b$.

Here the η-integral may be evaluated. It gives the Fourier transform $(\langle\eta\rangle^{-2N})^\wedge = (\langle\eta\rangle^{-2N})^\vee = E_N$, the unique fundamental solution in $\mathcal{S}'$ of $(1-\Delta)^{-N}$. An expression for E_N in terms of modified Hankel functions is known (cf. (0.24) for N=1; the method works for general $(1-\Delta)^{-s}$, s>0.) E_N decays exponentially, as $|x|\to\infty$. Without Hankel functions one confirms easily that E_N equals a function in S, for $|x|\ge 1$. (c.f. $[C_1]$,II). Hence (5.13) is continued as

$$(5.14)\qquad \begin{aligned} &=(-i)^{|\theta|}/\theta!\ a^{(\theta)}(x,\xi)\int đy((1-\Delta_y)^N b_{(\theta)}(x,\xi))E_N(x-y) \\ &(-i)^{|\theta|}/\theta!\ a^{(\theta)}(x,\xi)\ b_{(\theta)}(x,\xi)\ . \end{aligned}$$

confirming the first expression of (5.7), using (5.11), and for C_0^∞-functions only. For the remainder term, we look at

$$\text{p.f.}\int đyđ\eta\ e^{-iy\eta}\int_0^1(1-\tau)^N d\tau a^{(\theta)}(x,\xi-\tau\eta)(-\eta)^\theta b(x-y,\xi)$$

$$(5.15)\qquad = \int đyđ\eta\ e^{-iy\eta}\int_0^1(1-\tau)^N d\tau a^{(\theta)}(x,\xi-\tau\eta)(-\eta)^\theta\langle\eta\rangle^{-2N}b_N(x-y,\xi)$$

$$= \int_0^1(1-\tau)^N d\tau\int đyđ\eta\ e^{-iy\eta}a^{(\theta)}(x,\xi-\tau\eta)(-\eta)^\theta\langle\eta\rangle^{-2N}b_N(x-y,\xi)\ ,$$

as a matter of Fubini type integral exchanges. Write $(-\eta)^\theta e^{-iy\eta} = (-i)^{|\theta|}\partial_y^\theta e^{-iy\eta}$, and carry out a partial integration, for

$$(5.16)\qquad \begin{aligned} &=(-i)^{|\theta|}\int_0^1(1-\tau)^N d\tau\int đyđ\eta\ e^{-iy\eta}a^{(\theta)}(x,\xi-\tau\eta)b_{N(\theta)}(x-y,\xi)\langle\eta\rangle^{-2N} \\ &=(-i)^{|\theta|}\int_0^1(1-\tau)^N d\tau\int đyđ\eta\ e^{-iy\eta}a^{(\theta)}(x,\xi-\tau\eta)b_{(\theta)}(x-y,\xi)\ , \end{aligned}$$

completing the discussion of (5.7), for $a,b \in C_0^\infty$.

For general $a,b \in ST_1$ one now may use sequences a_j, $b_j \in C_0^\infty$, as in lemma 4.4, and arrive at (5.7) by passing to the limit $j\to\infty$, using lemma 4.4 and a formula similar to (4.25). Details are left to the reader. In particular the occurrance of the factor τ, in the argument of a , does not influence the derivation.

6. Calculus of ψdo's for symbols of Hoermander type.

The Leibniz' formulas with integral remainder of sec.5 may be of little use, unless we restrict symbols to the class SS of sec.3. For symbols in SS or ψh the formulas of thm.5.1 imply corresponding asymptotic expansions of a_{rl}, c_l, c_w, p_l, p_w, a^* . The key is an estimate of the remainders ρ_N of (5.4)-(5.9).

Theorem 6.1. Let $a \in \psi h_{m,\rho,\delta}$, $b \in \psi h_{m',\rho',\delta'}$, $c \in SS^{m'',\rho'',\delta''}$, with the 3 sets of paramaters satisfying (3.3). Then the remainders ρ_N of (5.4) through (5.9) satisfy the following estimates:

For (5.4): $\rho_N(x,\xi) = O\left(\langle\xi\rangle^{m_1-(N+1)(\rho_1-\delta)}\langle x\rangle^{m_2-(N+1)\rho_2}\right)$.

For (5.5): $\rho_N(x,\xi) = O\left(\langle\xi\rangle^{m_1''-(N+1)(\rho_1''-\delta'')}\langle x\rangle^{m_2''+m_3''-(N+1)\rho_3''}\right)$.

For (5.6): $\rho_N(x,\xi) = O\left(\langle\xi\rangle^{m_1''-(N+1)(\rho_1''-\delta'')}\langle x\rangle^{m_2''+m_3''-(N+1)\rho^{\sim}}\right)$,

$\rho^{\sim}=\mathrm{Min}\{\rho_2'',\rho_3''\}$.

For (5.7): $\rho_N(x,\xi) = O\left(\langle\xi\rangle^{m_1+m_1'-(N+1)(\rho_1-\delta')}\langle x\rangle^{m_2+m_2'-(N+1)\rho_2}\right)$.

For (5.8): $\rho_N(x,\xi) = O\left(\langle\xi\rangle^{m_1+m_1'-(N+1)(\rho_1^{\wedge}-\delta^{\wedge})}\langle x\rangle^{m_2+m_2'-(N+1)\rho_2^{\wedge}}\right)$,

$\rho_j^{\wedge} = \mathrm{Min}\{\rho_j,\rho_j'\}$, $\delta^{\wedge} = \mathrm{Max}\{\delta,\delta'\}$.

For (5.9): Same as for (5.4) .

Moreover, these remainders will be symbols of Hoermander type, of the order indicated by the above formulas. That is,

(6.1)
$\rho_N \in \psi h_{m-(N+1)(\rho-\delta e^1),\rho,\delta}$, $e^1=(1,0)$,for (5.4) and (5.9),

$\rho_N \in \psi h_{(m_1''-(N+1)(\rho_1''-\delta''),m_2''+m_3''-(N+1)\rho_3''),(\rho_1,\rho^{\sim}),\delta}$, for (5.5),

ρ_N as for (5.5), with ρ_3'' replaced by $\rho^{\sim}$, for (5.6) ,

$\rho_N \in \psi h_{m+m'-(N+1)(\rho-\delta' e^1),\rho^{\wedge},\delta^{\wedge}}$, for (5.7) ,and the same with δ' replaced by $\delta^{\wedge}$,for (5.8) .

Here we must require that $\delta^{\wedge} < \rho_1^{\wedge}$, in case of (5.7) and (5.8). Then both orders of ρ_N will tend to $-\infty$, as $N \to \infty$.

The proof proceeds similar as in sec.4. In each case we supply estimates, uniformly in τ, of the integrand $d\tau$ only. For example, to estimate ρ_N of (5.4), it suffices to consider

$$\text{p.f.}\int dy d\eta\, e^{-iy\eta} a_{(\theta)}^{(\theta)}(x-y,x-\tau\eta) \tag{6.2}$$

$$=\int dy d\eta\, e^{-iy\eta}\langle y\rangle^{-2M}(1-\Delta_\eta)^M(\langle\eta\rangle^{-2N} a_{N(\theta)}^{(\theta)}(x-y,\xi-\tau\eta)) ,$$

again with $a_N=(1-\Delta_x)^N a$, for sufficiently large M,N. Use (3.2) for

$$=O\left(\int dy\langle x-y\rangle^{m_2-\rho_2|\theta|}\langle y\rangle^{-2M}\int d\eta\langle\xi-\tau\eta\rangle^{m_1-\rho_1|\theta|+\delta|\theta|}\langle\eta\rangle^{-2N}\right) . \tag{6.3}$$

Here we also employed the estimate

$$(\langle\eta\rangle^{-R})^{(\alpha)} = O(\langle\eta\rangle^{-R-|\alpha|}) , \tag{6.4}$$

which is easily derived. Then apply the well known inequality

$$(6.5) \qquad \langle x\rangle^{-s}\langle x-y\rangle^{s} = O(\langle y\rangle^{|s|}) ,$$

valid for $s\in\mathbb{R}$, to estimate (6.3) by $\langle x\rangle^{m_2-\rho_2|\theta|}\langle\xi\rangle^{m_1-(\rho_1-\delta)|\theta|}$, confirming the estimate for the remainder ρ_N of (5.4).

Similarly for the other 5 estimates.Also we may apply Lemma 4.6 to the remainder of (5.4), for

$$(6.6) \quad \rho_{N(\beta)}^{(\alpha)}(x,\xi)=\int_0^1(1-\tau)^N/N!\int d\!\!{}^-y d\!\!{}^-\eta\, e^{-iy\eta}(-i\partial_x\cdot\partial_\xi)^{N+1}a_{(\beta)}^{(\alpha)}(x-y,\xi-\tau\eta).$$

Then the above estimate may be repeated for

$$(6.7) \quad \rho_{N(\beta)}^{(\alpha)} = O(\langle\xi\rangle^{m_1-(N+1)(\rho_1-\delta)-\rho_1|\alpha|+\delta|\beta|}\langle x\rangle^{m_2-(N+1)\rho_2-\rho_2|\beta|}),$$

which confirms (6.1) for the first remainder. Again similarly for the other five remainders, q.e.d.

<u>Theorem 6.2</u>. Under the assumptions of thm.6.1 the symbols a_{rl}, c_l, c_w, p_l, p_w, a^* of thm.5.1 are in ψh , provided that, for the product symbols p_l , p_w ,we still require that $\delta^\wedge < \rho_1^\wedge$, (which is automatically true if $(m,\rho,\delta) = (m',\rho',\delta')$) .

More precisely, we get a_{rl}, $a^*\in \psi h_{m,\rho,\delta}$, c_l, $c_w\in \psi h_{m^\sim,\rho^\sim,\delta}$, with $m^\sim=(m_1,m_2+m_3)$, $\rho^\sim=(\rho_1,\rho_2^\sim)$, $\rho_2^\sim=\mathrm{Min}\langle\rho_2,\rho_3\rangle$, p_l , $p_w \in \psi h_{m+m',\rho^\wedge,\delta^\wedge}$,with $\rho^\wedge$, $\delta^\wedge$ as in thm.6.1.

The proof of thm.6.2 is a consequence of (6.1).

Note that thm.6.2 contains the (so far unproven) thm.3.2 as a special case. Indeed, the symbol b of thm.3.2 is given by our c_l while b_1 may be obtained by writing $c(M_r,M_l,D)=(c(M_l,M_r,D)^* = \bar{b}_1(M_l,D)$, taking adjoints. b_2 coincides with c_w of thm.6.2. For uniqueness of b, b_1, b_2 note that a ψdo $A=a(M_l,M_r,D)$ vanishes if and only if its distribution kernel (2.16) vanishes. For $a(x,y,\xi)=b(x+y,\xi)$ set $w(x,y)=u(x+y)v(x-y)$, $u,v\in \mathcal{S}$. (2.16) yields $\int d\!\!{}^-\xi\int dsdt e^{i\xi t}a(s,\xi)u(s)v(t)=\int dsd\!\!{}^-\xi\, a(s,\xi)u(s)v^\wedge(\xi)=0$ for all $u,v \in \mathcal{S}$. This implies a=0. Thus the Weyl representation is unique. Similarly for the left (right) multiplying representations.

Similarly thm.6.2 implies the following result.

<u>Theorem 6.3.</u> Each set $\psi h_{\infty,\rho,\delta}$ forms an algebra under operator multiplication, containing its Hilbert space adjoints, whenever ρ,δ satisfy (3.3). The algebra product $a^\Delta b$ may be defined by anyone of the representations $a \leftrightarrow a(M_l,D)$, or, $a \leftrightarrow a(M_r,D)$, or $a \leftrightarrow a(M_w,D)$. That is $c=a^\Delta b$ satisfies $c(M_x,D)=a(M_x,D)b(M_x,D)$, x=l or x=r or x=w.

The proof, in essence, is an application of (5.7), (5.8), and (5.9), as modified by thm.6.2. Details are left to the reader.

In the following we tend to use the left multiplying representation $A=a(M_1,D)$, and then will use the more conventional notation $A=a(M_1,D)=a(x,D)$. For a class $\mathcal{X}$ of symbols let $Op\mathcal{X}=\{a(x,D): a\in \mathcal{X}\}$. For a class $\mathcal{Y}$ of ψdo's let $symb\mathcal{Y}$ be such that $Op(symb\mathcal{Y})=\mathcal{Y}$.

For a symbol $a\in \psi h_{m,\rho,\delta}$, and a sequence of symbols a_j, $j=0, 1,2,\ldots$, $a_j\in \psi h_{m^j,\rho,\delta}$, $m^j=(m_1^j,m_2^j)$, $m_k^j\to-\infty$, as $j\to\infty$, $k=1,2$, we will say that a has the <u>asymptotic expansion</u> $\sum_{j=0}^{\infty} a_j$, written as

$$(6.8)\qquad a = \sum_{j=0}^{\infty} a_j \ (\mathrm{mod}\ \psi h_{-\infty}) \quad , \text{ (or simply "(mod } \mathcal{S}\text{)") },$$

if both orders of the symbol $\rho^N=a-\sum_{j=0}^{N} a_j$ tend to $-\infty$, as $N\to\infty$ (that is, more precisely, if $r^N\in \psi h_{\mu^N,\rho,\delta}$, with $\mu^N=(\mu_N^1,\mu_N^2)$, where $\mu_j^k\to-\infty$, $j\to\infty$, $k=1,2$). It is seen at once that $\mu^j=m^{j+1}$ is the best possible choice for μ^j. Also we must have $m=m^0$.

Using this terminology we find that all formulas (5.4)-(5.9) imply corresponding asymptotic expansions For example,

$$(6.9)\qquad a_{1r}(x,\xi) = \sum_{j=0}^{\infty}(-i\partial_x\cdot\partial_\xi)^j a(x,\xi)/j! \quad (\mathrm{mod}\ \psi h_{-\infty}) \ ,$$

as $a\in \psi h_{m,\rho,\delta}$, $\delta<\rho$, and similarly for the other 5 formulas.

It is clear that the asymptotic sums in (6.8) or (6.9) are not in general convergent infinite series. For future applications it is important that, for <u>any</u> sequence a_j of symbols of order $m^j\to-\infty$, an asymptotic sum may be constructed, by the lemma, below.

<u>Lemma 6.4.</u> Let $a_j\in \psi h_{m,\rho,\delta}$, $j=0,1,\ldots$, where ρ,δ needs not satisfy (3.3), but is independent of j, and where $m^j=(m_1^j,m_2^j)$, with

$$(6.10)\qquad m_k^0 > m_k^1 > m_k^2 > \ldots \ , \ m_k^j \to -\infty \ , \text{ as } j\to\infty \ , \quad k=1,2 \ .$$

There exists a symbol $a \in \psi h_{m,\rho,\delta}$, $m = m^0$, such that

$$(6.11)\qquad s^N = a - \sum_{j=0}^{N-1} a_j \ \in \psi h_{m^N,\rho,\delta} \quad , \ N=0,1,2,\ldots \ .$$

<u>Proof.</u> Indicating first the idea of the proof, let

$$(6.12)\qquad \omega(x,\xi)=0 \ , \ |x|+|\xi| \le 1/2 \ , \ \omega=1 \ , \text{ as } |x|+|\xi| \ge 1 \ .$$

Assume $\omega \in C^{\infty}(\mathbb{R}^{2n})$, and let $\chi = 1-\omega \in C_0^{\infty}$. Assume $0 \le \omega, \chi \le 1$ in $\mathbb{R}^{2n}$. For an increasing seqence $\{t_j\}$, $t_j \in \mathbb{R}$, $t_j \to \infty$, $j \to \infty$, to be determined later, define

$$(6.13) \qquad a(x,\xi) = \sum_{j=0}^{\infty} a^j(x,\xi)\omega_j(x,\xi) \ , \ \omega_j(x,\xi) = \omega(x/t_j, \xi/t_j) \ .$$

Conclude that $\omega_j = 0$, as $|x|+|\xi| \le t_j/2$. Since $t_j \to \infty$ the sum in (6.13) is finite near x,ξ. Thus $a(x,\xi) \in C^{\infty}(\mathbb{R}^{2n})$ is well defined. We get

$$(6.14) \qquad a - \sum_{j=0}^{N-1} a_j = -\sum_{j=0}^{N-1} \chi_j a_j + \sum_{j=N}^{\infty} \omega_j a_j \ ,$$

where the first term at right is $C_0^{\infty}(\mathbb{R}^{2n})$, and satisfies (3.2) for any choice of parameters m,ρ,δ. For the second term we get

$$(6.15) \qquad |\omega_j a_j| \le c_j \langle\xi\rangle^{m_1} \quad \langle x\rangle^{m_2} \quad \sup\{\langle\xi\rangle^{\theta_1^j}\langle x\rangle^{\theta_2^j} : |x|+|\xi| \ge t_j/2\}$$

with $\theta_1^j = m_1^j - m_1^{j-1}$, $\theta_2^j = m_2^j - m_2^{j-1}$. By choosing t_j sufficiently large the supremum in (6.15) can be made less than $2^{-j}/c_j$. Using this in (6.15) , and then substituting (6.15) into (6.14) we get

$$(6.16) \qquad a - \sum_{j=0}^{N-1} a^j = o\left(\langle\xi\rangle^{m_1^{N-1}} \quad \langle x\rangle^{m_2^{N-1}}\right) .$$

Writing (6.16) for N+1, we notice that the highest term a_N , at left is of the same order as the right hand side. Thus we get the first estimate (3.2) contained in (6.11).

To get the other estimates we improve the choice of t_j. For a given α,β and N let $N_0 = \mathrm{Max}\{|\alpha|+|\beta|, N+1\}$. Write $s_N = a - \sum_{j=0}^{N} a_j$, and,

$$(6.17) \qquad \partial_x^{\alpha}\partial_{\xi}^{\beta} s_N = \sum_0^{N-1} \partial_x^{\alpha}\partial_{\xi}^{\beta}(a_j\chi_j) + \sum_N^{N_0} (\partial_x^{\alpha}\partial_{\xi}^{\beta}(a_j\omega_j) + \sum_{N_0+1}^{\infty} \partial_x^{\alpha}\partial_{\xi}^{\beta}(a_j\omega_j) \ .$$

Observe that $\omega_j \in \psi h_{0,\rho,\delta}$, for every ρ,δ with (3.3), and $0=(0,0)$. Thus also $a_j\omega_j \in \psi h_{m^j,\rho,\delta}$ and (6.15) follows for $\partial_x^{\alpha}\partial_{\xi}^{\beta}(\omega_j a_j)$ instead of $\omega_j a_j$,with c_j, m_1^{N-1} , m_2^{N-1} replaced by $c_{j,\alpha,\beta}$, $m_1^{N-1}+\delta|\alpha|-\rho_1|\beta|$, $m_2^{N-1}-\rho_2|\alpha|$, respectively, with a constant $c_{j,\alpha,\beta}$ depending on j, α,β only. The first sum in (6.17) is C_0^{∞} and satisfies (3.2). The second sum is finite; its terms satisfy the proper estimate. To control the last sum, t_j must be chosen according to

$$(6.18) \qquad c_{j,\alpha,\beta} \sup\{\langle\xi\rangle^{\theta_1^j} \ \langle x\rangle^{\theta_2^j} \ : |x|+|\xi| \ge t_j/2\} \le 2^{-j}, \text{ as } |\alpha|+|\beta| \le j.$$

This amounts to finitely many estimates, for each t_j, hence the selection is possible. In each term of the last sum we get $|\alpha|+|\beta| \le N_0 \le j$, by construction of N_0. Thus get the proper estimate,q.e.d.

<u>Lemma 6.5.</u> Let $\{b_{jl} : l=1,2,\ldots\}$ be sequences with the assumptions

of lemma 6.4, for j=1,2,... Also, let $b_{jl} \in \psi h_{m-(j+l)\mu,\rho,\delta}$, $\delta<\rho_1$, $\mu\geq 0$, so that the asymptotic sums $b_j = \sum_{k=0}^{\infty} b_{jk} \in \psi h_{m-j\mu,\rho,\delta}$ again form a sequence of symbols with the assumptions of lemma 6.4. Then

$$(6.19) \qquad \sum_{j=0}^{\infty}\sum_{k=0}^{\infty} b_{jk} = \sum_{j=0}^{\infty}\sum_{k=0}^{j} b_{k,j-k} \quad (\mathrm{mod}\ \psi h_{-\infty}),$$

where infinite sums are asymptotic. Equality in (6.19) is understood only up to a term in $\psi h_{-\infty}$.

Proof. All symbols b_{jl}, $j+l=k$, are in the same class $\psi h_{m-k\mu,\rho,\delta}$, hence their sum is in that class. Thus the asymptotic sum at right is meaningful. Denote the left and right hand side of (6.19) by b and b', and write $b^N = \sum_{j,k=0}^{N} b_{jk}$, $b'^N = \sum_{j+l\leq N} b_{jl}$. Note that $b^N - b'^N$ is a finite sum of symbols of order $m-N\mu$, hence itself is of that order. Moreover, $b' - b'^N$ is of order $m-N\mu$, while $b-b^N = b - \sum_{j=0}^{N} b_j + \sum_{j=0}^{N}(b_j - \sum_{k=0}^{N} b_{jk})$ again is of that order. Hence $b-b' = (b-b^N) + (b^N-b'^N) + (b'^N-b')$ is of order $m-N\mu$, for all $N = 0,1,2,\ldots$. It follows that $b-b' \in \psi h_{-\infty}$, q.e.d.

Lemma 6.6. An asymptotic expansion (mod S) of the form (6.8), with $a \in \psi h_{m,\rho,\delta}$, $a_j \in \psi h_{m^j,\rho,\delta}$, $m = m^0$, etc. may be differentiated term by term. That is, (6.8) implies

$$(6.20) \qquad a_{(\beta)}^{(\alpha)} = \sum_{j=0}^{\infty} a_{j(\beta)}^{(\alpha)} \quad (\mathrm{mod}\ S).$$

The proof is evident: All remainders r^N are symbols, their derivatives also are symbols, of orders tending to $-\infty$, as $N\to\infty$.

Problems. 1) A polycylinder $C=S^1\times\mathbb{R}^{n-1}$ is described analytically by its universal cover $\mathbb{R}^n=\mathbb{R}\times\mathbb{R}^{n-1}=\{(x_1,x^\Delta): x_1\in\mathbb{R},\ x^\Delta\in\mathbb{R}^{n-1}\}$: Write $C^\infty(C) = \{u\in C^\infty(\mathbb{R}^n):$ u periodic (2π) in $x_1\}$, $S(C)=\{u\in C^\infty(C): \partial_{x_1}^j u(x_1,.)\in S(\mathbb{R}^{n-1}), j=0,1,\ldots,$ uniformly in $x_1\}$. Show that a ψdo $A=a(x,D)\in Op\psi t_0$ with $a(x,\xi)$ periodic (2π) in x_1 may be regarded as a map $S(C)\to C^\infty(C)$, given by $Au(x)=\sum_{j=-\infty}^{\infty} e^{ijx_1}\int d\!\!{}^-\xi^\Delta e^{ix^\Delta\xi^\Delta} a(x_1,x^\Delta;j,\xi^\Delta)u_j^\wedge(\xi^\Delta)$, with $u_j(x^\Delta)=\frac{1}{2\pi}\int_0^{2\pi} e^{-ijx_1}u(x_1,x^\Delta)dx_1$, and the Fourier transform "$^\wedge$" with respect to x^Δ. 2) The operator $A=a(x,D)$ of pbm.1 may be regarded as a ψdo on $\mathbb{R}^{n-1}$ with operator valued symbol, acting on functions on $\mathbb{R}^{n-1}$ with values in $C^{\infty}(S^1)$: For $x^{\Delta},\xi^{\Delta}\in\mathbb{R}^{n-1}$ define $A(x^{\Delta},\xi^{\Delta}):C^{\infty}(S^1)\to C^{\infty}(S^1)$ by $A(x^{\Delta},\xi^{\Delta})u(\theta)=\sum e^{ij\theta}a(\theta,x^{\Delta},j,\xi^{\Delta})u_j$, where $u_j=\frac{1}{2\pi}\int_0^{2\pi}e^{-ij\theta}u(\theta)d\theta$. Then write $v(x^{\Delta})=u(.,x^{\Delta})$, and $Av(x^{\Delta})=\int d\!\!\!/\xi^{\Delta}e^{ix^{\Delta}\xi^{\Delta}}(A(x^{\Delta},\xi^{\Delta})v^{\wedge}(\xi^{\Delta}))$, for $v\in\mathcal{S}(R^{n-1},C^{\infty}(S^1))\cong\mathcal{S}(C)$. Show that, for a symbol $a\in\psi t_0$ (and even for $a\in ST_1$) this operator A maps $\mathcal{S}(\mathbb{R}^{n-1},C^{\infty}(S^1))\to\mathcal{S}(\mathbb{R}^{n-1},C^{\infty}(S^1))$, and continuously so when we equip $\mathcal{S}(C)=\mathcal{S}(\mathbb{R}^{n-1},C^{\infty}(S^1))$ with the Frechet topology of all norms $\|\langle x^{\Delta}\rangle^k u^{(\alpha)}\|_{L^{\infty}}$, $k=0,1,\ldots$, $\alpha\in\mathbb{N}^n$. (Recall that $a(x,\xi)$ must be 2π-periodic in x_1.) 3) Investigate linear operators $C^{\infty}(S^1)\to C^{\infty}(S^1)$ as $B=A(x^{\Delta},\xi^{\Delta})$ of pbm.1: Let $Bu(x_1)=\sum_j\int dy_1\,e^{ij(x_1-y_1)}b_j(x_1)u(y_1)$, with a sequence of functions $b_j(x_1)$, $j=0,\pm1,\pm2,\ldots$, bounded in j and 2π-periodic (and C^{∞}) in x_1. First assume that $b_j(x_1)$ is independent of j for j>>1 and j<<-1. Then $B=b_1(M)+b_2(M)H+K$, with the Hilbert transform H on S^1 (i.e., $Hu(x_1)=\int_0^{2\pi}dy_1\cot(x_1-y_1)/2\;u(y_1)$, with a principal value integral) and with an operator K of finite rank. Next assume that the limits $\lim_{j\to\pm\infty}b_j(x_1)=\beta_{\pm}(x_1)$ exist. Try for conditions still insuring the above form, whith more general K 4) Investigate the operators $B_1=(\langle\xi^{\Delta}\rangle^2-D_{x_1}^2)^{-1/2}$ and $B_2=D_{x_1}B_1$ in the sense of pbm.3.

7.Strictly classical symbols; some lemmata for application.

For many discussions, notably elliptic and hyperbolic theory it proves unnecessary to carry along the complicated classes of Hoermander developed for hypoelliptic theory. Let us introduce

$$(7.1)\qquad \psi c_m=\psi h_{m,(1,1),0}\ ,\ \psi c_{\infty}=\cup\,\psi c_m\ ,\ \psi c_{-\infty}=\cap\,\psi c_m\ ,$$

setting $\rho_1=\rho_2=1$, $\delta=0$. Symbols in ψc_{∞}, and corresponding operators in $Op\psi c_{\infty}$ will be called <u>strictly classical</u>. Note that

$$(7.2)\qquad \psi c_m\subset\psi l_m\subset\psi s_m\subset\psi t_m\ ,$$

with the classes ψt of (4.26) and ψl, ψs of ch.VIII.

The lemmata below are simple consequences of our calculus of

ψdo's in sec.6. We are preparing them for later application.

It is clear that the classes ψt_m and ψc_m , for finite m, are Frechet spaces under the collections of semi-norms, respectively,

$$(7.3) \qquad \sup \{|a^{(\alpha)}_{(\beta)}(x,\xi)|\langle x\rangle^{-m_2}\langle\xi\rangle^{-m_1} : x,\xi \in \mathbb{R}^n\} , \alpha,\beta \in \mathbb{N}^n ,$$

and

$$(7.4) \qquad \sup \{|a^{(\alpha)}_{(\beta)}(x,\xi)|\langle x\rangle^{-m_2+|\alpha|}\langle\xi\rangle^{-m_1+|\beta|} : x,\xi\in \mathbb{R}^n\} ,\alpha,\beta \in \mathbb{N}^n .$$

Similar sets of semi-norms for the classes $\psi h_{m,\rho,\delta}$ of sec.3 are suggested by the estimates (3.3),of course.

Recall that a subset *M* of a Frechet space is called <u>bounded</u> if each semi-norm of a defining set remains bounded on *M*. Clearly a bounded set of ψc_m is bounded in ψt_m. For a bounded set *M* of ψc_m the set $\{a^{(\alpha)}_{(\beta)}: a\in M\}$ is bounded in $\psi t_{m_1-|\alpha|,m_2-|\beta|}$. A bounded set in ψc_m (ψt_m) is bounded in $\psi c_{m'}$ ($\psi t_{m'}$) whenever $m_j \le m'_j$, j=1,2.

<u>Lemma 7.1.</u> For bounded sets $A=\{a\}\subset \psi c_m$, $B=\{b\}\subset \psi c_{m'}$ the collection

$$(7.5) \qquad C = \{c \in \psi c_{m+m'}: c(M_1,D)=a(M_1,D)b(M_1,D) , a \in A , b \in B\}$$

is bounded in $\psi c_{m+m'}$, and the sets $A^* =\{a^*\}$ of symbols of adjoints and $P = \{p \in \psi c_m : p(M_r,D) = a(M_1,D) ,a \in A \}$ are bounded in ψc_m . Similarly , if ψc is replaced by ψt , in all the above.

<u>Lemma 7.2.</u> Let $a \in \psi c_m$, $b \in \psi c_{m'}$. Then we have , with e = (1,1),

$$a(M_1,D)b(M_1,D) - (ab)(M_1,D) \in \mathrm{Op}\psi c_{m+m'-e} ,$$

$$(7.6) \qquad a^*(M_1,D) - \bar{a}(M_1,D) \in \mathrm{Op}\psi c_{m-e} ,$$

$$a(M_1,D) - a(M_r,D) \in \mathrm{Op}\psi c_{m-e} .$$

Also, if $A = a(M_1,D)$, $B = b(M_1,D)$ then

$$(7.7) \qquad [A,B] = AB - BA \in \mathrm{Op}\psi c_{m+m'-e} .$$

Moreover, if a, b range over bounded sets of ψc_m and $\psi c_{m'}$, respectively then the symbols of the expressions in (7.6) and (7.7) form bounded sets of $\psi c_{m+m'-e}$, ψc_{m-e}, ψc_{m-e}, $\psi c_{m+m'}$, respectively. In fact, this is correct if we do not necessarily require that $A\subset \psi c_m$, and $B \subset \psi c_{m'}$ are bounded, and $a \in A$, $b \in B$, but only that respectively, for the 4 relations in (7.6), and (7.7),

(i) $\{a^{(\theta)}\}$ bounded in ψc_{m-e^1} , and $\{b^{(\theta)}\}$ bounded in ψc_{m-e^2} ,

(ii) $\{a^{(\theta)}_{(\theta)}\}$ bounded in ψc_{m-e} ,

(iii) $\{a^{(\theta)}_{(\theta)}\}$ bounded in ψc_{m-e} ,

(iv) $\{a^{(\theta)}\}$ bounded in ψc_{m-e^1} , and $\{a_{(\theta)}\}$ bounded in ψc_{m-e^2} ,

and $\{b^{(\theta)}\}$ bounded in $\psi c_{m'-e^1}$, and $\{b_{(\theta)}\}$ bounded in $\psi c_{m'-e^2}$.
For (i) ,...,(iv) the conditions are required for all $|\theta|=1$.
Remark 7.3: Using the 1-1-correspondence $a \leftrightarrow A=a(M_1,D)$ between ψx_m and $Op\psi x_m$, for $x=c,t$, one may transfer the Frechet topology of the spaces ψx_m onto their corresponding sets $Op\psi x_m$. Then we may speak of bounded sets in $Op\psi c_m$, for example.
Proof. Clearly (7.6) and (7.7) are consequences of thm.6.1, noting that $\psi c_m = \psi h_{m,e,0}$. On the other hand all statements about boundedness of sets follow by just going again over the proofs leading to thm.6.1, keeping in mind the boundedness of sets in Frechet spaces. For example, regarding boundedness of the set C of (7.6) it is sufficient to show that the constant of O(.) in thm.6.1, regarding (5.5), as well as the corresponding constants for the derivatives $\rho^{(\alpha)}_{N(\beta)}$ of thm.6.2 of that remainder, all depend only on the bounds of the Frechet norms of a and b, in their bounded sets. This needs only be done for one N, say, for N=0. Or even, one could estimate the first expression (4.6) of the Beals formula for the product symbol, in its finite part integral form (4.19), of course, using the principles of the proof of thm.6.1.

Actually, going over that estimate again it becomes evident that indeed only bounds of expressions (6.2), for the symbols a and b, and corresponding bounds for the explicit symbols

$$(7.8) \qquad \pi_m(x,\xi) = \langle\xi\rangle^{m_1}\langle x\rangle^{m_2} \in \psi c_m ,$$

and existing definite integrals over $\mathbb{R}^n$

$$(7.9) \qquad \int\langle x\rangle^{-r}dx \quad , \quad r > n \quad ,$$

as well as binomial and multinomial coefficients occur.

We will not go over all the details again .

Chapter 2. ELLIPTIC OPERATORS AND PARAMETRICES IN $\mathbb{R}^n$.

0. Introduction.

In this chapter we focus at the main feature of our algebras $Op\psi h_{\rho,\delta}$: They contain (generalized) inverses of some of their operators - called (md-)(hypo-)elliptic. For historical reason such an inverse is called parametrix or Green inverse. Speaking algebraically, we deal with inverses modulo some 2-sided ideal, either the class $\mathcal{R}$ of integral operators with kernel in $\mathcal{S}(\mathbb{R}^{2n})$ (K-paramemetrix), or the operators of $\mathcal{R}$ with finite rank (Green inverse).

Integral operators with kernel in $\mathcal{S}(\mathbb{R}^{2n})$ are compact operators of $L^2(\mathbb{R}^n)$, as well as in every Sobolev space (cf.III,5). Thus a K-parametrix (Green inverse) of A will be an inverse mod $K(H)$ (a Fredholm inverse) as well, under proper assumptions, giving normal solvability of the equation Au=f (cf. [C_1],App.A1).

For $Op\psi h_{m,\rho,\delta}$ 'elliptic' usually will be 'md-elliptic', amounting to ellipticity 'at $|\xi|=\infty$' and 'at $|x|=\infty$'-i.e., "d-" and "m-" ellipticity. This is parallel to discussions in algebras of singular integral operators ([C_1],[C_2]): For operators of order 0 'md-elliptic' simply means that the symbol is $\neq 0$ at $|x|+|\xi|=\infty$.

An md-elliptic operator of order m will have a K-parametrix of order -m , and existence of a K-parametrix of order -m is in fact necessary and sufficient for md-ellipticity of $A\in Op\psi h_{m,\rho,\delta}$. However, it is possible for a non-md-elliptic operator to have a K-parametrix of order >-m. One such class of operators - the formally md-hypo-elliptic operators - is studied in sec.2.

The ellipticity concept can be localized, in the x-variable as well as in the ξ-variable, and 'local parametrices' may be constructed under various assumptions. In particular, (formal hypo-) ellipticity (for all ξ) over an open set $\Omega\subset\mathbb{R}^n$ implies hypo-ellipticity in the proper sense: All distribution solutions of Au=f are $C^\infty(\Omega)$ whenever $f\in\mathcal{D}'$ is $C^\infty(\Omega)$ (cf. sec.3 and 4). For proper discussion of such facts we introduce the wave front set WF(u) of $u\in\mathcal{D}'$ (sec.5), and show its invariance under spplication of ψdo's.

In sec.6 we discuss (left)(right) md-ellipticity of a matrix of ψdo's, together with certain partial inverses.

1. Elliptic and md-elliptic ψdo's.

Historically a parametrix is an integral kernel with a singularity similar to that of a Green's function, which can be used to invert a differential operator, up to an integral operator with continuous or smooth kernel (c.f. E.E.Levi [Lv_1], D.Hilbert [Hb_1]). Hoermander [Hr_2] used the term 'parametrix' for a local inverse up to an infinitely smoothening operator, for a ψdo. Here we will use a type of global parametrix with similar properties. Let A,B be ψdo's with symbol in *ST*. We shall say that B is a K-parametrix of A if we have $AB-1 \in Op\psi h_{-\infty}$, $BA-1 \in Op\psi h_{-\infty}$. Clearly then A and B are K-parametrices of each other. We speak of a left (right) K-parametrix B of A if only the second (first) condition holds.

A symbol $a \in \psi h_{m,\rho,\delta} \subset \psi h$ will be called md-elliptic (of order m) if (i) $a(x,\xi) \neq 0$, for all sufficiently large $|x|+|\xi|$, and (ii) the function $b=1/a$, well defined for large $|x|+|\xi|$, equals some symbol in $\psi h_{-m,\rho,\delta}$, for large $|x|+|\xi|$. Then the ψdo $A=a(x,D)$ also will be called md-elliptic. We will see that md-ellipticity is a property of the operator, insofar as, for an md-elliptic $A=a(M_l,D)$ $=b(M_r,D)=c(M_w,D)$ all 3 symbols a,b,c will be md-elliptic.

Proposition 1.1. A symbol $a \in \psi h_{m,\rho,\delta} \subset \psi h$ is md-elliptic if and only if, with constants $c,c'>0$, we have

$$|a(x,\xi)| \geq c\langle\xi\rangle^{m_1}\langle x\rangle^{m_2} \quad , \text{ as } |x|+|\xi| \geq c' , \tag{1.1}$$

Proof. If a is md-elliptic we get $b=1/a$ for large $|x|+|\xi|$, and $b\in \psi h_{-m,\rho,\delta}$, hence $b(x,\xi)=O(\langle\xi\rangle^{-m_1}\langle x\rangle^{m_2})$, using I,(3.2). This implies (1.1). Vice versa, if a symbol satisfies (1.1) then $b(x,\xi)=\frac{\omega(x,\xi)}{a(x,\xi)}$ with $\omega(x,\xi)\in C^\infty$, $\omega=1$, as $|x|+|\xi|\geq 3c'$, $\omega=0$, as $|x|+|\xi|\leq 2c'$, defines a $C^\infty(\mathbb{R}^{2n})$-function satisfying the first estimate I,(3.2). By lemma 1.2, below, it satisfies all estimates I,(3.2), q.e.d.

Lemma 1.2. Let $0\neq a\in C^\infty(\Omega)$. Then $(1/a)^{(\alpha)}_{(\beta)}$ is in the span of

$$\{[\Pi_j(a^{(\alpha^j)}_{(\beta^j)}/a)]/a \ : \ \sum\alpha^j = \alpha \ , \ \sum\beta^j = \beta \ \} \tag{1.2}$$

This lemma follows by induction.

Proposition 1.3. For $a\in \psi h_{m,\rho,\delta}\subset \psi h$, let $A=a(M_l,D)$. Let A also be given as $A=b(M_r,D)=c(M_w,D)=d(M_l,M_r,D)$, with symbols $b,c\in \psi h_{m,\rho,\delta}$, and $d\in SS^{m',\rho',\delta'}\subset SS$. The following conditions are equivalent: (i) b is md-elliptic ; (ii) c is md-elliptic ; (iii) $d(x,x,\xi)$ is md-elliptic (of order m). (iv) A is md-elliptic.

This follows from calculus of ψdo's: The symbol a may be expressed as an asymptotic sum of b or c or d and their derivatives. We get $b(x,\xi)=c(x,\xi)=d(x,x,\xi)=a(x,\xi)$ modulo terms of lower order. Thus (1.1) holds for all or for none of these functions, q.e.d.

Proposition 1.4.(a) $A\in \mathrm{Op}\psi h$ is md-elliptic if and only if A^* is.

(b) The product AB of two md-elliptic operators $A\in \mathrm{Op}\psi h_{m,\rho,\delta}$, $B\in \mathrm{Op}\psi h_{m',\rho,\delta}$, again is md-elliptic (of order m+m').

Again this follows from calculus of ψdo's.

Recall that a differential operator $A = \sum_{|\alpha|\le N} a_\alpha D^\alpha$ is called elliptic at x^0 if its principal symbol

$$(1.3) \qquad a_N(x,\xi) = \sum_{|\alpha|\le N} a_\alpha \xi^\alpha$$

is $\ne 0$ for $x=x^0$ and all $\xi\in \mathbb{R}^n$ with $|\xi|=1$. In that respect notice prop.1.5 below, which also motivates our terminology.

Proposition 1.5. A differential operator $A=a(x,D)\in \mathrm{Op}\psi h_{m,\rho,\delta}$, $m=(N,m_2)$ is md-elliptic (of order (N,m_2)) if and only if

$$(1.4) \qquad |a_N(x,\xi)| \ge p\langle x\rangle^{m_2}, \text{ for all } x, \text{ and all } |\xi| = 1,$$

and also, with suitable constants p, c>0,

$$(1.5) \qquad |a(x,\xi)| \ge p\langle \xi\rangle^N\langle x\rangle^{m_2}, \text{for all } x,\xi \text{ with } |x| \ge c.$$

Proof. Note that '$|x|+|\xi|$ large' means that 'either $|x|$ is large or $|\xi|$ is large' (or both). The first points to (1.5), the second, for a differential operator $A = a(M_1,D)$, translates into (1.4), using that the terms $a_\alpha\xi^\alpha$, with $|\alpha|< N$, all are $O(\langle\xi\rangle^{N-1}\langle x\rangle^{m_2})$.

By (1.5) md-elliptic differential operators are elliptic in $\mathbb{R}^n$. $\langle x\rangle^{-m_2}A$ even is 'uniformly elliptic' in $\mathbb{R}^n$. (1.4), in effect, is another ellipticity, with variables x, ξ reversed.

Theorem 1.6. An operator $A\in \mathrm{Op}\psi h_{m,\rho,\delta}$, admits a K-parametrix in $\mathrm{Op}\psi h$ of order -m if and only if A is md-elliptic.

Proof. If A and B are K-parametrices of each other then the relation AB=1 and the uniqueness of the left representation implies $1= a(x,\xi)b(x,\xi)$, modulo terms of order $-\varepsilon e$, $e=(1,1)$, $\varepsilon>0$. Thus it follows that $|ab| \ge 1/2$, as $|x|+|\xi| \ge c$, for large c. Accordingly, $|a| \ge \frac{1}{2|b|} \ge p\langle\xi\rangle^{m_1}\langle x\rangle^{m_2}$, $|x|+|\xi| \ge c$, $p>0$, using that b is of order -m hence, $b=O(\langle\xi\rangle^{-m_1}\langle x\rangle^{-m_2})$. Thus A is md-elliptic (order m). Vice versa, let $A=a(M_1,D)$ be md-elliptic. Using a recursion and I,lemma

6.4 we will construct a K-parametrix for A.

First, let b_1 be a symbol such that $ab_1=1$ for large $|x|+|\xi|$, by definition of md-ellipticity. If $B_1=b_1(M_1,D)$ then $AB_1-1=e_1(x,D)$ where $e_1(x,\xi)=\sum_{|\theta|\geq 1}(-i)^{|\theta|}/\theta! a^{(\theta)}b_{1(\theta)}$ (mod $\psi h_{-\infty}$). Here e_1 is of order $(-(\rho_1-\delta),-\rho_2)=\sigma$. Next set $b_2=e_1/a$, for large $|x|+|\xi|$, corrected to be $C^\infty(\mathbb{R}^{2n})$. Setting $B_2=b_2(M_1,D)$, get $AB_2+E_1=e_2(x,D)=E_2$ of order -2σ, where, again, we used ψdo-calculus for AB_2. Together, we get $A(B_1+B_2)=1+E_1-E_1+E_2=1+E_2$. It is clear now how to iterate: Next set $b_3=-e_2/a$, and $B_3=b_3(M_1,D)$, $AB_3+E_2=E_3=e_3(M_1,D)$, with e_3 of order -3σ, and $A(B_1+B_2+B_3)=1+E_3$, etc. Let $B=\sum B_j$ denote an asymptotic sum. Then $B-\sum_{j=1}^{N}B_j$ is of order $-(N+1)\sigma$, by definition. Also $A(\sum_j^{N=1}B_j)-1=\sum_j^{N=1}E_j$ is again of order $-N\sigma$. Thus AB-1 is of order $-N\sigma$, for every N = 1,2,.... It follows that AB-1 must be of order $-\infty$. Or, B is a right parametrix of A. Similarly construct a left parametrix $B^\sim$ of A. Get $B^\sim=B^\sim AB=B$ (mod $Op\psi h_{-\infty}$), q.e.d.

Finally, in this section, we show that $Op\psi h_{-\infty}=OpS(\mathbb{R}^{2n})=\mathcal{K}$:

<u>Lemma 1.7.</u> The class $Op\psi h_{-\infty}$ of ψdo's of order $-\infty$ coincides with the class $\mathcal{K}$ of all integral operators

$$(1.6)\qquad Ku(x)=\int k(x,y)u(y)đy\ ,\ u\in S\ ,$$

with rapidly decreasing kernel $k\in S(\mathbb{R}^{2n})$.

<u>Proof.</u> We have seen before (cf. I,(2.16)) that the distribution kernel k of a ψdo is given as

$$(1.7)\qquad k(x,y)=a^{\vee}(x,x-y)=\int đ\xi a(x,\xi)e^{i\xi(x-y)}\ .$$

For $a\in\psi h_{-\infty}=S(\mathbb{R}^{2n})$ (1.7) is a well defined integral. The Fourier transform and $b(x,y)\rightarrow b(x,x-y)$ are isomorphisms $S\leftrightarrow S$, q.e.d.

<u>2. Formally hypo-elliptic ψdo's.</u>

In sec.1 we found that a ψdo A of order m has a K-parametrix of order -m <u>if and only if</u> it is md-elliptic. It is possible, however, for $A\in Op\psi h_{m,\rho,\delta}\subset\psi h$ to have a K-parametrix in $Op\psi h_{m',\rho,\delta}$, with $m'_j\geq -m_j$ without being md-elliptic.

Let $a\in\psi h_{m,\rho,\delta}\subset\psi h$. We shall say that a (or A=a(x,D)) is

<u>formally md-hypo-elliptic</u> if for $|x|+|\xi|\geq\eta$ we have (i) $a(x,\xi)\neq 0$, and (ii) $\frac{1}{a}=O(\langle\xi\rangle^{m_1'}\langle x\rangle^{m_2'})$, with some $m'=(m_1',m_2')$, and (iii) for all α,β the functions $a^{(\alpha)}_{(\beta)}/a$ satisfy the estimates

$$(2.1)\qquad a^{(\alpha)}_{(\beta)}/a = O\left(\langle\xi\rangle^{-\rho_1|\alpha|+\delta|\beta|}\langle x\rangle^{-\rho_2|\beta|}\right) .$$

In this case we shall call $m'=(m_1',m_2')$ an <u>inverse</u> <u>order</u> of a .

<u>Lemma 2.1.</u> Let $a\in\psi h_{m,\rho,\delta}\subset\psi h$. The symbol a is formally md-hypo-elliptic if and only if (i) $\frac{1}{a}$ and $a^{(\alpha)}_{(\beta)}/a$ are $C^\infty(\{|x|+|\xi|\geq\eta\})$, (ii) the functions of (i), for all α,β, extend to symbols in $\psi h_{m',\rho,\delta}$, and $\psi h_{m^{\alpha,\beta},\rho,\delta}$, respectively, where

$$(2.2)\qquad m^{\alpha,\beta} = (-\rho_1|\alpha|+\delta|\beta|,-\rho_2|\beta|) \quad .$$

<u>Proof.</u> Trivially the condition of the lemma is stronger as that of the definition, because (2.1) is only the first I,(3.2), defining $\psi h_{m^{\alpha,\beta},\rho,\delta}$. We must show that also $\partial_x^\kappa\partial_\xi^\lambda(a^{(\alpha)}_{(\beta)}/a)$ satisfy I,(3.2). This follows from the lemma below easily proven by induction.

<u>Lemma 2.2.</u> For all $\alpha,\beta,\kappa,\lambda$ we have $\partial_x^\kappa\partial_\xi^\lambda(a^{(\alpha)}_{(\beta)}/a)=(a^{(\alpha)}_{(\beta)}/a)^{(\lambda)}_{(\kappa)}$ in the span of the products

$$(2.3)\quad \Pi_{j=1}^r(a^{(\theta^j)}_{(\vartheta^j)}/a) \ ,\ \Sigma_{j=1}^r\theta^j=\alpha+\lambda \ ,\ \Sigma_{j=1}^r\vartheta^j=\beta+\kappa \ ,\ r=1,2,\ldots \quad .$$

Again formal md-hypo-ellipticity is independent of the choice of left (right) multiplying representation.

<u>Proposition 2.3.</u> Let $A=a(M_l,D)=b(M_r,D)$,with $a,b\in\psi h_{m,\rho,\delta}\subset\psi h$. Then A (or a) is formally md-hypo-elliptic if and only if b is .
<u>Proof.</u> From formula I,(5.4) we conclude that

$$(2.4)\qquad b(x,\xi) = \sum_\theta(i)^{|\theta|}/\theta!\ a^{(\theta)}_{(\theta)}(x,\xi) \quad (\mathrm{mod}\ \psi h_{-\infty}) \quad .$$

If a is formally md-hypo-elliptic then the asymptotic sum

$$(2.5)\qquad c = \sum_\theta(i)^{|\theta|}/\theta!\ a^{(\theta)}_{(\theta)}/a$$

is well defined, assuming that 1/a has been suitably modified for $|x|+|\xi|<\eta$, to be $C^\infty(\mathbb{R}^{2n})$. This follows from lemma 2.1. Moreover, the highest term in the sum (2.5) is identically 1, implying that the symbol c is md-elliptic of order 0 , by prop.1.1. One confirms easily that ca = b (mod $\psi h_{-\infty}$). Indeed, from (2.5) we get

$$(2.6)\qquad c - \Sigma_j^{N=1}(i)^{|\theta|}/\theta!\ a^{(\theta)}_{(\theta)}/a \ \in\psi h_{-(N(\rho_1-\delta),N\rho_2),\rho,\delta} \quad .$$

This may be multiplied by the symbol a , using lemma 2.1, for

$$(2.7) \qquad ca - \sum_{j=1}^{N} (i)^{|\theta|}/\theta!\ a^{(\theta)}_{(\theta)} \in \psi h_{m-N\rho+N\delta e^1,\rho,\delta}\ ,\ N=0,1,2,\ldots.$$

In other words, we get ca = b (mod $\psi h_{-\infty}$) , using (2.4), as stated.

Using I,lemma 6.6, (2.4) may be differentiated, for

$$(2.8) \qquad b^{(\alpha)}_{(\beta)} = \sum_{\theta} (i)^{|\theta|}/\theta!\ a^{(\theta+\alpha)}_{(\theta+\beta)} \quad (\mathrm{mod}\ \psi h_{-\infty}) .$$

Then, repeating the above conclusion we see that

$$(2.9) \qquad b/a\ ,\ a/b \in \psi h_{0,\rho,\delta}\quad ,\ b^{(\alpha)}_{(\beta)}/a \ \in \psi h_{m^{\alpha,\beta},\rho,\delta} .$$

Again, the functions (2.9) must be modified on some $K \subset\subset \mathbb{R}^{2n}$ independent of α,β, before they are symbols as stated. (2.9) implies $1/b = (1/a)(a/b) \in \psi h_{m',\rho,\delta}$, $b^{(\alpha)}_{(\beta)}/b=(b^{(\alpha)}_{(\beta)}/a)(a,b) \in \psi h_{m^{\alpha,\beta},\rho,\delta}$, so that b is formally md-hypo-elliptic. Vice versa, if b is formally md-hypo-elliptic then a may be obtained as asymptotic expansion in terms of the derivatives of b. The above conclusion may be repeated to show that also a is formally md-hypo-elliptic, q.e.d.

<u>Proposition 2.4.</u> (a) $A \in \mathrm{Op}\psi h_{m,\rho,\delta} \subset \mathrm{Op}\psi h$ is formally md-hypo-elliptic if and only if A^*, its adjoint, is. (b) If $A=a(M_1,D)$,$B=b(M_1,D)$ $\in \bigcup_m \mathrm{Op}\psi h_{m,\rho,\delta} \subset \mathrm{Op}\psi h$ are formally md-hypo-elliptic then so is AB. (c) The formally md-hypo-elliptic symbols of (b) form a group under pointwise multiplication, for each ρ,δ. (d) An md-elliptic symbol of $\psi h_{m,\rho,\delta} \subset \psi h$ is formally md-hypo-elliptic, inverse order -m.

<u>Theorem 2.5.</u> A formally md-hypoelliptic $A=a(M_1,D) \in \mathrm{Op}\psi h_{m,\rho,\delta}$ admits a K-parametrix $B=b(M_1,D) \in \mathrm{Op}\psi h_{m',\rho,\delta}$, for suitable $m' \geq -m$.

<u>Proof.</u> As in the proof of thm.1.6 we focus on construction of a right K-parametrix. First set $b_1=\frac{1}{a}$, $|x|+|\xi|\geq\eta$, and $B_1=b_1(x,D)$. By assumption $b_1 \in \psi h_{m',\rho,\delta}$, and the product formula yields

$$(2.10) \quad AB_1-1=E_1=e_1(M_1,D)\ ,\ e_1= \sum_{|\theta|>0} \frac{(-i)^{|\theta|}}{\theta!} a^{(\theta)} b_{1(\theta)} \ (\mathrm{mod}\ \psi h_{-\infty}).$$

By lemma 1.2 the term $a^{(\theta)}b_{1(\theta)}=a^{(\theta)}(1/a)_{(\theta)}$ is in the span of $\{(a^{(\theta)}/a)\Pi_j(a_{(\alpha^j)}/a) : \sum\alpha^j=\theta\}$. Lemma 2.1 yields $a^{(\theta)}b_{1(\theta)} \in \psi h_{-N\rho+N\delta e^1,\rho,\delta}$, $|\theta|=N$. Accordingly $e_1 \in \psi h_{-\rho+\delta e^1,\rho,\delta}$ has order less than (0,0), as in thm.1.6.

Next, in the recursion, define $b_2=-e_1/a \in \psi h_{m'-\rho+\delta e^1,\rho,\delta}$. Try to verify that $B_2=b_2(M_1,D)$ satisfies $AB_2=-E_1+E_2$, with $E_2=e_2(x,D)$, $e_2 \in \psi h_{-2\rho+\delta e^1,\rho,\delta}$. Indeed we get

(2.11) $AB_2+E_1=e_2(M_1,D)$, $e_2 = \sum_{|\theta|>0} a^{(\theta)}(e_1/a)_{(\theta)}$ (mod $\psi h_{-\infty}$) .

A term $a^{(\theta)}(e_1/a)_{(\theta)}$, for $|\theta| = N$, is in the span of

(2.12) $\{ (a^{(\theta)}/a)\ e_{1(\alpha)}\ \Pi_j(a_{(\beta^j)}/a) : \alpha + \sum \beta^j = N \}$,

again by lemma 1.2. Using lemma 2.1, confirm the proper conditions e_2. Evidently the recursion of thm.1.6 works again, q.e.d.

3. Local md-ellipticity, and local md-hypo-ellipticity.

Recall that (formal) md-(hypo-) ellipticity of a symbol $a\in \psi h$, is a condition for large $|x|+|\xi|$. If a is modified in a set $K\subset\subset \mathbb{R}^{2n}$ these conditions are not disturbed.

A parallel, though slightly different concept of md-ellipticity was introduced in [C_1], for the operators of some C^*-subalgebras of $L(L^2(\mathbb{R}^n))$. Similar algebras are studied in [C_2] for general noncompact manifolds Ω. Their operators also have symbols now defined over the boundary of a compactification of the cotangent bundle $T^*\Omega$ - for $\Omega=\mathbb{R}^n$ for $|x|+|\xi|=\infty$, at the 'boundary' of $\mathbb{R}^{2n}$. The generators of such an algebra are ψdo's as studied here.

In [C_1], ch.IV, a compactification $\mathbb{P}^n\times\mathbb{P}^n$ of $\mathbb{R}^n\times\mathbb{R}^n=R^{2n}$ appears. The symbol of an operator A in the algebra, called A, for a moment is only defined at the boundary $\mathbb{M}=\partial(\mathbb{P}^n\times\mathbb{P}^n)=\mathbb{P}^n\times\mathbb{P}^n\setminus\mathbb{R}^{2n}$ of $\mathbb{P}^n\times\mathbb{P}^n$. Then $A\in A$ is called md-elliptic if its symbol is $\neq 0$ on $\mathbb{M}$. md-ellipticity is necessary and sufficient for $A\in A$ to be invertible modulo operators of finite rank. The algebra A is generated by the special ψdo's a(M) and b(D), where $a,b\in C^\infty(\mathbb{R}^{2n})$ are bounded and have $a^{(\alpha)}=o(1)$, $b^{(\beta)}=o(1)$, as $|x|\to\infty$, for all α. $\mathbb{P}^n$ just is the smallest compactification such that all such a,b extend to $C(\mathbb{P})$.

This result on C^*-algebras of ψdo's offers some guide lines on the question of local md-ellipticity. Within A we speak of md-ellipticity of $A\in A$ in a subset $M\subset \mathbb{M}$ if its symbol is $\neq 0$ near M . For fixed ρ,δ consider $A_{\rho,\delta}$, the closure of $\psi h_{0,\rho,\delta}$ within $CB(\mathbb{R}^{2n})$. The maximal ideal space $\mathbb{P}_{\rho,\delta}$ of this function algebra is a compactification of $\mathbb{R}^{2n}$ - again the smallest to which all functions of $\psi h_{0,\rho,\delta}$ extend continuously. We set $\mathbb{M}_{\rho,\delta}=\partial\mathbb{P}_{\rho,\delta}=\mathbb{P}_{\rho,\delta}\setminus\mathbb{R}^{2n}$ again.

Proposition 3.1. A symbol $a\in \psi h_{m,\rho,\delta}$ is md-elliptic if and only if (the continuous extension to $\mathbb{P}_{\rho,\delta}$ of) the function $b(x,\xi) = \langle\xi\rangle^{-m_1}\langle x\rangle^{-m_2} a(x,\xi)\in \psi h_{0,\rho,\delta}$ does never vanish on the set $\mathbb{M}_{\rho,\delta}$.

The proof follows from (1.1) and prop.1.4, and md-ellipticity

of $\langle\xi\rangle^{m_1}\langle x\rangle^{m_2} = \pi_m(x,\xi)\in \psi h_{m,\rho,\delta}$, using $0\le\rho_j\le 1$ and $\delta\ge 0$.
Remark: A result of III,1 below implies that $Op\psi h_{0,\rho,\delta}$, for $\delta\le\rho_1$, is an adjoint invariant subalgebra of $L(L^2(\mathbb{R}^n))$. The results on A sketched above, may be also derived for the norm closure $A_{\rho,\delta}$ of $Op\psi h_{0,\rho,\delta}$, assuming $\rho_j>0$, $0<\delta<\rho_1$ (cf. V,10, for $\rho_j=1$, $\delta=0$).

Following the above lead, a symbol $a\in \psi h_{m,\rho,\delta}$ will be called md-elliptic in a set $M\subset \mathbb{M}_{\rho,\delta}$ if (the continuous extension to $\mathbb{P}_{\rho,\delta}$) of $b(x,\xi)=a(x,\xi)\pi_{-m}(x,\xi)$ is $\neq 0$ on M.
We reformulate this definition, avoiding reference to values of b:

Definition 3.2. $a(x,\xi)\in \psi h_{m,\rho,\delta}$ is called md-elliptic in $M\subset \mathbb{M}_{\rho,\delta}$, if there exists a neighbourhood N of M in $\mathbb{P}_{\rho,\delta}$ with (1.1) in $K\cap \mathbb{R}^{2n}$ for every $K\subset\subset N$. A symbol $a\in \psi h_{m,\rho,\delta}$ is called formally md-hypoelliptic in M if (i),(ii),(iii) of sec.2 are valid in $K\cap \mathbb{R}^{2n}$, $K\subset\subset N$, with N as above. Again we call m' the inverse order of a .

The first part of def.3.2 is equivalent to our first definition of local md-ellipticity, because a continuous function is $\neq 0$ at a point p if and only if it is bounded away from 0 near p.

Def.3.2 still contains a reference to $M\subset \mathbb{M}_{\rho,\delta}\subset \mathbb{P}_{\rho,\delta}$. Also, $\mathbb{P}_{\rho,\delta}$ is of Stone-Cech type - its subsets are too general for our purpose. It seems practical to get restricted to a special kind of subsets M : The functions $s_j(x)$, $s_1(\xi)\in \psi h_{0,\rho,\delta}$ generate a subalgebra B of $A_{\rho,\delta}$, for $\rho_j\le 1$, $\delta\ge 0$. The maximal ideal space of B is a compactification of $\mathbb{R}^{2n}$ again- it equals $\mathbb{B}^n\times\mathbb{B}^n$, with the directional compactification $\mathbb{B}^n$ of $\mathbb{R}^n$ (cf. sec.5 or IV,1, below). $\mathbb{B}^n$ consists of $\mathbb{R}^n$ and the 'infinite' set $\{\infty x^0 : x^0\in \mathbb{R}^n\ |x^0|=1\}$. The function $s=(s_1\ldots s_n)(x)$ provides a homeomorphism $\mathbb{R}^n\leftrightarrow B_1=\{x\in \mathbb{R}^n, |x|<1\}$. $\mathbb{B}^n$ is characterized by the property that s(x) extends to a homeomorphism $\mathbb{B}^n\leftrightarrow B_1^c=\{|x|\le 1\}$. The inclusion $B\to A_{r,\delta}$ is an isomorphism, its dual is a surjective map $\iota:\mathbb{P}_{\rho,\delta}\to \mathbb{B}^n\times\mathbb{B}^n$.

Henceforth we will apply def.3.2 only for sets $M\subset \mathbb{M}_{\rho,\delta}$ of the form $M=\iota^{-1}M'$ with $M'\in \partial(\mathbb{B}^n\times\mathbb{B}^n)$. Then we may choose the neighbourhood N of def.3.2 in the form $N=\iota^{-1}N'$ where N' is a neighbourhood of M' in $\mathbb{B}^n\times\mathbb{B}^n$. $\mathbb{M}_{r,d}$ does not enter the estimates of def.3.2. Thus we refer to M', N' only, i.e., in def 3.2 replace $\mathbb{M}_{\rho,\delta}$ by $\mathbb{M}^0=\partial(\mathbb{B}^n\times\mathbb{B}^n)$, changing notation from M',N' to M,N. An $a\in \psi h_{0,\rho,\delta}$ needs not to extend to $C(\mathbb{M}_0)$, but for symbols of differential operators and their parametrices we have that property in $\mathbb{W}$ below:

$$(3.1)\qquad \mathbb{M}^0 = \partial(\mathbb{B}^n\times\mathbb{B}^n) = \mathbb{R}^n\times\partial\mathbb{B}^n \cup \partial\mathbb{B}^n\times\partial\mathbb{B}^n \cup \partial\mathbb{B}^n\times\mathbb{R}^n \quad ,$$

the first term at right denoting $\{(x,\xi): x\in \mathbb{R}^n,\ \xi\in \partial\mathbb{B}^n\}=\mathbb{W}$. The set $\mathbb{W}\subset \mathbb{M}^0$ will be called the wave front space, in view of an application in sec.5. If M is a subset of $\mathbb{W}$ then we will drop 'md-' and refer to a symbol (formally hypo-) elliptic in M. For M of the form $M=X\times\partial\mathbb{B}^n$, with $X\subset\mathbb{R}^n$ the operator $A=a(x,D)$ and its symbol a are called (formally hypo-) elliptic in X, in agreement with conventions for differential operators. For more general $M\subset\mathbb{W}$ another notation is in common use: Such sets are called non-characteristic sets of the operator A and symbol a (cf. [Tl_2], [Tr_1], [Hr_3]).

The union $\mathbb{M}_p=\mathbb{B}^n\times\partial\mathbb{B}^n$ of the first two terms in (3.1) and $\mathbb{M}_s=\partial\mathbb{B}^n\times\mathbb{R}^n$, the third term, are called the principal and secondary symbol space, resp. For $M\subset\mathbb{M}_s$ we speak of m-(hypo-)ellipticity. In [CP] a symbol $a\in\psi h_{m,\rho,\delta}\subset\psi h$ (formally) md-(hypo-) elliptic over $\mathbb{M}_p=\mathbb{B}^n\times\partial\mathbb{B}^n$ was called (formally) d-(hypo-) elliptic. "md-(hypo-) elliptic over $\mathbb{M}_p$ "just amounts to the conditions of (formal hypo-) ellipticity in some $|\xi|\geq\eta$, instead of a set $|x|+|\xi|\geq\eta$.

If a symbol is md-(hypo-) elliptic over $M\subset\mathbb{M}^\circ=\partial(\mathbb{B}^n\times\mathbb{B}^n)$, for an open set M, then one may expect a local parametrix for the corresponding ψdo $A=a(x,D)$. We mainly will consider the case $M\subset\mathbb{W}$, noting that the case $M\subset\mathbb{M}_s$ may be treated similarly, reversing x, ξ. Instead of a local parametrix as in [Hr_2] we construct a global ψdo inverting mod C^∞ in some set M of (formal hypo) ellipticity.

Let $\varphi=\varphi(x,\xi)$ be a symbol in $\psi h_{\nu,\rho,\delta}\subset\psi h$, for some $\nu\in\mathbb{R}^2$. A symbol $a\in\psi h_{m,\rho,\delta}$ will be called formally md-hypo-elliptic with respect to φ if for $|x|+|\xi|\geq\eta$, some $\eta>0$, we have (i) $a\neq0$ in supp φ, (ii) $\frac{\varphi}{a}\in\psi h_{\nu+m',\rho,\delta}$, (iii) the restriction of $a^{(\alpha)}_{(\beta)}/a$ to the set $\{(x,\xi)\in\text{supp }\varphi,\ |x|+|\xi|\geq\eta\}$ extends to a symbol in $\psi h_{m^{\alpha,\beta},\rho,\delta}$, with $m^{\alpha,\beta}$ of (2.2). If $m'=-m$ we will call a md-elliptic with respect to φ. Clearly a (formally) md-(hypo-) elliptic symbol is just a (formally) md-(hypo-)elliptic symbol with respect to $\varphi\equiv1$.

Theorem 3.3. Let $\varphi\in\psi h_{\nu,\rho,\delta}$, and let $a\in\psi h_{m,\rho,\delta}$ be (formally) md-(hypo-)elliptic with respect to φ. Then we have $e_j\in\psi h_{\nu-m',\rho,\delta}$, $j=1,2$, where $m'=-m$ in the elliptic case such that $A=a(x,D)$, $E_j=e_j(x,D)$, $\Phi_1=\varphi(x,D)=\varphi(M_l,D)$, $\Phi_2=\varphi(M_r,D)$ satisfy the relations

$$(3.2)\qquad AE_1=\Phi_1+K_1\ ,\quad E_2A=\Phi_2+K_2\ ,\ K_1,K_2\in\mathrm{Op}\psi h_{-\infty}\,.$$

Proof. 'Elliptic' just means $m'=-m$, hence needs no special consideration. We look at E_1 only, taking adjoints for E_2. With $E_2=B$ we once more go through the construction of the right parametrix of thm.1.6 (or 2.5). Set $b_0=\frac{\varphi}{a}$, getting $b_0\in\psi h_{\nu+m',\rho,\delta}$, by (ii). Write

$$(3.3)\qquad AB_0=\Phi_1+c_0(x,D)\ ,\ c_0=\sum_{|\theta|\geq1}(-i)^{|\theta|}/\theta!\,a^{(\theta)}\left(\tfrac{\varphi}{a}\right)_{(\theta)}\ ,$$

where the asymptotic sum $c_0 \in \psi h_{m+\nu+m'-\rho+\delta e',\rho,\delta}$. The next correction will be $b_1=-c_0/a$. Dividing (3.3) by a we get b_1 as an asymptotic sum of terms in the span of

$$(3.4) \qquad (a^{(\theta)}/a)(\tfrac{\varphi}{a})_{(\theta)} \ , \ |\theta| \geq 1 \ .$$

It follows that $b_1 \in \psi h_{\nu+m'-\rho+\delta e'}$, and that

$$(3.5) \quad AB_1=-c_0(x,D)+c_1(x,D) \ , \ c_1=-\sum_{|\theta|\geq 1}(-i)^{|\theta|}/\theta! a^{(\theta)}(c_0/a)_{(\theta)} \ .$$

This then leads into the recursion $b_N=-c_{N-1}/a$, $B_N = b_N(x,D)$,

$$(3.6) \quad AB_N=-c_{N-1}(x,D)+c_N(x,D) \ , \ c_N=\sum_{|\theta|\geq 1}(-i)^{|\theta|}/\theta! a^{(\theta)}b_{N(\theta)} \ ,$$

$N=1,2,\ldots$. We show that, successively, b_N, c_N are of proper order with $\sum_{j=0}^{\infty}B_j$ well defined, while the orders of c_N tend to $-\infty$.

As instrument in this proof we use the lemma below.

<u>Proposition 3.4</u>. Each symbol b_N ,$N=1,2,\ldots$, is an asymptotic sum of terms in the span of the expressions

$$(3.7) \qquad (\tfrac{\varphi}{a})^{(\alpha^0)}_{(\beta^0)} \Pi_{j=1}^{r}(a^{(\alpha^j)}_{(\beta^j)}/a) \ : \ \sum_{j=0}^{r}\alpha^j=\sum_{j=0}^{r}\beta^j=|\theta| \ ,$$

where $|\theta| \geq N$, for b_N .

Clearly prop.3.4 yields $b_N \in \psi h_{\nu+m'-N(\rho-\delta e')}$, hence $c_N=ab_{N+1} \in \psi h_{\nu+m+m'-N(\rho-\delta e')}$ has orders tending to $-\infty$. The asymptotic sum $\sum_{N=0}^{\infty}b_N=b \in \psi h_{\nu+m',\rho,\delta}$ is well defined and gives the desired $e_1=b$.

Thus thm.3.3 is reduced to prop.3.4.

We prove prop.3.4 by induction: It is true for N=1. (3.6) gives

$$(3.8) \qquad b_{N+1} = -\sum_{|\theta|\geq 1}(-i)^{|\theta|}/\theta!(a^{(\theta)}/a)b_{N(\theta)} \ .$$

If b_N is as in prop.3.4 we may reorder the double series by I, lemma 6.6, and get an asymptotic sum of terms in the span of

$$(3.9) \qquad (a_{(\theta')}/a)\bar{b}_N^{(\theta')} \ , \ |\theta'|\geq 1 \ ,$$

where $\bar{b}_N$ runs through all terms (3.7) , for $|\theta|=N$. Each term (3.9) is of the form (3.7) , for $|\theta|\geq N+1$, q.e.d.

We now apply thm.3.3 to a locally (hypo-)elliptic symbol:

<u>Proposition 3.5</u>. Let the symbol $a \in \psi h_{m,\rho,\delta} \subset \psi h$ be (formally hypo-) elliptic in an open set $M \subset \mathbb{W}$, and let $\varphi \in \psi h_{\nu,\rho,\delta}$, $\varphi=0$ for $(x,\xi) \notin K$ with $K \subset\subset N \subset \mathbb{B}^n \times \mathbb{B}^n$, as in def.3.2. Then the symbol a is (formally)

md-(hypo-) elliptic with respect to φ (so that thm.3.3 applies).

Also apply thm.3.3 to the (formally) d-(hypo-) elliptic case:

Proposition 3.6. If $a \in \psi h_{m,\rho,\delta}$ is (formally) d-(hypo-)elliptic, then it also is (formally) md-(hypo-)elliptic with respect to any function $\varphi(x,\xi)=\chi(\xi)$, where $\chi \in C^\infty(\mathbb{R}^n)$, $\chi=1$ for sufficiently large $|\xi|$, $\chi=0$ in a spere $|\xi| \le \eta$, for sufficiently large $\eta>0$.

The proof for both prop's is immediate. It already was seen that (formal) d-(hypo-)ellipticity amounts to (formal) md-(hypo-) ellipticity for large $|x|$ only, not for $|x|+|\xi|\ge\eta$.

Problems. 1) Discuss md-ellipticity on a cylinder in the setting of I,6, pbm's 1-4. Require the operator family $A(x^\Delta,\xi^\Delta):C^\infty(S^1) \to C^\infty(S^1)$ invertible for $|x^\Delta|+|\xi^\Delta|$ large; also, in the norm $\|.\|$ of $L^2(S^1)$, that $\|A^{(\alpha)}_{(\beta)}(x^\Delta,\xi^\Delta)\| = O(\langle x^\Delta\rangle^{m_2-\rho_2|\beta|}\langle\xi^\Delta\rangle^{m_1-\rho_1|\alpha|+\delta|\beta|})$, $\alpha,\beta \in \mathbb{N}^n$, as in I,(3.2)-(3.3). Discuss construction of a K-parametrix B with 1-AB, 1-BA$\in \mathcal{K}$ ($\mathcal{K}$ as in II,0, but with $\mathcal{L}(L^2(S^1))$-valued kernels). 2) Show that the above $\mathcal{K}$ (with operator-valued kernel) contains noncompact operators of $L^2(S^1\times\mathbb{R}^{n-1})$, but that $K\in\mathcal{K}$ with kernel in $\mathcal{K}(L^2(S^1))$ is compact. 3) For mutual K-parametrices A,B, as in pbm.1, a $\mathcal{K}(L^2(\mathbb{R}^{n-1}))$-valued symbol may be defined for each, $AB=1+K_1$ and $BA=1+K_2$, $K_j\in\mathcal{K}$. If and only if these symbols are invertible the operators A and B are Fredholm operators of $L^2(S^1\times\mathbb{R}^{n-1})$. 4) The generators [C2],VIII,(2.2) of the C^*-algebra $\mathcal{C}$ there -i.e., on our polycylinder $C=S^1\times R^{n-1}$, the multiplications by $a(x)\in C^\infty(\mathbb{R}^n)$ (2π-periodic in x_1), and the operators Λ, $\partial_{x_1}\Lambda$, $\partial_{x_2}\Lambda$, ..., $\Lambda = (1-\Delta)^{-1/2}$ (on C) all are ψdo's as in pbm.1 (verify!). Discuss the relation of pbm.3 above with [C2],VIII,thm.2.6.

4. Formally hypo-elliptic differential expressions.

Let Ω be a domain of $\mathbb{R}^n$, and let

$$L = a(x,D) = \sum_{|\alpha|\le N} a_\alpha(x)D^\alpha \tag{4.1}$$

be a differential expression with C^∞-coefficients : $a_\alpha \in C^\infty(\Omega)$. We shall call L (or its symbol $a(x,\xi)$)formally hypo-elliptic if for every compact set $K \subset \Omega$ there exist constants M_1, η, ρ, such that

$$\begin{aligned}&\text{(i) } a(x,\xi) \ne 0 \text{ , as } x \in K \text{ , } |\xi| \ge \eta \text{ ,}\\ &\text{(ii) } 1/a(x,\xi) = O(\langle\xi\rangle^{M_1}) \text{ , as } x \in K \text{ , } |\xi| \ge \eta \text{ ,}\\ &\text{(iii) } a^{(\alpha)}_{(\beta)}/a = O(\langle\xi\rangle^{-\rho|\alpha|}) \text{ , as } x \in K \text{ , } |\xi| \ge \eta \text{ .}\end{aligned} \tag{4.2}$$

(ii) is superfluous if L is locally of order >0 - i.e., L is of order N_x near x, with $0<N_x\le N$, $a_\alpha(x)\neq 0$ for some $|\alpha|=N_x$. Then

$$a^{(\alpha)}(x,\xi)/a(x,\xi) = \alpha! a_\alpha(x)/a(x,\xi) = O(\langle\xi\rangle^{-N_x\rho}) \quad , \text{ by (iii).}$$

This yields (ii) locally, with $M_1=-N_x\rho$, because $a_\alpha(x)$ is bounded away from zero, locally, for suitable α. This also shows that $M_1 = -\rho \operatorname{Min} N_x \le -\rho$ usually may be assumed negative :

For a constant coefficient L=a(D) of order >0 formal hypoellipticity is equivalent to hypo-ellipticity in the sense of the definition in ch.0, sec.5. This was shown by Hoermander [Hr_1],p.101.

On the other hand there exist hypo-elliptic expressions with variable coefficients which are not formally hypo-elliptic.

Here we use our parametrices of sec.3 for the following:

<u>Theorem 4.1</u>. A formally hypo-elliptic differential expression is hypo-elliptic. If M_1 of (ii) can be chosen <0 (as true for L of locally positive order) then $L-\lambda$ is hypoelliptic for all $\lambda\in\mathbb{C}$.
<u>Proof</u>. The second statement follows from the first, because for $M_1<0$ we get $\lim_{|\xi|\to\infty} a(x,\xi)=\infty$, uniformly on $K\subset\subset\Omega$. Accordingly, $\lim_{|\xi|\to\infty}(a(x,\xi)-\lambda)/a(x,\xi)=1$, uniformly on $K\subset\subset\Omega$, so that (i),(ii) (iii), for a , are equivalent to the same conditions for $a-\lambda$.

The first part of the theorem is best looked at in the light of invariance of the wave front set of sec.5, below. We actually will use thm.5.4, below. Assume $u\in\mathcal{D}'(\Omega')$, $f\in C^\infty(\Omega')$, for some open $\Omega'\subset\Omega$, and let Lu=f in Ω'. We must show that $u\in C^\infty(\Omega')$. Fix a point $x^0\in\Omega'$, and balls $B(x^0,\delta)$, $B(x^0,2\delta)$ with center x^0, radius δ, 2δ, closures $\subset\Omega'$. Let $\varphi\in C_0^\infty$, $\varphi=1$ near x^0, supp $\varphi\subset B(x^0,\delta)$. Define a global extension $a_0(x,\xi)$ of the symbol a as follows: Let $\chi\in C_0^\infty([0,2\delta))$, $\chi=1$ in $[0,\delta]$, $\chi=0$ in $[2\delta,\infty)$, $0\le t\chi(t)\le 2\delta$. Define

$$(4.3) \qquad a_0(x,\xi) = a(x^0+(x-x^0)\chi(|x-x^0|),\xi) \ , \text{ for } x,\xi\in\mathbb{R}^n .$$

Note that $a_0(x,\xi)=a(x^0,\xi)$, for x outside $B(x^0,2\delta)$. It follows that

$$(4.4) \qquad a_0\in\psi c_{N,0}\subset\psi h_{(N,0),(\rho,1),0} \quad ,$$

and that a_0 is formally d-hypo-elliptic in the latter space, with inverse order $M=(M_1,0)$. By prop.3.6 a_0 is formally md-hypoelliptic with respect to a suitable $\chi(\xi)$, $\chi=0$ near 0 , $\chi=1$ near ∞. By thm. 3.3 construct $E\in \mathrm{Op}\psi h_{M_1,0}$ with $Ea_0(x,D)=\chi(D)+K$, $K\in\mathrm{Op}\psi h_{-\infty}$. Clearly $a=a_0$, as $x\in B(x^0,\delta)$, hence $\varphi a=\varphi a_0$, for all x,ξ, assuming the left hand side zero whenever a is undefined. Accordingly

$$(4.5) \qquad \varphi f = \varphi Lu = L\varphi u + [\varphi,L]u = L_0(\varphi u) + [\varphi,L_0]\psi u \quad ,$$

with another cut-off ψ, supp $\psi\subset B(x^0,2\delta)$, $\psi=1$ in supp $[\varphi,L_0]$.

Note that $\varphi f \in C_0^\infty(\mathbb{R}^n) \subset S$, while $\varphi u \in E'(\mathbb{R}^n) \subset S'$. Thus $\varphi u \in H_s$, for some s, and $h=[\varphi, L_0]\psi u \in H_t$, for some t. Clearly, h=0 near x^0, since $\varphi=1$ near x^0, causing the commutator to vanish. We get x^0 in the complement of sing supp h . Hence $x^0 \notin$ sing supp Eh , using thm.5.4, below. Applying the operator E to (4.5) we get

$$(4.6) \qquad \varphi u + (1-\chi)(D)\varphi u + K\varphi u = E\varphi f - Eh .$$

Here $E\varphi f \in S$, since $\varphi f \in S$. Also $Kfu \in S$, since $\varphi u \in S'$, and K has order $(-\infty,-\infty)$. Also, $\omega = 1-\chi \in C_0^\infty(\mathbb{R}^n) \subset S$, thus $\omega(D)\varphi u = \int d\!\!\!{}^-y\,\omega^\vee(x-y)(\varphi u)(y) \in S$, the distribution integral existing, due to $\omega^\vee \in S$, $\varphi u \in S'$. Also Eh is C^∞ near x^0, hence φu is C^∞ near x^0. Since $\varphi=1$ near x^0 we find that also u is C^∞ near x^0 . Since this construction may be done for every $x^0 \in \Omega'$, the theorem is proven.

We made extensive use of hypo-ellipticity in [C_1],II,IV. Hypo-ellipticity of $L-\lambda$, especially of $L\pm i$ is required to make the Carleman alternative work, as well as in general the construction of e.s.a.-realizations in [C_2],II.

5. The wave front set and its invariance under ψdo's.

We define the <u>wave front set</u> WF(u) of a distribution $u \in D'(\Omega)$ over a domain $\Omega \subset \mathbb{R}^n$ as a subset of the <u>wave front space</u> $\mathbb{W}=\mathbb{W}(\Omega)$, already discussed in sec.3, for the special case $\Omega=\mathbb{R}^n$. For a general $\Omega \subset \mathbb{R}^n$ we define $\mathbb{W}(\Omega)=\Omega\times\partial\mathbb{B}^n \subset \mathbb{R}^n\times\partial\mathbb{B}^n=\mathbb{W}(\mathbb{R}^n)$.

It is common practice to interpret this space as $\mathbb{W}=\Omega\times S^n$, i.e.,

$$(5.1) \qquad \mathbb{W} = \mathbb{W}(\Omega) = \{(x,\xi) : x \in \Omega , \xi \in \mathbb{R}^n , |\xi| = 1 \} .$$

Similarly, for distributions on a differentiable manifold Ω one defines $\mathbb{W}(\Omega)$ as the "co-sphere bundle", i.e. the bundle of unit spheres in the cotangent space, with respect to a suitable smooth Riemannian metric. It is clear then that the unit vector ξ of a pair (x,ξ) with $x \in \Omega$ just indicates the direction tx , as $t \to \infty$, where the real point $(x,\infty\xi)$ is to be found. In other words, for a manifold Ω the wave front space is properly defined as the subset $\mathbb{W}$ of the symbol space $\mathbb{M}(\Omega)$, as defined in [C_2],VI, p.161.

One defines WF(u) as a subset of $\mathbb{W}$ by its complement $WF(u)^c = \mathbb{W}\setminus WF(u)$, as follows. Specify two types of cut-off functions called $\varphi(x)$, and $\psi(\xi)$, respectively. For $x^0 \in \Omega$ a cut-off φ is a C_0^∞-function, $\varphi\ge 0$, $\varphi=1$ near x^0. For a "direction" $\infty\xi^0 \in \partial\mathbb{B}^n$ given by $\xi^0 \in \mathbb{R}^n$ with $|\xi^0|=1$ a cut-off $\psi(\xi)$ is a $C(\mathbb{B}^n)$-function, $\psi\ge 0$, $\chi=1$ near $\infty\xi^0$, $\chi=0$ outside some neighbourhood of ξ^0, and, $\psi|\mathbb{R}^n \in C^\infty(\mathbb{R}^n)$. Also, χ

equals some homogeneous function of degree 0 for large $|\xi|$. One may think of a $C^\infty(\mathbb{R}^n)$-function $\psi=\lambda(|\xi|)\mu(\frac{\xi}{|\xi|})$ where $\lambda=0$ near 0, $=1$ near ∞, $\mu=1$ near ξ°.

We define: $(x^0,\xi^0)\in WF(u)^c$ (i.e., $(x^\circ,\alpha\xi^\circ)$ is not in $WF(u)$) if there exist φ, ψ, such that, with the Fourier transform "$\hat{}$",

$$(5.2) \qquad (\psi(\xi)(\varphi u)^\wedge(\xi)) \in \mathcal{S}.$$

Note that φu admits a natural extension to $\mathbb{R}^n$, zero outside Ω, a distribution in $\mathcal{S}'$, hence has a Fourier transform in $\mathcal{S}'$.

<u>Proposition 5.1</u>. We have $(x,\xi)\in WF(u)^c$ if and only if there exist cut-offs φ,ψ, near x and ξ, resp., such that the ψdo $C=\psi(D)\varphi(x)$ $\in Op\psi c_{0,-\infty}$ takes u into $\mathcal{S}$: $\psi(D)\varphi(x)u \in \mathcal{S}$.

The proof is evident, since $C=F^{-1}\psi F\varphi$, and because the inverse Fourier transform F^{-1} is an isomorphism of $\mathcal{S}$ onto itself.

From the definitions of WF(u) and s.s.u we conclude:

<u>Proposition 5.2</u>. The singular support s.s.u of $u\in \mathcal{D}'(\Omega)$ is the projection of WF(u) to the first component Ω of $\mathbb{W}=\Omega\times\partial\mathbb{B}^n$:

$$(5.3) \qquad \text{s.s.u} = \{x: \text{there exists } \xi \text{ such that } (x,\xi)\in WF(u)\} .$$

Indeed, if $x\in (\text{s.s.u})^c$, then a cut-off φ can be found with $\varphi u\in C_0^\infty\subset \mathcal{S}$, so that $(\varphi u)^\wedge\in \mathcal{S}$, -i.e. every $(x,\xi)\in WF(u)^c$. Vice versa if for some x every $(x,\xi)\in Wf(u)^c$, then $(\varphi u)^\wedge\in \mathcal{S}$ (i.e., $\varphi u\in \mathcal{S}$) follows for suitable φ: The unit sphere is compact, and for $|\xi|=1$ we can find $(\varphi^\xi u)^\wedge\psi^\xi\in \mathcal{S}$. A finite collection $\psi_j=\psi^{\xi^j}$ has supports covering the sphere, so that $\tilde\psi_j=\psi_j/\sum\psi_j$ is a partition of unity. Let $\varphi^\sim$ be a cut-off near x with $\varphi^\sim\varphi^{\xi^j}=\varphi^\sim$, for all j (i.e. $\varphi^{\xi^j}=1$ on supp $\varphi^\sim$), then $(\varphi^\sim u)^\wedge\psi_j \in \mathcal{S}$, by prop.5.3. Hence $(\varphi^\sim u)^\wedge(\sum\psi_j)\in \mathcal{S}$. But it is possible to arrange for $\omega=\sum\psi_j\geq 1$ outside a large sphere, and have ω homogeneous of degree zero there. Adding a suitable $C_0^\infty(\mathbb{R}^n)$-functions χ we get $\omega+\chi\geq 1$ in $\mathbb{R}^n$, and $(\omega+\chi)^{-1}\in \mathcal{T}$, the space of $C^\infty(\mathbb{R}^n)$-functions with derivatives of polynomial increase. Thus $(\varphi^\sim u)^\wedge\frac{\omega}{\omega+\chi}\in \mathcal{S}$, hence $(\varphi^\sim u)^\wedge\in \mathcal{S}$, since $1-\omega/(\omega+\chi)=\chi/(\chi+\omega)\in C_0^\infty$, and $(\varphi^\sim u)^\wedge\in C^\infty$. Hence $\varphi^\sim u\in \mathcal{S}$, $u\in C^\infty$ near x, or $x\in (\text{s.s.u})^c$, q.e.d.

We have used the proposition below, also useful later on:

Proposition 5.3. If we have $(\varphi u)^\wedge\psi\in S$, with a pair of cut-offs φ, ψ and a distribution u, and if $\varphi^\sim\varphi=\varphi$, for a cut-off $\varphi^\sim$, then we also get $(\varphi^\sim u)^\wedge\psi\in S$.

Proof. We know that $(\varphi u)^\wedge\psi\in S$ is equivalent to $\psi(D)\varphi(x)u\in S$, and similarly for $\varphi^\sim$ instead of φ. Write

$$(5.4)\qquad \psi(D)\varphi^\sim u=\psi(D)\varphi(x)\varphi^\sim u=\varphi^\sim\psi(D)\varphi(x)u+[\psi(D),\varphi^\sim(x)]\varphi u\ ,$$

where the first term, at right, is in S. For the second term we use the asymptotic expansion of the calculus of ψdo's : We get

$$(5.5)\quad \varphi[\varphi^\sim(x),\psi(D)]=p(x,D),\ p=-\sum_{|\theta|>0}\kappa_\theta\psi^{(\theta)}(\xi)\varphi(x)\varphi^{\sim}{}_{(\theta)}(x)\ (\mathrm{mod}\ S)$$

Each term of p is $\equiv 0$, hence $p(x,D)\in O(-\infty)$. The last term of (5.4) also is in S, due to $p(x,D)^*=-[\psi(D),\varphi^\sim(x)]\varphi(x)\in O(-\infty)$. Q.E.D.

Theorem 5.4. Let $u\in S'$, and let $A=a(x,D)$ be a ψdo with symbol in some $\psi h_{m,\rho,\delta}$, with the usual inequalities $0\le\delta<\rho_1$, $0<\rho_2$. Then

$$(5.6)\qquad WF(Au)\subset WF(u)\ ,\ \text{and s.s.}Au\subset\text{s.s.}u\ .$$

Proof. We are using H_s and $O(s)$ of ch.III, here, for convenience. It is sufficient to prove the first inclusion (5.6), by prop.5.2. We must show that $(x,\xi)\in WF(u)^c$ implies $(x,\xi)\in WF(Au)^c$. For $(x,\xi)\in WF(u)^c$ we have $\psi(D)\varphi(x)u\in S$ with suitable φ,ψ. It is found that

$$(5.7)\qquad \psi(D)\varphi(x)Au=A\psi(D)\varphi(x)u+[C,A]u\ ,\ C=\psi(D)\varphi(x)\ ,$$

where the first term at right is in S, since A takes S to S. $u\in S'$ is in some H_s, $s\in\mathbb{R}^2$, since $S'=\cup H_s$. $[C,A]$ is of lower differentiation order than A (and of multiplication order $-\infty$).

In order to get the desired property for Au, we will iterate the proceedure as follows. Construct sequences of cut-off functions $\varphi_0,\varphi_1,\dots,(\psi_0,\psi_1,\dots,)$ such that $\varphi\varphi_0=\varphi_0$, $\varphi_0\varphi_1=\varphi_1,\dots,$ $\varphi_j\varphi_{j+1}=\varphi_{j+1},\dots.$ (and $\psi\psi_0=\psi_0$, $\psi_0\psi_1=\psi_1,\dots,$). Prop.5.3. implies

$$(5.8)\qquad C_ju\in S\ ,\ j=0,1,\ \dots\ ,\ \text{where } C_j=\psi_j(D)\varphi_j(x)\ .$$

Proposition 5.5. We have

$$(5.9)\quad C_0C=CC_0=C_0\ (\mathrm{mod}\ O(-\infty)),\ C_jC_{j+1}=C_{j+1}C_j=C_{j+1}\ (\mathrm{mod}\ O(-\infty)).$$

Proof. We have $CC_0=\psi(D)\varphi(x)\psi_0(D)\varphi_0(x)=\psi_0(D)\varphi_0(x)+$ $+\ \psi(D)[\varphi(x),\psi_0(D)]\varphi_0(x)$ .Hence it is enough to show that the last term, called V for a moment, is in $O(-\infty)$. But we have

(5.10) $V^*=z(x,D)$, $z(x,\xi)= \sum_{|\theta|>0} \kappa_\theta \varphi_0(x)\psi_0^{(\theta)}(\xi)\varphi^{(\theta)}(x)\varphi_0(x) \pmod{\mathcal{S}}$,

where again all terms of the asymptotic sum vanish identically . Accordingly $V^* \in O(-\infty)$, hence $V \in O(-\infty)$. This shows that $CC_0 = C \pmod{O(-\infty)}$. Similarly for C_jC_{j+1} . On the other hand,

$$C_0C = \psi_0(D)\varphi_0(x)\psi(D)\varphi(x) = \psi_0(D)\varphi_0(x) + \psi_0(D)[\varphi_0(x),\psi(D)]\varphi(x) .$$

Let W be the last term. Consider the asymptotic expansion of W^* :

$$W^* = p(x,D) \ , \ p(x,\xi) = -\sum_{|\theta|>0} \kappa_\theta \varphi(x)\psi_0(\xi)\psi^{(\theta)}(\xi)\varphi_0^{(\theta)}(x) \pmod{\mathcal{S}} ,$$

which again implies $W^* \in O(-\infty)$ since all terms vanish, q.e.d.

Using prop.5.5, we may write

$$C_NAu = C_{N-1}C_NAu \pmod{\mathcal{S}} = C_{N-1}AC_Nu + C_{N-1}[C_N,A]u \pmod{\mathcal{S}}$$

(5.11) $$=C_{N-2}C_{N-1}[C_N,A]u \pmod{\mathcal{S}} = C_{N-2}[C_{N-1},[C_N,A]]u \pmod{\mathcal{S}}$$

$$= \dots = [C,[C_0,[C_1,\dots.[C_{N-1},[C_N,A]]\ \dots]]u \pmod{\mathcal{S}} .$$

Here the N+2-fold commutator has multiplication (differentiation) order $-\infty$ $(m_1-(N+2)(\rho_1-\delta))$. Accordingly the right hand vector of (5.11) is in $\mathcal{H}_{M,\infty}$, for every M, as N gets large. Thus $C_NAu \in \mathcal{H}_{M,\infty}$, as $N>N_0(M)$. The φ_k,ψ_k may be constructed such that $\varphi_k=1$ in a fixed neighbourhood $\mathcal{N}$ of x (and similarly for ψ_k and a neighbourhood $\mathcal{N}'$ of ξ). Thus we can find cut-offs $\varphi^\sim$, $\psi^\sim$ with $\varphi^\sim\varphi_k=f^\sim$, $\psi^\sim\psi_k=\psi^\sim$, for all k . Using prop.5.5 we thus get

(5.12) $C^\sim Au = C^\sim C_NAu \in \mathcal{H}_{M,\infty}$, for every $M \Rightarrow C^\sim Au \in \mathcal{S} = \cap \mathcal{H}_s$.

This completes the proof of thm.5.4.

<u>Corollary 5.6</u>. Let $A\in Op\psi h$ be formally md-hypo-elliptic with respect to a symbol φ . Let $u\in \mathcal{S}'$, and let $Au=f$, and $\Phi=\varphi(x,D)$. Then

(5.13) $WF(f) \subset WF(u)$, and $WF(\Phi u) \subset WF(f)$.

Morever, if $\varphi \equiv 1$ (i.e., if A is formally md-hypo-elliptic),then

(5.14) $WF(u) = WF(f)$.

The proof is an evident consequence of thm.5.4 and thm.3.3.

6. Systems of ψdo's.

It is often convenient to consider a $\nu\times\mu$=matrix

(6.1) $A = ((A_{jl}))_{j=1,..,\nu,l=1,..,\mu}$, $A_{jl}= a_{jl}(M_l,M_r,D)$,

of ψdo's where we generally assume that $a_{jl} \in SS^{m,\rho,\delta} \subset SS$, m,ρ,δ independent of j,l. Introduce $a(x,y,\xi) = ((a_{jk}(x,y,\xi)))$ as a matrix-valued symbol and write $a\in SS^{m,\rho,\delta}$, and $A=a(M_l,M_r,D)$. In the present section, we will use all symbol class notations, such as ST, ψh, $\psi h_{m,\rho,d}$ also for $\nu\times\mu$-matrices to indicate that the entries are in the corresponding symbol class. The operator A of (6.1) will act on spaces of μ-vector-valued functions or distributions.

All formulas of the ψdo-calculus of ch.I remain valid, except those involving a commutator, such as I,(7.5).

For md-ellipticity we focus on square ($\mu\times\mu$-) matrices of ψdo's We call $A=a(x,D)\in Op\psi h_{m,\rho,\delta}$ (and its symbol a) md-elliptic if

(6.2) $c(x,\xi) = \det((a_{jl}(x,\xi))) \in \psi h_{\mu m,\rho,\delta}$

is md-elliptic. A symbol $a\in SS$ (and its operator A) will be called md-elliptic if $b(x,\xi)=a(x,x,\xi)$ is md-elliptic.

Lemma 6.1. A square-matrix-valued symbol $a\in \psi h_{m,\rho,\delta}$ is md-elliptic if and only if for $|x|+|\xi|\geq\eta$, with some $\eta>0$, the matrix $a(x,\xi)$ is nonsingular, and there exists a square-matrix-valued symbol $b\in \psi h_{-m,\rho,\delta}$ such that $a^{-1}(x,\xi) = b(x,\xi)$ as $|x|+|\xi|\geq\eta$.

Trivially the Hilbert space adjoint of an elliptic ψdo and the product of 2 elliptic ψdo's are elliptic again.

We can expect a theorem like thm.1.6 for square matrices, and even a more general result for rectangular matrices. For the latter let us define as follows: If for a pair of ($\nu\times\mu$- and $\mu\times\nu$-) symbols $a\in \psi h_{m,\rho,\delta}$, $b\in \psi h_{-m,\rho,\delta}$,resp., we have (for some $\eta>0$)

(6.3) $a(x,\xi)b(x,\xi) = 1 = ((\delta_{jl}))_{j,l=1,..\nu}$, for all $|x|+|\xi|\geq \eta$,

then we will call a right md-elliptic and b left md-elliptic . We then must have $\nu\leq\mu$. a is elliptic if and only if it is left and right elliptic. b is then called a right inverse symbol of a, etc.

Theorem 6.2. Let a $\nu\times\mu$-symbol $a\in \psi h_{m,\rho,\delta}$ be (right) (left) md-elliptic, and let $b\in \psi h_{-m,\rho,\delta}$ be a (right) (left) md-inverse symbol. There exists a (right) (left) K-parametrix $E=e(x,D)$, $e\in \psi h_{-m,\rho,\delta}$, of the ψdo $A=a(x,D)$ such that

(6.4) $e(x,\xi)-b(x,\xi) \in \psi h_{(-m_1-\rho_1+\delta,-m_2-\rho_2),\rho,\delta}$.

Vice versa, if a (right) (left) parametrix of A of order $-m$ exists then A (and a) will be (right) (left) md-elliptic.

We prove thm.6.2 like thm.1.6, constructing a right or left K-parametrix. If both exist they coincide mod $O(-\infty)$, as usual.

Many 'hypo-elliptic' arguments work as in the scalar case.

For matrix-valued symbols there is another interesting case to consider: Assume the matrix $a(x,\xi)$ of constant rank for large $|x|+|\xi|$, and that, moreover, $a\in \psi h_{m,\rho,\delta}$ allows a decomposition

$$a = a_0+a_1 \ , \ a_1\in \psi h_{m',\rho,\delta} \ , \ m'<m \ , \tag{6.5}$$

where a_0 allows a 'partially md-inverse symbol' $b_0\in \psi h_{-m,\rho,\delta}$, in the sense that $b_0(x,\xi)$ is a $\mu\times\nu$-matrix with

$$a_0(x,\xi)b_0(x,\xi)=p_0(x,\xi) \text{ for large } |x|+|\xi| \ , \tag{6.6}$$

where $p_0(x,\xi)$ is a $\nu\times\nu$-projection matrix of the same rank as $a_0(x,\xi)$ and $b_0(x,\xi)$ (i.e., $p_0(x,\xi)^2=p_0(x,\xi)$).

Clearly (6.6) and rank p_0=rank a_0 imply im $p_0(x,\xi)$=im $a_0(x,\xi)$ so that p_0 projects onto im a_0, hence $1-p_0$ vanishes on im a_0, i.e, $(1-p_0)a_0=0$. Introduce $q_0(x,\xi)=b_0(x,\xi)a_0(x,\xi)$. Then get $q_0-q_0^2=b_0a_0-b_0a_0b_0a_0=b_0a_0-b_0p_0a_0=b_0(1-p_0)a_0=0$, so that $q_0(x,\xi)$ also is a projection. Evidently rank $q_0(x,\xi)$=rank $p_0(x,\xi)$, im p_0 = im b_0 .

Generally, if $b=b_0+b_1$, where $b_1\in \psi h_{-m',\rho,\delta}$, with $-m'<-m$, we will call the symbols a and b <u>partially md-inverse to each other</u>.

<u>Theorem 6.3</u>. Assume that $a\in \psi h_{m,\rho,\delta}$ and $b\in \psi h_{-m,\rho,\delta}$ are partially md-inverse symbols of each other (where $\rho>0$, $\rho_1>\delta$). Let $p=ab$, $q=ba$, and let $A=a(x,D)$, $B=b(x,D)$, $P=p(x,D)$, $Q=q(x,D)$. Then

$$AB - P = r(x,D) \ , \ BA - P = s(x,D) \ , \tag{6.7}$$

where the symbols $r(x,\xi)$, $s(x,\xi)$ are $\nu\times\nu$-matrix-valued and $\mu\times\mu$-matrix-valued, respectively, and are in $\psi h_{-\varepsilon e,\rho,\delta}$, for some $\varepsilon>0$.

The proof is calculus of ψdo's. We cannot expect better para metrices making r or s of order $-\infty$, with improved B, etc.

Matrix-symbols also occur in VI,IX,X, under different aspects

Chapter 3. L^2-SOBOLEV THEORY AND APPLICATIONS.

0. Introduction.

In this chapter we consider ψdo's as linear operators of a class of weighted L^2-Sobolev spaces over $\mathbb{R}^n$. We specialize on L^2-spaces and neglect L^p-theory, because ψdo's in general are not continuous operators on L^p-Sobolev spaces, for $p\neq 2$. To be more precise, general L^2-boundedness theorems for ψdo's are true for $A=a(x,D)\in Op\psi h_{0,\rho,\delta}$, assuming $\rho\geq 0$, $0\leq\delta\leq\rho_1$, $\delta\neq 1$, but corresponding L^p-boundedness statements are false, except for $\rho_1=1$. There is an extensive theory in L^p-spaces of Sobolev and other types (cf. Beals [B_4], Coifman-Meyer [CM], Marshall [Mr_1], Muramatu [Mm_1], Nagase [Ng_1], Yamazaki [Ym_1]).

In sec's 1, 2 we prove the L^2-boundedness theorem, for $\delta=0$, and $0<\delta\leq 1$, respectively. This result often is quoted as Calderon-Vaillancourt theorem. In sec.3 we look at weighted L^2-Sobolev norms. Our class of spaces $H_s=H_{s_1,s_2}$ is left invariant by the Fourier transform, just as many of our ψdo-classes. A ψdo of order $m=(m_1,m_2)$ is a bounded map $H_s\to H_{s-m}$, for every s. For every $m\in\mathbb{R}^2$ an order class $O(m)$ is introduced - the operators $S\to S$ extending to operators in $L(H_s,H_{s-m})$ for all s. $O(m)$ is a Frechet space under the norms of $L(H_s,H_{s-m})$; $O(0)$ and $O(\infty)=\cup O(m)$ are algebras. A ψdo of order m belongs to $O(m)$.

A refined Fredholm theory holds for (formally) md-(hypo-) elliptic ψdo's. Such an operator admits a Green inverse- the equivalent of the integral operator of the generalized Green's function. This is discussed in sec.4. In sec.5 we prove that ψdo's of negative order ($m_1<0$, $m_2<0$) are compact operators $H_s\to H_s$, for all s.

1. L^2-boundedness of zero-order ψdo's.

We refer to the class $\psi t_0=\psi h_{(0,0),(0,0),0}$ of symbols here, as introduced in I,(4.26), and discuss the following result.

Theorem 1.1. An operator $A=a(M_1,D)\in Op\psi t_0$ is bounded in $H=L^2(\mathbb{R}^n)$. More precisely: The restriction $A|H$ maps $H\to H$ and belongs to $L(H)$.

This result is quoted as the Calderon-Vaillancourt theorem (c.f.[CV_1]). For independent proofs c.f. [CC], and VIII,thm.2.2, below. The very short approach below is due to Beals [B_3].

<u>Proof.</u> Construct a partition of unity for $\mathbb{R}^n$, as follows. With the cube $Q_r=\{|x_j|<r, j=1,\dots,n\}$ choose $0\le\varphi\in C_0^\infty(Q_2)$, $\varphi>0$ in $Q_{3/2}$. Define $\varphi_\alpha(x)=\varphi(x-\alpha)$, $\alpha\in\mathbb{Z}^n$, and $\psi_\alpha=\varphi_\alpha(\sum_{\beta\in\mathbb{Z}^n}\varphi_\beta^2)^{-1/2}$. Clearly then

$$(1.1)\qquad \psi_\alpha\in C_0^\infty(\{x\in\mathbb{R}^n: |x-\alpha|<2, j=1,\dots,n\}) \ ,\ \sum_\alpha\psi_\alpha{}^2(x)=1 \ ,\ \psi_\alpha\ge 0 \ .$$

Given some symbol $a\in\psi t_0$ we define

$$(1.2)\qquad a_\alpha(x,\xi)=\psi_\alpha(x)a(x,\xi) \ ,\ A_\alpha=a_\alpha(x,D) \ .$$

We first show that $A_\alpha\in L(H)$. Since the Fourier transform is unitary we may show $B_\alpha=FA_\alpha F^{-1}\in L(H)$. But for $a\in\psi t_0$, supp $a\subset\{|x|\le p\}$,

$$(1.3)\qquad \begin{aligned} FAF^{-1}u(\eta) &= \int đx e^{-ix\eta}\int đ\xi\ a(x,\xi)\ e^{ix\xi}u(\xi) \\ &= \int đ\xi\ u(\xi)\int đx\ a(x,\xi)\ e^{ix(\xi-\eta)} = \int đ\xi\ a^{\vee}(\xi-\eta,\xi)\ u(\xi) \ , \end{aligned}$$

with "$^\vee$" with respect to x written as "$^{\downarrow}$", and trivial integral interchanges. The integral operator at right of (1.3) is L^2-bounded by Schur's lemma ([C_1],I,1.1.1). We get the operator bound

$$(1.4)\qquad \sup\{\int đ\zeta|a^{\vee}(\xi-\zeta,\xi)| \ ,\int d\zeta|a^{\vee}(\zeta-\eta,\zeta)| \ :\xi,\eta\in\mathbb{R}^n\} \ .$$

In particular,

$$(1.5)\qquad \begin{aligned} \langle\xi-\eta\rangle^{2N}a^{\vee}(\xi-\eta,\xi) &= \int đx\ a(x,\xi)(1-\Delta_x)^N e^{ix(\xi-\eta)} \\ &= \int đx a_N(x,\xi)e^{ix(\xi-\eta)} \ ,\ \text{with } a_N=(1-\Delta_x)^N a \ , \end{aligned}$$

which implies

$$(1.6)\qquad a^{\vee}(\xi-\eta,\xi) = O(\langle\xi-\eta\rangle^{-2N}\|a_N\|_{L^\infty}) \ ,$$

hence

$$(1.7)\qquad \|A\|_{L^2} \le c\ \|a_N\| \ ,\ 2N>n \ ,$$

where c depends only on N , n and p .

We need an improved estimate for the operators

$$(1.8)\qquad A_{\alpha\beta}=A_\alpha\psi_\beta(M)=a_{\alpha\beta}(M_1,M_r,D) \ ,\ a_{\alpha\beta}=\psi_\alpha(x)\psi_\beta(y)a(x,\xi) \ .$$

Taking the Fourier transformed operator $A^\wedge$ for $A=a(M_l,M_r,D)$ again, with $a(x,y,\xi)$ vanishing for (x,y) outside $K\subset\subset \mathbb{R}^{2n}$, we get

$$(1.9)\quad A^\wedge u=FAF^{-1}u(\eta)=\int e^{-ix\eta}đx\int đ\xi\int đy e^{i\xi(x-y)}a(x,y,\xi))\int đ\kappa e^{iy\kappa}u(\kappa)\ .$$

Integrals interchange readily, as $u\in\mathcal{S}$. Therefore,

$$(1.10)\qquad A^\wedge u(\eta) = \int đ\kappa\ u(\kappa)\int đ\xi\ a^{\wedge\wedge}(\xi-\eta,\kappa-\xi,\xi)\quad .$$

With the methods used to derive (1.6) one finds that

$$(1.11)\qquad \begin{aligned} a^{\wedge\wedge}(\xi-\eta,\kappa-\xi,\xi) &= \|a_{NN}\|_{L^\infty}\, O(\langle\xi-\eta\rangle^{-2N}\langle\xi-\kappa\rangle^{-2N}) \\ a_{NN} &= (1-\Delta_x)^N(1-\Delta_y)^N a \quad . \end{aligned}$$

Note that (1.11) implies

$$(1.12)\qquad \int a^{\wedge\wedge}(\xi-\eta,\kappa-\xi,\xi)đ\xi = \|a_{NN}\|_{L^\infty}\, O(\langle\kappa-\eta\rangle^{-2N})\quad .$$

Here it is important to observe that the "O(.)-constant" depends only on n,N, and the volume of the compact set K, the latter since an L^1-norm has to be estimated by the corresponding L^∞-norm. Also, $\int d\xi\langle\xi-\eta\rangle^{-2N}\langle\xi-\kappa\rangle^{-2N}=\int d\xi\langle\xi-(\eta-\kappa)\rangle^{-2N}\langle\xi\rangle^{-2N}$, and $\int(\langle t\rangle\langle 2s-t\rangle)^{-2N}dt$

$$=\int_{|t|\le|s|}+\int_{|t|\ge|s|} = I_1+I_2\ ,\ \text{where } I_2\le c\langle s\rangle^{-2N},\ c=\int\langle t\rangle^{-2N}dt\ .$$

For the integral I_1 the inequality implies $|2s-t|\ge 2|s|-|t|\ge|s|$,

so that the integrand is bounded by $\langle s\rangle^{-2N}\langle t\rangle^{-2N}$, i.e., we get the same estimate as for I_2. This shows that indeed (1.12) holds.

Next we transform the operator $A_{\alpha\beta}$, as follows.

$$(1.13)\qquad \begin{aligned} A_{\alpha\beta}u(x) &= \int đ\xi\int đy((1-\Delta_\xi)^M a_{\alpha\beta}(x,y,\xi))\langle x-y\rangle^{-2M} e^{i\xi(x-y)}u(y) \\ &= a^M_{\alpha\beta}(M_l,M_r,D)u(x)\ ,\ a^M_{\alpha\beta}(x,y,\xi)=\langle x-y\rangle^{-2M}(1-\Delta_\xi)^M \alpha_{\alpha\beta}(x,y,\xi)\ , \end{aligned}$$

where the partial integration is justified as in sec.1.

<u>Proposition 1.2</u>. We have supp $a^M_{\alpha\beta}$ contained in the set

$$(1.14)\qquad \{(x,y,\xi)\ :\ |x-\alpha|\le 2^n\ ,\ |y-\beta|\le 2^n\}\ .$$

Moreover,

$$(1.15)\quad \|a^M_{\alpha\beta NN}\|_{L^\infty}\le c\langle\alpha-\beta\rangle^{-2M}\max\{\|a^{(\sigma)}_{(\tau)}\|_{L^\infty(\mathbb{R}^2)}\ :|\sigma|\le 2M,|\tau|\le 2N\}\quad ,$$

where the constant c depends only on n, M, and N .

Indeed, (1.14) is trivial, by construction of the ψ_α , while (1.15) is a calculation, using that $\langle x\rangle^{(\kappa)}/\langle x\rangle = O(1)$ for all κ .

Combining (1.12) and (1.15) we get

$$(1.16)\qquad \|A_{\alpha\beta}\|_{L^2} \le c\ \langle\alpha-\beta\rangle^{-2M}[[a]]_{2N,2M}\ ,\quad 2N > n\ ,$$

with

$$(1.17)\qquad [[a]]_{k,l} = \max\ \{\|a^{(\sigma)}_{(\tau)}\|_{L^\infty} : |\tau|\le k\ ,\ |\sigma|\le l\}\ ,$$

where c depends only on n,N,M. Choose the set K as product of the balls with center α,β, radius 2^n, its volume independent of α,β.

For $u,v\in C_0^\infty(\mathbb{R}^n)$, $u_\alpha=\psi_\alpha u$, $v_\alpha=\psi_\alpha v$, 2M>n, we get

$$(1.18)\qquad \begin{aligned} |(u,Av)| &= |\sum_{\alpha,\beta}(u_\alpha,A_{\alpha\beta}v_\beta)| \le c_1 \sum_{\alpha,\beta}\|u_\alpha\|\,\|v_\beta\|\langle\alpha-\beta\rangle^{-2M} \\ &\le c_2\{\textstyle\sum_\alpha\|u_\alpha\|^2\ \sum_\beta\|v_\beta\|^2\}^{1/2} = c_2\|u\|\ \|v\|\ , \end{aligned}$$

with $\|.\|=\|.\|_{L^2}$, inner product (.,.) of $\mathcal{H}$, where $c_j=c_j'[[a]]_{2M,2N}$, j=1,2, with c_j' depending only on n,M,N. In (1.18) we used $\|u\|^2= \sum_\alpha\|u_\alpha\|^2$, by construction of ψ_α, and Schur's lemma in discrete form for l^2-boundedness of $((\langle\alpha-\beta\rangle^{-2M}))$, 2M>n. Clearly (1.18) implies thm.1.1: Set $u=u_k\in C_0^\infty(\mathbb{R}^n)$ where $u_k\to Av\in\mathcal{S}$ in $\mathcal{H}$.

Moreover, we also have proven the useful corollary, below.

<u>Corollary 1.3</u>. There exists c>0 only depending on n such that

$$(1.19)\qquad \|a(x,D)\|_{L^2(\mathbb{R}^n)} \le c[[a]]_{NN},\ N=[n/2]+1 \text{ with } [[.]] \text{ of } (1.17).$$

Revisiting our above proof we notice that the same arguments apply for an $A=a(M_l,M_r,D)$ with $a\in \mathcal{SS}_{0,0,0}$, (i.e., m=0, ρ=0, δ=0). The $A_{\alpha\beta}$ are of this form anyway. Instead of (1.16) , (1.17) we get

$$(1.20)\qquad \|A_{\alpha\beta}\|_{L^2} \le c\ \langle\alpha-\beta\rangle^{-2M}\ [[[a]]]_{2N,2N,2M}\ ,$$

with

$$(1.21)\qquad [[[a]]]_{j,k,l} = \max\ \{\|\partial_x^\sigma\partial_y^\tau\partial_\xi^\kappa a\|_{L^\infty} : |\sigma|\le j, |\tau|\le k, |k|\le l\}\ .$$

Thus we have

<u>Corollary 1.4</u>. All operators $A=a(M_l,M_r,D)$,with $a\in\mathcal{SS}_{0,0,0}$ are $L^2(\mathbb{R}^n)$-bounded. Moreover, we have

$$(1.22)\qquad \|A\|_{L^2} \le c[[[a]]]_{N,N,N}\ ,\ N = [n/2]+1\ .$$

2. L^2-boundedness for the case of $\delta>0$.

Note that thm 1.1 gives L^2-boundedness of all ψdo's in the Hoermander classes $\psi h_{0,\rho,0}$, with $\rho_j \geq 0$, but that it does not apply to the case $\delta>0$. The theorem below asserts L^2-boundedness for general m=0, and $\rho_j\geq 0$, $\rho_1\geq\delta$, $0\leq\delta<1$. It was first proven in [CV_2]. Other proofs may be found in [Ka_2], [CM]. We discuss a proof after a scheme in [B_3] which again is relatively short.

Theorem 2.1. A ψdo $A=a(x,D)$ with symbol $a \in \psi h_{0,\delta e^1,\delta}$, $0<\delta<1$, induces a bounded operator of $H=L^2(\mathbb{R}^n)$.

Proof. We use cor.1.4 and the following decomposition. For a partition ω_j, $j=0,\pm1,\pm2,\ldots$, like ψ_α but for n=1, i.e.,

$$(2.1)\qquad \omega_j(t) = \omega(t-j)\ ,\ \operatorname{supp}\omega \subset \{|t|\leq 2\}\ ,\ \sum_j\omega_j^2 = 1\ ,\ \omega_j\geq 0\ ,$$

set $\chi_j=\omega_j(\log_2(\langle\xi\rangle^\delta))$. Clearly χ_j has support near $\log_2\langle\xi\rangle^\delta=j$, i.e, $\langle\xi\rangle^\delta = 2^j$. Since $\langle\xi\rangle^\delta\geq 1$ we get $\chi_j=0$, as $j<-1$. The partition of unity $\Sigma\chi_j^2(\xi)=1$ has supports in concentric spherical shells:

$$(2.2)\qquad \operatorname{supp}\chi_j \subset \{2^{(j-2)/\delta} \leq \langle\xi\rangle \leq 2^{(j+2)/\delta}\}\ .$$

For $A=a(x,D)$ with a satisfying our assumptions, we write

$$(2.3)\quad (u,Av)=\sum_{jl}(u_j,A_{jl}v_l),\ A_{jl}=\chi_j(D)A\chi_l(D),\ u_j=\chi_j(D)u,\ v_j=\chi_j(D)v.$$

We apply Schur's lemma once more to show that the matrix $((2^{-\varepsilon|j-l|}))$ is l^2-bounded, and prove an estimate of the form

$$(2.4)\qquad \|A_{jl}\|_{L^2} \leq c2^{-\varepsilon|j-l|}\ ,\ j,l \in \mathbb{Z}\ ,\ \text{for some } \varepsilon>0\ .$$

Then the argument following (1.18) will give boundedness of A .

We prove (2.3) for $A_{jl}{}^\wedge=FA_{jl}F^{-1}$ instead of A_{jl} again. Here $A_{jl}{}^\wedge=P_{jl}=p_{jl}(M_l,M_r,D)$ with $p_{jl}=\chi_j(x)\chi_l(y)a(-\xi,y)$. Get $p_{jl}=0$ unless

$$(2.5)\qquad 2^{j-2}\leq\langle x\rangle^\delta\leq 2^{j+2}\ ,\ 2^{l-2}\leq\langle y\rangle^\delta\leq 2^{l+2}\ .$$

Repeating once more the partial integration of (1.13) we may write

$$P_{jl}=p^M_{jl}(M_l,M_r,D)\ ,\ p^M_{jl}=\langle x-y\rangle^{-2M}((1-\Delta_\xi)^M p_{jl})(x,y,\xi)\ .$$

$$(2.6)\qquad = \Theta^M(x,y)\chi_j(x)\chi_l(y)b_M(-\xi,y)\ ,\ \text{where } \Theta(x,y)=\langle y\rangle^{2\delta}/\langle x-y\rangle^2\ ,$$

$$\text{and } b_M(x,\xi) = \langle\xi\rangle^{-2M\delta}(1-\Delta_x)^M a(x,\xi)\ .$$

Notice that $b_M \in \psi h_{0,\delta e',\delta}$ again, for every M=0,1,..., .

Proposition 2.2. We have $\|b_M(-D,M_r)\chi_l(M_r)\| \le c$, with c independent of l , for each fixed M=0,1,2,.... .

Applying the Fourier transform we may prove $\|b_M(x,D)\chi_l(D)\| \le c$ instead. Note also that $\chi_l(\xi) \in \psi h_{0,\delta e',\delta}$, so that $c_l(x,\xi) = b_M(x,\xi)\chi_l(\xi) \in \psi h_{0,\delta e',\delta}$.

Proposition 2.3. The family $q_l(x,\xi)=c_l(2^{-l}x,2^l\xi)$ is bounded in ψt_0

Indeed,

$$
\begin{aligned}
q_{l(\beta)}^{(\alpha)}(x,\xi) &= 2^{l(|\alpha|-|\beta|)} c_{l(\beta)}^{(\alpha)}(2^{-l}x,2^l\xi) \\
&= 2^{l(|\alpha|-|\beta|)} O(\langle 2^l\xi\rangle^{\delta(|\beta|-|\alpha|)}) ,
\end{aligned}
\tag{2.7}
$$

with O(.) independent of l. Also, $\chi_l(2^l\xi)=0$ except if $\langle 2^l\xi\rangle^\delta$ differs from 2^l by a factor between 1/4 and 4. Since q_l contains the factor $\chi_l(2^l\xi)$ we conclude from (2.7) that $q_{l(\beta)}^{(\alpha)}=O(1)$, with O(.) independent of l,x,ξ. Thus cor.1.3 gives the uniform L^2-bound for $q_l(x,D)=Q_l$. Note that $Q_l=V_l b_M(x,D)\chi_l(D)V_l^{-1}$, with unitary $V_l:u(x) \to 2^{-ln/2}u(2^{-l}x)$, so that $\|Q_l\|=\|b_M(x,D)\chi_l(D)\|$, proving prop.2.2.

Applying prop.2.2 for M=0 get $\|A_{jl}\| \le c$, c independent of j,l. For (2.4) it suffices to look at $|j-l| \ge k_0$ with suitable fixed k_0.

Proposition 2.4. For $|j-l| \ge k_0=5\delta$, and $\varepsilon=\frac{2}{\delta}(1-\delta)$ we have

$$
\Theta(x,y) \le c\, 2^{-\varepsilon|j-l|} \text{ , as } (x,y) \text{ belong to the set (2.5) ,} \tag{2.8}
$$

with c independent of j,l.

For the proof we use the estimate

$$
\langle x-y\rangle \ge |\langle x\rangle - \langle y\rangle| , \tag{2.9}
$$

easily derived by an elementary calculation. We get

$$
\Theta(x,y) \le \langle y\rangle^{2\delta}/|\langle x\rangle-\langle y\rangle|^2 = \upsilon_1 4^l/|\upsilon_2 2^{j\eta}-\upsilon_3 2^{l\eta}|^2 , \tag{2.10}
$$

with υ_j satisfying $\frac{1}{4} \le \upsilon_j \le 4$ and $\eta=1/\delta>1$. For $2^{\eta|j-l|} \ge 32$, i.e., $|j-l| \ge 5\delta=k_0$ the right hand side of (2.10) is bounded by $c4/|2^{\eta j}-2^{\eta l}|^2 \le c4^l/4^{\eta j} \le c4^{\eta(l-j)}=c4^{-\eta|j-l|}$, as j>l, and by $c4^l/4^{\eta l}=c4^{-(\eta-1)l|j-l|}$, as j<l. Hence we get the statement with $\varepsilon=2\eta-2=2(1-\delta)/\delta$.

Regarding the proof of thm.2.1 note that P_{jl} has distribution kernel $\Theta^M(x,y)k(x,y)$, with the kernel k of $\chi_j(x)\chi_l(y)b_m(-\xi,y)$. The latter symbol defines a uniformly bounded operator family, while, in supp k , we have (2.8), hence it seems that (2.4) should follow. For a real proof however we argue as follows: First show that

$$(2.11)\qquad \partial_x^\alpha\partial_y^\beta\Theta/\Theta \le c \sum_{k+j=|\beta|} \langle y\rangle^{-j}\langle x-y\rangle^{-k-|\alpha|} \ .$$

Indeed, for $|\alpha|+|\beta|=1$, writing $\partial_x^\alpha\partial_y^\beta\upsilon=\upsilon_{|\alpha\beta}$ for a moment, we get $\frac{\Theta}{\Theta}{}_{|\alpha\beta}=(\log\Theta)_{|\alpha\beta}=2\delta(\log\langle y\rangle)_{|\alpha\beta}-2(\log\langle x-y\rangle)_{|\alpha\beta}$ in the span of $\frac{T}{T}{}_{|\alpha\beta}$ and $\frac{S}{S}{}_{|\alpha\beta}$, with $T=\langle y\rangle$, $S=\langle x-y\rangle$. For any α,β, and $|\varepsilon|+|\eta|=1$ we get $(\frac{R}{R}{}_{|\alpha\beta})_{|\varepsilon\eta} = \frac{R_{|\alpha+\varepsilon,\beta+\eta}}{R} + (\frac{R_{|\alpha\beta}}{R})(\frac{R_{|\varepsilon\eta}}{R})$, where we may set $R=S,T,\Theta$. An induction argument gives $\frac{\Theta_{|\alpha\beta}}{\Theta}$ in the span of $\Pi K(\alpha^j,\beta^j)\ \Pi L(\gamma^j,\delta^j)$, where $K(\alpha,\beta)=\frac{S_{|\alpha\beta}}{S}$, $L(\alpha,\beta)=\frac{T_{|\alpha\beta}}{T}$, with product over all partitions $\alpha=\sum\alpha^j+\sum\gamma^j$, $\beta=\sum\beta^j+\sum\delta^j$, with multiindices $\alpha^j,\beta^j,\gamma^j,\delta^j$. Finally note that $\partial_y^\alpha T=0$, as $\alpha\ne0$. Thus we may assume $\sum\gamma^j=0$, $\sum|\alpha^j|=|\alpha|$. Let $\sum|\beta^j|=k$, $\sum|\delta^j|=l$ and note that $\langle x\rangle^{(\alpha)}/\langle x\rangle=O(\langle x\rangle^{-|\alpha|})$, as used earlier. Hence the above products are $O(S^{-|\alpha|-k}T^{-l})$, where $k+l=|\beta|$, implying (2.11).

Now we again recall the above unitary map V_l. Look at (2.6), for M=1 , describing a symbol of P_{jl} , rewritten as

$$(2.12)\qquad p_{jl}^1(x,y,\xi) = \Theta(x,y)\chi_j(x)b_l(-\xi,y)\chi_l(y)$$

The operator $W_{jl}=V_l^{-1}P_{jl}V_l$ then has symbol

$$(2.13)\qquad w_{jl}(x,y,\xi) = \Theta(2^lx,2^ly)\chi_j(2^lx)b_l(-2^{-l}\xi,2^ly)\chi_l(2^ly) \ .$$

From prop.2.3 we get $b^l(-2^{-l}\xi,2^ly)\chi_l(2^ly)$ describing a bounded set in ψt_0. Notice that $w_{jl}(x,y,\xi)$ of (2.13) describes a bounded set with respect to the norm (1.22), while $[[[w_{jl}]]]_{N,N,N}\le c2^{-\varepsilon|j-l|}$, so that cor.1.4 gives (2.3). Indeed, $\partial_x^\alpha\partial_y^\beta\partial_\xi^\gamma w_{jl}$ is in the span of

$$(2.14)\qquad ((\partial_x^\kappa\partial_y^\lambda\Xi)/\Xi)(\partial_x^\mu\omega_l)\ \Xi\ \psi_{(\nu)}^{(\gamma)} \quad , \ \kappa+\mu=\alpha \ , \ \lambda+\nu=\beta \ ,$$

where $\Xi(x,y)=\Theta(2^lx,2^ly)$, $\omega_j(x)=\chi(2^lx)$, $\psi(x,\xi)=(\chi_lb_l)(-2^{-l}\xi,2^lx)$.

Note that $\partial_x^\alpha\partial_y^\beta\Xi/\Xi=2^{l(|\alpha|+|\beta|)}(\partial_x^\alpha\partial_y^\beta\Theta/\Theta)(2^lx,2^ly)$. For boundedness in supp w_{jl} we show that $2^l/\langle 2^ly\rangle$ and $2^l\langle 2^lx-2^ly\rangle$ are bounded, by (2.11). In supp w_{jl} get $2^l/\langle 2^ly\rangle\le c2^l/2^{l/\delta}\le c$. Also, $2^l/\langle 2^lx-2^ly\rangle\le c2^l/|2^{j/\delta}-2^{l/\delta}|$, as near (2.10). With similar argument (and $\eta=1/\delta$) we get $\le c2^{l(1-\eta)}$ for $l>j$, and $\le c2^{l-\eta j}$, as $l<j$, so $\le c$ in both cases.

In other words, the first factor of (2.14) is bounded. We already know the last factor bounded, since $\psi\in\psi t_0$. For the second we get $\delta_x^\mu \omega = 2^{|\mu| l} O(\langle 2^l x\rangle^{-|\mu|}) \le c 2^{|\mu|(l-\eta j)}$, bounded for l>j, but <u>not</u> for l<j. The third factor is bounded by $c2^{-\varepsilon|j-l|}$. Together we get the estimates for (2.4), via cor.1.4 , but <u>only for</u> l>j .

For j<l observe that prop.2.2 does not involve the support restriction $\langle 2^l x\rangle^\delta \approx 2^j$, while prop.2.3 remains intact if that restriction is replaced by $\langle 2^l x\rangle \le 2^{l-5} \Leftrightarrow |x| \le \sqrt{4^{-5}-4^{-l}}$. This means that we may replace the factor $\omega_j(x)$ in w_{jl} by $\omega_j(x)\theta(x)$, using a cut-off $\tau=0$ for $|x|\ge 2^{-6}$, =1 near 0. For j<<l we get $\omega_j = \omega_j\theta$. Replacing ω_j in (2.13) by $\omega_j\theta$ we get a new w_{jl} with also the third factor of (2.14) bounded for j<l. The old w_{jl} will now be $\omega_j(x)w_{jl}$ giving the ψdo $\omega_j(M)W_{jl}$, i.e., the correct old W_{jl}. The estimates after (2.11) also remain intact with the new x-restriction.

This completes the proof of thm.2.1. We also get

<u>Corollary 2.5</u>. Under the assumptions of thm.2.1 there exists a constant c depending on n and δ only such that for 0<δ<1 we have

$$(2.15)\qquad \|A\| \le c \max\{\|a^{(\alpha)}_{(\beta)}(x,\xi)\langle\xi\rangle^{\delta(|\alpha|-|\beta|)}\|_{L^\infty} : |\alpha|\le 2, |\beta| \le [\tfrac{n}{2}]+1\} .$$

with the L^2-operator norm $\|A\|$.

The proof is evident. Note that a result like cor.1.4 is easily derived for the case δ>0, following the above guide lines.

<u>3. Weighted Sobolev spaces; K-parametrix and Green-inverse</u>.

Define the spaces $H_s = H_{s_1,s_2}$, for $s=(s_1,s_2)$, as the classes

$$(3.1)\qquad H_s = \{u \in S' : \pi_s(x,D)u \in H\} ,$$

where $H = L^2(\mathbb{R}^n)$, and $\pi_s(x,\xi) = \langle x\rangle^{s_2}\langle\xi\rangle^{s_1} \in \psi c_s$. For $u,v \in H_s$ we introduce a norm and an inner product by

$$(3.2)\qquad (u,v)_s = (\Pi_s u, \Pi_s v)_{L^2} ,\ \|u\|_s = \|\Pi_s u\|_{L^2} ,\ \Pi_s = \pi_s(x,D) .$$

Note that the strictly classical ψdo Π_s is invertible, as an operator $S \to S$, or $S' \to S'$, with inverse

$$(3.3)\qquad \Pi_s^{-1} = \Pi_{-s}^* = \pi_{-s}(M_r,D) = \langle D\rangle^{-s_1}\langle M\rangle^{-s_2} \in Op\psi c_{-s} .$$

In particular Π_s is md-elliptic of order s , and,

$$(3.4)\qquad \Pi_s^{-1} - \Pi_{-s} \in Op\psi c_{-s-e} ,$$

by calculus of ψdo's. From (3.2) conclude that $\Pi_s : H_s \to H$ is an isometry $H_s \to H$. In particular, H_s is a Hilbert space, for all $s \in \mathbb{R}^2$.

We refer to the spaces H_s as (weighted L^2-) Sobolev spaces over $\mathbb{R}^n$. In the special case $s_2=0$ we obtain the ordinary (unweighted) L^2-Sobolev spaces, commonly known as H_s - now with $s\in\mathbb{R}$.

First we discuss properties of unweighted Sobolev spaces, for the moment also called H_s, $s\in\mathbb{R}$: For details (and a more general discussion involving L^p-norms) cf.[C1],III, or [GT]. Write

(3.5) $\Lambda_s=\langle D\rangle^s$, $H_s=\{u\in S':\Lambda_s u\in H\}$, $(u,v)_s=(\Lambda_s u,\Lambda_s v)$, $\|u\|_s=\|\Lambda_s u\|$.

1) Using Parseval's relation (0,(1.8)) we may write

$$\|u\|_s^2 = \int d\!\!\!{}^-\xi\, |u^\wedge(\xi)|^2\langle\xi\rangle^{2s} , \tag{3.6}$$

with the Fourier transform $u^\wedge=Fu\in L^1_{loc}(\mathbb{R}^n)$, where $\int$ exists as a Lebesgue integral. In particular,

$$H_s = \{u\in S' : u^\wedge\in L^2(\mathbb{R}^n,\langle\xi\rangle^{2s}d\xi)\} . \tag{3.7}$$

2) The spaces H_s form a decreasing set of subspaces of S', containing S, as s increases from $-\infty$ to $+\infty$. Moreover, $\|u\|_s$ is an increasing function of s, either on $\mathbb{R}$, or on a half-line $(-\infty,\sigma)$.

3) For s=k , a nonnegative integer, we get

$$H_s = \{u\in S' : u^{(\alpha)}\in H \text{ ,for all } |\alpha|\le k\} , \tag{3.8}$$

with distribution derivatives $u^{(\alpha)}$. The norm $\|u\|_s$ is equivalent to

$$\|u\|_k^{\sim} = \{\sum_{|\alpha|\le k}\|u^{(\alpha)}\|^2\}^{1/2} . \tag{3.9}$$

This follows from (3.6) :

$$\begin{aligned}\|u\|_k^2 &= \int d\xi\,|u^\wedge|^2(1+\textstyle\sum\xi_j^2)^{2k} = \sum_{|\alpha|\le k}\binom{k}{\alpha}\int d\xi\,\xi^{2\alpha}|u^\wedge(\xi)|^2 \\ &= \sum_{|\alpha|\le k}\binom{k}{\alpha}\|u^{(\alpha)}\|^2\end{aligned} \tag{3.10}$$

with the multinomial coefficients $\binom{k}{\alpha} = k!/(\alpha!(k-|\alpha|)!)$.)

4) We have Sobolev's lemma : For $s > n/2$ the space H_s is (compactly) imbedded into the space $CO(\mathbb{R}^n)$ of all continuous functions over $\mathbb{R}^n$ vanishing at infinity. We have the inequalities

$$\|u\|_{L^\infty} \le c_s\|u\|_s , \quad c_s = \left\{ \int\langle\xi\rangle^{-2s}d\xi \right\}^{1/2} , \tag{3.11}$$

and

$$|u(x)-u(y)|\le\delta_s(|x-y|)\|u\|_s , \quad \delta_s=\left\{2\int\sin^2(\xi_1 t/2)\langle\xi\rangle^{-2s}d\xi\right\}^{1/2}. \tag{3.12}$$

Moreover, if even $s > n/2+j$, for some integer $j > 0$, then we have a compact imbedding $H_s \to CO^j(\mathbb{R}^n)$, defined as the space of all $u \in CO$ such that also $u^{(\alpha)} \in CO$, for all $|\alpha| \le j$, and inequalities of the form (3.11), (3.12) are valid for all $u^{(\alpha)}$, $|\alpha| \le j$.

Sobolev's inequality (3.11) follows by showing that the Fourier transform $u^\wedge$ is L^1 for $u \in H_s$, $s > n/2$ (a matter of Schwarz' inequality). (3.12) is derived by estimating the Fourier transform (cf.[C_1],III). The compactness of the imbedding is a consequence of the Ascoli theorem: Equicontinuity of H_s-bounded sets follows. The imbedding $H_s \to CO_j$, as $s > n/2+j$, follows from (5) below.

5) For $u \in H_s$ we have $u^{(\alpha)} \in H_{s-|\alpha|}$, and

$$\|u^{(\alpha)}\|_{s-|\alpha|} \le \|u\|_s \quad . \tag{3.13}$$

This is a consequence of the inequality

$$\|u^{(\alpha)}\|^2_{s-|\alpha|} = \int |u^\wedge|^2 \xi^{2\alpha} \langle\xi\rangle^{2s-2|\alpha|} d\xi \le \int |u^\wedge|^2 \langle\xi\rangle^{2s} d\xi = \|u\|_s^2 \ , \tag{3.14}$$

using that $|\xi^{2\alpha}| \le \langle\xi\rangle^{2|\alpha|}$.

6)The spaces

$$H_\infty = \cap\{H_s : s \in \mathbb{R}\} \ , \ H_{-\infty} = \cup\{H_s : s \in \mathbb{R}\} \ , \tag{3.15}$$

have simple locally convex topologies (cf.[C_1], ch.III). The space H_∞ is contained in $CO^\infty(\mathbb{R}^n)$, but contains S , while $H_{-\infty}$ contains E', but is contained in S'. All inclusions mentionned are proper.

7) H_s and H_{-s} are mutually adjoint under the pairing

$$(u,v) = (\Lambda^s u, \Lambda^{-s} v) = \int \bar{u} v dx \quad , \ u \in H_s \ , \ v \in H_{-s} \ , \tag{3.16}$$

defined by the second expression, involving the inner product of $H = L^2$, using that $\Lambda^s : H_s \to H$ and $\Lambda^{-s} : H_{-s} \to H$ define isometries.

Returning to weighted Sobolev spaces $H_s = H_{s_1,s_2}$ of (3.1) we note analogously the following properties of H_s .

(i) We have

$$H_s = \{u \in S' : \|\Pi_s u\| < \infty\} = \{u \in S' : \|\Pi_s^* u\| < \infty\} \ , \tag{3.17}$$

with the operator $\Pi_s^* = \langle D\rangle^{s_1} \langle M\rangle^{s_2} = \pi_s(M_r, D)$. Moreover, the norms (3.2) and (3.18) below are equivalent.

$$\|u\|_s^\cdot = \|\Pi_s^* u\| = \|\langle D\rangle^{s_1} \langle M\rangle^{s_2} u\| \quad . \tag{3.18}$$

This follows by calculus of ψdo's : $\Pi_s \Pi_s^{*-1}$ and $\Pi_s^* \Pi_s^{-1}$ belong to $\mathrm{Op}\psi c_0 \subset \mathrm{Op}\psi t_0$, They are L^2-bounded, by thm.1.1, so that

(3.19) $$c_s\|u\|_s \le \|u\|_s^{\bullet} \le c_s'\|u\|_s$$

(ii) We have

(3.20) $$\mathcal{H}_s \supset \mathcal{H}_t \ , \text{ as } s_j \le t_j \ , \ j=1,2 \ ,$$

although the norm $\|u\|_s$ not necessarily increases, as s_j increases. Moreover, the imbedding $\mathcal{H}_t \to \mathcal{H}_s$ is compact, whenever $t_j > s_j$, $j=1,2$.

(3.20) follows from the estimate

(3.21) $$\|u\|_s \le \|u\|_{s_1,t_2} \le 1/c_{s_1,t_2}\|u\|^{\bullet}_{s_1,t_2} \le c_t/c_{s_1,t_2}\|u\|_t \ ,$$

using (3.2) and (3.19). Or, $\|u\|_s=\|\Pi_s u\|=\|\Pi_t(\Pi_t^{-1}\Pi_s)u\|=\|\Pi_t^{-1}\Pi_s u\|_t$, where $K_{st}=\Pi_t^{-1}\Pi_s \in Op\psi c_{s-t}$ is a bounded (compact) operator of $\mathcal{H}_t$, due to $\Pi_t K_{st}\Pi_t^{-1}=\Pi_s\Pi_t^{-1}=\Pi_{s-t}(\Pi_{s-t}^{-1}\Pi_s\Pi_t^{-1})=\Pi_{s-t}L_{st}$, where $L_{st}\in Op\psi c_0 \subset L(\mathcal{H})$, $\Pi_r=\langle x\rangle^{r_1}\langle D\rangle^{r_2}\in K(\mathcal{H})$, as $r_j<0$ (cf. thm.5.1).

(iii) For $s_1=k\ge 0$, k an integer, we have

(3.22) $$\mathcal{H}_s = \{u\in \mathcal{S}' : u^{(\alpha)} \in L^2(\mathbb{R}^n,\langle x\rangle^{2s_2}dx), \ |\alpha|\le k\} \ ,$$

and a norm equivalent to $\|u\|_s$ is given by

(3.23) $$\sum_{|\alpha|\le k} \|\langle x\rangle^{s_2} u^{(\alpha)}\| \quad .$$

For an integer $s_2=j\ge 0$ $\mathcal{H}_s$ consists precisely of all $u\in S'$ such that $x^\alpha u\in \mathcal{H}_{s_1}=H_{s_1,0}$ (with the unweighted $\mathcal{H}_{s_1}$) for all $|\alpha|\le j$. A norm equivalent to $\|u\|_s$ is given by

(3.24) $$\sum_{|\alpha|\le j} \|x^\alpha u\|_{s_1} \quad .$$

(iv) For $s\in \mathbb{R}^2$, and $k=(k_1,k_2)$, with integers k_1, k_2,

(3.25) $$\mathcal{H}_{s+k} = \{u\in \mathcal{H}_s : x^\alpha u^{(\beta)} \in \mathcal{H}_s \ , \text{ as } |\alpha|\le k_1 \ , \ |\beta|\le k_2\} \ ,$$

and $\|u\|_{s+k}$ is equivalent to

(3.26) $$\sum_{|\beta|\le k_1, |\alpha|\le k_2} \|x^\alpha u^{(\beta)}\|_s \quad .$$

(v)(Sobolev's lemma): For $s_1>n/2+j$ we get a compact imbedding

(3.27) $$\mathcal{H}_s \to \{u\in C^j(\mathbb{R}^n) : \langle x\rangle^{s_2} u \in CO^k(\mathbb{R}^n)\} \ .$$

(vi) $\mathcal{H}_s$ and $\mathcal{H}_{-s}$ are mutually adjoint, under the pairing

(3.28) $$(u,v) = (\Pi_s u,\Pi_{-s}v) = \int \bar{u}v\,dx \ , \ u\in \mathcal{H}_s \ , \ v\in \mathcal{H}_{-s}$$

where $(\Pi_s u,\Pi_{-s}v)$ (with (.,.) of $\mathcal{H}$) defines (u,v), $u\in \mathcal{H}_s$, $v\in \mathcal{H}_{-s}$.

Note that $\Pi_s^* = \Pi_{-s}^{-1}$, and that (3.28) is independent of s .

That is, the continuous linear functionals on H_s precisely are given as $v \to (u,v)$, with some $u \in H_{-s}$ and $(.,.)$ of (3.28).

This follows because $\Pi_s : H_s \to H$ and $\Pi_{-s} : H_{-s} \to H$ are isometries.

(vii) We have

$$S = \cap \{H_s : s \in \mathbb{R}^2\} \ , \ S' = \cup \{H_s : E\ \mathbb{R}^2\} \ .$$

Indeed, (v) gives $\cap H_s \subset C^\infty(\mathbb{R}^n)$. Combined with (3.25) we get $\cap H_s \subset S$, the opposite inclusion being evident from (3.1). On the other hand, (3.28) implies that $\cup H_s \subset S'$, with $\langle u,\varphi\rangle = (\bar{u},\varphi)$, $\varphi \in S$, $u \in H_s$. The opposite follows from [C₁],I,(6.3) (or [Schw_1]).

(viii) The space S is dense in every space H_s .
Indeed, a sequence $u_m \to u$ may be generated, for $u \in H_s$ by setting $u_m = \chi_m * (\omega_m u)$, $\chi_m = m^{-n}\chi(mx)$, $\omega_m = \omega(\frac{x}{m})$, with suitable cut-off's χ,ω .

The spaces H_s will play a central role. They allow a description of topologies on ψdo-algebras in terms of L^2-norms.

<u>Theorem 3.1</u>. For a symbol $a \in \psi t_m$ the corresponding ψdo $A = a(x,D)$ is a continuous operator $H_s \to H_{s-m}$, for every $s \in \mathbb{R}^2$. Similarly, for $a \in \psi h_{m,\delta e^1,\delta}$, $0<\delta<1$, A is continuous from H_s to H_{s-m} .

The proof is simple, due to our preparations: Since Π_s is an isometry $H_s \to H$, one must show that $\Pi_{s-m} A \Pi_s^{-1}$ is L^2-bounded. By I,thm. 4.8 this is an operator in $Op\psi t_0$, due to $\Pi_{s-m} \in Op\psi c_{s-m} \subset Op\psi t_{s-m}$, $\Pi_s^{-1} \in Op\psi c_{-s} \subset Op\psi t_{-s}$. Hence the desired L^2-boundedness follows from thm.1.1. Similarly for $a \in \psi h_{m,\delta e^1,\delta}$, using thm.2.1 instead.

We now define a type of <u>order classes</u>: An operator $A: S \to S$, with the property of thm.3.1 (i.e., A admits continuous extensions $A_s : H_s \to H_{s-m}$, for all $s \in \mathbb{R}^2$, and given $m \in \mathbb{R}^2$) is said to have order m. The class of all operators of order m is denoted by $O(m)$.

Trivially $A \in O(m)$ admits a continuous extension $A_\infty : S' \to S'$ with $A_s = A_\infty | H_s$. Thm.3.1 now may be expressed by stating that

$$\text{(3.29)} \qquad Op\psi h_{m,\rho,\delta} \subset O(m) \ , \text{ for all } m \in \mathbb{R}^2 \ , \ \rho \ge 0 \ , \ \delta \le \rho_1 \ , \ 0 \le \delta < 1 \ .$$

We also define

$$\text{(3.30)} \qquad O(\infty) = \cup\{O(m) : m \in \mathbb{R}^2\} \ , \ O(-\infty) = \cap\{O(m) : m \in \mathbb{R}^2\} \ .$$

The proposition below is trivial.

<u>Proposition 3.2</u>. The classes $O(\infty)$ and $O(0)$ are algebras under operator multiplication and $O(-\infty)$ is an ideal of both, $O(0)$, $O(\infty)$. Moreover, we have

(3.31) $$O(m)O(m') \subset O(m+m') .$$

The class $O(0)$ has the natural locally convex topology, induced by the collection $\{ \|A\|_s : s\in \mathbb{R}^2\}$ of all operator norms $\|A\|_s = \sup\{\|Au\| : \|u\|_s \le 1\}$.

Proposition 3.3. The above topology is a Frechet topology. An equivalent countable system of semi-norms is given by

(3.32) $$\{ \|A\|_k : k\in \mathbb{Z}^2\} .$$

Moreover, $O(0)$, under this topology, is a Frechet algebra, i.e., it is complete, and the algebra operations are continuous.

This follows from the Calderon interpolation theorem (in $[C_1]$,III the corresponding is shown for unweighted spaces H_s) .

Proposition 3.4. The class $O(-\infty)$ of all operators of order $-\infty$ coincides with the class $\psi h_{-\infty}$, and, again, with the class $\mathcal{K}$ of all integral operators with kernel in $S(\mathbb{R}^{2n})$.

Proof. We have seen in II, lemma 1.7 that $\psi h_{-\infty}$ is the class of all integral operators with rapidly decreasing kernel. Also, as a consequence of thm.3.1, get $\mathrm{Op}\psi h_{-\infty}=\mathrm{Op}\psi t_{-\infty}\subset O(-\infty)$. For the converse let $A\in O(-\infty)$. Observe that $A_{\alpha\beta\alpha'\beta'}=M^{\alpha}D^{\beta}AD^{\beta'}M^{\alpha'}\in O(-\infty)$ for all $\alpha,\beta,\alpha',\beta'$. We must have $A:S'\to S$, since A takes each H_s into $\cap H_t=S$, and $S'=\cup H_t$. In particular $A:H\to S$, hence $Au(x)=\int\bar{k}_x(y)u(y)dy$, $u\in H$, with some $k_x\in H$, for $x\in \mathbb{R}^n$, by the Frechet-Riesz theorem. Similarly,

(3.33) $$Au(x) = (k_x^s,u)_s \ , \ s\in \mathbb{R}^2 \ , \ x\in \mathbb{R}^n ,$$

with some $k_x^s \in H_s$. Using (3.2) one concludes that

(3.34) $$Au(x) = \int\Pi_s^*\Pi_s\bar{k}_x^s(y)u(y)dy \quad , u \in S ,$$

with a distribution integral, where the temperate distribution

(3.35) $$k_x = \Pi_s^*\Pi_s k_x^s \in H_{-s} \ , \ s \in \mathbb{R}^2 ,$$

must be independent of s , hence will be in $H_\infty=S$. It follows that

(3.36) $$Au(x) = \int\bar{k}(x,y)u(y)dy \ , \ k(x,y) = k_x(y) .$$

But the same conclusion also applies to all $A_{\alpha\beta\alpha'\beta'}$. It follows by integration of (3.36) for large α, β, α', β' that k must have derivatives of all orders, for x, y , and that

$$(3.37)\qquad A_{\alpha\beta\alpha'\beta'}u(x) = \pm i\int x^{\alpha}y^{\alpha'}\partial_x^{\beta}\partial_y^{\beta'}k(x,y)u(y)dy \quad ,$$

where the kernel has the same properties. Thus $k\in C^{\infty}(\mathbb{R}^{2n})$, and $x^{\alpha}y^{\alpha'}\partial_x^{\beta}\partial_y^{\beta'}k(x,y)=O(1)$, i.e., we conclude that $k\in \mathcal{S}(\mathbb{R}^{2n})$, q.e.d.

Recall the concept of K-parametrix of a ψdo of II,1. We now recognize that a K-parametrix B of A is an inverse modulo $\mathcal{O}(-\infty)$. The concept is meaningful for arbitrary $A\in \mathcal{O}(\infty)$.

Finally we introduce a stronger type of inverse: Two operators $A,B\in \mathcal{O}(\infty)$ will be called <u>Green</u> <u>inverses</u> of each other if

$$(3.38)\qquad AB-1 \ ,\ BA-1 \ \in \mathcal{F} \ ,$$

where $\mathcal{F} \subset \mathcal{O}(-\infty)$ denotes the class of operators with finite rank.

Notice that $\mathcal{F}$ is a 2-sided ideal of the algebras $\mathcal{O}(\infty)$, $\mathcal{O}(0)$. Clearly a Green inverse is a special kind of K-parametrix. In sec. 4 we will show that an operator admitting a K-parametrix also admits a Green inverse. Clearly, if $A \in \mathcal{O}(\infty)$ admits a Green inverse then it is a Fredholm operator of $\mathcal{S}'$ as well as of $\mathcal{S}$, since the Green inverse acts as a Fredholm inverse in both spaces (cf. $[C_1]$,app.A1, thm.2.1).

More generally, if $A\in \mathcal{O}(s)$ and $B\in\mathcal{O}(t)$ are Green inverses of each other then $r=s+t\geq 0$ follows. In case of $r=0$ - i.e., $s=-t$ - we have $A:\mathcal{H}_m\to\mathcal{H}_{m-s}$, $B:\mathcal{H}_{m-s}\to\mathcal{H}_m$ continuous, for every $m\in \mathbb{R}^2$, so that again A and B are Fredholm inverses of each other, as maps $\mathcal{H}_m \leftrightarrow \mathcal{H}_{m-s}$. Hence then also A is Fredholm as a map $\mathcal{H}_m\to\mathcal{H}_{m-s}$.

For general $r=s+t>0$ we may regard A as an unbounded operator from $\mathcal{H}_m$ to $\mathcal{H}_{m+t}$ with domain dom A = $\mathcal{H}_{m+r}$. Then $B:\mathcal{H}_{m+t}\to\mathcal{H}_m$ is bounded, and we get $BA \subset 1+F$, $AB \subset 1+F'$, with $F,F' \in \mathcal{F}$. This implies that the closure A^c of $A:\mathcal{H}_{m+r}=\text{dom } A \to\mathcal{H}_{m+t}$ is a closed unbounded Fredholm operator from $\mathcal{H}_m$ to $\mathcal{H}_{m+t}$, for every m. We shall discuss details in sec.4, below (cf. $[C_1]$,app.A1, sec.6).

<u>Remark 3.5</u>. Note that A must have the same Fredholm index as a map $\mathcal{S}\to\mathcal{S}$, $\mathcal{S}'\to\mathcal{S}'$, $\mathcal{H}_m\to\mathcal{H}_{m-s}$ (or dom A = $\mathcal{H}_{m+r}\subset \mathcal{H}_m \to \mathcal{H}_t$) .

<u>Problems</u>. 1) For the ψdo's on the cylinder of I,6-II,3,pbms, if $\|A^{(\alpha)}_{(\beta)}(x^{\Delta},\xi^{\Delta})\|_{L^2(S^1)}=O(1)$, all α,β, show $L^2(S^1\times\mathbb{R}^{n-1})$-boundedness of A. 2) Define weighted L^2-Sobolev spaces on the cylinder, using the norms $\int_0^{2\pi}dx_1\int dx^{\Delta}\langle x^{\Delta}\rangle^{2s_2}|(1-\Delta)^{s_1}u|^2=\|u\|_s$, with properly defined power of $1-\Delta=1-\partial_{x_1}^2-\Delta^{\Delta}$. 3) Discuss $\mathcal{H}_s$-boundedness of ψdo's as in (1).

4. Existence of a Green inverse.

In this section we will show that also a Green inverse exists if only an operator $A\in O(\infty)$ admits a K-parametrix. We have introduced the concept of Green inverse in sec.3 as that of an inverse modulo the ideal $F \subset O(-\infty)$ of operators of finite rank.

All results, below, (especially thm.4.1, thm.4.2, cor.4.3) are valid not only for a ψdo with complex-valued symbol but even for a matrix-valued ψdo $A =((a_{jk}(x,D)))_{j,k=1,\dots,\nu}$, although we only discuss the scalar cases. Proofs extend literally.

Actually, it is practical to prove a stronger result: An A with Green inverse has a finite dimensional null space $\ker A \subset S$. There is a complement T of $\ker A$ in S' and an operator $B_0\in O(\infty)$ such that $B_0|\operatorname{im} A$ inverts $A|T$. Here the order of B_0 may be chosen equal to the order of B. Also we may choose T as the ortho-complement $T=\{u\in S':(u,\varphi)=0 \text{ for all } \varphi\in \ker A\}$, with the pairing (u,φ) of (3.28). With this construction B_0 has the properties of a special Fredholm inverse ([C_1], p.259). In particular,

$$1 - BA = P\ ,\ 1 - AB = Q \quad , \tag{4.1}$$

with projections $P,Q\in F$ onto ker A and ker A^*, respectively. Also, $P,\ Q \in O(0)$ are orthogonal projections in $H=H_0$.

For a proper setting of (4.1) let A^* be the adjoint of $A\in O(\infty)$ under the pairing (3.28). That is, $A^*:H_{m-s}\to H_s$ is the operator satisfying $(Au,v)=(u,A^*v)$ for all $u\in H_s$, assuming that $A\in O(m) \subset O(\infty)$, with $(.,.)$ of (3.28). This defines A^* for each H_s. All of them agree on S. By continuity they must agree wherever they are jointly defined. Thus we obtain A^* as a map $S'\to S'$ with continuous restrictions $S\to S$, $H_{m-s}\to H_{-s}$. Then $A\in O(m)$ amounts to the condition

$$\Pi_{s-m} A \Pi_s^{-1} \in L(H)\ ,\ \text{for all } s\in \mathbb{R}^2\ . \tag{4.2}$$

For the adjoint A^* this translates into

$$\Pi_s^{-1*} A^* \Pi_{s-m}^* = \Pi_{-s} A^* \Pi_{m-s}^{-1} \in L(H)\ ,\ s \in \mathbb{R}^2\ . \tag{4.3}$$

Or, replacing s by m-s, $\Pi_{s-m}A^*\Pi_s^{-1}\in L(H)$, i.e., $A^*\in O(m)$. It follows that $A\in O(m) \Leftrightarrow A^*\in O(m)$, including the cases $m=\pm\infty$. Clearly, "*" has the properties of an involution $O(\infty)\to O(\infty)$. For a ψdo A this adjoint is the formal Hilbert space adjoint A^* of I,(2.3).

Theorem 4.1. Let $A\in O(m)$ have a K-parametrix $B\in O(m')$. Then the operator A , as a map $A:S' \to S'$ has the following properties.

(i) $\ker A \subset S$, $\ker A^* \subset S$;

(ii) $\dim \ker A < \infty$, $\dim \ker A^* < \infty$;

(iii) $\mathrm{im}\, A = \{f \in S' : (f,\varphi) = 0 \text{ for all } \varphi \in \ker A^*\}$.

(iv) For $f\in H_s$ with $(f,\varphi)=0$ for $\varphi\in \ker A^*$ all solutions u of $Au=f$ are contained in $H_{s-m'}$. There exists $p_s>0$ such that

(4.4) $\|f\|_s=\|Au\|_s\ge p_s\|u\|_{s-m'}$, whenever $u\in H_{s-m'}$, $(u,\varphi)=0$ for $\varphi\in \ker A$

with p_s independent of u and f.

<u>Proof</u>. For $u\in S$ with $Au=0$ get $(1+K)u=BAu=0$, where $K\in O(-\infty)$. Thus $u=-Ku\in S$, since an integral operator with kernel in $S(\mathbb{R}^{2n})$ maps $S'\to S$. Similarly $B^*A^*=1+K'^*$, hence $\ker A^*\subset S$, confirming (i). To show (ii) note that $\ker A \subset \ker(1+K)$. But $\ker(1+K)$ is closed in H_s since $u\in H_s$, $(1+K)u=0$ implies $u\in S$. For an infinite orthonormal set $u_j\in \ker(1+K)$ - with respect to (u,v) of H - the sequence Ku_j has a convergent subsequence, in $L(H)$, K being compact in H, its kernel in $L^2(\mathbb{R}^{2n})$ makes it a Schmidt operator. But $u_j=-Ku_j$ cannot have a convergent subsequence, as an orthonormal set. Therefore $\dim(1+K)<\infty$, implying $\dim A <\infty$. Similarly for A^* , proving (ii).

For (iii) we first observe that, trivially, every $f \in S'$, $f=Av$, satisfies $(f,\varphi) = (Av,\varphi) = (v,A^*\varphi)=0$, for all $\varphi \in \ker A^*$. In order to show the converse we will first establish (iv) and show that a solution $u \in H_{s-m'}$ exists for $f\in H_s$, satisfying (4.4).

To clarify the relation with the inner product of H we assume that $m'=0$. This is no restriction: For general $A\in O(m)$, $B\in O(m')$ observe that $(A\Pi^*_{m})(\Pi_{-m'}B)=1+K$, $(\Pi_{-m'}B)(A\Pi^*_{m'})=1+K''$, with $K''=\Pi_{-m'}K'\Pi_{m'}^{-1}\in O(-\infty)$. Also $Au=f \Leftrightarrow (A\Pi^*_{m'})(\Pi_{-m'}u)=f$, and $\ker A^* = \ker (A\Pi^*_{m'})^*$, and $\|u\|_{s-m'}\approx\|\Pi_{-m'}u\|_s$. Thus all terms may be translated to the case $m'=0$: Replacing $A\Pi^*_{m'}\in O(r)$, $\Pi_{-m'}B\in O(0)$ with A,B we get a pair A,B of mutual K-parametrices with $A\in O(r)$, $r=m+m'\ge 0$, and must solve $Au=f$ under the corresponding conditions.

Now let $f\in S'$ be 'orthogonal' to $\ker A^*$. We must have $f\in H_s$ for some s, and now will proceed to show that $u\in H_s$ orthogonal to $\ker A$ exists solving $Au=f$, where u and f satisfy (4.4) (with $m'=0$)

A further reduction of the statement is useful: we may assume $s=0$: $Au=f \Leftrightarrow (\Pi_s A\Pi_s^{-1})(\Pi_s u)=\Pi_s f$, where $\Pi_s u$, $\Pi_s f\in H=H_0$, while $\Pi_s A\Pi_s^{-1} \in O(r)$ has K-parametrix $\Pi_s B\Pi_s^{-1}\in O(0)$. Also, $(f,\varphi)=0$ for $\varphi\in \ker A^*$ $\Leftrightarrow (\Pi_s f,\psi)=0$ for $\psi=\Pi_{-s}\varphi$ satisfying $(\Pi_s A\Pi_s^{-1})^*\psi=\Pi_{-s}A^*\Pi_{-s}^{-1}\Pi_{-s}\varphi = 0$.

Thus we now are reduced to an operator equation $Au=f$ in the Hilbert space H , where we may regard A as an unbounded operator with domain dom $A = \mathcal{H}_r$, $r=m+m'\geq 0$. The operator A satisfies $AB=1+K$, $BA=1+K'$ with $K,K'\in O(-\infty)$. Let us next prove (4.4), or,

$$(4.5) \qquad \|Au\| \geq p\|u\| \text{ for all } u\in \mathcal{H}_r= \text{dom } A \text{ with } u \perp \ker A^* .$$

Suppose this inequality is false. Then there exists $u_j\in \mathcal{H}_r$, $u_j \perp \ker A^*$, $\|u_j\|=1$, with $\|Au_j\| \to 0$, $j\to\infty$. But we get

$$(4.6) \qquad \|u\| \leq \|(1+K')u\| + \|K'u\| = \|BAu\| + \|K'u\| \leq c\|Au\| +\|K'u\| ,$$

for all $u\in \mathcal{H}_r$. Since $K'\in K(\mathcal{H})$, as used before, there must be a convergent subsequence of $K'u_j$ (we assume $K'u_j$ convergent itself). It follows that $\|u_j-u_l\|\leq c\|Au_j\|+c\|Au_l\|+\|K'u_j-K'u_l\|\to 0$, $j,l\to\infty$. Thus, $u_j\in \mathcal{H}_r$, $u_j\to u$ (in $\mathcal{H}$), $\|u\|=1$, $Au_j\to 0$. Hence $u\in$ dom A^c , with the closure A^c of the unbounded operator with domain $\mathcal{H}_r$, as discussed. But we also have $\|Au_j\|_{s+r}\to 0$, $\|u-u_j\|_s\to 0$ for $s\leq 0$, $s+r\leq 0$, where now $A:\mathcal{H}_{s+r}\to\mathcal{H}_s$ is continuous, so that u is in the null space of $A:\mathcal{H}_{s+r} \to \mathcal{H}_s$. It was seen that this null space is independent of s, hence =ker $A\subset S$. Since $u_j\to u$ in $\mathcal{H}$ and $u_j \perp$ ker A in $\mathcal{H}$ it follows that $u \perp$ ker A, hence $1=(u,u)=\|u\|_0^2=0$, a contradiction, proving (4.4).

The adjoint $A^*\in O(r)$ also may be interpreted as unbounded $A^*:\text{dom } A^*=\mathcal{H}_r\to \mathcal{H}$. For this operator we have

$$(4.7) \qquad (u,Av) = (A^*u,v) \text{ for all } u,v \in \text{dom } A = \text{dom } A^* = \mathcal{H}_r ,$$

by definition of A^*. In other words, A and A^* are in adjoint relation in the Hilbert space $\mathcal{H}$: The Hilbert space adjoint A^Δ of A extends A^*. We get ker A^Δ =ker A^*. Indeed, ">" is trivial. Let $f\in$ ker A^Δ : $f\in \mathcal{H}$, $(f,Av)=0$ for all $v\in \mathcal{H}_r$. Then $A^*f=0$ wth $A^*:\mathcal{H}\to\mathcal{H}_{\mp r}$. Or $u\in$ ker $A^*\subset S$, ker A^* independent of s, so that ker $A^\Delta\subset$ ker A^*.

We conclude that A^c is an unbounded closed Fredholm operator of the Hilbert space $\mathcal{H}$ since (4.6) implies that im A^c is closed. It then is well known that im A^c =(ker $A)^\perp$ (cf.[C1] app.A_1,6,7).

Notice finally that A^c equals the restriction to dom A^c of the operator $A:\mathcal{H}\to\mathcal{H}_{-r}$. Indeed, $u_j\in \mathcal{H}_r=$ dom A, $u_j\to u$ (in $\mathcal{H}$), $Au_j\to A^cu$ (in $\mathcal{H}$) implies $u_j\in \mathcal{H}$, $u_j\to u$ (in $\mathcal{H}$) , $Au_j\to A^cu$ (in $\mathcal{H}_{-r}$), hence $A^cu = Au$, since $A:\mathcal{H} \to \mathcal{H}_{-r}$ is continuous. This completes the proof of (iii) and (iv) of thm.4.1. Q.E.D.

<u>Theorem 4.2</u>. An operator $A \in O(m)$ admits a Green inverse B_0 if and only if it has a K-parametrix B. If $B\in O(m')$, then also $B_0\in O(m')$. Also, B_0 may be chosen in such a way that (4.1) is valid.

<u>Proof</u>. A Green inverse is a K-parametrix. Thus we must construct

a Green inverse whenever a K-parametrix exists. Applying thm.4.1 we define an operator B_0 by setting $B_0u=0$ for $u\in \ker A^*$, and $B_0=(A|(\ker A^0)^{\perp})^{-1}$ in $\operatorname{im} A=(\ker A^*)^{\perp}$. Every $u\in S'$ is in some H_s, and we have a unique decomposition $u=v+w$, $v\in \ker A$, $w\perp \ker A$. Thus the above defines a linear operator $S'\to S'$. Clearly $B_0Au=u$ for $u\perp \ker A$, and $B_0Au=0$ for $u\in \ker A$. But $AB_0u=u$ for $u\in \operatorname{im} A=(\ker A^*)^{\perp}$, and $AB_0=0$ for $u\in \ker A^*$. In other words, (4.1) is valid.

On the other hand, (4.4) implies

$$(4.8)\qquad \|A^{-1}u\|_{s-m'}\le \frac{1}{p_s}\|u\|_s \text{ for all } u\in H_s\ ,\ u\perp \ker A^*,$$

for the abstract inverse A^{-1} of $A:\{u\in H_{s+m}: u\perp \ker A\}\to H_s$. Actually, we get (4.8) for A^{c-1} with the closure A^c of the unbounded operator $A:\operatorname{dom} A=H_{s+m-r}=H_{s-m'}\to H_{m'}$, and A^{c-1} is defined in all of $\{u\in H_s : u\perp \ker A^*\}$. By our definition of B_0 (4.8) implies $\|B_0u\|_{s-m'}=\|B_0v\|_{s-m'}\le\frac{1}{p_s}\|v\|_s\le\frac{c}{p_s}\|u\|_s$, where $u=v+w$, $w\in(\ker A^*)^{\perp}$ is the above unique decomposition, and where we used that $\|v\|_s\le c\|u\|_s$ $u\in H_s$, c independent of u. Indeed, with an orthonormal base $\varphi_1,\dots,\varphi_k$ (in H) of $\ker A^*$ we have $w=\sum\varphi_j(\varphi_j,u)$, $v=u-w$, hence $\|v\|_s$ $\le \|u\|_s+\sum\|\varphi_j\|_s\|\varphi_j\|_{-s}\|u\|_s$, so that $c=1+\sum\|\varphi_j\|_s\|\varphi_j\|_{-s}$ may be chosen.

Clearly $Q=\sum\varphi_j\rangle\langle\varphi_j$ and $P=\sum\psi_j\rangle\langle\psi_j$ (with an orthonormal base $\{\psi_j\}$ (in H) of ker A) belong to $O(0)$. The proof is complete.

Corollary 4.3. The operators B and B_0 of thm.4.2 differ by an operator in $O(-\infty)=\mathrm{Op}\psi h_{-\infty}$ only. Thus a Green inverse of a (formally) md-(hypo-)elliptic ψdo always is in the same symbol class $\psi h_{m,\rho,\delta}$ of ch.II as the K-parametrix constructed there.

Proof. Indeed, that Green inverse is a special K-parametrix, and we observed earlier that, as inverse modulo the ideal $O(-\infty)$, a K-parametrix is unique up to an additive term in $O(-\infty)$.

We conclude this section with the remark that a Green inverse of a ψdo in essence has properties very similar to the integral operator with kernel equal to the generalized Greens function of a boundary value problem: It constitutes an inverse modulo operators of finite rank; its distribution kernel has singular support at the diagonal x=y only (as a consequence of I, thm.3.3).

Problems. 1) The operator $B: C^\infty(S^1)\to C^\infty(S^1)$ given by $u\to\sum e^{ij\theta}b_j(\theta)u_j$

with $u_j=\frac{1}{2\pi}\int e^{-ij\varphi}u(\varphi)d\varphi$, as in I,6,pbm.3, is $L^2(S^1)$-bounded if we assume that $b_j^{(k)}(\theta)=O(1)$ (in j and θ , for each k=0,1,... . [Derivatives of all orders are not required, but how many are needed for a proof similar to that of thm.1.1?] 2) Investigate the C^*-subalgebra of $L(L^2(S^1))$ generated by all operators B as in (1). Show that its finite Fourier transform is the subalgebra X of $l^2(\mathbb{Z})$ generated by the shift operator $(u_j)\to(u_{j+1})$ and all diagonal matrices with bounded coefficients. 3) Does X have compact commutators? 4) Answer the same questions if the condition of pbm.1 is modified by requesting that $\nabla^l a_j^{(k)} = O(\langle j\rangle^{-\rho l})$ in θ and j , with the finite difference $\nabla a_j=a_j-a_{j-1}$, and some ρ>0 , for all (some?) l and k.

5. H_s-compactness of ψdo's of negative order.

We shortly discuss an often used compactness result.

Theorem 5.1. A (matrix of) ψdo's $A=a(M_l,M_r,D)$, with $a\in SS$ and orders m_j satisfying $m_1<0$, $m_2+m_3<0$, is a compact operator $H_s\to H_s$, for every $s\in\mathbb{R}^2$. Moreover, for general m_j, A: $H_s\to H_t$ is compact whenever $m_1<s_1-t_1$, $m_2+m_3<s_2-t_2$. Especially, $A=a(x,D)\in Op\psi h_{m,\rho,\delta}$, $\rho_1>0$, $\rho_2>\delta$, is $\in K(H_s,H_t)$ whenever $m_j<s_j-t_j$, j=1,2.

Proof. Observe that $Q_\varepsilon=\langle x\rangle^{-\varepsilon}\langle D\rangle^{-\varepsilon}$ is compact $H\to H$, as ε>0 (cf.[C_1] III,lemma 8.1, for example - but other proofs are known. Details: The operator $\langle D\rangle^{-\varepsilon}$ may be written as convolution $\langle x\rangle^{-\varepsilon\vee}*$, where

(5.1) $$p_\varepsilon(x) = \langle x\rangle^{-\varepsilon\vee}(x) = c_{\varepsilon,\nu}|x|^{(\varepsilon-n)/2}K_{(\varepsilon-n)/2}(|x|) ,$$

with a constant $c_{\varepsilon,n}$, and the modified Hankel function K_σ. (5.1) is verified using techniques as in ch.0, sec.4, for construction of the fundamental solution of $\Delta+k^2$. Note the function (5.1) is $L^1(\mathbb{R}^n)$. The kernel $q_\varepsilon(x,y)=\langle x\rangle^{-\varepsilon}p^\varepsilon(x-y)$ may be approximated in the sense of Schur's lemma by Schmidt kernels $q_\varepsilon^j\in L^2(\mathbb{R}^{2n})$. Then (i) the operators Q_ε^j, $Q_\varepsilon^j u(x)=\int q_\varepsilon^j(x,y)u(y)dy$, are compact, from H to H . Second, we have $\|Q_\varepsilon-Q_\varepsilon^j\|\to 0$, $j\to\infty$, by Schur's lemma, so that $Q_\varepsilon\in K(H)$, since $K(H)$ is closed under uniform operator convergence.

With L^2-compactness of Q_ε we get the statement at once: Compactness $A:H_s\to H_t$ means compactness of $\Pi_s^{-1}A\Pi_t=(\Pi_s^{-1}A\Pi_{t+\varepsilon e})Q_\varepsilon:H\to H$, where the first factor is bounded, by III,thm.1.1 (or thm.2.1), assuming the inequalities of thm.5.1, by calculus of ψdo's. Thus indeed A: $H_s\to H_t$ is compact if these inequalities hold. Q.E.D.

Chapter 4. PSEUDO-DIFFERENTIAL OPERATORS

ON MANIFOLDS WITH CONICAL ENDS

0. Introduction.

In the present chapter we will focus on pseudo-differential operators on differentiable manifolds. We assume either that Ω is compact -then our theory will not differ from others - or that Ω is a noncompact Riemannian space with conical ends.

While the Fourier transform and, correspondingly, the concept of Fourier multiplier $a(D) = F^{-1}a(M)F$ is meaningfull only for functions or distributions defined on $\mathbb{R}^n$, the kind of ψdo's we introduced has a natural environment on a type of differentiable manifold, to be studied. The reason: Our ψdo's of ch.2 are invariant under a type of coordinate transform (discussed in sec.3) while Fourier multipliers do not have this property.

In sec.1 we discuss distributions on manifolds. A special type of 'S-manifolds' is preferred, allowing the definition of a class $\mathcal{S}(\Omega)$ of rapidly decreasing functions. The linear functionals on $\mathcal{S}(\Omega)$ will be our <u>temperate distributions</u>. For simplicity we will consider only manifolds Ω allowing a compactification Ω^0 to which the C^∞-structure can be extended - making Ω^0 a compact manifold with boundary. In essence then $\mathcal{S}(\Omega)$ will be the class of functions over Ω vanishing of all orders on $\partial\Omega$.

In sec.2 we will introduce 'admissible' charts, cut-off's, partitions, as well as admissible coordinate transforms, all designed to give $\mathcal{S}(\Omega)$ similar properties than $\mathcal{S}(\mathbb{R}^n)$. In particular a Riemannian metric is introduced, making Ω a space with conical ends. In sec.3 we prove invariance of pseudo-differential operators under admissible coordinate transforms.

Sec's 4 and 5 generalize the calculus of ψdo's to spaces with conical ends. In particular, we again get md-elliptic and formally md-hypo-elliptic operators, defined by their <u>local</u> <u>symbols</u>.Results of ch.'s II and III regarding K-parametrix, Sobolev spaces, order classes, Green inverse, etc., all generalize almost literally.

Similar results, on coordinate invariance as well as md-elliptic operators on manifolds, were discussed by Schrohe [$Schr_j$] in somewhat different setting. Regarding application to differential operators we point to [CDg_1] , [C_2] where similar C^*-algebras of singular integral operators on $L^2(\Omega)$ are considered, with corresponding results, but abstract proofs.

1. Distributions and temperate distributions on manifolds.

In ch.0 we were discussing distributions on an open subset Ω of $\mathbb{R}^n$. Presently, we first will extend the distribution concept to differentiable manifolds, compact or not.

For a C^∞-manifold Ω of dimension n (always assumed paracompact -even with a countable atlas) the space $\mathcal{D}=\mathcal{D}(\Omega)=C_0^\infty(\Omega)$ is well defined, and the convergence $\varphi_j\to 0$ in $\mathcal{D}(\mathcal{O})$ just as well: One requires that (i) supp $\varphi_j\subset K$, with a set $K\subset\subset\Omega$ independent of j.

We get $K\subset\cup\Omega_l$, a finite union of charts Ω_l, and $\varphi_j=\sum\omega_l\varphi_j$, with a

partition of unity $\sum\omega_l = 1$ in K , supp $\omega_l\subset\Omega_l$, $\omega_l\ge 0$, $\omega_l\in C_0^\infty(\Omega)$.

One also requires (ii) that $\omega_l\varphi_j\to 0$ in $\mathcal{D}(\Omega_l)$, as $j\to 0$,in the coordinates of the chart Ω_l , for $l=1,\dots,N$.

We define a distribution $u\in\mathcal{D}'(\Omega)$ to be a continuous linear functional over $\mathcal{D}(\Omega)$, where 'continuous' means that

$$(1.1)\qquad \langle u,\varphi_j\rangle \to 0 \text{ whenever } \varphi_j \to 0 \text{ in } \mathcal{D}(\Omega) .$$

With this definition a conceptual difficulty arises if we attempt to interpret a function $f\in L^1_{loc}(\Omega)$ as a distribution. For $\Omega\subset\mathbb{R}^n$ we defined $\langle f,\varphi\rangle=\int f\varphi dx$, taking advantage of the existence of a <u>distinguished</u> <u>measure</u> on $\mathbb{R}^n$ - the Lebesgue measure as Haar measure of the group $\mathbb{R}^n$. For a manifold Ω there no longer is such a distinction, but we may construct a positive C^∞-measure $d\mu$, locally, $d\mu=\kappa dx$, $0<\kappa\in C^\infty$, on a paracompact manifold Ω , and then define

$$(1.2)\qquad \langle f,\varphi\rangle=\int_\Omega f\varphi d\mu \ , \ \varphi\in\mathcal{D}(\Omega) ,$$

for $f\in L^1_{loc}(\Omega)$, establishing the analogous imbedding $L^1_{loc}\to\mathcal{D}'$.

This dependence on prior choice of a measure may be avoided by defining distributions as linear functionals on a properly topologized space of signed $C_0^\infty(\Omega)$-measures - expressions of the form $\varphi d\mu$, $\varphi\in\mathcal{D}(\Omega)$, in our terminology. But we are tied to the use of Sobolev norms anyway, where we use a distinguished measure. Hence we follow our habit of [C2], and always assume given a pair $\{\Omega,d\mu\}$ of a manifold Ω and positive C^∞-measure $d\mu$ on Ω, so that $d\mu$, distinguished by other uses in our theory always is on hand. Then we may use (1.2) to define the imbedding $L^1_{loc} \to \mathcal{D}'(\Omega)$. Often $d\mu=dS$ will be the surface measure of a Riemannian metric on Ω .

Most concepts obviously extend, such as restriction of a distribution to an open subset, (singular) support, definition of Lu, for a differentials expressions with smooth coefficients, etc.

A differential expression L on Ω (with C^∞-coefficients) is defined as a linear map $L:C^\infty(\Omega) \to C^\infty(\Omega)$ with $Lu=\sum a_\alpha \partial_x^\alpha u$ in derivatives ∂_x^α of the local coordinates and C^∞-coefficients a_α for u with support in a single chart. The adjoint differential expression is defined as the expression $L^*:C^\infty(\Omega) \to C^\infty(\Omega)$ satisfying

$$(1.3) \qquad \int \overline{Lu}v d\mu = \int \overline{u} L^* v d\mu \ , \text{ for all } u,v \in C^\infty(\Omega) \text{ with } uv \in C_0^\infty(\Omega).$$

The dual expression $L^\sim$ of L is defined by setting $L^\sim f=\overline{L^*\overline{f}}$, $f\in \mathcal{D}$. For a general distribution $u\in \mathcal{D}'(\Omega)$ one then defines Lu by setting

$$(1.4) \qquad \langle Lu,\varphi\rangle = \langle u,L^\sim \varphi\rangle \ , \quad \text{for all } \varphi \in \mathcal{D}(\Omega) \ .$$

One derives the familiar facts on $\mathcal{E}'(\Omega)$, the space of compactly supported distributions, on regularization, and once again may introduce the standard topologies on $\mathcal{D}'$ and $\mathcal{E}'$ (cf. $[C_1]$,I,6).

For an extension of $\mathcal{S}(\Omega)$ and $\mathcal{S}'(\Omega)$ we introduce a type of manifold looking conical at ∞ .

The main feature of $f\in \mathcal{S}=\mathcal{S}(\mathbb{R}^n)$ is the behaviour of f for large $|x|$. In II,5, looking at wave front sets, we introduced a type ψ of directional cut-off and regarded f belonging to $\mathcal{S}$ in a conical sector only if $\psi f\in \mathcal{S}$ for suitable such ψ .

A general change of coordinates can destroy the property of a function u to belong to $\mathcal{S}$ (or $\psi u\in \mathcal{S}$). Accordingly, a special structure on Ω, allowing only special C^∞-coordinate transforms as admissible coordinate changes is required before a space $\mathcal{S}=\mathcal{S}(\Omega)$ can be defined, with general properties of $\mathcal{S}(\mathbb{R}^n)$. Such structures were introduced by Schrohe $[Schr_{1,2}]$ and were called *SG*-structures

We only look at a restricted class of such Ω. First define the diffeomorphism $s:\mathbb{R}^n \to B_1$, $B_1=B_1(0)=\{y\in \mathbb{R}^n: |y|<1\}$, by setting

$$(1.5) \qquad y = s(x) = x/\langle x\rangle \quad , \ x = s^{-1}(y) = y/\sqrt{1-|y|^2} = t(x) \ .$$

Introduce $[y] = \sqrt{1-|y|^2}$, then,

(1.6) $y = x/\langle x\rangle \quad , \quad x = y/[y] \quad , \quad [y] = 1/\langle x\rangle$.

Proposition 1.1. The map

(1.7) $u(x) \to v(y) = u(s^{-1}(y)) = (u\circ s^{-1})(y)$

defines a bijection $\mathcal{S}(\mathbb{R}^n) \to CO^\infty(B_1)$.
In prop.1.1 $CO^\infty(B_1)$ denotes the class of $C^\infty(B_1)$-functions vanishing with all their derivatives at ∂B_1.

Remark 1.2. A function $f\in CO^\infty(B_1)$ has a natural extension to $\mathbb{R}^n$ - $f=0$ there. The extension v is $C_0^\infty(\mathbb{R}^n)$, with supp $v\subset B_1^c=(B_1)^{\text{clos.}}$. Vice versa, for $v\in C_0^\infty(\mathbb{R}^n)$ with supp $w\in B_1^c$ we have $w|B_1\in CO^\infty(B_1)$.

Proof. If $v\in CO^\infty(B_1)$ then v and all its derivatives vanish of infinite order at $|y|=1$, by Taylor's formula. Thus we get

(1.8) $v^{(\alpha)}(y) = O(((1-|y|^2)^k)$, for all $k=0,1,\dots$, and all α .

For $\alpha=0$ this implies $u(x)=v(y)=O([y]^k)=O(\langle x\rangle^{-k})$, $k=0,1,\dots$.
But $u^{(\alpha)}(x)=\partial^\alpha(v(x/\langle x\rangle))$ is a linear combination of terms

(1.9) $v^{(\beta)}(\frac{x}{\langle x\rangle})\Pi\, \partial_x^{\gamma^k}(x_{j_k}/\langle x\rangle) \quad , \quad |\beta|\le|\alpha| \quad , \quad \sum\gamma^k = \alpha$,

by induction. Get $\partial_x^\gamma(x_j/\langle x\rangle)=O(\langle x\rangle^{-|\gamma|})$, easily confirmed. Each term (1.9), hence $u^{(\alpha)}$ is $O(\langle x\rangle^{-k})$, for all k, i.e., $u\in\mathcal{S}$.

Vice versa, let $u\in\mathcal{S}$,and $v(y) = u(y/[y])$. Note that $\langle y/[y]\rangle^2=1+|y|^2/(1-|y|^2)=1/(1-|y|^2)=[y]^{-2}$, hence $\langle y/[y]\rangle=[y]^{-1}$,

(1.10) $u^{(\alpha)}(y/[y]) = O(\langle y/[y]\rangle^{-k}) = O([y]^k)$, for all k .

For $\alpha=0$ it follows that $v(y)\in CO(B_1)$. For arbitrary α one again uses the chain rule to express $v^{(\alpha)}(y)$ as a linear combination of

(1.11) $u^{(\beta)}(y/[y])\Pi\, \partial_y^{\gamma^j}(y_{l_j}/[y])$, $|\beta|\le|\alpha|$, $\sum\gamma^j=\alpha$,

similarly as (1.9). Here we get $\partial_y{}^\gamma(y_l/|y|) = O([y]^{-2|\gamma|-1})$, and (1.10) still implies all terms (1.11) to vanish, hence $v^{(\alpha)}=0$ at $|y|=1$. Thus $v\in CO^\infty(B_1)$ follows and prop.1.1 is proven.

The above suggests the concept of an $\mathcal{S}$-manifold as the interior Ω of a smooth compact manifolds Ω^0 with boundary. Then define the space $\mathcal{S}=\mathcal{S}(\Omega)$ of rapidly decreasing functions on Ω by setting $\mathcal{S}(\Omega)=CO^\infty(\Omega)=\{$all $u\in C^\infty(\Omega^0)$ vanishing of infinite order on $\partial\Omega\}$.

Ω^0 is a compactification of Ω, just as $\mathbb{B}^n$ of $\mathbb{R}^n$ in II,(3.1).

$\partial\Omega^0$ is called the infinite boundary of Ω; its points at infinity.

We will insist, in the following, that Ω^0 be a compact C^∞-manifold with boundary. Ω^0 will be useful for function classes other than $S(\Omega)$. On the other hand, for defining $S(\Omega)$ this is not required. For example we might allow Ω^0 as a cartesian product of finitely many manifolds with boundary, or allow Ω^0 to have corners vertices -i.e., at some boundary points we have charts in such cartesian products $\Pi_{j=1}^n I_j$, with $I_j=\mathbb{R}$ or $I_j=\mathbb{R}^+=[0,\infty)$, depending on j.

Notice that $\mathbb{R}^n\times\mathbb{R}^n=\mathbb{R}^{2n}$ is (should be) an S=manifold, but the product $\Omega^0\times\Omega^0=B_1^c\times B_1^c$ no longer is a C^∞-manifold with boundary. As in prop.1.1 one shows that we get a bijection $S(\mathbb{R}^{2n}) \leftrightarrow CO^\infty(B_1^c\times B_1^c)$ as well, induced by the map $s\times s:\mathbb{R}^{2n} \to B_1^c\times B_1^c$.

Thus, for $\Omega=\mathbb{R}^{2n}$ we could use as compactification $\Omega^0=B_1^c$ (in 2n dimensions) or $\Omega^0=B_1^c\times B_1^c$. We choose the first and keep it fixed, for the following (note the exception at the beginning of sec.4).

Before discussing temperate distributions on a general S-manifold Ω we look at the transfer of the Frechet topology of $S(\mathbb{R}^n)$ onto $CO(B_1)$ under the bijection of prop.1.1. The seminorms

$$(1.12)\qquad \|u\|_k = \sup\{\|\langle x\rangle^k u^{(\alpha)}\|_{L^\infty(R^n)} : |\alpha|\le k\},\ k=0,1,2,\dots,$$

generating the topology of S, are equivalent to the seminorms

$$(1.13)\qquad \nu_k(v) = \sup\{\|[y]^{-k} v^{(\beta)}\|_{L^\infty(B_1)} : |\beta|\le k\},\ k=0,1,2,\dots,$$

in the sense that

$$(1.14)\qquad \|u\|_k\le c_k\nu_{l(k)}(v),\ \text{and } \nu_k(v)\le c_k\|u\|_{l(k)},\ k=0,1,2,\dots,$$

with c_k, $l(k)$ independent of u,v. This is confirmed looking at the proof of prop.1.1. On the other hand, since all $v^{(\beta)}(y^0)=0$ as $|y^0|=1$, Taylor's formula with integral remainder (I,(5.20)) implies

$$(1.15)\qquad v(y) = (N+1)\sum_{|\theta|=N+1}\frac{(y-y^0)^\theta}{\theta!}\int_0^1(1-\tau)^N v^{(\theta)}(y^0+\tau(y-y^0))d\tau .$$

For $0\ne y\in B_1^c$ let $y^0=\frac{y}{|y|}$, so that $|(y-y^0)^\theta|\le|y-y_0|^{|\theta|}\le[y]^N$. Then

$$(1.16)\qquad v(y)=\mathrm{Max}\{\|v^{(\theta)}\|_{L^\infty(B_1)} :|\theta|=N\}.O([y]^{N+1}),\ N=0,1,2,\dots .$$

Similarly for derivatives $v^{(\alpha)}$. Combining (1.14) and (1.16) we get

<u>Proposition 1.2</u>. Under the bijection $S(\mathbb{R}^n) \leftrightarrow CO^\infty(B_1)$ of prop.1.1 the Frechet topology of S (induced by (1.12)) is equilent to that of $CO^\infty(B_1)$ as a subspace of $C^\infty(B_1^c)$, induced by (with $C^k=C^k(B_1^c)$)

$$(1.17)\qquad \|v\|_{C^k} = \mathrm{Max}\{\|v^{(\alpha)}\|_{L^\infty(B_1)} : |\alpha|\le k\},\ k=0,1,2,\dots .$$

As a consequence of prop.1.2 a temperate distribution $u \in \mathcal{S}' = \mathcal{S}'(\mathbb{R}^n)$ defines a distribution $v \in \mathcal{D}'(B_1)$, by $\langle v,\psi\rangle = \langle u,\varphi\rangle$, where $\varphi \in \mathcal{D}(\mathbb{R}^n)$, $\psi(y)=\varphi(x)$. We get an injective map $\mathcal{S}'(\mathbb{R}^n) \to \mathcal{D}'(B_1)$. Actually, v defines a functional on the subspace $CO^\infty(B_1) \subset C^\infty(B_1^c)$, extending to $w: C^\infty(B_1^c) \to \mathbb{C}$, by the Hahn-Banach theorem. More precisely, $v: CO^\infty(B_1) \to \mathbb{C}$ is bounded with respect to some $\|v\|_{C^k}$. Then Hahn-Banach, applied for the B-space $C^k(B_1^c)$ gives a continuous functional on C^k whose restriction to $C^\infty(B_1^c)$ gives the desired w .

In turn we may interpret w as a distribution $z \in \mathcal{D}'(\mathbb{R}^n)$ with supp $z \subset B_1^c$ by setting $\langle z,\varphi\rangle = \langle w, \varphi | B_1^c\rangle$, $\varphi \in \mathcal{D}(\mathbb{R}^n)$. Thus v may be obtained as restriction $v = z|B_1$ of a distribution $z \in \mathcal{D}'(\mathbb{R}^n)$. Vice versa, a restriction $v = z|B_1$ of $z \in \mathcal{D}'(\mathbb{R}^n)$ clearly transforms to $u \in \mathcal{S}'$.

The last arguments may be repeated for a general $v \in CO^\infty(\Omega) \subset C^\infty(\Omega^0)$: <u>The space of continuous linear functionals on</u> $CO^\infty(\Omega)$ <u>under the Frechet topology of</u> $C^\infty(\Omega^0)$ <u>coincides with the space</u> $\mathcal{D}^\Delta(\Omega)$ <u>of restrictions</u> $z|\Omega$ <u>of</u> $z \in \mathcal{D}'(\Omega_1)$, with any C^∞-manifold $\Omega^1 \supset\supset \Omega^0$.

Then we define $\mathcal{S}'(\Omega)$, the space of <u>temperate distributions</u> on Ω by setting $\mathcal{S}'(\Omega) = \mathcal{D}^\Delta(\Omega)$. We will work on some details in sec.2. Later on we will introduce ψdo's on $\mathcal{S}$-manifolds.

2. <u>Distributions on $\mathcal{S}$-manifolds; manifolds with conical ends.</u>

We return to the discussion of $\mathcal{S}$-manifolds as interior Ω of a manifold with boundary Ω^0 . In order to make the space $CO^\infty(\Omega)$ look like $\mathcal{S}(\Omega)$ we will use a special kind of chart only. First, instead of relating the charts of Ω^0 to coordinates defined in open subsets of a half space, we use charts (for Ω^0) of the form

$$\text{(2.1)} \qquad \omega^0 : U^0 \to \omega^0(U^0) \subset B_1^c ,$$

with a homeomorphism ω^0 between the open sets $U^0 \subset \Omega^0$ and $\omega^0(U^0)$. Then an $\mathcal{S}$-<u>admissible chart</u> of Ω is of the form

$$\text{(2.2)} \qquad \omega: U \to \omega(U) \subset \mathbb{R}^n , \text{ where } U = U^0 \cap \Omega , \ \omega = s^{-1} \circ (\omega^0 | U) ,$$

with an Ω^0-chart $\{U^0, \omega^0\}$ as in (2.1), and $s^{-1} = t$ of (1.5) . The $\mathcal{S}$-<u>structure</u> on Ω is induced by an atlas of $\mathcal{S}$-admissible charts.

An interior chart $U \subset\subset \Omega$ of the noncompact manifold Ω is $\mathcal{S}$-admissible, since we may define $U^0 = U$, $\omega^0 = s \circ \omega$. A general chart $\omega: U \to \omega(U) \subset \mathbb{R}^n$ is $\mathcal{S}$-admissible if and only if the map $s \circ \omega: U \to B_1$ extends to a diffeomorphism $\omega^0 : U^0 \to V^0$ between open subsets where $U \subset U^0 \subset \Omega^0$, $s \circ \omega(U) \subset V^0 \subset B_1^c$, and where U and $s \circ \omega(U)$ are dense in U^0 , V^0 , respectively, and $U^\circ\cap\Omega=U$. For a given *S*-admissible $U\subset\Omega$ the chart $U^\circ\subset\Omega^\circ$ is given as $U^\circ=U\cup(U^c\cap\partial\Omega^\circ)$, with the closure U^c in Ω° of U. In particular, $\{U^\circ,\omega^\circ\}$ is uniquely determined by $\{U,\omega\}$.

Note the 'shape' of a neighbourhood of an infinite boundary point $p^\circ\in\partial\Omega^\circ$ in 'admissible coordinates': It contains a 'lense' $\{|y-y^\circ|<\varepsilon,\ |y|<1\}$, with $|y^\circ|=1$, y° image of p°, in B_1^c-coordinates of Ω°. With s^{-1} this goes to $N_\varepsilon=\{|\frac{x}{\langle x\rangle}-y^\circ|\le\varepsilon\}$. After a rotation assume $y^\circ=(1,0,\dots,0)$. The neighbourhood system $N_\varepsilon=\{(1-\frac{x_1}{\langle x\rangle})^2+\frac{|x'|^2}{\langle x\rangle}\le\varepsilon^2\}$ is equivalent to $M_\varepsilon=\{1-\frac{x_1}{\langle x\rangle}\le\varepsilon,\frac{|x'|}{\langle x\rangle}\le\varepsilon\}$, where $x=(x_1,x')$. M_ε is the intersection of the solid hyperboloids

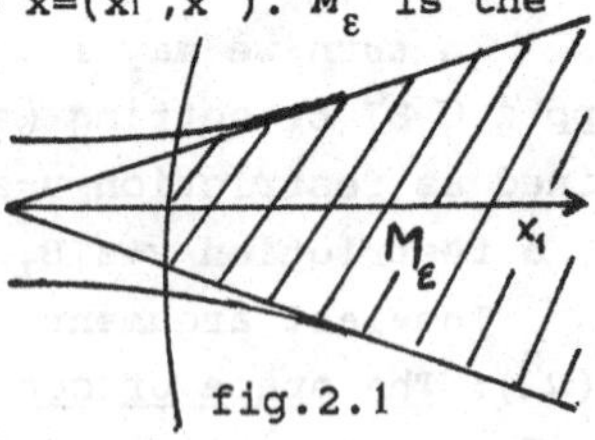

(2.3) $\quad x_1^2\ge\lambda^2(1+\rho^2)\ ,\ x_1>0\ ,\ \lambda=\frac{1-\varepsilon}{\sqrt{2\varepsilon-\varepsilon^2}}$,

with $\rho=|x'|$, and,

(2.4) $\quad \rho^2\le\delta^2(1+x_1^2)\ ,\ \delta=\frac{\varepsilon}{\sqrt{1-\varepsilon^2}}$,

fig.2.1

(fig.2.1). In other words, the set N_ε is contained in cut-off cone

(2.5) $$CC_{y^\circ,\eta}=\{x\in\mathbb{R}^n:\ |x|>\frac{1}{\eta}\ ,\ |\frac{x}{|x|}-y^\circ|\le\eta\}\ ,\ \eta>0\ ,$$

where we may choose $\eta=\eta(\varepsilon)\to0$ as $\varepsilon\to0$. Vice versa, any set N_ε contains a $CC_{y^\circ,\eta}$, for a smaller $\eta>0$. This describes the neighbourhoods of our infinite points. The sets (2.5) form a base at p°.

An <u>*S*-admissible</u> <u>coordinate</u> <u>transform</u> $\omega\circ\chi^{-1}:\chi(U\cap V)\to\omega(U\cap V)$, for *S*-admissible charts $\omega:U\to\mathbb{R}^n$, $\chi:V\to\mathbb{R}^n$, must be of the form

(2.6) $$\kappa=\omega\circ\chi^{-1}=s^{-1}\circ\kappa^\circ\circ s,\ \text{with}\ \kappa^\circ=\omega^\circ\circ\chi^{\circ\,-1}:\chi^\circ(U^\circ\cap V^\circ)\to\omega^\circ(U^\circ\cap V^\circ)\ ,$$

a diffeomorphism between open subsets of B_1^c.

For investigation of maps of this form we focus on the Jacobian matrix $P(x)=\partial\varphi/\partial x=((\partial\varphi_j/\partial x_l))_{j,l=1,\dots,n}$ of a global coordinate transform $\varphi=t\circ\psi\circ s$, $t=s^{-1}$ with some diffeomorphism $\psi:B_1^c\leftrightarrow B_1^c$. Write $I=((\delta_{jl}))$, $pq^T=((p_jq_l))$, for $p,q\in\mathbb{R}^n$, (j=row, l=col.-index):

(2.7) $$\partial t/\partial x=\frac{1}{[x]}(I+t(x)t(x)^T)\ ,\ \partial s/\partial x=\frac{1}{\langle x\rangle}(I-s(x)s(x)^T)\ .$$

It follows that

$$P(x)=(\langle x\rangle[\psi(s(x))])^{-1}(I+t(\psi(s(x)))t(\psi(s(x)))^T)\Psi(I-s(x)s(x)^T)\ ,$$

with $\Psi=\partial\psi/\partial s(s(x))$. Here $\lambda(x)=\frac{1}{\langle x\rangle}=\sqrt{1-s(x)^2}$, hence

(2.8) $$P(x)=(\frac{1-s^2}{1-\psi(s)^2})^{1/2}(I+t(\psi(s))t(\psi(s))^T)\Psi(s)(I-ss^T)\ .$$

Taking determinants we get

$$(2.9)\qquad J_\varphi(x) = |\det P(x)| = \left(\frac{1-s^2}{1-\psi(s)^2}\right)^{3/2}|\det \Psi(s)| ,$$

using that $\det(I+t(x)t(x)^T)=1+t(x)^2=\frac{1}{1-x^2}$. Therefore $J_\varphi \circ s^{-1}$ is (extends to) a smooth positive function on B_1^c. Moreover, $\frac{1-s^2}{1-\psi(s)^2} = \frac{1+|s|}{1+|\psi(s)|}\frac{\text{distance}(s,\Gamma)}{\text{distance}(\psi(s),\Gamma)} \in C^\infty(B_1^c)$ (with $\Gamma=\partial B_1^c$) follows, because $\psi: B_1^c \leftrightarrow B_1^c$ must map Γ to Γ . (Near s^0 with $|s^0|=1$ introduce local coordinates $\upsilon=(\upsilon_1,\ldots,\upsilon_n)$, setting $\upsilon_1=\text{dist}(s,\Gamma)$, with coordinates $\upsilon_2,\ldots,\upsilon_n$ of $\frac{s}{|s|}$ on Γ. In such coordinates the diffeomorphism ψ is represented by an n-tuple of funtions $\psi_j^{\sim}$, where $\text{dist}(\psi(s(\upsilon)),\Gamma)=$ $=\psi_1^{\sim}(\upsilon)$. Clearly $\psi_1^{\sim}=0$ as $\upsilon_1(s)=0$. Thus $\psi_1^{\sim}/\upsilon_1 \in C^\infty$, and it is clear that this quotient is >0 , since ψ is a diffeomorphism.)

This result may be extended, as follows.

Proposition 2.1. The function $Q(s)=P\circ s^{-1}=P(\frac{s}{[s]})=(\partial\varphi/\partial x)(\frac{s}{[s]})$, $s\in B_1$, extends to a $C^\infty(B_1^c)$-function, nonsingular for all $s\in B_1^c$.

Proof. First of all, the scalar factor up front in (2.8) is $C^\infty(B_1^c)$ and the matrix $\Psi(s)(1-ss^T)$ has entries in $C^\infty(B_1^c)$. Introducing an orthonormal base $s^0,\ldots,s^{n-1}$ of $\mathbb{R}^n$ varying smoothly with $s\in B_1$ near some point of Γ such that $s^0=\frac{s}{|s|}$, we may write

$$(2.10)\qquad I- ss^T = (1-|s|^2)s^0 s^{0\,T} + \textstyle\sum_{j=1}^{n-1} s^j s^{jT} ,$$

hence

$$(2.11)\qquad \Psi(s)(I- ss^T) = (1-|s|^2)(\Psi(s)s^0)s^{0\,T} + \textstyle\sum_{j=1}^{n-1}(\Psi(s)s^j)s^{jT} .$$

Applying the matrix $t(\psi(s))t(\psi(s))^T=\frac{1}{1-\psi(s)^2}\psi(s)\psi(s)^T$ to the terms at right of (2.11), the first term gives a $C^\infty(B_1^c)$-matrix, using that $\frac{1-s^2}{1-\psi(s)^2} \in C^\infty(B_1^c)$. The same is true for the other terms, since $\Psi(s)$ must map tangential vectors at $s\in\Gamma$ to tangential vectors at $\psi(s)\in\Gamma$. Or, $\psi(s).\Psi(s)s^j(s)=0$ as $|s|=1$, since $\psi(s)$ is normal to Γ at $\psi(s)$. Also, $q(s) = (\psi(s).\Psi(s)s^j(s))\psi(s)$ is C^∞ near $|s|=1$. Thus the quotient $\frac{q(s)}{1-\psi(s)^2}$ also is C^∞ near $|s|=1$. Q.E.D.

There exists a finite S-admissible atlas. Indeed we may choose a finite atlas $\{U_j^0,\omega_j^0\}$ of the compact Ω^0 and construct the corresponding charts $\{U_j,\omega_j\}$ of (2.2). Moreover, for a partition of unity $\sum\lambda_j^0=1$ subordinated to $\{U_j^0,\omega_j^0\}$ (with supp $\lambda_j^0\subset U_j^0$) a corresponding partition subordinated to $\{U_j,\omega_j\}$ is defined by $\lambda_j=(\lambda_j^0|\Omega)\circ s$. Notice that, in admissible coordinates, we have $\lambda_j\in SS_{0,(1,1,1),0}$, i.e., $\lambda_j^{(\alpha)}=O(\langle x\rangle^{-|\alpha|})$ for all α, (λ_j extended 0 outside $\omega_j(U_j)$ in $\mathbb{R}^n$). The same is true for the restriction $\chi=\chi^0|\Omega$ of a $C^\infty(\Omega^0)$-function χ° with support in a single chart U^υ. We will call such $\{\lambda_j\}$ and χ an *S*-admissible partition of unity, and cut-off function, respectively. (Often we drop "*S*-admissible" for "admissible").

With an admissible partition $\{\lambda_j\}$ and finite atlas $\{U_j,\omega_j\}$ we define a Riemannian metric on Ω by setting

$$ds^2 = \sum_j \lambda_j dx^2 , \tag{2.12}$$

with the Euclidian metric dx^2 of $\omega_j(U_j) \subset \mathbb{R}^n$, in each summand.

From prop.2.1 it is evident that an admissible coordinate change $\varphi:\mathbb{R}^n \leftrightarrow \mathbb{R}^n$ converts the Euclidean metric dx^2 of $\mathbb{R}^n$ into a metric of the form $\sum h_{jl}((s(x))dx_j dx_l$ where $h_{jl} \in C^\infty(B_1^c)$, and the matrix $((h_{jk}))$ is defined and positive definite still at $\partial\Omega^0$. Thus in admissible coordinates, the metric tensor $((g_{jl}))$ of (2.12) has the same properties - $g_{jk}\circ s^{-1}$ extends as a C^∞-function to the infinite boundary of the chart, and the matrix is >0 there.

Proposition 2.2 For each infinite boundary point p^0 and admissible chart $p^0 \in U \to \omega(U) \subset \mathbb{R}^n$ there exists a cut-off-cone $CC_{y^0,\eta} \subset \omega(U)$, $y^0 = \omega^0(p^0)$ such that (i) the Riemannian distance $d(p,q)$ of the metric (2.12) is equivalent to the Euclidean metric in $CC_{y^0,\eta}$, in the sense that $c_1|x-x^\Delta| \le d(p,p^\Delta) \le c_2|x-x^\Delta|$ for all $x,x^\Delta \in CC_{y^0,\eta}$, and corresponding $p,p^\Delta \in \Omega$; (ii) the Riemann space Ω (with metric (2.12)) is complete; (iii) after a linear transform the metric is approximately Euclidean, insofar as $g_{jl}(x)=g^0_{jl}+\varepsilon_{jl}$, $g^0_{jl}=g_{jl}(t(y^0))$ =const. in $CC_{y^0,\eta'}$ with $\varepsilon_{jl}\to 0$ as $\eta\le\eta'\to\infty$; (iv) $g_{jl}\circ s^{-1}$ is C^∞, and the metric is positive definite, even at $\partial\Omega^0$.

A complete Riemannian space Ω with above properties - e.g., Ω admits a compactification Ω^0, a compact C^∞-manifold with boundary; a finite atlas of Ω exists with charts derived from charts of Ω^0 , in the sense of (2.1),(2.2); in a neighbourhood of a point of $\partial\Omega^0 = \Omega^0 \setminus \Omega$ the metric of Ω has the properties (i),(iii),(iv) of prop. 2.2 - will be referred to as a space with conical ends. Note that this term is used synonymously to 'manifold with *S*-structure' or '*S*-manifold', '*S*-space', in the following sense: An *S*-manifold becomes a space with conical ends by introducing a metric of the form (2.12) (or any other Riemannian metric with properties (i)-(iv)of prop.2.2 - called a conical metric). Vice versa, a space with conical ends is just an *S*-manifold together with a conical metric, not necessarily of the form (2.12). In the following we always assume a space with conical ends, and then choose the surface measure $d\mu=dS$ $(=\det((g_{jk}))^{1/2})dx)$ of the conical metric $ds^2=$ $=\sum g_{jk}dx^j dx^k$ as C^∞-measure for our distributions and Sobolev spaces In particular we introduce the Hilbert space $\mathbf{H}=L^2(\Omega,d\mu)$.

As a side remark it may be mentioned that the quasi-Euclidean metric ds^2 has a corresponding Beltrami-Laplace operator Δ . Using the comparison triple $\{\Omega,1-\Delta,d\mu\}$ in the sense of $[C_2]$,V one may generate L^2-comparison algebras with complex-valued symbol, containing a core of ψdo's of the kind to be studied below.

Remark. For a Riemannian manifold Ω introduce $d(x)=d(x,x^\Delta)$, $x\in\Omega$, and $d(\mathbf{X})=\sup\{d(x):x\in\mathbf{X}\}$, $\mathbf{X}$ a subset of Ω , with the geodesic distance $d(x,x^\Delta)$ and a fixed point x^Δ . An end Ω' of Ω is defined as a subdomain of Ω with compact $\partial\Omega'$ and $d(\Omega')=\infty$ such that no decomposition $\Omega'=\Omega_1\cup\Omega_2$ exists with $\Omega_1\cap\Omega_2$ bounded and $d(\Omega_1)=d(\Omega_2)=\infty$ (think of the two "ends" of a cylinder). (Actually, two such Ω' , called Ω' , Ω'' , define the same end if $\Omega'\backslash\Omega''$ and $\Omega''\backslash\Omega'$ are bounded.) Such an end will be called conical if (a suitable) Ω' is isometric to an end of an $\mathbf{S}$-manifold, as above.

It is possible that some ends of a Riemannian space Ω are conical, others are not. Also, one may consider cases like the cone $x_1=\sqrt{(x_2^2+\ldots+x_n^2)}$, $x_1>0$, having the concial end $\{x_1>1\}$, but a conical tip $\{x_1<1\}$. Our theory, below, will not apply then. But the conical end and conical tip may be separated, using algebra surgery, similar to that in $[C_2]$,VIII, as not to be discussed here (cf. the problems of chapters I, II, III, IV).

We summarize, below, also discussing some trivial additions:

a)Temperate distributions and rapidly decreasing functions are introduced for a noncompact manifold Ω which is the interior of a compact manifold Ω^0 with boundary, where $\partial\Omega^0=0$ is permitted, giving the special case of a compact Ω . We define $\mathbf{S}=\mathbf{S}(\Omega)=CO^\infty(\Omega^0)$, $\mathbf{S}'=\mathbf{S}'(\Omega)=\mathbf{D}^\Delta(\Omega)=\{u\in\mathbf{D}'(\Omega):u=v|\Omega$ with $v\in\mathbf{D}'(\Omega^1)$, with an $\Omega^1\supset\supset\Omega^0\}$.

b) $\mathbf{S}(\Omega)$ and $\mathbf{S}'(\Omega)$ assume their conventional looks only in special coordinates: An $\mathbf{S}$-structure, making Ω an $\mathbf{S}$-manifold is introduced by declaring certain charts, coordinates, coordinate changes etc. as admissible. Admissible charts, atlantes, cut-off's, partitions of unity, all are obtained from charts,..., of Ω^0 , using the map $s(x)=\frac{x}{\langle x\rangle}$: $\mathbb{R}^n\to B_1$ of (2.1), (2.2), ...

c) A neighbourhood base of an infinite boundary point in admissible coordinates is of the form (2.5).

d) An $\mathbf{S}$-manifold Ω possesses a distinguished type of Riemannian metric ds^2 , called conical metric. In admissible coordinates

ds^2 has properties (i)-(iv) listed in prop.2.2. A manifold with conical ends is defined as an S-manifold Ω with distinguished conical metric. Write d(p,q) for the geodesic distance, and d(p)= =d(p,p°) with a fixed point p° and variable p in the following. The distinguished measure dμ on Ω is introduced as the surface measure of the distinguished conical metric. We have $d\mu=\kappa(s(x))dx$ in admissible charts, with dx in local coordinates and $\kappa\in C^\infty(U^0)$. Also, globally, we have $d\mu=\langle d(p)\rangle^{n-1}d\nu$ with a positive C^∞-measure dν on the compact Ω^0, and $\langle d(p)\rangle=\sqrt{1+d(p)^2}$.

e) We have

$$(2.13)\qquad S(\Omega)=\{u\in C^\infty(\Omega):\ (\lambda u)\circ\omega^{-1}\in S(\mathbb{R}^n)\text{ for all }\omega,\lambda\}\ ,$$

and

$$(2.14)\qquad S'(\Omega)=\{u\in D'(\Omega):\ (\lambda u)\circ\omega^{-1}\in S'(\mathbb{R}^n)\text{ for all }\omega,\lambda\}\ ,$$

where, in each case, $\omega:U\to\omega(U)\subset\mathbb{R}^n$ and λ denote an admissible chart and subordinated cut-off function (e.g., supp $\lambda\subset U$) .

f) The space

$$(2.15)\qquad L^1_{pol}(\Omega)=\{u\in L^1_{loc}(\Omega):\ d(p)^{-k}u\in L^1(\Omega,d\mu)\text{ for some }k\}$$

is naturally imbedded in $S'(\Omega)$ by

$$(2.16)\qquad \langle u,\varphi\rangle=\int_\Omega u\varphi d\mu\ ,\quad \varphi\in S(\Omega)\ .$$

g) Define $L(\Omega)$ as the space of differential expressions $L:C^\infty(\Omega)\to C^\infty(\Omega)$ such that in every distinguished chart $\omega:U\to\omega(U)\subset\mathbb{R}^n$ L is represented by a PDE $\Sigma a_\alpha\partial_x^\alpha$ with a_x^α of polynomial growth -that is, $\lambda a_\alpha\circ\omega^{-1}\in T(\mathbb{R}^n)$, with $T(\mathbb{R}^n)$ of [C1],p.28, the space of S'-multipliers (cf. [Schw1]) for every admissible λ, , supp $\lambda\subset U$. Then $L:S'(\Omega)\to S'(\Omega)$ for all $L\in L$, and such L is a contionuous map.

h) We have

$$(2.17)\qquad S(\Omega)=\{u\in C^\infty(\Omega):\ Lu=O(\langle d(p)\rangle^{-k})\text{ for all }L\in L\ ,\ k\ge 0\}\ .$$

i) The topology of $S(\Omega)$ is the Frechet topology of $C^\infty(\Omega^0)$. it may be generated by all seminorms $\|\langle d(p)\rangle^k Lu\|_{L^\infty}$ where k=0,1,.., and where L ranges over $L(\Omega)$ (or a suitable countable subset). In $S'(\Omega)$ we use the topology of weak convergence. (Note that the inductive limit topology based on $S'=\cup_s H_s$ is available as well.)

3. Coordinate invariance of pseudodifferential operators.

It is trivial that ψdo's on $\mathbb{R}^n$ transform into ψdo's on $\mathbb{R}^n$ under a linear change of coordinates, of the form $x=mx'+p$, where $p\in\mathbb{R}^n$, and $m=((m_{jk}))_{j,k=1,..,n}$ is invertible, $m_{jk}\in\mathbb{R}$. That is, if $Au=a(M_1,M_r,D)u=f$, where $u,f\in\mathcal{S}(\mathbb{R}^n)$, $a\in ST$, and $u(mx'+p)=v(x')$, $f(mx'+p)=g(x')$, then we get

$$(3.1)\qquad g(x')=|m|\int đ\xi\int đy'e^{i\xi.m(x'-y')}a(mx'+p,my'+p,\xi)v(y')\ ,$$

where $|m|=|\det m|$, for a moment. With $\xi=m^{-t}\xi'$, $m^{-t}=(m^{-1})^t$, we get

$$(3.2)\qquad g(x')=\int đx'\int đ\xi' e^{i\xi'(x'-y')}a(mx'+p,my'+p,m^{-t}\xi')v(y')\ .$$

Or, $g=b(M_1,M_r,D)v$, where $b(x,y,\xi)=a(mx+p,my+p,m^{-t}\xi)$ generally is in the same symbol class as $a(x,y,\xi)$.

For more general local coordinate changes and local ψdo's a we refer to Hoermander [Hr_2]. A very elegant proof may be found in [Fr_3]; its idea seems due to Kuranishi (unpublished?). Using this technique Schrohe [$Schr_3$] proved a result for a class of global transforms and ψdo's on $\mathbb{R}^n$ of our general kind.

Here we will use the same 'Kuranishi trick' again, for a subcollection of Op*ST*. We only admit coordinate transforms $\varphi:\mathbb{R}^n\to\mathbb{R}^n$ with the property that $s\circ\varphi\circ s^{-1}=\psi$ extends to a diffeomorphism $B_1^c\leftrightarrow B_1^c$. Clearly then the diffeomorphism $\varphi:\mathbb{R}^n\to\mathbb{R}^n$ extends to a homeomorphism $\mathbb{B}^n\leftrightarrow\mathbb{B}^n$. The homeomorphism may be used to carry over the manifold structure of B_1^c to the compactification $\mathbb{B}^n$ of $\mathbb{R}^n$. Thus we may regard $\mathbb{B}^n$ as a compact manifold with boundary. The above type of coordinate transform is precisely the class of homeomorphisms $\mathbb{B}^n\leftrightarrow\mathbb{B}^n$ preserving this manifold structure.

Given such $\varphi:\mathbb{R}^n\to\mathbb{R}^n$ and a ψdo $A=a(M_1,M_r,D)$, $a\in SS_{m,\rho,\delta}$ let again $f=Au$, with $u,f\in\mathcal{S}$. Let $g(x)=f(\varphi(x))$, $v(x)=u(\varphi(x))$. We get

$$(3.3)\qquad g(x)=\int đ\xi\int đye^{i\xi(\varphi(x)-\varphi(y))}a(\varphi(x),\varphi(y),\xi)J_\varphi(y)v(y)\quad ,$$

where $J_\varphi(y)=|\det((\partial\varphi_j/\partial y_k))|$.

Let us split the expression at right of (3.3) into two parts by inserting a partition $1=\chi(s(x)-s(y))+\omega(s(x)-s(y))$ under the integral signs, where $\omega(z)=0$ for $|z|\le\varepsilon$, $\omega(z)=1$ for $|z|\ge 2\varepsilon>0$. Writing $g=g_1+g_2$, correspondingly, we first consider g_2. First look at g_2 in the old coordinates. With $f_2(x)=g_2(\theta(x))$, $\theta(x)=\varphi^{-1}(x)$, we get

$$(3.4)\qquad f_2(x)=\int đ\xi\int đye^{i\xi(x-y)}\omega(s(\theta(x))-s(\theta(y)))a(x,y,\xi)u(y)\quad .$$

With the inverse function $\zeta=\psi^{-1}=s\circ\theta\circ s^{-1}$ we have $s\circ\theta=\zeta\circ s$, hence

$$(3.5)\qquad f_2(x)=\int d\!\!{}^-\xi\int d\!\!{}^-y e^{i\xi(x-y)}\omega(\zeta(s(x))-\zeta(s(y)))a(x,y,\xi)u(y) \ .$$

We use the identity $e^{i\xi(x-y)}=|x-y|^{-2N}(-\Delta_\xi)^N(e^{i\xi(x-y)})$. The integrand vanishes near x=y. An N-fold partial integration gives

$$(3.6)\qquad f_2(x)=\int d\!\!{}^-\xi\int d\!\!{}^-y e^{i\xi(x-y)}\frac{\omega(\zeta(s(x))-\zeta(s(y)))}{|x-y|^{2N}}a_N(x,y,\xi)u(y) \ ,$$

$$\text{with } a_N(x,y,\xi) = (-\Delta_\xi)^N a(x,y,\xi) \ .$$

It is clear that this partial integration is legal, similar as the ones performed earlier. The function ζ, as a diffeomorphism $B_1^c\to B_1^c$, satisfies an inequality $|\zeta(x)-\zeta(y)|\ge p|x-y|$, with some $p>0$. Hence the integrand in (3.6) vanishes for $|s(x)-s(y)|\le\frac{1}{p}|\zeta(s(x)-\zeta(s(y))|$ $\le\frac{\varepsilon}{p}$. In other words, we may assume that $|s(x)-s(y)|\ge\frac{\varepsilon}{p}$.

Observe that $s^{-1}(x)=t(x)=\frac{x}{[x]}$, where $[x]=\sqrt{1-x^2}$. We get

$$(3.7)\qquad T(x) = ((t_{j|x_1})) = \frac{1}{[x]}((\delta_{j1}+t_jt_1)) = \frac{1}{[x]}(I+t)\langle t\rangle \ ,$$

implying T(x) to be real symmetric, $T(x)\ge\frac{1}{[x]}$, $x\in B_1$. Notice that

$$t(x)-t(y) = \int_0^1 d\tau \frac{d}{d\tau}(t(x+\tau(y-x))) =\{\int_0^1 d\tau T(x+\tau(y-x)\}(y-x) = S(y-x),$$

where $S\ge\mathrm{Min}\{\frac{1}{[x]},\frac{1}{[y]}\}$. We get $\frac{1}{[x]}=\langle t(x)\rangle$, hence

$$(3.8)\qquad |t(x)-t(y)|\ge(\mathrm{Min}\{\langle t(x)\rangle,\langle t(y)\rangle\})|x-y|,\ x,y\in B_1 \ , \text{ or,}$$

$$\frac{1}{|x-y|}\mathrm{Min}\{\langle x\rangle,\langle y\rangle\} \le \frac{1}{|s(x)-s(y)|} \ ,\ x,y\in\mathbb{R}^n \ .$$

Thus we may assume that

$$(3.9)\qquad |x-y|^{-2N} \le (\tfrac{p}{\varepsilon})^{2N}\mathrm{Max}\{\lambda^{2N}(x),\lambda^{2N}(y)\} \ ,\ \lambda(t)=\frac{1}{\langle t\rangle} \ .$$

Assuming $a\in SS_{m,\rho,\delta}$, $\rho_1>0$, and N sufficiently large the integrals $\int d\xi\int dy$ in (3.6) may be interchanged, for

$$f_2(x) = \int d\!\!{}^-y u(y)k_2(x,y) \ , \text{ with integral kernel}$$

$$(3.10)\qquad k_2(x,y)=\frac{\omega(\zeta(s(x))-\zeta(s(y)))}{|x-y|^{2N}}\int d\!\!{}^-\xi e^{i\xi(x-y)}a_N(x,y,\xi) \ ,$$

where N is arbitrary (sufficiently large), and where $\int d\xi$ exists, since the ξ-order of a_N is $m_1-2N\rho_1 < -n$, as N gets large.

Proposition 3.1. The kernel $k_2(x,y)$ of (3.10) as well as its transform $k_2(\varphi(x),\varphi(y))J_\varphi(y) = k_{2,\varphi}(x,y)$ under the coordinate change $x \to \varphi(x)$, are in $S(\mathbb{R}^{2n})$. Hence the operator K_2 defined by $K_2u=f_2$, and its transform under φ are ψdo's of order $-\infty$.

Proof. The integral in (3.10)₂ may be written as $\overset{3}{a}_N(x,y,x-y)$. Clearly $\overset{3}{a}_N(x,y,z)\in C^\infty(\mathbb{R}^{3n}\setminus\{z=0\})$. Indeed, $\overset{3}{a}_N(x,y,z)=(\frac{1}{|z|}2M)\overset{3}{a}_{N+M}(x,y,z)$ where a_{N+M} decays better as M increases, so $\overset{3}{a}_{N+M}$ admits more and more z-derivatives, as $M\to\infty$. Similarly,

$$(3.11)\quad \partial_x^\alpha\partial_y^\beta\partial_z^\gamma\overset{3}{a}_N(x,y,z)=O(\langle x\rangle^{m_2}\langle y\rangle^{m_3})\ ,\ \text{as } |z|\ge\varepsilon>0,\ x,y,z\in\mathbb{R}^n,$$

with 'O(.)-constant' depending on N, but not on the m_j. Therefore,

$$(3.12)\quad \partial_x^\alpha\partial_y^\beta\overset{3}{a}_N(x,y,x-y)=O(\langle x\rangle^{m_2}\langle y\rangle^{m_3})\ ,\ \text{as } x,y\in\mathbb{R}^n\ ,\ |x-y|\ge\varepsilon>0\ .$$

For the factor k_3 in front of the integral in (3.10)₂ we get

$$(3.13)\quad \partial_x^\alpha\partial_y^\beta k_3(x,y) = O(\lambda^{2\eta}(x-y)\ \mathrm{Max}\{\lambda^\eta(x),\lambda^\eta(y)\})\ ,$$

where, of course, $\eta=\frac{2}{3}N$ may be choosen arbitrarily large. Here

$$(3.14)\quad \mathrm{Max}\{\lambda^\eta(x),\lambda^\eta(y)\}=\lambda^\eta(x)\mathrm{Max}\{1,(\tfrac{\lambda(y)}{\lambda(x)})^\eta\}\le\lambda^\eta(x)\langle x-y\rangle^\eta\ ,$$

using the well known inequality. Accordingly,

$$(3.15)\quad \partial_x^\alpha\partial_y^\beta k_3(x,y) = O(\lambda^\eta(x)\lambda^\eta(x-y)),\ \text{for all } \eta>0\ .$$

Combining (3.12) and (3.15) we indeed get $k_2\in S(\mathbb{R}^{2n})$. The remainder of prop.3.1 then is a consequence of prop.1.1 and prop.2.1 : The map $(x,y)\to(s(x),s(y))$ takes $S(\mathbb{R}^{2n})$ to $CO^\infty(B_1^c\times B_1^c)$, which is preserved by $(x,y)\to(\psi(x),\psi(y))$. The Jacobian determinant defines a $C^\infty(\Omega^0)$-function $J_\varphi\circ s^{-1}$, by prop.2.1. Q.E.D.

After dealing with the part of $A=a(M_l,M_r,D)$ belonging to $\omega(s(x)-s(y))$ we now turn to the other part,

$$(3.16)\quad g_1(x) = \int d\!\!{}^-\xi\int d\!\!{}^-y e^{i\xi(\varphi(x)-\varphi(y))}\chi a(\varphi(x),\varphi(y),\xi)J_\varphi(y)v(y)\ ,$$

where $\chi=\chi(s(x)-s(y))=0$ for $|s(x)-s(y)|\ge2\varepsilon$. Write

$$(3.17)\quad \varphi(x)-\varphi(y)=\int_0^1 d\tau\{\tfrac{d}{d\tau}\varphi(y+\tau(x-y))\}= M(x,y)(x-y)\ ,$$
$$\text{where}\quad M(x,y) = \int_0^1 d\tau(\partial\varphi/\partial x)(y+\tau(x-y))\ .$$

If x and y are close then clearly $M(x,y)\approx\partial\varphi/\partial x$ which is an invertible matrix. We have $\xi(\varphi(x)-\varphi(y))= (M^t(x,y)\xi)(x-y)$. Then an integral substitution $\xi^{\sim}=M^{t}(x,y)\xi$, for fixed x,y, close together, may be used to convert (3.16) into the form of a ψdo-integral again.

Having employed our partition we may assume now that $|s(x)-s(y)|\leq 2\varepsilon$, where $\varepsilon>0$ is arbitrary. The question is whether this condition is sufficient to guarantee a global such integral substitution, valid for all x,y,ξ with $|s(x)-s(y)|\leq 2\varepsilon$.

Write M of (3.17) as $M(x,y)=\int_0^1 d\tau Q(s(y+\tau(x-y))$ with $Q(s)=\partial\varphi/\partial x(\frac{s}{[s]})\in C^{\infty}(B_1^c)$ (by prop.2.1). The curve $\{s(y+\tau(x-y)):\tau\in\mathbb{R}\}=\Gamma$ is a half-ellipse with center 0 and vertices at $\pm\frac{x-y}{|x-y|}$ and $\frac{z}{\langle z\rangle}$, z= vector to the point of the line $\{y+t(x-y)\}$ closest to 0, as easily seen. The arc $\{s(y+\tau(x-y)):0\leq\tau\leq 1\}$ connects s(y) and s(x) on Γ. It is contained in the ball $|s-\frac{1}{2}(s(x)+s(y))|\leq\frac{1}{2}|s(x)-s(y)|\leq\varepsilon$. The $C^{\infty}(B_1^c)$-function Q(s) is uniformly continuous and invertible . Thus indeed ε may be chosen small to insure invertibility of M for all (x,y) as $|s(x)-s(y)|\leq 2\varepsilon$.

Carrying out the integral substitution as indicated yields

$$g_1(x)=\int đ\xi\int đy e^{i\xi(x-y)}a^{\sim}(x,y,\xi)v(y) \text{ , with} \tag{3.18}$$

$$a^{\sim}(x,y,\xi)=\frac{\det M(y,y)}{\det M(x,y)}\,a(\varphi(x),\varphi(y),M^{-t}(x,y)\xi)\ \chi(s(x)-s(y)) \ .$$

The following observation about the new symbol $a^{\sim}$ is useful.

<u>Proposition 3.2</u>. For any function $b(s)\in C^{\infty}(B_1^c)$ the composition $c=b\circ s$ is a symbol in $SS_{0,(1,1,1),0}$. More precisely, we have

$$c^{(\alpha)}(x) = O(\langle x\rangle^{-|\alpha|}) \text{ , for all } \alpha \ . \tag{3.19}$$

Indeed, this is a consequence of the chain rule, and the fact that $s^{(\alpha)}(x)=O(\langle x\rangle^{-|\alpha|})$, for all $|\alpha|$.

Looking at the ξ-derivatives of $a^{\sim}$ of (3.18) it appears that the parameter ρ_1 remains unchanged under our coordinate transform. An x- (or y-) derivative may land on the $M^{-t}(x,y)\xi$ inside the ξ-argument of a , or anywhere else. In the first case we get a factor $O(\langle x\rangle^{-1}\langle\xi\rangle^{1-\rho_1})$, in addition to the already existing ones. In the other cases we get a factor $O(\langle x\rangle^{-\rho_2}\langle\xi\rangle^{\delta})$ or $O(\langle x\rangle^{-1})$. (All this for an x-derivative).

It follows that $a^{\sim}\in SS_{m,\rho,\delta^{\Delta}}$ with $\delta^{\Delta}=\mathrm{Max}\{\delta,1-\rho_1\}$. To verify this we use prop.2.1 when differentiating for the x in φ(x): ∂φ/∂x is of the form needed in prop.3.2. Also, bounds of the form $0<c\leq\langle\varphi(x)\rangle/\langle x\rangle\leq C$ are easily derived. Summarizing:

<u>Theorem 3.3</u>. Let $a\in SS_{m,\rho,\delta}$, with $\rho_1>0$, and $\rho_j\leq 1$, j=1,2,3.

Assume that $\varphi:\mathbb{R}^n\to\mathbb{R}^n$ with inverse $\theta:\mathbb{R}^n\to\mathbb{R}^n$ has the property that $\psi= s\circ\varphi\circ s^{-1}$ (with $s(x)=\frac{x}{\langle x\rangle}=\frac{x}{\sqrt{1+x^2}}$), a map $B_1\to B_1$, extends to a diffeomorphism $B_1^c\leftrightarrow B_1^c$ of the closed unit ball B_1^c. Then the linear operator $A^\sim = T_\varphi^{-1}AT_\varphi$, with $A=a(M_l,M_r,D)$ and $T_\varphi u(x)=(u\circ\varphi)(x)=u(\varphi(x))$, $u \in \mathcal{S}$, is a ψdo again with symbol $a^\sim \in \mathcal{SS}_{m,\rho,\delta^\Delta}$, $\delta^\Delta=\max\{\delta,1-\rho_1\}$. Up to an additional term in $\mathcal{S}(\mathbb{R}^{2n})$ the symbol $a^\sim$ is given by (3.18).

Usually we will tend to apply thm.3.3 for operators of the form $a(x,D)$, $a(M_r,D)$, $a(M_w,D)$, with $a\in \psi h$. These are special cases of $A=a(M_l,M_r,D)$ in thm.3.3. However, thm.3.3 then will give an $A^\sim=a^\sim(M_l,M_r,D)$ which must be converted to $b(x,D)$, $c(M_r,D)$, $e(M_w,D)$, using I,thm.6.2, assuming that $\delta^\Delta<\rho_1$ - i.e., $\delta<\rho_1$ and $1-\rho_1<\rho_1 \Leftrightarrow \rho_1>\frac{1}{2}$. We therefore have

<u>Corollary 3.4</u>. Let $a\in \psi h_{m,\rho,\delta}$ with $\rho_2>0$, $\rho_1>\frac{1}{2}$, $\delta<\rho_1$. Then the coordinate transform of thm.3.3 takes each of the operators $a(x,D)$ $a(M_r,D)$, $a(M_w,D)$ into an operator of the same form with symbol in $\psi h_{m,\rho,\delta^\Delta}$, $\delta^\Delta=\text{Max}\{\delta,1-\rho_1\}$. The new symbol, up to terms in $\psi h_{m',\rho,\delta^\Delta}$, $m'<m$, is given by $a^\sim(x,x,\xi)$ with $a^\sim$ of (3.18).

<u>Problems</u>. 1) For $n=1$, consider the transform $t\to x=\varphi(t)=e^t$, a diffeomorphism $\mathbb{R}\leftrightarrow\mathbb{R}^+=(0,\infty)$ (and a group isomorphism). $\varphi(t)$ and its inverse $t=\theta(x)=\log x$) do not satisfy the cdn's of thm.3.3, but they take $A=a(M_l,M_r,D)$ to a ψdo $A^\sim$ if only $\text{supp}\, a(x,y,\xi)\subset K\times K\times\mathbb{R}$, $K\subset\subset\mathbb{R}$ ($K\subset\subset\mathbb{R}^+$ in case φ). 2) The Mellin transform is defined as $M=FT_\varphi$ with the Fourier transform F and $T_\varphi u(x)=u(\varphi(x))$, φ of pbm.1. Express M and M^{-1} like Fourier integrals. Show that M is the Fourier-Plancherel transform of the group $\mathbb{R}^+$: Define $c=a\circledast b=\int a(\frac{x}{y})b(y)\frac{dy}{y}$, for $a,b\in L^1(\mathbb{R},\frac{dx}{x})$, and get $Mc=(Ma)(Mb)$. 3) Consider $\mathbb{R}^{2*}=\mathbb{R}^2\setminus\{0\}$ as Riemann space with metric $ds^2=\sum(\delta_{jk}+n_jn_k)dx_jdx_k$, $n_j=x_j/|x|$ (a cone $\mathbf{Z}$ isometric to $\{y=|x|\}\subset\mathbb{R}^3=\{(x_1,x_2,y)\}$). Show that this cone is mapped conformally onto the cylinder $\{-\infty<t<\infty,\ 0\le\theta\le 2\pi\}$ with metric $2dt^2+d\theta^2$ by the map $x=(x_1,x_2)\to(t,\theta)$, where $t=\log|x|$, $\theta=\arg x$ $\arg x = \arctan(x_2/x_1)$. 4) Use the diffeomorphism between cone and cylinder of pbm.3 to install a natural class of ψdo's on $\mathbf{Z}$. In particular, the ψdo's should be the global coordinate transforms of the "cylinder ψdo's" introduced in II,3,pbms 1,2. They should be "local ψdo's", in the sense $\lambda A\mu$ is a ψdo for $\lambda,\mu\in C_0^\infty(\mathbf{Z})$. Try for a concept of K-parametrix, for md-elliptic operators to be defined especially with an operator-valued symbol at the conical tip.

4. Pseudodifferential operators on $\mathcal{S}$-manifolds.

We start our discussion with a generalization of I, thm.3.3.

Proposition 4.1. For the special $\mathcal{S}$-manifold $\Omega=\mathbb{R}^n$ let θ,λ be admissible cut-off functions such that $\theta^\circ=\theta\circ s^{-1}$ and $\lambda^\circ=\lambda\circ s^{-1}$, extended to B_1^c, have disjoint supports. Then, for any $A\in \mathrm{Op}\psi h_m$ we have $\theta A\lambda \in \mathrm{Op}\psi h_{-\infty}$, an integral operator with kernel in $\mathcal{S}(\mathbb{R}^{2n})$.

Proof. By I, thm.3.3 the distribution kernel $k(x,y)$ of A has singular support at $x=y$ only, thus the kernel $\kappa(x,y)=k(x,y)\theta(x)\lambda(y)$ of $\theta A\lambda$ is $C^\infty(\mathbb{R}^{2n})$. In fact, k equals a function in $\mathcal{S}$ for $x\in K\subset\subset \mathbb{R}^n$ and large $|y|$, and vice versa. Moreover, looking at I,(3.15), using that $|x|\approx|x-y|$, $|y|\approx|x-y|$, as $x\in \mathrm{supp}\,\theta$, $y\in \mathrm{supp}\,\lambda$, $|x|$, $|y|>>1$, we indeed find that $x^\gamma y^\delta \partial_x^\alpha \partial_y^\beta \kappa(x,y) = O(1)$, for all $\alpha,\beta,\gamma,\delta$ - i.e., $\kappa\in \mathcal{S}(\mathbb{R}^{2n})$, q.e.d.

Now assume that Ω is a Riemann space with conical ends, interior of Ω° , a compact manifold with boundary, as in sec.2. We will use admissible cut-off functions now, defined as $\theta=\theta^\circ\circ s$, with $\mathrm{supp}\,\theta^\circ \subset U^\circ$ with a chart $\omega^\circ:U^\circ\rightarrow\omega^\circ(U^\circ)\subset B_1^c$ of Ω°, and s of (1.5).

The above relation between the cut-off θ and the chart U will be expressed by writing $\theta \lessdot U$. The same notation will be used for another relation: Writing

$$\theta \gtrdot \lambda \ , \text{ or } , \ \lambda \lessdot \theta \tag{4.1}$$

indicates that $\theta=\theta^\circ\circ s$, $\lambda=\lambda^\circ\circ s$, where θ° , λ° are cut-off's of Ω° with support in a chart U° , and $\theta^\circ=1$ near $\mathrm{supp}\,\lambda^\circ$.

The following is evident: Just look at Ω° instead of Ω :

Proposition 4.2. For a given admissible cut-off function $\theta\lessdot U$ we may construct an infinite sequence $\theta=\theta_0\lessdot\theta_1\lessdot\theta_2\lessdot\ldots\lessdot U$.

First define the class $\mathcal{O}\!\!\mathcal{R}=\mathcal{LS}_{-\infty}=\mathcal{LS}_{-\infty}(\Omega)$ of integral operators

$$Ku(x) = \int_\Omega k(x,y)u(y)d\mu(y) \ , \ u\in \mathcal{S}(\Omega) \ , \tag{4.2}$$

(or $u\in \mathcal{S}'(\Omega)$, with a distribution integral) with kernel $k(x,y)\in \mathcal{S}(\Omega\times\Omega) = CO^\infty(\Omega\times\Omega)$ (note the remark above (2.12)). We will regard $\mathcal{O}\!\!\mathcal{R}$ as the class of ψdo's of order $-\infty$, as in case of $\Omega=\mathbb{R}^n$.

A pseudodifferential operator A on Ω then is defined as a continuous linear operator $A:\mathcal{S}(\Omega)\rightarrow \mathcal{S}(\Omega)$ such that, given any admissible cut-off function λ and chart $\omega:U\rightarrow\mathbb{R}^n$, with $\lambda\lessdot U$, there exists $A_\lambda=a_\lambda(x,D)\in \mathrm{Op}\psi h_{m,\rho,\delta}$, $K_\lambda\in \mathcal{O}\!\!\mathcal{R}$ with $a_\lambda(x,\xi)=0$ as $x\notin\omega(U)$, and

$$A(\lambda u) = K_\lambda u + (A_\lambda((\lambda u)\circ\omega^{-1}))\circ\omega \ , \ u\in \mathcal{S}(\Omega) \ , \tag{4.3}$$

where we assume m,ρ,δ independent of λ and $\omega:U\rightarrow\mathbb{R}^n$.

Note, we assume $m=(m_1, m_2)$, $\rho=(\rho_1, \rho_2)$, δ <u>independent of</u> λ, U. As usual, $m_j \in \mathbb{R}$, $0 \le \rho_j \le 1$, $0 \le \delta \le \rho_1$, but for a <u>local calculus</u> of ψdo's <u>and coordinate invariance</u> we require $\rho_j > 0$, $1-\rho_1 \le \delta < \rho_1$. <u>The class of such</u> ψdo's <u>for a given</u> m, ρ, δ <u>will be denoted by</u> $LS_{m,\rho,\delta}$. Write $LS_m = \cup LS_{m,\rho,\delta}$, union over ρ, δ with $\rho > 0$, $1-\rho_1 \le \delta < \rho_1$, enabling local calculus and coordinate change. Write $LS_{\infty,\rho,\delta} = \cup LS_{m,\rho,\delta}$, $\cup LS_m = LS_\infty = LS$. We shall see: $= LS_{-\infty} = \cap LS_m$. The special case $\rho = e = (1,1)$, $\delta = 0$ will be focus of interest. We thus define $LC_m = LS_{m,e,0}$, $LC_\infty = \cup LC_m$.

Note that (4.3) is of strictly local nature, insofar as λu as well as $(A_\lambda((\lambda u)\circ\omega^{-1})\circ\omega$ have support in a single admissible chart U. Thus it is practical to regard ω in (4.3) as an identification of the points of U with those of $\omega(U)$ - i.e., as the identity map, which may be omitted in writing. Then (4.3) reads

$$A(\lambda u) = K_\lambda u + A_\lambda(\lambda u) . \tag{4.3'}$$

Note, we also get

$$A(\lambda u) = K_\lambda u + \lambda^0 A_\lambda(\lambda u) , \tag{4.3"}$$

with any other admissible cut-off λ_0 , $\lambda \lhd \lambda_0 \lhd U$. Indeed, we get $\lambda \lhd U_0$ with the admissible chart $U_0 = \{\lambda_0 = 1\}^{\mathrm{int}}$, $\omega_0 = \omega|U_0$. By definition we get (4.3'), possibly with another $a_\lambda(x,\xi)$, but $a_\lambda = 0$ for $x \notin U_0$, hence (4.3") follows. On the other hand, if (4.3") holds for some $\lambda \lhd \lambda_0 \lhd U$ we may use the symbol $b_\lambda(x\xi) = \lambda_0(x) a_\lambda(x,\xi)$ in place of $a_\lambda(x,\xi)$, to get (4.3'), for equivalence (4.3') $\Leftrightarrow$ (4.3") .

From now on we always assume $A \in LS$ - i.e., $0 < \rho_j$, $1-\rho_1 \le \delta < \rho_1$.

If we have both $\lambda \lhd U$, $\theta \lhd U$, for two cut-off's λ, θ , then $A\lambda\theta = a_\lambda(x,D)\lambda\theta + K_\lambda\theta = a_\theta(x,D)\lambda\theta + K_\theta\lambda$, showing that $(a_\lambda - a_\theta)(x,D)\theta\lambda \in \mathcal{O}$. If $\lambda(p) \ne 0$, $\theta(p) \ne 0$, for some $p \in \Omega^0$, then prop.4.1 implies existence of an admissible cut-off κ with $\kappa = 1$ near p such that

$$\kappa(x)(a_\lambda(x,\xi) - a_\theta(x,\xi)) \in S(\mathbb{R}^{2n}) . \tag{4.4}$$

Clearly (4.4) expresses a uniqueness property of the local symbol.

Note that $LS_{m,\rho,\delta}(\mathbb{R}^n) = \mathrm{Op}\psi h_{m,\rho,\delta}$, assuming $\rho_j > 0$, $1-\rho_1 \le \delta < \rho_1$: First let $A = a(x,D) \in \mathrm{Op}\psi h$. For $\lambda \lhd \lambda_0 \lhd U$, $\omega = \mathrm{id}|U$, construct $\lambda \lhd \lambda_0 \lhd U$. Then $(1-\lambda_0)A\lambda \in \mathcal{O}$,by prop.4.1, hence (4.3"). Similarly for a general ω extending to an admissible coordinate transform $\mathbb{R}^n \Leftrightarrow \mathbb{R}^n$. One then must use cor.3.4. For general admissible $\omega: U \to \mathbb{R}^n$ first cover $U \subset \mathbb{R}^n$ by a finite collection $\cup U_j$ such that $\omega^{\circ}|U_j^0$ is approximately linear, hence $\omega|U_j$ extends to an admissible map $\mathbb{R}^n \Leftrightarrow \mathbb{R}^n$. Then, with an admissible partition $1 = \Sigma\mu_j$, $\mu_j \lhd U_j$, let $\lambda_j = \lambda\mu_j$. Get $A\lambda_j = a_{\lambda_j}(x,D)\lambda_j + K_{\lambda_j}$, and sum over j, for (3.3). For

$A\in LS_{m,\rho,\delta}(\mathbb{R}^n)$ choose $U=\mathbb{R}^n$, $\omega=id.$ to show that $A\in \psi h_{m,\rho,\delta}$.

For general Ω we clearly get $LS_{-\infty}(\Omega)\subset LS_{m,\rho,\delta}(\Omega)$. For general $A\in LS_m$ and finite admissible partition of unity $\Sigma\lambda_j=1$, $\lambda_j<\varepsilon)U_j$ get

(4.5) $\quad Au=\sum A\lambda_j u=\sum \lambda_j^1 A_{\lambda_j}(\lambda_j u)+Ku \ , \ u\in S \ , \ K\in LS_{-\infty} \ , \ \lambda_j<\varepsilon)\lambda_j^1<\varepsilon)U_j \ ,$

just by repeated application of (4.3). Vice versa, a ψdo $A\in LS_m$ may be constructed by assuming $A_{\lambda_j}=A_j\in Op\psi h_{m,\rho,\delta}$ in (4.5) as arbitrary ψdo's. Before confirming this let us prove the following.

<u>Proposition 4.3</u>. For admissible cut-off's $\theta=\theta^\circ\circ s$, $\lambda=\lambda^\circ\circ s$, if $(\mathrm{supp}\ \theta^\circ)\cap(\mathrm{supp}\ \lambda^\circ)=\emptyset$, then $\theta A\lambda\in LS_{-\infty}$, for every $A\in LS_m$.

<u>Proof</u>. Construct a finite admissible atlas and partition of unity $\Sigma\lambda_j=1$, $\lambda_j\lessdot U_j$, $\omega_j:U_j\to\omega_j(U_j)\subset\mathbb{R}^n$ such that for each j either $U_j^0\cap\ \mathrm{supp}\ \lambda^\circ=\emptyset$ or $U_j^0\cap\ \mathrm{supp}\ \theta^\circ=\emptyset$. With this partition write

$$\theta A\lambda = \sum\theta(A\lambda_j)\lambda = \sum\theta(\lambda_j^0 A_{\lambda_j}\lambda_j+K_j)\lambda \ .$$

The sum may be extended over j with $U_j^0\cap\ \mathrm{supp}\ \theta^\circ\neq 0$, else the factor $\theta\lambda_j^0$ vanishes. In the left-over terms we get $\theta\lambda_j^0 A_{\lambda_j}\lambda_j\lambda=0$, since then $\lambda_j\lambda=0$, so that $\theta A\lambda=\Sigma\theta K_j\lambda\in LS_{-\infty}$, q.e.d.

For $\{U_j,\omega_j\}$, λ_j, as above, and general $A_j\in Op\psi h_{m,\rho,\delta}$ write

(4.6) $$A=\sum_{j=1}^N\lambda_j^1 A_j\lambda_j \ , \ \lambda_j\lessdot\lambda_j^1\lessdot U_j \ .$$

Let us prove that $A\in LS_{m,\rho,\delta}$. We trivially get $A:S(\Omega)\to S(\Omega)$ continuous. To confirm (4.3) let λ, $\lambda_\circ$ be as in (4.3"). For each j the product $\lambda\lambda_j$ is an admissible cut-off with $\mathrm{supp}(\lambda\lambda_j)\subset U\cap U_j$. The coordinate transform $\omega\circ\omega_j^{-1}:\omega_j(U\cap U_j)\to\omega(U\cap U_j)$ is of the form $\kappa=\omega\circ\omega_j^{-1}=s^{-1}\kappa^\circ\circ s$ with a diffeomorphism $\kappa^\circ:\omega_j^0(U^\circ\cap U_j^0)\to\omega^0(U^\circ\cap U_j^0)$. We may assume the partition ω_j^0 refined so far that κ° differs little from a linear map. Thus we may assume κ° extendable to a diffeomorphism $B_1^c\to B_1^c$. Hence κ extends to a diffeomorphism $\mathbb{R}^n\leftrightarrow\mathbb{R}^n$. Thm.3.3 (or cor.3.4) may be applied to transform A_j to the coordinates of the chart U , resulting in a ψdo $A_j^\Delta\in Op\psi h_{m,\rho,\delta}$. To be precise, in the formulation of (4.2) we have

(4.7) $$(A_j((\lambda\lambda_j u)_{\,\circ}\omega_j^{-1}))\circ\omega_j=(A_j^\Delta((\lambda\lambda_j u)\circ\omega^{-1}))\circ\omega \ ,$$

with properly extended diffeomorphism $\omega\circ\omega_j^{-1}$, but $(\lambda\lambda_j u)\circ\omega_j^{-1}$ has its support within $\omega_j(U\cap U_j)$. Using prop.4.1, write the right hand side of (4.7) as $(\theta_j A_j^\Delta((\lambda\lambda_j u)\circ\omega^{-1}))\circ\omega+K_j$, $K_j\in LS_{-\infty}(\mathbb{R}^n)$, with an admissible cut-off θ_j, $\mathrm{supp}\ \theta_j\subset U$. By prop.4.2 the integral operator K_j has a kernel of the form $\upsilon_1(x)\upsilon_2(y)k_j(x,y)$, with admissible cut-off's υ_j, supp $\upsilon_j \subset U$. Returning to the notation of (4.3) -(4.3") it follows that indeed each $(\lambda_j^0 A_j \lambda_j \lambda)$ transforms to a term of the form (4.3"), proving that A of (4.6) belongs to $LS_{m,\rho,\delta}$.

Clearly a differential expression $L \in L(\Omega)$, $L(\Omega)$ of sec.2,(g) defines a ψdo $L \in LS_{(N,m_2),(1,m_2),0}$, for some N (the order of the expression) provided that the local coefficients a_α^j in admissible coordinates satisfy $a_\alpha^{j(\beta)}(x)=O(\langle x\rangle^{m_2-\rho_2|\beta|})$ for all α,β.

As for differential operators a global symbol in general is not defined, for an $A \in LS_{m,\rho,\delta}$, although we have a well defined <u>local symbol</u> $a_\lambda(x,\xi)$ as above. Note that a_λ is not unique, since (i) any term $c(x,\xi) \in S(\mathbb{R}^{2n})$ may be added, and (ii) $a_\lambda(x,\xi)$ is more or less arbitrary for x outside supp λ. For a differential operator $L \in L(\Omega)$ the local symbol coincides with the polynomial $a(x,\xi)= \Sigma a_\alpha(x)\xi^\alpha$, where in local coordinates $L=\Sigma a_\alpha(x)D^\alpha$, $D=\frac{1}{i}\partial_x$.

It is clear that we will seek properties of operators in LS similar to those of Opψh in earlier chapters. We summarize corresponding facts in thm.4.4. Proofs are straight extensions, and will not be discussed in detail.

<u>Theorem 4.4</u>. The classes $LS_{m,\rho,\delta}$ for $m=0,\pm\infty$, form algebras for each given $\rho_j>0$, $1-\rho_1<\delta<\rho_1$. More generally, we have

$$(4.8) \qquad LS_{m,\rho,\delta} \cdot LS_{m',\rho,\delta} \subset LS_{m+m',\rho,\delta}\ ,\ LC_m \cdot LC_{m'} \subset LC_{m+m'}\ .$$

One finds that $LS_{-\infty}$ is an ideal of LS_0, LS_∞, $LS_{m,\rho,\delta}$, $m=0,\infty$. All spaces are invariant under the involution "*" defined by the Hilbert space adjoint (with respect to $(u,v)=\int \bar{u}v d\mu$).

Calculus of ψdo's holds locally: For $A \in LS_{m,\rho,\delta} \subset LS_m$, $B \in LS_{m',\rho,\delta}$ the local symbol c_λ of $C=AB \in LS_{m+m',\rho,\delta}$ is expressed by the asymptotic Leibniz formula (of I,$(5.1)_1$), near any point $p \in \Omega^0$ with $\lambda(p)\neq 0$. The symbols a_λ, a_λ^* of $A \in LS_{m,\rho,\delta} \subset LS_m$ and its adjoint A^* are related by an asymptotic I,$(5.1)_2$ (with $\kappa=\frac{d\mu}{dx}$). In detail,for admissible $\lambda \Subset \lambda_0 \Subset U$, near $p \in \Omega^0$ with $\lambda(p)\neq 0$, a_λ,.....,of (4.3"),get

$$(4.9) \quad c_\lambda = \sum_\theta \frac{(-i)^{|\theta|}}{\theta!} a_\lambda^{(\theta)} b_{\lambda(\theta)}\ ,\ a_\lambda^* = \sum_\theta \frac{(-i)^{|\theta|}}{\kappa\theta!} \{\kappa\bar{a}_\lambda\}_{(\theta)}^{(\theta)} \ (\mathrm{mod}\ \psi h_{-\infty})\ ,$$

where "f=g (mod $\mathfrak{K}$) near p" means $(f-g)\chi \in \mathfrak{K}$, with a cut-off χ near p. Local symbols are unique: $a_\lambda = a_\theta$ (mod$\mathfrak{K}$)near $\lambda(p)\neq 0 \neq \theta(p)$, cf. (4.4).

The left multiplying representation of (4.3") may be replaced by Weyl (or right multiplying) representation without changing LS.

As next important point we shall look into existence of a parametrix, for $A \in LS$. Here $A,B \in LS$ are said to be parametrices of each other if $AB-1$, $BA-1 \in \boldsymbol{LS}_{-\infty}$. This problem is quickly resolved, using our local parametrices of II,3. For a finite admissible partition $1=\Sigma\lambda_j$ let $\lambda_j \prec \lambda_j^1 \prec \lambda_j^2 \prec \lambda_j^3 \prec U_j$, as in prop.4.2. Given $A \in \boldsymbol{LS}$ write $A\mu = K_j + A_\mu \mu$, $\mu=\lambda_j^3$. Assume A_μ to admit a left K-parametrix B_j with respect to the symbol λ_j , so that $B_j A_\mu = \lambda_j + K_j$. Write $A_\mu = A_j$. Notice we also have $\lambda_j^1 B_j \lambda_j^2 A_j = \lambda_j + K_j$, since $\lambda_j^1 B_j (1-\lambda_j^2) \in \mathrm{Op}\psi\boldsymbol{h}_{-\infty}$, by prop.1.1. As a consequence, $\lambda_j^1 B_j \lambda_j^2 A =$

$= \lambda_j^1 B_j \lambda_j^2 A_j \lambda_j^3 + \lambda_j^1 B_j \lambda_j^2 A (1-\lambda_j^3) = \lambda_j + K_j$, using prop.4.3 with $\theta=\lambda_j^2$ $\lambda=1-\lambda_j^3$. Assuming all left parametrices B_j to belong to the same $\psi\boldsymbol{h}_{m',\rho,\delta}$, $m' \geq -m$, it follows that $B=\Sigma\lambda_j^1 B_j \lambda_j^2 \in \Lambda\Sigma_{m',\rho,\delta}$ is a left K-parametrix of the operator $A \in \boldsymbol{LS}$. Involving adjoints to possibly convert right into left parametrices, we have proven:

<u>Theorem 4.5</u>. Let $A \in \boldsymbol{LS}_{m,\rho,\delta} \subset \boldsymbol{LS}_m$. For an admissible partition $\Sigma\lambda_j = 1$, $\lambda_j \prec \theta_j \prec U_j$ with admissible cut-offs θ_j and charts $\omega: U_j \to \mathbb{R}^n$, let the local operators A_{θ_j} of (4.3') admit left (right) K-parametrix in $\psi\boldsymbol{h}_{m',\rho,\delta}$, (with inverse order m') with respect to λ_j , for all j. Then A admits a left (right) K-parametrix of order m' .

A ψdo $A \in \boldsymbol{LS}_{m,\rho,\delta} \subset \boldsymbol{LS}$ is said to <u>be md-elliptic</u> (of order m) if for every admissible cut-off $\theta \prec U$, $\omega: U \to \mathbb{R}^n$ an admissible chart, there exists $\theta_0 \succ \theta$, $\theta_0 \prec U$, such that the local symbol a_{θ_0} is md-elliptic (of order m) with respect to θ .

<u>Theorem 4.6</u>. An md-elliptic ψdo $A \in \boldsymbol{LS}_{m,\rho,\delta} \subset \boldsymbol{LS}_m$ admits a parametrix $B \in \boldsymbol{LS}_{-m,\rho,\delta}$. Vice versa, if $A \in \boldsymbol{LS}_{m,\rho,\delta} \subset \boldsymbol{LS}_m$ admits a K-parametrix in $\boldsymbol{LS}_{-m,\rho,\delta}$ then it is md-elliptic.

This theorem is an immediate consequence of thm.4.5, as far as sufficiency of the condition is concerned. For necessity one must use local calculus of ψdo's .

<u>Theorem 4.7</u>. Suppose $A \in \boldsymbol{LS}_{m,\rho,\delta} \subset \boldsymbol{LS}_m$ has the property that, for some $m' \geq -m$ and some admissible finite partition $1=\Sigma\lambda_j$, $\lambda_j \prec \theta_j \prec U_j$, with admissible θ_j , $\omega_j: U_j \to \mathbb{R}^n$, the local operator A_{θ_j} of (3.3) is formally md-hypo-elliptic (of inverse order m') with respect to λ_j. Then A admits a K-parametrix $B \in \boldsymbol{LS}_{m',\rho,\delta}$ of order m' .

This theorem is an evident consequence of thm.4.5.

<u>Problems</u>. 1) The operators $B: C^\infty(S^1) \to C^\infty(S^1)$ of III,4,pbm.1 are ψdos in $\boldsymbol{LS}(S^1)$, under proper assumptions on the sequence $b_j(\theta)$ of periodic functions. 2) Consider ψdo's A with operator-valued symbol $A(x,\xi)$, where $B=A(x,\xi)$, for fixed x,ξ , is a ψdo in $\boldsymbol{LC}_0(\boldsymbol{B})$, for a

smooth compact manifold B. As in I,6, pbm's 1-4 and II,3, pbm's 1-4 such operator should act on $S(C)$, $C=B\times\mathbb{R}^n$, properly defined. We should get a ψdo-calculus, a K-parametrix construction, and, generally, results analogous to those in the problems mentioned, including those of $\mathcal{C}$ in [C₂],VIII. 3) Reflect on a theory of ψdo's on a more general Riemannian manifold Ω which has conical amd cylindrical ends both, as well as conical tips: In a subdomain Ω'_c containing a single cylindrical end, Ω_c should identify with a neighbourhood of the right end of the cylinder of I,6, pbms.1,2,3 (or of a more general cylinder $M\times\mathbb{R}^k$, M a compact manifold). Near a conical tip Ω_z should identify with a neighbourhood of the tip x=0 of the cone Z of sec 3,pbms.3,4 (or a more general such cone $(0,\infty)\times M$, M compact). There should be an operator-valued symbol at a cylindrical end as well as at a conical tip, but not at a conical end (cf. [CDg]). 4) Show that the operators A of pbm.2 are ψdo's on the non compact manifold $C=B\times\mathbb{R}^n$ insofar as, for cut-off's ω,χ with support in a chart $U\subset\subset\Omega$, we have $\omega A\chi$ a ψdo on $\mathbb{R}^n$. 5) For a distribution $u\in S'(\Omega)$, Ω a manifold with conical ends, define the concepts of WF(u) - the wave front set - and ZF(u), looking at $\genfrac{}{}{0pt}{}{WF}{ZF}(\lambda_j u)$ of II,6 and VI,7, with an admissible $1=\Sigma\lambda_j$, $\lambda_j<\varepsilon)U_j$, $\omega_j:U_j\to\mathbb{R}^n$. A wave fron space $\mathbb{W}(\Omega)$ may be introduced such that $WF(u)\subset\mathbb{W}(\Omega)$. Show that $\mathbb{W}(\Omega)$ may be interpreted as the cosphere bundle $S^*(\Omega)$, i.e., the bundle of unit spheres in the cotangent space, with respect to any (conical) Riemannian metric on Ω . What would be the corresponding space $\mathbb{Z}(\Omega)$, for ZF(u) ? .

5. Order classes and Green inverses on S-manifolds.

On an S-manifold Ω introduce the weighted Sobolev norms

$$(5.1)\qquad \|u\|_s = \{\sum\|\lambda_j u\|_s^2\}^{1/2} , \quad u\in S'(\Omega) , \ s=(s_1,s_2)\in\mathbb{R}^2 ,$$

with $1=\sum\lambda_j$ a given fixed finite admissible partition of unity subordinated to an admissible atlas $\{U_j\}$. To be precise, we abbreviated $\|\lambda_j u\|_s=\|(\lambda_j u)\circ\omega_j^{-1}\|_s= \|\Pi_s((\lambda_j u)\circ\omega_j^{-1})\|_{L^2(R^n)}$. The ($L^2$-Sobolev-) space $H_s=H_s(\Omega)$ consists of all $u\in S'(\Omega)$ with finite norms $\|u\|_s$. Generally, when writing $\|u\|_s$ we imply $\|u\|_s<\infty$, i.e., $u\in H_s$.

Again we regard the maps ω_j as identifications, hence $\lambda_j u$ as functions on $\mathbb{R}^n$ (extended 0 outside $\omega_j(U_j)$). Notice that

$$(5.2)\qquad \|\lambda_j u\|_s^2 = (u,\lambda_j\Pi_s^*\Pi_s\lambda_j u) , \text{ as } u\in H_s ,$$

with the pairing III,(3.28), or, with the inner product of $L^2(\mathbb{R}^n)$, as $u\in \mathcal{H}_{2s}$, where $Q_j=\lambda_j\Pi_s^*\Pi_s \in Op\psi c_{2s}$. Hence we get

(5.3) $\|u\|_s=(u,P_{2s}u)$, $P_{2s}=\sum\kappa_j Q_j\lambda_j$, $\kappa_j=d\mu/dx$, for all $u\in \mathcal{H}_{2s}$,

with the inner product $(u,v)=\int \bar{u}v d\mu$ of $\mathcal{H}=L^2(\Omega,d\mu)$. Clearly P_{2s} is a ψdo , $P_{2s}\in \mathcal{LS}_{2s,e,0} = \mathcal{LC}_{2s}$, as follows by comparing (5.3) and (4.6). Moreover, P_{2s} is md-elliptic of order 2s, as confirmed by looking at the local symbol: At each point it is a finite sum of transforms of the non-vanishing symbols of $Q_j\lambda_j$, all of them locally md-elliptic and nonnegative, so that they cannot cancel each other, looking at formula (3.18) . It follows that $P_{2s}P_{-2t}$ and $P_{-2t}P_{2s}$ both are md-elliptic of order 2s-2t, by calculus of ψdo's.

From estimates III,3,(3.19),(3.21) we conclude at once that

(5.4) $\mathcal{H}_t\subset \mathcal{H}_s$, and $\|u\|_s\le c_{s,t}\|u\|_t$, as $s \le t$, $u\in \mathcal{H}_t$.

Also one at once confirms III,(vii), i.e.,

(5.5) $\mathcal{S}(\Omega)= \bigcap_s \mathcal{H}_s(\Omega)$, $\mathcal{S}'(\Omega) = \bigcup_s \mathcal{H}_s(\Omega)$,

as well as formulas similar to (3.22) through (3.27) in ch.III .

In detail, we have

(5.6) $\mathcal{H}_s=\{u\in \mathcal{S}' : \lambda_j u \in \mathcal{H}_s(\mathbb{R}^n) , j=1,\dots,n\}$.

This validates III,2,(i)-(v),(vii) for $\mathcal{H}_s(\Omega)$, just as for $\Omega=\mathbb{R}^n$. Similar arguments imply that, again, $\mathcal{S}$ is dense in every $\mathcal{H}_s$.

It should be desirable to obtain an isometry $\mathcal{H}_s\to\mathcal{H}$, just as Π_s of (3.2) for $\Omega=\mathbb{R}^n$. Note that P_{2s}:dom $P_{2s} = \mathcal{H}_{2s}\to \mathcal{H}=L^2(\Omega,d\mu)$ may be regarded as an unbounded hermitian positive definite operator of $\mathcal{H}$, assuming $s_j\ge 0$. It is found that the positive square root of the Friedrichs extension of P_{2s} (or its continuous extension to $\mathcal{H}_s$) provides this isometry. Instead of engaging in an argument to show that this operator is a ψdo in $\mathcal{LC}_s$ we prefer to use the operator $P_s\in \mathcal{LC}_s$ instead. While P_s generally is not an isometry we will show that it is an isomorphism $\mathcal{H}_s \leftrightarrow \mathcal{H}$ at least.

Indeed, it already was seen that $P_s:\mathcal{H}_s \to \mathcal{H}$. In particular,

(5.7)
$$\|P_s u\|^2 = \|\sum\lambda_j\Pi_{s/2}^*\Pi_{s/2}\lambda_j u\|^2\le c\sum\|\Pi_{s/2}^*\Pi_{s/2}\lambda_j u\|^2$$
$$\le c\sum\|\Pi_s\lambda_j u\|^2 = c\|u\|_s^2 , \quad u\in \mathcal{H}_s ,$$

using III, thm.1.1 on $\Pi_{s/2}^*\Pi_{s/2}\Pi_s^{-1} \in Op\psi c_0$ to show that $P_s:\mathcal{H}_s\to\mathcal{H}$ is continuous for every s. As md-elliptic operator P_s admits a K-parametrix Q_{-s}, a ψdo in LC_{-s}. If $u\in H$, $P_s u=0$, then $0=Q_{-s}P_s u=u-K_s u$, $K_s \in LS_{-\infty}$, $\Rightarrow u=K_s u\in S \Rightarrow (u,P_s u)=\sum\|\Pi_{s/2}\lambda_j u\|^2=0$, $\Rightarrow \lambda_j u=0 \Rightarrow u=0$, showing that $P_s:H_s\to H$ is 1-1. In H let $f\in H$ be ⊥ to im P_s. Then $(f,P_s\varphi)=0$, $\varphi\in S \Rightarrow 0=(f,P_sQ_{-s}\psi)=(f,(1-L_s)\psi)$, $\psi\in S$, where $L_s\in LS_{-\infty}$. Hence $(f-L_s^*\varphi,\psi)=0$, $\psi\in S \Rightarrow f=L_s^*\varphi\in S \Rightarrow 0=(f,P_s f) =\sum \|\Pi_{s/2}\lambda_j f\|^2 \Rightarrow f=0$. It follows that $P_s:H_s\to H$ has dense image P_s. Then $P_s^{-1}:P_s\to H_s$ defines an unbounded operator from H to H_s with dense domain.

Proposition 5.1. For $s\in \mathbb{R}^2$ we have $T_s=P_{s_1}P_{-s}\in LC_0$. T_s is bounded in $H=L^2(\Omega,d\mu)$ and has a bounded inverse $T_s^{-1}:H\to H$. Moreover, $T_s^{-1}\in LC_0$ is a ψdo as well, and thus is a special K-parametrix of T_s.

Proof. We already noted that $T_s\in LC_0$ is md-elliptic, hence has a K-parametrix $R_s\in LC_0$. As ψdo's of order 0 , T_s and R_s are L^2-bounded. For example, $\|T_s u\|=\|\sum T_{s,j}\lambda_j u+Ku\|\le c\|u\|+\sum\|T_{s,j}u\|$, where III,thm.1.1, or III,thm.2.1 may be used to show $\|T_{s,j}u\|\le c\|u\|$.

Next confirm that ker T_s =0, and that im T_s is dense in H, where we mean the map $T_s:H\to H$. Indeed $T_s u=P_s P_{-s}u=0$ yields $u\in S$, using the K-parametrix R_s, then $P_{-s}u\in S\Rightarrow P_{-s}u=0\Rightarrow u=0$, so T_s is 1-1. Similarly, using R_s, show that $(f,T_s\varphi)=0$ for $\varphi\in S$ implies f=0, hence im T_s is dense. Existence of the (L^2-bounded) K-parametrix R_s amounts to existence of an inverse mod $K(H)$. Thus T_s is a Fredholm operator (cf.[C_1],App.A1,thm.4.8), it has closed range. Since ker $T_s=0$ and im T_s is dense in H , T_s is invertible, $T_s^{-1}\in L(H)$. But we have $T_s R_s=1-K_s$, $R_s T_s=1-L_s$, with K_s, $L_s\in LS_{-\infty}$. This implies $R_s= T_s^{-1}-T_s^{-1}K_s=T_s^{-1}-L_sT_s^{-1}$. Or, $T_s^{-1}=R_s+T_s^{-1}K_s$, and $T_s^{-1}=R_s+L_sT_s^{-1}$. Substitute the second into the right hand side of the first: $T_s^{-1}=R_s+R_sK_s +L_sT_s^{-1}K_s$. Here $R_s\in LS_0$, while $R_sK_s\in LS_{-\infty}$. The third term also belongs to $LS_{-\infty}$, since we get $\|L_sT_s^{-1}K_s u\|_m\le c_1\|T_s^{-1}K_s u\|\le c_2\|K_s u\|\le c_3\|u\|_{m'}$ for all m,m' . It follows that $T_s^{-1}\in LC_0$, q.e.d.

Using prop.5.1 we show that $P_s\in L(H_s,H)$ admits a continuous inverse. Indeed, $T_s=P_{s_1}P_{-s}$ implies $P_s^{-1} = P_{-s}T_s^{-1}$ where we regard the right hand side as a composition of $T_s^{-1}\in L(H)$, and $P_{-s}\in L(H,H_s)$. To confirm the latter note that $\|P_{-s}u\|_s^2=\sum\|\Pi_{-s}(\lambda_j P_{-s}u)\|^2= \sum\|\Pi_{-s}(\sum_k\kappa_k\lambda_k\Pi_{s/2}^*\Pi_{s/2}\lambda_k u)\|^2$. The point is that, for fixed j, all terms $\Pi^*_{s/2}P_{s/2}\lambda_k u$ must be taken to the coordinates of the j-th chart, using thm.3.3. and only within a neighbourhood of supp $\lambda_j \subset U_j$. They will give operators of order s , insuring L^2-boundedness of each term of the sum. Hence $\|P_{-s}u\|_s \le c\|u\|$, i.e., $P_{-s} \in L(H,H_s)$, implying $P_s^{-1} \in L(H,H_s)$. So, indeed, we have the result, below.

Proposition 5.2. The operator P_s of (5.3) are ψdo's in LS_s . There exists a ψdo $Q_{-s} \in LS_{-s}$, acting as an inverse -i.e.,

$$P_sQ_{-s} = Q_{-s}P_s = 1 \quad . \tag{5.8}$$

We have $Q_{-s}: H \to H_s$ continuous, i.e., $P_s: H_s \leftrightarrow H$ and $Q_{-s}: H \leftrightarrow H_s$ define isomorphisms between H_s and H . With constants c_s , $c'_s > 0$ we have

$$c_s\|P_su\| \le \|u\|_s \le c'_s\|P_su\| \ , \ u \in H_s \ . \tag{5.9}$$

Using the isomorphism P_s we can prove H_s-boundedness of ψdo's:

Theorem 5.3. A ψdo $A \in LS_m$ satisfies

$$\|Au\|_{s-m} \le c_s\|u\|_s \ , \text{ for all } u \in H_s \ , \ c_s \text{ independent of } u. \tag{5.10}$$

Proof. We have $\|Au\|_{s-m} \le c\|(P_{s-m}AQ_{-s})P_su\|$, where $P_{s-m}AP_s \in LS_0$ is L^2-bounded, as seen above. Thus $\|Au\|_{s-m} \le c\|P_su\| \le c'\|u\|_s$, q.e.d.

In particular thm.5.3 insures that the ψdo's $P_s \in LS_s$ and $Q_{-s} \in LS_{-s}$ are operators of $L(H_t,H_{t-s})$ and $L(H_t,H_{t+s})$ respectively, for every $t \in \mathbb{R}^2$, not only for t=s or t=0 , as known earlier.

Next we again introduce a pairing between H_s and H_{-s} by

$$(u,v) = (P_su,Q_{-s}v) = \int_\Omega \bar{u}v d\mu \ , \ u \in H_s \ , \ v \in H_{-s} \ . \tag{5.11}$$

Here P_su, $Q_{-s} \in H$, hence the middle term in (5.11) is well defined, as inner product of elements of H. It is clear also that $(u,P_su) = (P_su,u)$, $(u,Q_{-s}u) = (Q_{-s}u,u)$, as $u \in S$, so that (5.11) is meaningful.

With such preparations we reintroduce order classes, setting

$$O(m) = \{A \in L(S(\Omega)) : P_sAQ_{-s} \in L(H_m,H) \ , \text{ for all } s \in \mathbb{R}^2\} \ . \tag{5.12}$$

Or, $O(m)$ consists of all $A \in L(S)$ such that for all $s \in \mathbb{R}^2$ A extends to a continuous map $H_s \to H_{s-m}$. Then, clearly, $LS_m \subset O(m)$ for all m. We again define $O(\pm\infty)$ by III,(3.30), and get $LS_{-\infty} = O(-\infty)$. The concept of K-parametrix is meaningfull not only for ψdo's, but also for general $A \in O(\infty)$. A Green inverse of $A \in O(\infty)$ is defined as a K-parametrix $B \in O(\infty)$ with $1-AB$, $1-BA \in O(-\infty)$ of finite rank.

We now can repeat every line of argument of III,4, showing

that a Green inverse - and even a special Green inverse, with all properties III,(4.1), exists if and only if a K-parametrix exists, for $A \in O(\infty)$. This holds, because all arguments there are abstract, using only the system of Hilbert spaces H_s, the pairing (5.12), and the operator Π_s, here represented by P_s :

<u>Corollary 5.4</u>. All statements of III,thm.4.1, thm.4.2, and cor.4.3 are also valid for operators of our present order classes $O(m)$. Moreover, they still hold for operators between sections of admissible vector bundles (the latter denoting the restriction to Ω of a C^{∞}-vector bundle over Ω^0 , where evidently H_s-norms and H_s-spaces and ψdo's may be introduced for sections, in the same way).

As a final remark, we mention without detailed proof that a system of equivalent norms is given by

$$(5.13) \qquad \|u\|_s^{\cdot} = \|\langle d(x)\rangle^{s_2}(1-\Delta)^{-s_1/2}u\| \quad , \; u \in S(\Omega) \; ,$$

with the Beltrami-Laplace operator Δ of the distinguished conical metric (or any conical metric): With constants c_s, c_s' we have

$$(5.14) \qquad c_s\|u\|_s \le \|u\|_s^{\cdot} \le c_s'\|u\|_s \; , \; u \in S \; .$$

In particular, the system of spaces H_s and the order classes $O(m)$ are independent of the partition of unity and atlas chosen.

As an argument leading into (5.13), (5.14): For $s_1 \in \mathbb{Z}$, $s_1 > 0$, (5.14) follows immediately. For other $s_1 > 0$ use an interpolation argument, as for the proof of III,prop.3.3. For $s_1 < 0$ one uses calculus of ψdo's in combination with the above.

Let us not forget to mention that $O(0)$ is a Frechet algebra again, with topology induced by the operator norms in $L(H_s)$. Use Calderon interpolation as in III,prop.3.3 to show this.

<u>Problems</u>. 1) Connecting to the problems of IV,4, introduce order classes and Green inverses for ψdo's on manifolds with all three, conical and cylindrical ends, and conical tips. Note, there will be a difference in technique of constructing a Green iverse, once a K-parametrix exists: md-ellipticity is not enough ([C₃],VIII).

Chapter 5. ELLIPTIC AND PARABOLIC PROBLEMS

In this chapter we take up the lead of ch.0, sec.4, with regard to elliptic and parabolic problems. There we applied the Fourier-Laplace method to free-space problems of elliptic equations, and to evolutionary half-space problems of the (parabolic) heat equation, all with constant coefficients. We covered Dirichlet and Neumann problems in a half-space, for elliptic equations.

With the tools developed in I, II, III, IV we now can give a similar "Fourier-Laplace treatment" to much more general variable coefficients elliptic and parabolic problems. This may be done in "free space" (that is, in $\mathbb{R}^n$, or on a smooth compact manifold Ω or on a noncompact Ω with conical ends - but without the presence of boundary points). Such results are special cases of theorems on Green inverses of ψdo's already discussed, but they will be summarized (in more general form) in sec.1, below. If Ω is compact we need ellipticity, else md-ellipticity of the operator. Not only (md-) elliptic operators on a complex-valued function but even maps between crosssections of vector bundles are considered.

Note that there is a different approach - a functional analysis approach - to these theorems, not using ψdo's at all. Elliptic theory, in its beginnings, was developed for 2-nd order equations. Such 2-nd order theory is of dominating importance for many physical applications. The Laplace, Helmholtz and Schroedinger operator each has its own well developed theory.

One finds that virtually <u>all</u> results we state for $\mathbb{R}^n$ can be reached by focusing on the C^*-subalgebra A of $L(L^2(\mathbb{R}^n))$ generated by the multiplications $a(M):u\rightarrow au$, $a\in C(\mathbb{B}^n)$ (cf.II,3) and the operators $D_j(1-\Delta)^{-1/2}=S_j$, defining $S_j=s_j(D)=F^{-1}s_j(M)F$ as Fourier multiplier or else $(1-\Delta)^{-1/2}$ as inverse positive square root of the unique self-adjoint realization of $1-\Delta\geq 0$, $\Delta=\sum\partial^2_{x_j}$ ([C₁],III,IV).

Similarly, in case of a general Riemannian manifold with conical and cylindrical ends, and conical tips (cf. IV,4,5, and the problems of IV,4) we may generate such C^*-algebra from $D(1-\Delta)^{-1/2}$ with the Beltrami-Laplace operator Δ of Ω under the metric discussed for such manifolds, with $D\in \mathcal{D}^{\#}$, and a(M) . $a\in \mathcal{A}^{\#}$, with suitable classes $\mathcal{A}^{\#}$ of functions and $\mathcal{D}^{\#}$ of FOLPDE's on Ω . Such approach to elliptic theory is discussed in detail in [C_1] and [C_2]. Here we look at this in sec.10, only to clarify some questions comparing both the ψdo- and the C^*-algebra approach.

The C^*-algebra approach is of importance not only for elliptic but also to hyperbolic theory (of ch's VI and VII):Conjugation with the evolution operator e^{iLt} of a first order hyperbolic equation defines an automorphism of the above algebra $\mathcal{A}$. The dual of this automorphism will be a Hamiltonean flow in symbol space determining propagation of singularities, as in Egorov's theorem.

In sec.2 we start focusing on the general elliptic boundary problem for smooth boundaries under Lopatinski-Shapiro type boundary conditions, but only for compact $\Omega\cup\Gamma \subset\subset \mathbb{R}^n$, for simplicity. First we discuss common facts, such as reduction to the case of a homogeneous boundary conditions or a homogeneous PDE.

In ch.0 we used a reflection principle to convert a Dirichlet or Neumann problem over a half-space into a free-space problem. Similarly, for a general boundary problem, here we will extend all functions from $\Omega\cup\Gamma$ into all $\mathbb{R}^n$, and the elliptic operator to an md-elliptic operator on $\mathbb{R}^n$ (sec.3). The boundary problem will become a 'Riemann-Hilbert type problem' involving (genuine) distributions. We get a type of distribution best classified as 'multi-layer potentials'. In potential theory one uses a single- or multi-layer ansatz, using the same kind of distribution for solving a boundary problem. Here we find that <u>every</u> solution necessarily is a sum of a C^∞-function and certain multilayer potentials.

Actually, in sec. 4, we discuss a result we call <u>boundary hypoellipticity</u>: Just as $u\in C^\infty$ for $f\in C^\infty$ follows for a distribution u solving Lu=f for a hypo-elliptic L , inside Ω , we will show that, if f admits a certain asymptotic expansion near the boundary so must u satisfy the same kind of expansion, under proper assumptions on L (i.e., hypo-ellipticity, and that Γ is non-characteristic for L).This will be very useful not only for the elliptic problem but also for parabolic problems, later on.

In sec.6 we then discuss existence and uniqueness (again in finite dimensional degeneration- i.e., normal solvability) of the elliptic boundary problem, if the boundary conditions are of 'Lopatinski-Shapiro type'. Of course we use the "multi-layer ansatz". We only look at the simplest nontrivial case: A single even order equation. But the generalization to operators (of even or odd order) between vector bundles should be fairly evident; also extension to boundary problems for subdomains of an *S*-manifolds should involve no new ideas, under proper assumptions on Ω and Γ.

The multi-layer ansatz (for a homogeneous PDE) will convert the boundary conditions into a system of ψdo's on Γ. Ellipticity of that system will amount precisely to the L.-S.-conditions. Actually, for existence and uniqueness considerations we need two different such elliptic systems, one $n/2\times n/2$-system, the other $n\times n$.

In sec.7 we turn to the parabolic initial-boundary problem. A parabolic problem is evolutionary - we assume an equation $\partial u/\partial t= Lu$ with a differential operator L under initial conditions at t=0. The evolution operator $U(t)=e^{itL}$ is well defined as a semigroup. For a parabolic problem the operator $\partial/\partial t-L$ is hypo-elliptic. For existence of the semigroup we assume standard results, such as the Hille-Yosida theorem. Boundary-hypo-ellipticity (sec.4) will be usefull to secure C^{∞}-solutions instead of distribution solutions.

An efficient use of results such as the Hille-Yosida theorem in sec.7, will require investigation of $R(\lambda)=(L-\lambda)^{-1}$ of an elliptic operator, as will spectral theory of elliptic PDO's. Thus, in sec's 8-9, we look at spectral theory and $R(\lambda)$, for elliptic L.

We only consider the cases $\Omega=\mathbb{R}^n$, Ω = compact manifold , Ω = manifold with conical ends, $\Omega\subset\subset\mathbb{R}^n$ a domain with smooth boundary. Extensions to (i) systems of equations (ii) differential operators between sections of vector bundles of equal dimension, (iii) exterior boundary problems, (iv) Riemann-Hilbert type problems are fairly evident, but are left for the reader to deploy.

Existence of a Green inverse for operators defined with boundary conditions is discussed as auxiliary result (thm.8.3), extending results of sec.6. Mainly we focus on compactness of the resolvent, under proper assumptions. Results on selfadjointness and dissipativity of differential operators (making thm's 7.3, 7.5 applicable) are discussed under various assumptions (thm.8.5 for Ω without boundary, thm.9.1 for $\Omega\subset\subset\mathbb{R}^n$ with boundary). In all cases proper deployment of the results is left to functional analysis - insofar as the spectral theorem , the Hille-Yosida theorem, or existence of an orthonormal base of eigenfunctions for self-adjoint operators with compact resolvent is not discussed in detail.

In sec.10, finally, we discuss the C^*-subalgebra $\mathcal{A}$ of $\mathcal{L}(\mathcal{H})$, $\mathcal{H}=L^2(\mathbb{R}^n)$, generated by our algebra Opψc₀ of ch.I. The point: $\mathcal{A}^{\vee}$ = $\mathcal{A}/\mathcal{K}(\mathcal{H})$ is a commutative C^*-algebra isometrically isomorphic to $C(\partial\mathbb{P})$, with the boundary $\partial\mathbb{P}$ of a certain compactification $\mathbb{P}$ of $\mathbb{R}^{2n}$ A corresponding result holds for the norm closures in $L^2(\Omega)$ of $\mathcal{LS}_{0,\rho,\delta}$, $\delta<\rho_2$, of IV,4, with similar proof, not discussed here.

1. Elliptic problems in free space; a summary.

In the present section we summarize (and trivially extend) results on elliptic equations, previously discussed in the more general context of pseudodifferential equations. "Free space" is interpreted as "no boundary" - a compact manifold is a free space.

Theorem 1.1. Given an elliptic differential operator $A:C^{\infty}(\Omega)\to C^{\infty}(\Omega)$ of order N, on a compact C^{∞}-manifold Ω (of dimension n) without boundary. That is, in coordinates of a chart $U\subset\Omega$, for $u\in C_0^{\infty}(U)$,

$$(1.1)\qquad Au(x) = \sum_{|\alpha|\le N} a_{\alpha}(x)D^{\alpha} \ , \ a_{\alpha}\in C^{\infty}(U) \ ,$$

with

$$(1.2)\qquad a_N(x,\xi) = \sum_{|\alpha|=N} a_{\alpha}(x)\xi^{\alpha} \ne 0 \ , \text{ as } |\xi|=1 \ , \ x\in U \ .$$

Assertion: (1) The differential equation

$$(1.3)\qquad Au=f \quad , \ f \in C^{\infty}(\Omega) \ ,$$

is normally solvable, for $u\in C^{\infty}(\Omega)$. That is, (i) the homogeneous equation Au=0 admits at most finitely many linearly independent solutions; (ii) there exists a solution $u\in C^{\infty}(\Omega)$ of Au=f if and only if f satisfies finitely many linear conditions of the form

$$(1.4)\qquad \int_{\Omega} d\mu\psi_j u = 0 \quad , \ j=1,\dots,j_0 \ .$$

with any positive C^{∞}-measure $d\mu$ on Ω and certain $C^{\infty}(\Omega)$-functions ψ_j [In fact, the ψ_j may be chosen as a basis of ker A^* , with the adjoint A^* of A with respect to the inner product $\int_{\Omega}\bar{u}vd\mu$] .

(2) The operator $A:C^{\infty}(\Omega)\to C^{\infty}(\Omega)$ admits a special Green inverse $G \in LS_{-N} = LS_{(-N,0),(1,0),0}$, the latter as in IV,4, with

$$(1.5)\qquad GA=1-P \ , \ AP=1-Q \ ,$$

where P is a projection onto ker A, annihilating some (arbitrarily chosen) complement of ker A , and Q a projection onto an (arbitrarily chosen) complement of im A annihilating im A .

Proof. The compact manifold Ω is a special case of S-manifold studied in ch.IV - there are no conical ends at all. Since there is no infinity, the classes $LS_{m,\rho,\delta}$ are independent of m_2 , ρ_2 ; We thus write $LS_{m_1,\rho_1,\delta}$, and LC_{m_1} in case of $\rho_1=1$, $\delta=0$.

Also, since there is no infinity, md-elliptic means the same as 'elliptic' - i.e., cdn.(1.2) above. Thus the existence of a K-parametrix in LS_{-N} follows from IV,thm.4.6. The special Green inverse G may be constructed as in III,4 (cf.IV,cor.5.4). Once we have G , the normal solvability (1) is immediate.

Thm.1.1 generalizes immediately to the case of an elliptic differential operator $A:\mathcal{X}\to\mathcal{Y}$, with $\mathcal{X},\mathcal{Y}$ the spaces of C^{∞}-sections of two vector bundles E,F over Ω with $\dim E = \dim F = r_0$. We will not discuss details of the proof of thm.1.2, below: They are just formal extensions of earlier discussions.

Theorem 1.2. Given an elliptic differential operator $A:\mathcal{X}\to\mathcal{Y}$, with $\mathcal{X},\mathcal{Y}$ as described above, A of order N , on a compact C^{∞}-manifold Ω of dimension n without boundary. That is, in local coordinates of a chart $U\subset\Omega$, for $u\in\mathcal{X}$ with $\operatorname{supp} u\subset U$, we have (1.1) , where now $a_{\alpha}(x)$ are $r_0\times r_0$-matrix-valued, while (1.2) is replaced by

(1.6) $$\det a_N(x,\xi)\neq 0 \quad , x\in U \, , |\xi|=1 \; .$$

Then the assertions of thm.1.1 hold again, as follows:

(1) The equation

(1.7) $$Au=f \;, f\in\mathcal{Y} \;,$$

is normally solvable , as $u\in\mathcal{X}$ - that is, $\dim\ker A<\infty$, $\operatorname{codim}\operatorname{im} A<\infty$.

(2) There exists a special Green inverse G , locally an $r_0\times r_0$-matrix of ψdo's in LC_{-N} , such that (1.5) holds.

Proof: See the remarks above.

If the manifold Ω without boundary no longer is compact, but, at infinity, still is conical - in the sense of ch.IV , then our theory of earlier chapters yields results of similar structure but only if (i) the space $C^{\infty}(\Omega)$ is replaced by a smaller space - $\mathcal{X}(\Omega)$ still containing all of $C_0^{\infty}(\Omega)$; (ii) the coefficients a_{α} of (1.1) and their derivatives satisfy growth restrictions at ∞ ; (iii) we supplement the ellipticity condition (1.2) (or 1.6)) by a condition at infinity (m-ellipticity).

A useful choice for $\mathcal{X}(\Omega)$ is the space $H_{\infty,\sigma}=\cap\{H_s(\Omega):s_2=\sigma\}$, for any $\sigma\in\mathbb{R}$. Then, to get a normally solvable operator $A:\mathcal{X}\to\mathcal{X}$, we may choose $a_{\alpha}(x)$ as restrictions to U of functions in the symbol class $\psi h_{0,e,0}$, whenever the chart U gets near a point of Ω_{∞} .

(That is, the full local symbol $a(x,\xi)=\sum_{|\alpha|\le N} a_{\alpha}(x)\xi^{\alpha}$ belongs to

$\psi c_{(N,0)}$.) Finally, instead of (1.2) we ask for

(1.8) $$|a(x,\xi)| \geq c\langle\xi\rangle^N \ , \text{ as } |x|+|\xi|\geq \eta \ , \text{ with some } c,\eta>0 \ .$$

That is, A must be md-elliptic of order (N,0) .

More generally we may regard operators $A:X\to Y$, $X=H_{\infty,\sigma}(\Omega)$, $Y=H_{\infty,\tau}(\Omega)$, or even let X,Y be spaces of smooth sections of S-admissible vector bundles of the same dimension r_0 , with components in $H_{\infty,\kappa}$, $\kappa=\sigma,\tau$, again. Let us summarize this in thm.1.3, below, again a consequence of IV,thm.4.6, and III,4 (i.e., IV,cor.5.1).

<u>Theorem 1.3</u>. Given an md-elliptic differential operator $A:X\to Y$, with the spaces X, Y of C^∞-sections on two S-admissible vector bundles E, F, dim E =dim F =r_0 , with components in $H_{\infty,\sigma}$ and $H_{\infty,\tau}$, respectively. Let A be a differential operator , $A \in LC_{(N,\kappa)}$, where $\kappa=\sigma-\tau$, and let A be md-elliptic of order (N,κ) . That is, in the coordinates of an S-admissible chart $U\subset\Omega$, we have (1.1),

(1.9) $$a_\alpha^{(\beta)}(x) = O(\langle x\rangle^{\kappa-|\beta|}) \ , \text{ for all } \beta \ ,$$

and (with the $r_0\times r_0$-matrix norm $|.|$, and some $c,\eta>0$)

(1.10) $$|a(x,\xi)|=|\sum_{|\alpha|\leq N} a_\alpha(x)\xi^\alpha| \geq c\langle x\rangle^\kappa\langle\xi\rangle^N \ , \text{ as } |x|+|\xi|\geq\eta \ .$$

Then the differential equation

(1.11) $$Au=f \ , \ f\in Y \ ,$$

is normally solvable, for $u\in X$ - i.e., dim ker A $<\infty$, codim im A $<\infty$. There exists a special Green inverse $G\in LS_{(-N,-\kappa)}$ such that (1.5) holds, with projections P,Q as described.

<u>Problems</u>. Consider the paraboloid $P:x_1=x_2^{\ 2}+x_3^{\ 2}$ and the one- and two-shell hyperboloids $B_\pm:x_1^{\ 2}-x_2^{\ 2}-x_3^{\ 2}=\pm1$ in $\mathbb{R}^3$. Let Δ be the Beltrami-Laplace operator of the Riemannian metric induced in P and $B_\pm$ by the Euclidean metric of $\mathbb{R}^3$. 1) Show that the natural S-structure of $\mathbb{R}^3$ discussed in IV,1 also induces an S-structure in $B_\pm$:An atlas of admissible charts may be obtained by restricting admissible charts of $\mathbb{R}^3$ to $B_\pm$.) With this S-structure, $1-\Delta$ is md-elliptic. 2) Discuss the corresponding facts for the paraboloid P .

<u>2. The elliptic boundary problem</u>.

In this section we start a discussion of the elliptic boundary problem on compact domains with smooth boundary.

Consider a compact subdomain $\Omega\subset\mathbb{R}^n$ with boundary $\Gamma=\partial\Omega\subset\mathbb{R}^n$, where $\Gamma=\cup\Gamma_j$ is a finite disjoint union of compact smooth n-1-dimensional submanifolds of $\mathbb{R}^n$. Let $A=a(x,D)$, $B_j=b_j(x,D)$, j=1,.. ..M, be differential operators of orders N, N_j, with smooth coefficients, defined near $\Omega\cup\Gamma$ and near Γ, respectively. Assume complex (globally defined) coefficents, for the moment, - i.e.,

$$A=\sum_{|\alpha|\le N} a_\alpha(x)D^\alpha \ , \ B_j=\sum_{|\alpha|\le Nj} b^j_\alpha(x)D^\alpha \ , \tag{2.1}$$

where $a_\alpha\in C^\infty(\mathcal{N})$, $b^j_\alpha\in C^\infty(\mathcal{N}_j)$, $\Omega\cup\Gamma\subset\subset\mathcal{N}$, $\Gamma\subset\subset\mathcal{N}_j$, $\mathcal{N},\mathcal{N}_j\subset\mathbb{R}^n$.

For given functions $f\in C^\infty(\Omega\cup\Gamma)$, $\varphi_j\in C^\infty(\Gamma)$ one seeks to find a function $u\in C^\infty(\Omega\cup\Gamma)$ satisfying the equations

$$Au=f \ , \ x\in\Omega\cup\Gamma \ , \quad B_ju=\varphi_j \ , \ x\in\Gamma \ , \ j=1,\dots,M \ . \tag{2.2}$$

The problem of finding u for given f, φ_j (or the discussion of existence and uniqueness of such u) is called a boundary problem; the first relation Au=f is called the differential equation, while the relations $B_ju=\varphi_j$, required for $x\in\Gamma$ only, are called the <u>boundary conditions</u> of the problem.

We recall the definition of $C^\infty(\Omega\cup\Gamma)$ and $C^\infty(\Gamma)$: $u\in C^\infty(\Omega\cup\Gamma)$ <u>extends</u> to a $C^\infty(\mathcal{N})$-function for some $\Omega\cup\Gamma\subset\subset\mathcal{N}$. Similarly for Γ.

<u>Example 2.1</u>. Consider the Laplace operator $A=\Delta=\sum_{j=1}^n\partial^2_{x_j}$ (i.e., N=2), with M=1, $B_1=1$ (i.e., $B_1(u)=u$). The boundary problem (2.2) with this choice of A and B_j is called the <u>Dirichlet problem</u> of the Laplace equation. It is well known in potential theory, and has many physical applications.

<u>Example 2.2</u>. Problem (2.2) with $A=\Delta$ as in Ex'le 5.1, but $B_1=B_M=\partial_\nu=\sum\nu_j\partial_{x_j}$ (where $\nu=(\nu_1,\dots,\nu_n)$ denotes the exterior unit normal of Γ) is called the <u>Neumann problem</u> of the Laplace equation. It has many applications as well. More generally, for

$$B_1=B_M=\partial_\nu+h \ , \ h\in C^\infty(\Gamma) \ , \tag{2.3}$$

(one often assumes h(x)>0) one gets Hilberts boundary problem <u>of the third kind</u>, known for problems of heat conduction for example.

Physical applications are not restricted to 2-nd order equation problems. For example, the case of N=4, $A=\Delta^2$, under various choices of B_j (such as $B_1=1$, $B_2=B_M=\partial^2_\nu$) describes a loaded elastic plate over Ω, under various restraints at the boundary Γ.

Often a different problem is of interest, of the form

$$(2.4)\qquad Au=f\ ,\ x\in \mathbb{R}^n\setminus\Omega\ ,\ B_j u=\varphi_j\ ,\ x\in\Gamma\ ,\ j=1,\dots,M,\ u\in \mathcal{B}_\infty\ ,$$

where now A and f are defined (and C^∞) in the noncompact closed set $\mathbb{R}^n\setminus\Omega$. Here $\mathcal{B}_\infty$ denotes a linear set of functions defined near infinity, containing all $u\in C_0^\infty(\mathbb{R}^n)$, so that "$u\in \mathcal{B}_\infty$" amounts to a (set of) conditions at infinity. Also, to arrive at a well posed (or normally solvable) problem one must replace the condition $f\in C^\infty(\mathbb{R}^n\setminus\Omega)$ by a stronger one "$f\in C^\infty(\mathbb{R}^n\setminus\Omega)\cap\mathcal{X}_\infty$ with some space $\mathcal{X}_\infty$ satisfying the above conditions for $\mathcal{B}_\infty$.

(2.4) is often called an exterior boundary problem. As example consider $A=\Delta$, $B_1=B_M=1$, $\mathcal{B}_\infty=\{u\in C_\infty(\mathbb{R}_n): u(x)=O(\frac{1}{|x|})\}$, leading to the <u>exterior</u> <u>Dirichlet</u> <u>problem</u> of the Laplace equation.

The examples given all use an elliptic expression $A=a(x,D)$. Here we used $A=D$ and $A=D^2$. Other examples, using a general elliptic $A=\sum a_{jk}(x)\partial_{x_j}\partial_{x_k}+A_1$ with A_1 of first order, $a_{jk}(x)$ real, C^∞, $((a_{jk}(x)))$ a positive definite n×n-matrix, are easily quoted.

For the moment we shall focus on (2.2), also referred to as <u>interior</u> <u>problem</u>, assuming that A is elliptic in $\Omega\cup\Gamma$. A fortiori, we assume that A extends to an md-elliptic expression on $\mathbb{R}^n$:

$$(2.5)\qquad A=a(x,D)\ ,\ a=\{c|N\}\ ,\ \text{where } c=\sum_{|\alpha|\le N} c_\alpha\xi^\alpha\in\psi c_{N,m_2}\ ,$$

is md-elliptic of order (N,m_2), for some $m_2\in\mathbb{R}$.

(2.5) will be convenient but is not essential. It trivially holds after restricting A to any small ball, hence will disappear if the theory is drawn up for general manifolds with boundary, as we shall not do here. Note that, under (2.5), interior <u>and</u> exterior boundary problem both are meaningful, once $\mathcal{B}_\infty$ and $\mathcal{X}_\infty$ are fixed.

If, for a moment, we admit an empty domain $\Omega=\emptyset$, then the exterior problem fits into our earlier theory of md-elliptic problems on $\mathbb{R}^n$: Choosing $\mathcal{B}_\infty=\mathcal{H}_s$ - for any given fixed $s=(s_1,s_2)$, problem (2.4) becomes normally solvable if we require in addition that $f\in\mathcal{H}_{s-(N,m_2)}=\mathcal{X}_\infty$. Accordingly we tend to think of similar spaces $\mathcal{B}_\infty$, $\mathcal{X}_\infty$, also for non-empty Ω and Γ.

One will expect that the B_j must satisfy certain conditions before solution of (2.2) or (2.4) can be attempted. A (somewhat complicated) set of conditions called <u>Lopatinski</u>-<u>Shapiro</u> <u>conditions</u> - will be introduced below. Then we will prove normal solvability of (2.2) : u exists for all f, φ_j satisfying finitely many

For $f=\varphi_j=0$ the space of solutions is finite dimensional.

In other words, with an elliptic A and a system $\{A,B_1,\dots B_M\}$ satisfying the Lopatinskij-Shapiro conditions, the linear map

$$(2.6)\qquad A\times B_1\times\dots\times B_M : C^\infty(\Omega\cup\Gamma) \to C^\infty(\Omega\cup\Gamma)\times C^\infty(\Gamma)\times\dots\times C^\infty(\Gamma) ,$$

defined by $u \to (Au,B_1u,\dots,B_Mu)$ is a Fredholm map.

Apart from existence and uniqueness we will pursue a question parallel to hypo-ellipticity, as in II,thm.4.1. For open $\Omega\subset\mathbb{R}^n$ if $u\in\mathcal{D}'(\Omega)$ has $Cu\in C^\infty(\Omega)$ then $u\in C^\infty(\Omega)$, by hypo-ellipticity of C.

In the line of solving boundary problems, a similar result, with $\Omega\cup\Gamma$ instead of Ω would be useful. Of course, $\mathcal{D}'(\Omega\cup\Gamma)$ is not defined. Using ideas of Melrose [Me_1] we replace $\mathcal{D}'(\Omega\cup\Gamma)$ by a space of extendable distributions. Then this result indeed extends, since existence of certain asymptotic expansions is preserved.

We shall refer to this as <u>boundary hypo-ellipticity</u>.

As already mentioned, a boundary problem (2.2), in its abstract setting, amounts to the problem of inverting (or Fredholm inverting) a linear map of the form (2.6). However, (2.2) proves equivalent to either of the two simpler problems arising if one either assumes $f=0$ on Ω or $\varphi_j=0$, $j=1,\dots,M$, on Γ. That is, either

$$(2.2')\qquad Au=0 \ , \ x\in\Omega\cup\Gamma \ , \ B_ju=\varphi_j \ , \ x\in\Gamma \ , \ j=1,\dots,M \ .$$

or,

$$(2.2'')\qquad Au=f \ , \ x\in\Omega\cup\Gamma \ , \ B_ju=0 \ , \ x\in\Gamma \ , \ j=1,\dots,M \ .$$

Clearly, if (2.2) is (normally) solvable then so are (2.2') and (2.2") each. Vice versa, let (2.2') be (normally) solvable. Assume <u>Condition</u> S: The differential equation $Av=f$ admits a solution $v\in C^\infty(\Omega\cup\Gamma)$ for each $f\in C^\infty(\Omega\cup\Gamma)$. Moreover, a unique such v can be assigned to every f, such that $v=Xf$ becomes a linear operator (In other words, $A:C^\infty(\Omega\cup\Gamma)\to C^\infty(\Omega\cup\Gamma)$ admits a (linear) right inverse).

Given $f\in C^\infty(\Omega\cup\Gamma)$ and a solution u of (2.2) we define $w=u-v$. Then w solves (2.2') with φ_j replaced by φ_j-B_jv. Since (2.2') is normal ly solvable we get a solution w if only φ_j-B_jXf satisfies finally many linear conditions, translating into finitely many linear conditions for f,φ_j . Clearly also the null space of (2.2) coincides with the null space of (2.2'), hence it is finite dimensional.

Next let (2.2") be normally solvable. Assume <u>Condition</u> B_S: For an arbitrary selection of $\varphi_j\in C^\infty(\Gamma)$ there exists a function $v\in C^\infty(\Omega\cup\Gamma)$ such that $B_jv=\varphi_j$ on Γ , and, again, a linear right inverse of $v \to (B_j v)$ may be constructed.

Then, for given f,φ_j , and a solution u of (2.2) set w=u-v again. Then w satisfies (2.2") with f replaced by f-Av. A solution w exists if f-Av satisfies finitely many linear conditions, etc.

When are cdn.S or cdn.B_S satisfied? For cdn B_S there is a simple answer: Observe that the normal derivatives of order 0,1,.. ... of a $C^\infty(\Omega\cup\Gamma)$-function v may be arbitrarly prescribed, and that the equations $B_j v=\varphi_j$ translate into a system of differential equations for $(v_0,\ldots,v_{N-1})$, where $v_j=\partial_\nu^j v|\Gamma$ on the boundary Γ . Cdn. B_S simply says that this linear system can be solved for every φ_j, and, moreover, that a linear right inverse can be constructed. One may check with the examples and find this trivially satisfied.

For cdn.S we use that C is md-elliptic hence has a special Green inverse G (III,thm.4.2): CG=1-P, P a finite dimensional projection onto some complement of im C, annihilating im $C=(\ker C^{\sim})^{\perp}$. Let $g\in C_0^\infty(\mathbb{R}^n)$ extend f to $\mathbb{R}^n$: g=f near $\Omega\cup\Gamma$, $g\in C_0^\infty$. Clearly

$$Pz=\sum\beta_j\langle\alpha_j,z\rangle \quad , z\in \mathcal{S}' , \tag{2.7}$$

with a basis $\{\alpha_j\}$ of ker $C^{\sim}$, $C^{\sim}$ the distribution adjoint of I,3, and a basis $\{\beta_j\}$ of the complement of im C, bi-orthogonal to $\{\alpha_j\}$. Try to select β_j such that $\beta_j=0$ near $\Omega\cup\Gamma$. We only must construct a set $\{\gamma_j\}$ of linear independent functions $\gamma_j\in \mathcal{S}$ such that $\gamma_j=0$ in $\mathcal{B}_\eta=\{|x|\le\eta\}$, for some sufficiently large η , and that no linear com bination γ of γ_j satisfies $\langle\gamma,\alpha_l\rangle=0$ for all l.

Let us first assume that we have "unique continuation" of solutions of $C^{\sim}u=0$, a property common to all second order (and much more general classes of) elliptic equations (cf. [As_1], [CU], [Hr_3],ch.28). We will not discuss such results here. In particular our present use is not essential for discussion of L-S-theory here

Under unique continuation no solution α of $C^{\sim}\alpha=0$ may vanish outside $\mathcal{B}_\eta$, except $\alpha\equiv0$. Thus a basis of ker $C^{\sim}$ is linearly independent in $X_\eta=\mathbb{R}^n\setminus\mathcal{B}_\eta$: The matrix $((\int_{X_\eta}\alpha_j\alpha_l dx))=Z$ is nonsingular.

For a suitable cut-off χ with supp $\chi\subset X_{\eta-\varepsilon}$ $\chi=1$ in X_η and $\gamma_j=\chi\alpha_j$ we get $((\langle\gamma_j,\alpha_l\rangle))$ close to Z , thus nonsingular as well. Thus γ_j give the desired functions.

For P of (2.7) with α_j , β_j as constructed and the corresponding Green inverse G let w=Gg. Then $Cw=CGg=g-\sum\beta_j\langle\alpha_j,g\rangle=f$ for x near $\Omega\cup\Gamma$. Thus a v solving Cv=f near $\Omega\cup\Gamma$ exists: Set v=w in $\mathcal{N}(\Omega\cup\Gamma)$. Moreover, the above construction has given a right inverse of A , as required, and cdn. S always holds, under above general assumptions, assuming that $C^{\sim}u=0$ has the unique continuation property.

Remark 2.3. Note that the last argument is superfluous if C has a fundamental solution -i.e., a special Green inverse which is also a right inverse. Examples 1 and 2 above satisfy this condition - the fundamental solution $c_n|x-y|^{2-n}$ (or $c_2\log|x-y|$) is known

A transformation of the general (2.2) to a form similar to (2.2') will be essential, in the following. Since the above discussion undesirably depends on unique continuation we note the proposition, below, tailored to fit the proof of thm.6.3.

Proposition 2.4. Pick any basis $\gamma_1,\ldots,\gamma_R$ of a complement of im C (We may choose the γ_j as $C_0^\infty(\mathbb{R}^n)$-functions.). With the above Green inverse G of C let $z=Gf^{\backprime}$, $f^{\backprime}$ an extension of $f\in C\infty(\Omega\cup\Gamma)$ to $C_0^\infty(\mathbb{R}^n)$. Then $u\to\psi=u-z|\Omega\cup\Gamma$ transforms (2.2) to the following problem:

$$(2.8)\qquad Av=\gamma\ ,\ x\in\Omega\cup\Gamma,\ B_j\psi=\psi_j=\varphi_j-B_jz\ ,\ x\in\Gamma\ ,$$

with $\gamma\in M=\mathrm{span}\ \{\beta_j|\Omega\cup\Gamma\}$, a fixed finite dimensional space, with β_j of (2.7). We may arrange $f\to f^{\backprime}$ such that $f\to v$ is a linear operator.

The proof follows our discussion around (2.7).

3. Conversion to an $\mathbb{R}^n$-problem of Riemann-Hilbert type.

We enter the discussion of solving (2.2) (or (2.4)) by setting up relations for the extended functions v , g of u and f :

$$(3.1)\qquad v=u \text{ in } \Omega\cup\Gamma\ ,\ g=f \text{ in } \Omega\cup\Gamma\ v=g=0 \text{ in } \mathbb{R}^n\backslash(\Omega\cup\Gamma)\ .$$

For a function $z\in L^1_{loc}(\Omega\cup\Gamma)$ we denote the "zero-extension to $\mathbb{R}^n$" by $z^{\sim}$, so that $z^{\sim}=z$ in $\Omega\cup\Gamma$, $z^{\sim}=0$ in $\mathbb{R}^n\backslash\Omega\backslash\Gamma$. Thus $g=f^{\sim}$, $v=u^{\sim}$.

We have $Cv-g=0$ in Ω and in $\mathbb{R}^n\backslash\Omega$. In general v,g are discontinuous on Γ, although clearly $v,g\in L^1_{pol}(\mathbb{R}^n)\subset S'$. Looking at $Cv-g=h$ (interpreting the derivatives of C as distribution derivatives),

$$(3.2)\qquad Cv-g = h \in E'\subset S'\ ,\ \text{where supp } h\subset\Gamma\ .$$

A distribution with support in a single point $a\in\mathbb{R}^n$ is known to be a finite sum of derivatives $\delta^{(\alpha)}(x-a)$ (Schwartz, [Schw1],thm.35) Similarly, a distribution with support on a smooth hypersurface locally is a finite sum of distributions $\delta^k_{\psi,\Gamma}\in D'(\mathbb{R}^n)$ of the form

$$(3.3)\qquad \langle\delta^k_{\psi,\Gamma},\varphi\rangle = \langle\psi,\partial^k_\nu\varphi|\Gamma\rangle\ \ \varphi\in D(\mathbb{R}^n)\ ,\ \text{with some } \psi\in D'(\Gamma)\ ,$$

where ∂_ν^k denotes the k-th normal derivative at Γ (l.c.,thm.37). Therefore h of (3.2) must be such a sum.

Actually, for an $u\in C^\infty(\Omega\cup\Gamma)\cap\mathbf{H}$, we get such a representation for h explicitly, as a result of Green's formula. To be precise we introduce (τ,ν) as new coordinates on Ω near Γ , τ denoting the foot point of the perpendicular from x to Γ , and ν the distance from x to Γ , with negative sign in the interior of Ω . For a sufficiently small neighbourhood $\mathbf{N}_\Gamma$ of Γ the footpoint τ of x is unique, so that (τ,ν) are useful coordinates. Then ∂_ν^k simply denotes the k-th partial derivative for ν, - derivative along τ=const.

Let $\varphi\in\mathbf{D}(\mathbb{R}^n)$ be a testing function, and let $u\in C^\infty(\Omega\cup\Gamma)$, Lu=f, and v, as above. For the distribution derivative Cv=c(x,D)v we get

$$\langle c(x,D)v,\varphi\rangle = \langle v,c(M_r,-D)f\rangle = \int_\Omega uc(M_r,-D)\varphi dx \ .$$

The right hand side may be integrated by parts, for

$$(3.4)\qquad \langle Cu^{\sim},\varphi\rangle = \int_\Omega f\varphi dx + \sum_{|\alpha|+|\beta|\le N-1}\int_\Gamma \gamma_{\alpha\beta}(x)u^{(\alpha)}\varphi^{(\beta)}dS \ .$$

Here $\gamma_{\alpha\beta}$ are $C^\infty(\Gamma)$-functions depending on the order of partial integration, dS denotes the surface element on Γ . Perhaps it is useful to introduce the coordinates (ν,τ) in the second integral:

$$(3.5)\qquad \langle h,\varphi\rangle = \langle Cv-g,\varphi\rangle = \sum_{|\alpha|+|\beta|+j+k\le N-1}\int_\Gamma \gamma_{\alpha\beta jk}\partial_\tau^\alpha\partial_\nu^j u\partial_\tau^\beta\partial_\nu^k\varphi dS \ .$$

We still may integrate by parts along Γ , using that $\varphi\in C^\infty(\Gamma)$. Accordingly, all τ-derivatives may be assumed on $\partial_\nu^p u$. We proved:

<u>Proposition 3.1</u>. Given $u\in C^\infty(\Omega\cup\Gamma)$, $f=a(x,D)u=Cu$. Let $v=u^{\sim}$, $g=f^{\sim}$, as above. Then, with differential operators P_l of order l on Γ ,

$$(3.6)\qquad \langle h,\varphi\rangle = \sum_{j+k\le N-1}\int_\Gamma (P_{N-1-k-j}\partial_\nu^j u)\partial_\nu^k\varphi dS \ , \quad \varphi\in\mathbf{D}(\mathbb{R}^n) \ ,$$

where the P_l have smooth coefficients.

Note that (3.6) amounts to

$$(3.7)\qquad h = \textstyle\sum_{k=0}^{N-1}\delta^k_{\psi_k,\Gamma} \ , \quad \psi_k=\sum P_{N-1-k-j}\partial_\nu^j u \ , \quad \delta^k_{\psi,\Gamma} \text{ of } (3.3) \ ,$$

hence it gives the desired decomposition of h explicitly.

<u>Remark 3.2</u>: For a distribution $T\in\mathbf{D}'(\mathbb{R}^n)$ of the form $T=\sum\delta^j_{\psi_j,\Gamma}$, with $\delta^j_{\psi_j,\Gamma}$ of the form (3.3) and a finite sum, the distributions $\delta^j_{\nu_j,\Gamma}\in\mathbf{D}'(\Gamma)$ are uniquely determined (as long as we define ∂_ν^k by means of the above coordinates (ν,τ)). Indeed, for any given inte-

ger $k\geq 0$, and $\varphi\in D(\Gamma)$, a function $\varphi^\circ \in D(\mathbb{R}^n)$ may be constructed such that $\partial_\nu^j\varphi^\circ = \varphi\delta_{jk}$ on Γ , $j=0,1,\ldots$. Then clearly

$$(3.8) \qquad \langle \psi_k,\varphi\rangle = \langle T,\varphi^\circ\rangle \ ,$$

showing that indeed the ψ_k , hence the δ_k are fully determined.

Proposition 3.3. Suppose for any u, $f\in C^\infty(\Omega\cup\Gamma)$ and their zero-extensions u^- , f^- we have (in the sense of $D'(\mathbb{R}^n)$)

$$(3.9) \qquad Cu^- = f^- + h \ , \ \mathrm{supp}\, h \in \Gamma \ .$$

Thus Au=f in Ω, and h coincides with the distribution (3.7).

Indeed, we trivially have $Cu^-=f$ in the open set Ω . Since u, $f\in C^\infty(\Omega)$ derivatives in $Cu^-|\Omega$ are classical, hence we may write Au =f in Ω. Taking closure in $\Omega\cup\Gamma$ get Au=f in $\Omega\cup\Gamma$. By above construction, get (3.7) for h, with the precisely determined ψ_k, q.e.d.

The discussion shows that, for $u,f\in C^\infty(\Omega\cup\Gamma)$, the PDE Au=f is exactly equivalent to (3.9) with h of (3.7). For a boundary problem (2.2) we must solve Au=f, hence (3.9), for given f, where $u\in C^\infty(\Omega\cup\Gamma)$, $\psi_0, \psi_1, \ldots \psi_{N-1} \in C^\infty(\Gamma)$ are to be determined.

The right hand side of (3.9) belongs to $S'(\mathbb{R}^n)$. Therefore (3.9), for given ψ_j , may be solved with the methods of ch.II (or V,thm.1.1). The md-elliptic operator C admits a Green inverse G . We get CG=1-P where P may be chosen as in (2.7). The distribution

$$(3.10) \qquad v = Gf^- + Gh$$

will solve $Cv=f^-+h$ if and only if $P(f^-+h)=0$ (i.e., $\langle\alpha_j,f^-+h\rangle=0$, for a basis α_j of ker C^-). Then the general solution (in S') is

$$(3.11) \qquad v = Gf^- + \textstyle\sum_{j=0}^{N-1} G\delta^j_{\psi_j} + \sum\lambda_j\omega_j \ , \ \psi_j\in C^\infty(\Gamma) \ ,$$

with $\lambda_j\in\mathbb{C}$, and a basis ω_j of ker C. For arbitrarily given f, $\psi_0,\ldots,\psi_{N-1}$ one will not expect such v to vanish outside $\Omega\cup\Gamma$ - i.e., v is not a zero-extension of some u. In view of the above our problem of solving (2.2) thus appears reduced to the following:

Problem 3.4. For given $f\in C^\infty(\Omega\cup\Gamma)$, $\varphi_j\in C^\infty(\Gamma)$ determine $\psi_0,\ldots,\psi_{N-1} \in C^\infty(\Gamma)$ such that (i) $\langle f^-+h,\alpha\rangle=0$ for all $\alpha\in$ ker C^- , and that, for suitable λ_j the function v of (3.11) satisfies (ii) v=0 in $\mathbb{R}^n\backslash\Omega\backslash\Gamma$; (iii) $v|\Omega$ extends to a function $u\in C^\infty(\Omega\cup\Gamma)$, so that $v=u^-$; (iv) we get $B_ju=\varphi_j$, $x\in\Gamma$, $j=1,\ldots,M$ - i.e., u satisfies the cdn's of (2.2)

Actually, v of (3.11) is a temperate distribution only, now to be examined for its properties. Evidently we have $\omega_j\in S$. The other two terms involve the operator $G\in O\psi c_{-N,-m_2}$, of differentiation order $-N<0$. G sends H_s to $H_{s+(N,m_2)}$, as we know. From II, 2.5 we know that the ψdo G leaves wave front sets and singular supports invariant. Also, the distribution kernel $k(x,y)=g^{\vee}(x,x-y)$ g=symb(G) is equal to some function in $S(\mathbb{R}^n)$, for fixed y and large $|x|$ (and uniformly so for $y\in K\subset\subset\mathbb{R}^n$). We have proven:

Proposition 3.5. v of (3.11) is $C^\infty(\mathbb{R}^n\backslash\Gamma)$. Also, v is equal to a function in $S(\mathbb{R}^n)$ for sufficiently large $|x|$.

Indeed, sing supp$(f^{\sim}+h)\subset\Gamma$, and $Gf^{\sim}=\int_{\Omega\cup\Gamma}k(f^{\sim}+h)dy$, $\Omega\cup\Gamma\subset\subset\mathbb{R}^n$.

We will make a detailed discussion of $Gf^{\sim}$ and $G\delta^j_{\psi,\Gamma}=\upsilon_j$ near their singular support Γ in sec.4, below, and then return to pbm. 3.4. In essence one finds a system of ψde's for the ψ_j of pbm.3.4. The υ_j will be called multilayer potentials (induced by G). This notation is suggested by considering examples 2.1, 2.2 again. For $C=\Delta$ use of the fundamental solution G of 0,(4.12) instead of the Green inverse is natural. The two distributions υ_0, υ_1 are known in potential theory as single and double layer potentials. They are used to solve the Dirichlet and Neumann problem for $C=\Delta$.

With a result of sec.4, we get an improvement of prop.3.5:

Proposition 3.6. 1) For $f\in C^\infty(\Omega\cup\Gamma)$ we have $w=G\varphi^{\sim}\in C^{N-1}(\mathbb{R}^n)$. Moreover the N-th derivatives $w^{(\alpha)}$, $|\alpha|=N$, jump accross Γ, but are otherwise smooth: For every α the restrictions of $w^{(\alpha)}$ to Ω and to $\mathbb{R}^n\backslash(\Omega\cup\Gamma)$ extend to functions in $C^\infty(\Omega\cup\Gamma)$ and $C^\infty(\mathbb{R}^n\backslash\Omega)$, respectively

2) For $\psi\in C^\infty(\Gamma)$, and $j=0,\dots,N-2$ the multilayer potential $\upsilon_j=G\delta^j_\psi$ is a function in $C^{N-2-j}(\mathbb{R}^n)$. For all derivatives $\upsilon_j^{(\alpha)}$ the restrictions to Ω and $\mathbb{R}^n\backslash(\Omega\cup\Gamma)$ extend to functions in $C^\infty(\Omega\cup\Gamma)$ and in $C^\infty(\mathbb{R}^n\backslash\Omega)$, respectively, $j=0,\dots,N-1$. (That is, all derivatives of order $\geq N-1-j$ may be expected to jump on Γ).

Proof. We will get $C(Gf^{\sim})=f^{\sim}+s'$, $C\upsilon^j=\delta^j_\psi+s''$, $s',s''\in S$. Then thm.4.4 implies the statement; details are obtained by comparing the particular asymptotic expansions of $f^{\sim}$ and δ^j_ψ.

4. Boundary hypo-ellipticity; asymptotic expansion mod ∂_ν.

In this section we focus on matters of boundary hypoellipticity, as mentioned in sec.2, and used in prop.3.6. With notations of sec's 2 and 3 focus on PDE's of the form (3.9), i.e.,

$$Cv=f^{\sim}+\sum_{j=0}^{p}\delta^j_{\psi_j}, \tag{4.1}$$

where $v\in D'(\mathbb{R}^n)$. We will mainly be interested in the case $p=N-1$,

$f \in C^\infty(\Omega \cup \Gamma)$, $\psi_j \in C^\infty(\Gamma)$, attempting to solve pbm.3.4, but results will be interesting for general p and $\psi_j \in \mathcal{D}'(\Gamma)$ and $f \in C^\infty(\Omega)$ with $f^\sim \in L^1_{loc}(\mathbb{R}^n)$, as will be assumed henceforth. Define the spaces

$$(4.2) \quad \mathcal{X}_\Omega = \{w \in \mathcal{D}'(\mathbb{R}^n) : \operatorname{supp} w \subset \Omega \cup \Gamma \ , \ w \in C^\infty(\Omega)\} \ , \ \mathcal{Y}_\Omega = \mathcal{X}_\Omega \cap L^1_{loc}(\mathbb{R}^n) \ .$$

Let us use the coordinates (ν,τ) introduced in sec.3, where $\nu \in \mathbb{R}$, $\tau \in \Gamma$. They are valid only in some neighbourhood $\mathcal{N}$ of Γ . Since Γ is compact $\mathcal{N}$ contains some set $\mathcal{N}_\varepsilon = I_\varepsilon \times \Gamma$, $I_\varepsilon = \{|\nu| \le \varepsilon\}$, $\varepsilon > 0$. Here the neighbourhood $\mathcal{N}_\varepsilon$ of Γ may be regarded as a subset of the cylinder $\Omega' = \mathbb{R} \times \Gamma$ as well. With a cut-off function κ , $\operatorname{supp} \kappa \subset I_\varepsilon$, $\kappa = 1$ near 0 , and $w = \kappa v$, equation (4.1) takes the form

$$(4.3) \quad Cw = f^\sim \kappa + \textstyle\sum_{j=0}^{p} (-)^j \partial_\nu^j \delta(\nu) \otimes \psi_j + \lambda \ , \ \lambda \in C_0^\infty(\{\nu \ge 0\}) , \ \lambda = 0 \text{ as } \nu < \varepsilon/2,$$

with the Dirac measure δ on $\mathbb{R} = \{-\infty < \nu < \infty\}$, and the distribution tensor product (We were using the fact that $v \in C^\infty(\mathbb{R}^n \setminus \Gamma)$, by hypoellipticity of C). We get $v = w$ near $\Gamma = \{\nu = 0\}$. The behaviour of v near Γ may be studied by looking at w of (4.3), while (4.3) may be regarded as a PDE on the cylinder Ω' . (Assume C extended to Ω' , for example by freezing the coefficients: In these coordinates we get

$$(4.4) \qquad C = \textstyle\sum_{j=0}^{N} C_{N-j}(\nu) \partial_\nu^j \ ,$$

with PDE's $\{C_k(\nu) : \nu \in [-\varepsilon,\varepsilon]\}$, of order $\le k$. Extend $C_k(\nu)$ to $\mathbb{R}$ by setting $C_k = C_k(\pm\varepsilon)$ as $\nu \ge \varepsilon$ and $\nu \le -\varepsilon$, resp., smoothened near $\pm\varepsilon$.) We work with (4.3) and $\Omega^0 = \{\nu > 0\} \subset \Omega' = \mathbb{R} \times \Gamma$ instead of (4.1) and Ω. Correspondingly, $\mathcal{X}_{\Omega^0}$ and $\mathcal{Y}_{\Omega^0}$ are defined using $\mathbb{R} \times \Gamma$ instead of $\mathbb{R}^n$.

Clearly $C_0(\nu) = c_0(\nu,\tau)$ of (4.5) is a function. By ellipticity of C we have $c_0(\nu,\tau) \ne 0$ on $\Omega' = \mathbb{R} \times \Gamma$. This condition alone - in fact only the property of Γ being noncharacteristic for C , i.e.,

$$(4.5) \qquad c_0(0,\tau) \ne 0 \qquad \text{(on all of } \Gamma) \ ,$$

is sufficient for our present discussion. We assume C hypo-elliptic (not necessarily elliptic) apart from (4.5). By assumption f is $C^\infty(\Omega \cup \Gamma)$. It admits a Taylor expansion in the variable ν :

$$(4.6) \quad \begin{aligned} f(\nu,\tau) &= \textstyle\sum_{j=0}^{q} (\partial_\nu^j f(0,\tau)) \frac{\nu^j}{j!} + r_q(\nu,\tau) \ , \\ r_q(\nu,\tau) &= \int_0^\nu (\partial_\nu^{q+1} f(0,\tau)) \frac{(\nu-\kappa)^q}{q!} d\kappa \ , \ q = 0,1,\dots \ , \ \nu \ge 0 \ . \end{aligned}$$

For the extended function $f^\sim$ we get an expansion valid in a 2-sided neighbourhood of $\Gamma = \{\nu = 0\}$ by introducing $\chi_j(\nu)$, $j = 0,1,\dots$,

$$(4.7) \qquad \chi_j(\nu) = \nu^j/j! \ , \text{ as } \nu \ge 0 \ , \ \chi_j(\nu) = 0 \ , \text{ as } \nu < 0 \ .$$

For N=0,1,2,...,

$$(4.8)\qquad f^{-}(\nu,\tau) = \sum_{j=0}^{q}\partial_\nu^j f(0,\tau)\chi_j(\nu) + r_q^{-}(\nu,\tau)\ ,\ \tau\in\Gamma\ ,\ |\psi|\le\varepsilon\ .$$

The $\chi_j(\nu)$ are homogeneous of degree j in ν . (4.8) may be expressed as an asymptotic expansion mod ∂_ν (at ν=0), in the sense of [C₁],II,3 : For fixed $\tau=\tau^0\in\Gamma$ we have

$$(4.9)\qquad f^{-}(\nu,\tau^0) = \sum_{j=0}^{\infty}\partial_\nu^j f(0,\tau^0)\chi_j(\nu)\ (\mathrm{mod}\ \partial_\nu)\ ,\ \text{at}\ \nu=0\ .$$

In details, the terms of (4.9) are homogeneous of degree $\rho_j=j\to\infty$ j=0,1,... , ($\chi_j\in H_j''$, speaking in terms of [C₁],II,(3.3)).

We have $f^{-}-\sum_{j=0}^{q}\partial_\nu^j\varphi(0,\tau^0)\chi_j=r_q^{-}\in C^q(\Omega')$, by well known properties of the remainder r_N . Equivalently, for N=0,1,2,... there exists M such that $r_k^{-}\in C^N$ for all k≥M . The terms of (4.9) are L^1_{pol} hence also are homogeneous distributions.

In [C₁],II,3 we were discussing more general such asymptotic expansions, allowing arbitrary homogeneity degree ρ_j - positive and negative, and even complex - but with certain restrictions for degrees $\rho_j\in\mathbb{Z}$, $\rho_j\le -n$, and allowing certain non-homogeneous terms for integers $\rho_j\ge 0$. Here it is practical to extend in a somewhat simpler way: For j=1,2,... define the distributions

$$(4.10)\qquad \chi_{-j} = \partial_\nu^j c_0 = \partial_\nu^{j-1}\delta\ .$$

Observe that then, for k=0,1,...,

$$(4.11)\qquad \begin{aligned}&\partial_\nu^k\chi_j = \chi_{j-k}\ ,\ j\in\mathbb{Z}\ ,\ \nu^k\chi_j=\frac{(k+j)!}{j!}\chi_{k+j}\ ,\ j\ge 0\ ,\\ &\nu^k\chi_{-j}=(-1)^k\frac{(j-1)!}{(j-1-k)!}\chi_{k-j}\ ,\ j>k\ ,\ =0\ ,\ j\le k\ .\end{aligned}$$

All χ_j , also for j<0 , are homogeneous distributions of degree j.

We shall say that a distribution $T\in\mathcal{X}_{\Omega'}$ allows an <u>extended Taylor expansion</u> (mod ∂_ν) at ν=0, (abbreviated "ETE") written as

$$(4.12)\qquad T = \sum_{j=p}^{\infty}\chi_j\otimes\gamma_j\ (\mathrm{mod}\ \partial_\nu)\ \text{at}\ \nu=0\ ,$$

where $p\in\mathbb{Z}$, and $\gamma_j\in\mathcal{D}'(\Gamma)$, j=p,p+1,.... , if for every M=0,1,... there exists $N_0=N_0(M)$ such that $T-\sum_{j=p}^{q}\chi_j\otimes\gamma_j\in C^M$, near ν=0, as N≥ N_0. The largest such p will be called the order of the expansion. Actually we always will require that $\gamma_j\in C^\infty(\Gamma)$, as j≥0 , while the γ_j , j<0 will be allowed as general distributions in $\mathcal{D}'(\Omega)$.

This definition proves useful at once: (4.3) may be written as

(4.13)
$$Cw = \sum_{j=-p-1}^{\infty} \chi_j \otimes \gamma_j \pmod{\partial_\nu} \text{ at } \nu=0 \text{ , with}$$
$$\gamma_j = \partial_\nu^j f(0,\tau) \text{ , } j \geq 0 \text{ , } \gamma_j = (-1)^{-j-1} \psi_{-j-1} \text{ , as } j<0 \text{ .}$$

It is evident that the 0-extensions of $C^\infty(\Omega \cup \Gamma)$-functions are characterized by the nature of their asymptotic expansions:

Proposition 4.1. A distribution $g \in \mathcal{D}'(\Omega^1)$ with supp $g \subset \{\nu \geq 0\}$ and sing supp $g \subset \{\nu=0\}$ is $C^\infty(\{\nu \geq 0\})$ if and only if there exists functions $\gamma_j(\tau) \in C^\infty(\Gamma)$ such that

(4.14)
$$g = \sum_{j=0}^{\infty} \chi_j \otimes \gamma_j \pmod{\partial_\nu} \text{ at } \nu=0 \text{ .}$$

Then we have $\gamma_j = \partial_\nu^j g(0+,\tau)$, $\tau \in \mathcal{G}$.

Proof. One direction was shown above. Vice versa, if the expansion (4.14) exists, then it is evident that $g|\{\nu>0\}$ extends to a $C^N(\{\nu \geq 0\})$-function, for every N , hence is $C^\infty(\{\nu \geq 0\})$, q.e.d.

Proposition 4.2. 1) An extended Taylor expansion may be differentiated term by term - for ν as well as τ , of arbitrary order.

2)The product ωf , where $\omega \in C^\infty(\Omega^1)$ and f allows an extended Taylor expansion, possesses an ETE with terms explicitly determined by Cauchy multiplication of the expansions of f and $\omega^0 = (\omega|\Omega^0)^-$ using (4.11). (Note that ω^0 has an ETE , by prop.4.1.)

The proof is left to the reader. A consequence of prop.4.2:

Proposition 4.3. If u^- allows an extended Taylor expansion then so does Cu^- , where C is any differential expression as in (4.4).

The matters of boundary hypo-ellipticity, as well as prop.3.6 now are settled by proving the following.

Theorem 4.4. Let C of (4.4) be hypo-elliptic with smooth coefficients. Assume $c_0(\nu,\tau)$ of order 0 (i.e., a function), satisfying (4.5), but the $C_k(\nu)$, $k>0$, are differential expressions in τ only, of arbitrary (finite) order, with coefficients depending smoothly on ψ and τ. If, for $u \in \mathcal{D}'(\Omega^1)$ we have $Cu=f \in \mathcal{X}_{\Omega^0}$, and if f allows an ETE of order p at $\nu=0$, with smooth negative coefficients, then so does u. Moreover the expansion of u is of order $p+N$.

Proof. Assume we have supp $u \subset \{\nu \geq 0\}$ and

(4.15)
$$Cu = \sum_{j=p}^{\infty} \chi_j \otimes \varphi_j \pmod{\partial_\nu} \text{ at } \nu=0 \text{ .}$$

As a first step we will construct a distribution of the form $u_0=$

$\chi_{p+N}\otimes\gamma_{p+N}$ such that the extended Taylor expansions of Cu and Cu_0 have the same term of order p. By (4.11) we get $\partial_v^j u_0 = \chi_{p+N-j}\otimes\gamma_{p+N}$. There will be a Taylor expansion $c_k(v) \sim \sum c_k^{(j)}(0)v^j/j!$. Hence

$$(4.16) \qquad c_{N-j}(v)\partial_v^j u_0 = \chi_{p+j}\otimes c_{N-j}(0)\gamma_{p+N} + \dots \ (\mathrm{mod}\ \partial_v) \text{ at } v=0 .$$

Taking the sum (4.4) one finds that

$$(4.17) \qquad Cu_0 = \chi_p\otimes(c_0(0,\tau)\gamma_{p+N}) + \dots \ (\mathrm{mod}\ \partial_v) \text{ at } v=0 .$$

To get the same lowest order terms of (4.15) and (4.17) we choose

$$(4.18) \qquad \gamma_{p+N} = \varphi_p/c_0(0,\tau) .$$

(Note (4.5) - i.e.,Γ is noncharacteristic for C).

After constructing u_0 we form $w_1=u-u_0$. Clearly the expansion of Cw_1 is of order p+1. We may repeat the process to construct $u_1=\chi_{p+N+1}\otimes\gamma_{p+N+1}$ such that Cw_1 and Cu_1 have the same lowest order term, so that $C(w_1-u_1)=C(u-u_0-u_1)$ has an expansion of order p+2. By iteration we get $u_0,u_1,\dots,$ such that $C(u-u_0-\dots-u_M)$, for each M, has an expansion of order p+M .

Finally notice that, formally, the infinite sum $u_0+u_1+\dots.$ is well defined. Write $u_k = \sum_{j=p+k+N}^{\infty}\chi_j\otimes\gamma_j^k$. Then

$$(4.19) \qquad \sum_{k=0}^{\infty}u_k = c_{p+N}\otimes\gamma^0_{p+N} + \chi_{p+N+1}\otimes(\gamma^1_{p+N}+\gamma^0_{p+N+1}) + \dots ,$$

the factors of χ_1 at right being well defined finite sums. Setting

$$(4.20) \qquad \gamma_{p+N+j} = \gamma^0_{p+N+j}+\gamma^1_{p+N+j-1} + \dots + \gamma^j_{p+N} , \ j=0,1,\dots .$$

we may write the formal sum (4.19) as

$$(4.21) \qquad \sum_{k=0}^{\infty}u_k = \sum_{j=p+N}^{\infty}\chi_j\otimes\gamma_j .$$

We show that (i) there exists $v\in \mathcal{D}'(\Omega')$, supp $v\subset\{v\geq 0\}$ with

$$(4.22) \qquad v = \sum_{j=p+N}^{\infty}\chi_j\otimes\gamma_j \ (\mathrm{mod}\ \partial_v) \text{ at } v=0 ,$$

and that (ii) we have

$$(4.23) \qquad C(u-v_q)\in C^q , \ q=p+N,p+N+1,\dots , \text{ where } v_q=\sum_{j=p+N}^{q}\chi_j\otimes\gamma_j .$$

Constructing v of (4.22) means writing $v=\sum\kappa_j\chi_j\otimes\gamma_j$, with suitable cut-off's $\kappa_j(\nu)$ near $\nu=0$. With a technique similar to that of I,lemma 6.4, we show that κ_j may be chosen to converge weakly in $\mathcal{D}'(\Omega')$, and to satisfy (4.22). Then (4.23) follows easily.

It follows that $C(u-v)$ has an ETE with all terms vanishing. This means that $C(u-v)\in C^\infty(\Omega')$. By hypo-ellipticity of C this implies $u-v\in C^\infty(\Omega^0)$, implying that $u=\sum\chi_j\otimes\gamma_j \pmod{\partial_\nu}$ at $\nu=0$. Q.E.D.

Corollary 4.5. Every solution $v\in\mathcal{D}'(\mathbb{R}^n)$ of equation (3.11), with $f\in C^\infty(\Omega\cup\Gamma)$, $\psi_j\in C^\infty(\Gamma)$, has $v|\Omega$ extending to a $C^\infty(\Omega\cup\Gamma)$-function.

The proof is evident.

Remark 4.6. Prop.3.6 follows from thm.4.4 and remarks following it

Remark 4.7. The technique may be used as well for nonsmooth $Cv=z=f^{\sim}+\sum_{j=0}^{N-1}\delta_{\nu_j}^j$, assuming that f,ψ_j belong to suitable Sobolev spaces.

Remark 4.8. Thm.4.4 is of local nature:If its conditions hold only for a subregion R of Γ, then (4.15), valid only for $(\nu,\tau)\in\mathbb{R}\times R$, implies a corresponding ETE for u , also valid only in $\mathbb{R}\times R$. Indeed, all discussions of the proof extend literally.

5. A system of ψde's for the ψ_j of problem 3.4.

We return to our pbm.3.4, the general elliptic boundary problem. After sec.4 we know that $v\in\mathcal{D}'(\mathbb{R}^n)$, satisfying (3.11) is $C^\infty(\mathbb{R}^n\setminus\Gamma)$ and has interior and exterior limits for all $\partial_x^\alpha v$ on Γ. If v is to satisfy cdn's (i)-(iv) of pbm.3.4, then the exterior (interior) limits must vanish (satisfy the boundary conditions.

In the present section we will analyze these conditions, and convert them into a more explicit form. We shall see that a system of ψde's on Γ results, which is normally solvable, under certain conditions - the Lopatinskij-Shapiro conditions already mentioned.

First of all, let us assume cdn.S (of sec.2) satisfied. In other words, then it suffices to focus on (2.2') - i.e., assume $f=0$ in $\Omega\cup\Gamma$. This simplifies (3.11): Up to the additional solution $\sum\lambda_j\omega_j$ of $Cu=0$, v of (3.11) must be a sum of multilayer potentials. Thus, clearly, the well known multi-layer Ansatz from potential theory appears to be justified. We have proven:

Proposition 5.1. Every solution u of the boundary problem (2.2')

is representable (mod ker C) as a sum $v=\sum_{j=1}^{N-1} G\delta_{\psi_j}^{j}$ of multi-layer potentials, with certain $\psi_j \in C^{\infty}(\Gamma)$. That is, u is the continuous extension of $v|\Omega$ to $\Omega\cup\Gamma$, up to an additional term in ker C.

Remark 5.2. Note that the behaviour of v in the outside of $\Omega\cup\Gamma$ is unessential. If z is a solution of an exterior boundary problem, using the same C and Γ, but possibly different boundary operators B_j, (but with f=0 in $\mathbb{R}^n\backslash(\Omega\cup\Gamma)$), then the $L^1_{loc}(\mathbb{R}^n)$-function w=u (in Ω), w=z (in $\mathbb{R}^n\backslash(\Omega\cup\Gamma)$), undefined on the null set Γ, solves $Cw=\sum_{j=1}^{N-1}\delta_{\psi_j}^{j}$, with certain $\psi_j \in C^{\infty}(\Omega)$, hence

$$(5.1) \qquad v = \sum_{j=0}^{N-1} G\delta_{\psi_j}^{j} + \omega \ , \ \omega \in \ker C \ , \ \psi_j \in C^{\infty}(\Gamma) \ .$$

This suggests that -possibly- a solution u of (2.2') has many representations as $u=v|\Omega$, with v of (5.1), and different sets of ψ_j.

To prove existence of u - a solution of (2.2') - it is enough to find ψ_j such that v of (5.1) (with certain ω) satisfies the boundary conditions (2.2'), using derivatives from the inside of Γ. In details the ψ_j must satisfy (with interior derivatives)

$$(5.2) \qquad \sum_{j=0}^{N-1} B_k G\delta_{\psi_j}^{j} + B_k\omega = \varphi_k \ , \ k=1,\dots,M \ ,$$

Then $u^{\sim}$, the zero extension of $u=v|\Omega$ will also satisfy (5.1) but with redefined ψ_j, as we have derived initially in sec.3.

Moreover, $u^{\sim}$, of course, will satisfy the additional boundary conditions (in addition to the interior conditions (2.2))

$$(2.2)_1 \qquad \partial_\nu^k u^{\sim} = \sum_{j=0}^{N-1}(-iD_\nu)^{k} G\delta_{\psi_j}^{j} + \partial_\nu^k\omega = 0 \ , \ x \in G \ , \ k=0,\dots,N-1 \ ,$$

with derivatives from the exterior - since $u^{\sim}=0$ outside $\Omega\cup\Gamma$. Thus, for uniqueness, we may show that the N+M conditions (2.2) (with $\varphi_j=0$) and $(5.2)_1$, imposed on N functions ψ_j, imply $\psi_j=0$.

We now will attempt to express terms of the form $BG\delta_{\psi}^{j}$, with a boundary operator B - such as B_j of (2.2) or $B=\partial_\nu^k$ of $(2.2)_1$, with interior or exterior derivatives, in the form $BG\delta_{\psi}^{j}=Q\psi$, where Q must be studied. Q will be a ψdo, and (2.2'),$(2.2)_1$, after modification, systems of ψdo's. Fixing some ψ_j, (2.2') alone is converted to an elliptic system with Green inverse. Thus we will settle existence and uniqueness (prove normal solvability of (2.2')).

The manifold structure of Γ, and the fact that $\Gamma=\partial\Omega$ implies existence of a coordinate transform, for each $x_0 \in \Gamma$, mapping a neighbourhood N_{x^0} of x^0 onto a neighbourhood of 0 in the halfspace $\mathbb{R}^n_+=\{x_1 \geq 0\}$. Γ is covered by a finite number of such neighbourhoods. Using a partiton of unity $\{\kappa_j\}$ subordinate to such covering $\{N_{x^j}\}$ we split: $\psi = \sum \kappa_j \psi$, supp $\kappa_j \psi \subset N_{x^j}$, and write $\upsilon=\sum \upsilon_k$, $\upsilon_k = B\delta^j_{\kappa_k \psi}$. Our above coordinate transform (for N_{x^j} chosen sufficiently small) may be extended to a global transform meeting the requirements of IV,3. In the new coordinates υ_k transforms into a linear combination of multilayer potentials with respect to the half space $\mathbb{R}^n_+$ with boundary γ given by $\{x_1=0\} = \mathbb{R}^{n-1}$. That boundary is noncompact. However, we now may assume the density ψ to have compact support. G, of course, will transform into a ψdo of the same properties. Returning to our old notations, we now will have

$$\Omega=\{x=(\nu,\tau)\in \mathbb{R}^n : \nu>0\} \;,\; \Gamma=\mathbb{R}^{n-1}=\{x=(\nu,\tau)\in \mathbb{R}^n : \nu=0\} \;,\; \psi\in C_0^\infty(\Gamma) \;,$$

(5.3)

$$\upsilon = G\delta^k_\psi = (-1)^k G(\delta^{(k)}(\nu)\otimes \psi(\tau)) = (-i)^k G D_1^k(\delta(\nu)\otimes \psi(\tau)) \;,$$

where G is a special Green inverse of $C\in \mathrm{Op}\psi c_{N,m_2}$, an md-elliptic differential operator of order N . For our boundary conditions (2.2) we also need derivatives $D^\alpha \upsilon$, $|\alpha|\leq N-1$, where

$$D^\alpha \upsilon = (-i)^k D^\alpha G D^k_\nu (\delta(\nu)\otimes\psi(\tau)) \;,\; \psi\in C_0^\infty(\Gamma) \;. \tag{5.4}$$

For the calculation of (5.4) we approach $\delta(\nu)$ by a δ-family: Choose any family $\delta_\varepsilon(\nu)\in C_0^\infty(\mathbb{R})$, $\varepsilon>0$, with $\delta_\varepsilon\to\delta$, as $\varepsilon\to 0$, weakly in $D'(\mathbb{R})$ - for example $\delta_\varepsilon(\nu)=\frac{1}{\varepsilon}\delta_1(\frac{\nu}{\varepsilon})$, $\varepsilon>0$, with $\delta_1\in C_0^\infty(\mathbb{R})$, $\int\delta_1 d\nu=1$. By continuity of the tensor product and the operator B we then get

$$D^\alpha \upsilon = (-i)^k D^\alpha G D^k_\nu (\delta(\nu)\otimes\psi(\tau) = (-i)^k \lim_{\varepsilon\to 0} T_\varepsilon \;,$$

(5.5)

$$T_\varepsilon = D^\alpha G D_1^k(\delta_\varepsilon(\nu)\otimes\psi(\tau)) = P(\delta_\varepsilon\otimes\psi) \;,\; P=D^\alpha G D_1^k \in \mathrm{Op}\psi c_{(|\alpha|+k)e^1-m} \,.$$

Clearly $\delta_\varepsilon\otimes\psi \in C_0^\infty(\mathbb{R}^n) \subset S(\mathbb{R}^n)$, hence (with $x=(\nu,\tau)=(x_1,x^\Delta)$)

$$T_\varepsilon(x_1,x^\Delta) = \int d\!\!{}^-\tau \int d\!\!{}^-\kappa e^{i\tau(x^\Delta-\kappa)}\psi(\tau)\int d\!\!{}^-\nu e^{i\nu x_1}\delta_\varepsilon^\wedge(\nu)p(x,(\nu,\tau)) \;, \tag{5.6}$$

with $p=\mathrm{symb}(P)$, $\delta_\varepsilon^\wedge$=Fourier transform of δ_ε . We have written $\xi=(\nu,\tau)=(\xi_1,\tau)$, $y=(y_1,\kappa)$, and used I,(1.5) with some Fubini-type integral exchanges, due to $\delta_\varepsilon^\wedge\in S$, p of polynomial growth.

Introducing the n-1-dimensional symbol (with parameter x_1)

$$(5.7)\qquad \beta_\varepsilon(x_1,x^\Delta,\xi^\Delta) = \int d\!\!\!/ v e^{ix_1 v}\delta_\varepsilon{}^\wedge(v)p(x_1,x^\Delta,v,\xi^\Delta)$$

we get $T_\varepsilon = \beta_\varepsilon(x_1,x^\Delta,D_{x^\Delta})\psi$. We get $\delta_\varepsilon{}^\wedge \to \delta^\wedge = \sqrt{2\pi}$, as $\varepsilon\to 0$, hence expect

$$(5.8)\quad \beta_\varepsilon\to\beta_0(x_1,x^\Delta,\xi^\Delta) = \int dv e^{ix_1 v}p(x_1,x^\Delta,v,\xi^\Delta) = \kappa_1^{-1}p^{\vee}(x_1,x^\Delta,x_1,\xi^\Delta),$$

possibly with a distribution integral. Actually, (5.7) amounts to $\beta_\varepsilon = p(x_1,x^\Delta,D_{x_1},\xi^\Delta)\delta_\varepsilon$, for fixed x^Δ,ξ^Δ, implying the convergence (5.8) for each fixed x^Δ,ξ^Δ, in $\mathcal{S}'(\mathbb{R})$. Formally we thus expect

$$(5.9)\quad i^k D^\alpha \upsilon(v,\tau) = \beta_0(v,\tau,D_\tau)\psi\ ,\ \beta_0(v,\tau,\xi^\Delta) = \kappa_1^{-1}p^{\vee}(v,\tau,v,\xi^\Delta)\ ,$$

with the inverse Fourier transform $^{\vee}$ for the 3-rd argument.

Let us analyze the special form of $p(x,\xi)=\mathrm{symb}(D^\alpha B D_{x_1}^k)$.

<u>Proposition 5.3</u>. $p(x,\xi)$ (and its derivatives for x,ξ of arbitrary order) may be written as homogeneous asymptotic sums

$$(5.10)\qquad p(x,\xi) = \sum_{j=0}^\infty p_j(x,\xi)\quad (\mathrm{mod}\ |\xi|)\ \text{at}\ \infty\ ,$$

in the sense of [C₁],II,def.5.2, and uniformly so for $x\in K \subset\subset \mathbb{R}^n$, any K. Here the p_j are rational functions of ξ with coefficients smooth in x, homogeneous of degree $-N-j+|\alpha|+k$ in ξ. Specifically,

$$(5.11)\qquad \begin{aligned} &p_0(x,\xi)=\xi^{\alpha+ke_1}/c_N(x,\xi)\ ,\ e_1=(1,\dots,0)\ ,\\ &p_j(x,\xi) = r_j(x,\xi)/c_N^{j+1}(x,\xi)\ ,\ r_j\in \psi c_{jm-je+(|\alpha|+k)e^1}\ ,\end{aligned}$$

where the r_j are polynomials in ξ of degree $\le Nj-j+|\alpha|+k$ with coefficients in $\psi c_{0,jm_2-j}$, and where $c_N(x,\xi)$ denotes the principal part polynomial of $c(x,\xi)$ (i.e. the homogeneous part of degree N). Asymptotic convergence mod $|\xi|$ at $\xi=\infty$ means that

$$(5.12)\qquad p(x,\xi)-\sum_{j=0}^R p_j(x,\xi) = O(|\xi|^{-q})\ \text{as}\ R\ge R_0(q),$$

uniformly for $x\in K\subset\subset \mathbb{R}^n$, all K, and with all (x,ξ)-derivatives.

<u>Proof</u>. G is the operator B constructed in II,thm.1.6, up to an additional term in $Op\mathcal{S}(\mathbb{R}^{2n})$. In the notations used there, $b(x,\xi)$ is an asymptotic sum - $b=\sum_0^\infty b_j\ (\mathrm{mod}\ \psi c_{-\infty})$ - where $b_0=\frac{1}{c}$ (large $|x|+|\xi|$) while, for $j=1,2,\dots$, the b_j are recursively defined :

$$(5.13)\ b_{j+1}=\sum \kappa_\theta b_{j(\theta)}\frac{c^{(\theta)}}{c}\ ,\ \kappa_\theta=(-i)^{|\theta|}/\theta!\ \text{for}\ 1\le|\theta|\le N,\ =0\ \text{otherwise}$$

for large $|x|+|\xi|$ again. At the moment we have x varying over a bounded set, so, 'large $|x|+|\xi|$' $\Leftrightarrow$ 'large $|\xi|$'. For our present notation we use g=symb(G) and replace b_j by g_j in (5.13).

<u>Proposition 5.4</u>. We have (for $x \in K \subset\subset \mathbb{R}^n$, $|\xi|$ large)

$$(5.14) \qquad g_j \in \operatorname{span}\{\tfrac{1}{c}\Pi(c_{(\beta^j)}^{(\alpha^j)}/c) : j \le \sum|\alpha^j| = \sum|\beta^j| \le jN\} .$$

The proof is immediate, using (5.13) and induction.

Notice that a term at right of (5.14) is of the form

$\frac{1}{c}\Pi(c_{(\beta^j)}^{(\alpha^j)}/c) : \sum|\alpha^j|=\sum|\beta^j|=k$. Such a term is a symbol in ψc_{-m-ke} , $m=(N,m_2)$, $e=(1,1)$. Asymptotic convergence of $\sum g_j$, in our case, means that $g-\sum_{j=0}^{R} g_j = O(|\xi|^{-q})$ as $R \ge R_0(q)$ and $x \in K$, hence coincides with asymptotic convergence mod $|\xi|$, as defined. Terms of (5.14) are rational functions of ξ with coefficients C^∞ in x. Denominators are powers of $c(x,\xi)$. By reordering the expansion we get

$$(5.15) \quad g = \sum_{j=0}^{\infty} p_j/c^{j+1} \ (\mathrm{mod}\ |\xi|) \text{ at } \infty , \ p_0=1 , \ p_j \in \psi c_{Nj-j,Nm_2-j} ,$$

where p_j are polynomials in ξ of degree $\le Nj-j$. Write $c=\sum_{j=0}^{N} c_j$, with homogeneous c_j of degree j. The md-ellipticity of c implies

$$(5.16) \qquad |c_N(x,\xi^0)| \ge p>0, \text{ for all } |\xi^0|=1 , \ x \in K .$$

We get

$$(5.17) \qquad \frac{1}{c} = \frac{1}{c_N(x,\xi)} \{1+\frac{c_{N-1}(x,\xi^0)}{c_N(x,\xi^0)}|\xi|^{-1}+\dots+\frac{c_0(x,\xi^0)}{c_N(x,\xi^0)}|\xi|^{-N}\}^{-1}$$
$$= \frac{1}{c_N(x,\xi)} \{1+\gamma_1(x,\xi)|\xi|^{-1} + \dots\} ,$$

where the γ_j are homogeneous of degree 0 . Uniform convergence of the series $1+\gamma_1|\xi|^{-1}+\dots$ in (5.17) for large $|\xi|$ and $x \in K$ is evident, using the geometric series $\frac{1}{1+y}=1-y+y^2-+\dots$. Again, a study of the coefficients γ_j, from expanding $y^j=(\frac{c_{N-1}}{c_N}+\dots\frac{c_0}{c_N})^j$, shows that

$$(5.18) \qquad \frac{1}{c} = \frac{1}{c_N}\sum_{k=0}^{\infty} q_k/c_N^k \quad , \text{ degree } q_k \le kN-k .$$

The series converges uniformly for $x \in K$ and large $|\xi|$ with all (term by term) derivatives. A fortiori we have asymptotic convergence mod $|\xi|$. Similar expansions result for all powers c^{-j} . Substituting (5.18), etc., into (5.15), and reordering we get the desired expansion (5.12), for the case $k=\alpha=0$.

For general $P=D^{\alpha}GD^{k}_{x_1}$ we use I,(5.7), giving the finite sum

$$P=\sum\frac{(-i)^{|\theta|}}{\theta!}\binom{\alpha}{\theta}g_{(\theta)}(x,D)D^{\theta}D_1^{j} \quad (5.19)$$

In other words, the symbol of $P=D\alpha GD^{k}_{x_1}$ is a finite combination of symbols $g_{(\beta)}(x,\xi)\xi^{\theta}$. Since we know that (5.12) may be differentiated term by term, and since multiplication by a power of ξ does not change the asymptotic convergence, nor the structure of the expansion terms, we get (5.12) in the general case. In particular the term $p_0(x,\xi)$ is as stated, as a comparison shows. Q.E.D.

Here we return to formula (5.9), so far instituted only formally. Substituting $p(x,\xi)$ from (5.10), taking the Fourier transform "^{3}" term by term, we must deal with the distribution integral

$$\sqrt{2\pi}p_j^{\vee}(\nu,\tau,\nu,\xi^{\Delta}) = \int d\kappa e^{i\nu\kappa}p_j(\nu,\tau,\kappa,\xi^{\Delta}) \quad , \quad (5.20)$$

where $p_j = r_j/c_N^{j+1}$ is a homogeneous rational function of degree $k+|\alpha|-N-j$ in the variables $\xi=(\kappa,\xi^{\Delta})$. For fixed ν,τ,ξ^{Δ}, the p_j are rational functions of the single variable κ . Their poles, in the complex κ-plane, are independent of j - just the roots of the denominators c_N^{j+1} i.e., of the polynomial equation

$$c_N(\nu,\tau,\kappa,\xi^{\Delta}) = 0 \ . \quad (5.21)$$

Since C is elliptic there are no real roots of this equation, by (5.16). Assuming first that (5.20) is a Lebesgue integral, one will tend to evaluate it by a complex method: Assume $\nu\neq 0$ - i.e., x is not on Γ. Depending on the sign of ν we will integrate over a semi-circle in the upper or lower half-plane: For $\nu>0$ ($\nu<0$) the function $e^{i\psi\kappa}=e^{i\nu\mathrm{Re}\kappa}.e^{-\nu\mathrm{Im}\kappa}$ will decay as $\mathrm{Im}\,\kappa \to \infty$ ($\mathrm{Im}\,\kappa \to -\infty$) . Let the roots of (5.21) be called κ_1 . Then it is clear that

$$(5.22)\quad \int d\kappa e^{i\nu\kappa}p_j(\nu,\tau,\kappa,\xi^{\Delta})=\sum_{\mathrm{Im}\kappa_1>0}\mathrm{Res}_{\kappa_1} e^{i\nu\kappa}p_j(\nu,\tau,\kappa,\xi^{\Delta}) \ , \text{ as } \nu>0,$$

$$= -\sum_{\mathrm{Im}\kappa_1<0} \mathrm{Res}_{\kappa_1} e^{i\nu\kappa}p_j(\nu,\tau,\kappa,\xi^{\Delta}) \ , \text{ as } \nu<0.$$

So far we assumed a Lebesgue integral. However, in general, the term $p_j=r_j/c_N^{j+1}$ is a linear combination of terms $\kappa^q\xi^{\Delta\,\theta}/c_N^{j+1}$, where $\xi^{\Delta\,\theta}/c_N^{j+1}$ is L^1, or at least improperly Riemann integrable, as $\nu\neq 0$. Thus (5.22) holds for these terms, while "$.\kappa^q$" amounts to "D_ν^q" , under Fourier transform. Differentiation for a parameter may be taken under the residue. Therefore we get (5.22) in the general case

Looking at (5.9) we will be interested in the two limits

(5.23) $\quad i^k D^\alpha \upsilon(\pm 0,\tau) = \lim_{\nu\to 0} i^k D^\alpha \upsilon(\nu,\tau)$, as $\psi>0$,and $\psi<0$, resp.

With (5.22) we get the following result.

<u>Proposition 5.5</u>. We have

(5.24) $\quad i^k D^\alpha \upsilon(\pm 0,\tau) = \pi^{\pm}(\tau,D_\tau)\psi$, $\quad \psi\in C_0^\infty(\mathbb{R}^{n-1})$,

where $\pi^{\pm}(\tau,\xi^{\Delta})$ is smooth in τ,ξ^{Δ} , for all τ,ξ^{Δ} , and satisfies

(5.25) $\quad \pi^{\pm}(\tau,\xi^{\Delta}) = \sum_{j=0}^{\infty} \pi^{j,\pm}(\tau,\xi^{\Delta})$ (mod $|\xi^{\Delta}|$) at $\xi^{\Delta}=\infty$,

with

(5.26)
$$\pi^{j,+}(\tau,\xi^{\Delta}) = \sum_{\operatorname{Im}\kappa_1>0} \operatorname{Res}_{\kappa_1} p_j(0,\tau,\kappa,\xi^{\Delta}) \quad ,$$
$$\pi^{j,-}(\tau,\xi^{\Delta}) = \sum_{\operatorname{Im}\kappa_1<0} \operatorname{Res}_{\kappa_1} p_j(0,\tau,\kappa,\xi^{\Delta}) \quad .$$

for $\tau\in K\subset\subset \mathbb{R}^{n-1}$, all K , uniformly in τ , and for partial derivatives of all orders. In particular, the $\pi^{j,\pm}(\tau,\xi^{\Delta})$ are smooth in τ,ξ^{Δ} , as $\xi^{\Delta}\neq 0$, and positive homogeneous of degree $-N+|\alpha|+k-j+1$,

(5.27) $\quad \operatorname{Res}_{\kappa_1} p_j(0,\tau,\kappa,\xi^{\Delta}) = \int_{|\kappa-\kappa_1|=\varepsilon} p_j(0,\tau,\kappa,\xi^{\Delta})d\kappa$,

with small $\varepsilon>0$.

<u>Proof</u>. Note that the roots $\kappa_1=\kappa_1(\nu,\tau,\xi^{\Delta})$ of (5.21) are continuous in ν,τ,ξ^{Δ}, as long as $\xi^{\Delta}\neq 0$. Moreover, these roots are homogeneous of degree 1 in ξ' : $\quad \kappa_1(\nu,\tau,\rho\xi^{\Delta}) = \rho\kappa(\nu,\tau,\xi^{\Delta})$, $\rho>0$. Using this, and that p_j are homogeneous of degree $-N-j+|\alpha|+k$ we find that the $p^{j,\pm}(\tau,\xi^{\Delta})$ are homogeneous of degree $-N-j+|\alpha|+k+1$ in ξ^{Δ} , by an integral substitution in (5.27). Furthermore, the $\pi^{j,\pm}(\tau,\xi^{\Delta})$ are smooth in τ,ξ^{Δ} as well, since they may be represented by an integral over a countour containing all roots in the given half-plane, where the countour may be locally kept constant.

As for the asymptotic convergence in (5.25), the remainder

$\rho_q^{\vee} = p-\sum_0^q p_j = O(|\xi|^{-R})$ as $q\geq q_0(R)$, and $|x|^{-R}=\{|\xi^{\Delta}|^2+\kappa^2\}^{-R/2}$,

hence $\rho_q=(p-\sum_0^q p_j)^{\vee} = O(\int d\kappa|\xi|^{-R}) = O(|\xi^{\Delta}|^{-R}+1)$. Similarly we may argue for partial derivatives.

Finally, regarding smoothness of $\pi^{\pm}(\tau,\xi^{\Delta})$ we look at the expansion (5.15) (which, in fact, is mod $\mathcal{S}(\mathbb{R}^{2n})$, not only mod $|\xi|$). That is, the remainders are $O(\langle\xi\rangle^{-R})$ not only $O(|\xi|^{-R})$, so that the "ϑ"-remainders are arbitrarily smooth. Regarding $(p_j/c^{j+1})^{\vartheta}$: The same residue calculus may be applied - but remember, the polynomial $c(x,\kappa,\xi^\Delta)$ may have real roots, in a bounded set $|\kappa|\le\eta$. But we really have $\chi p_j/c^{j+1}$ with $\chi=0$ as $|\kappa|\le\eta$. Thus the path of the Fourier integral may be laid clear of those roots , generating a remainder involving an integral over a C^∞-function on a compact countour - a C^∞-function. For $\int_\gamma p_j/c^{j+1}$, along a path γ clear of the roots of c -but coinciding with $\mathbb{R}$ for $|\kappa|>2\eta$ we may repeat the above residue argument to get smoothness of the term. Q.E.D.

6. Lopatinskij-Shapiro conditions; normal solvability of (2.2).

It is time now to return to the two systems of equations (5.2) and $(5.2)_1$ seen to govern our boundary problem. Writing

$$(6.1)\qquad B_k=b_k(\tau,D_x)=\sum_{|\alpha|\le N^k} b^k_\alpha(\tau)D_x^\alpha \ ,\ k=1,\dots,M\ ,\ N^k\le N-1\ ,$$

in our boundary coordinates of sec.3, using (5.9), (5.10), (5.11) and (5.24),(5.25),(5.27), equations (5.2) assume the form

$$(6.2)\qquad \sum_{k=0}^{N-1} g_{lk}(\tau,D_\tau)\psi_k = \varphi_l - B_l\omega\ ,\ l=1,\dots,M\ ,$$

$$\text{where } g_{lk}(\tau,\xi^\Delta)=(-i)^k \sum_{|\alpha|\le N^l} b^l_\alpha(\tau)\,\pi^{+,k,\alpha}(\tau,\xi^\Delta)\ .$$

(We write $\pi^\pm$ of (5.24) as $\pi^{\pm,k,\alpha}$.) In view of (5.25) the g_{lk} are smooth in τ,ξ^Δ , and have asymptotic expansions

$$(6.3)\qquad \begin{gathered} g_{lk} = \sum_{j=0}^\infty g_{lkj}(\tau,\xi^\Delta)\ (\mathrm{mod}\ |\xi^\Delta|)\ \text{at}\ \xi^\Delta=\infty\ ,\\ g_{lkj}=(-i)^k \sum_{|\alpha|\le N^l} b^l_\alpha(\tau)\pi^{j,+,k,\alpha}(\tau,\xi^\Delta)\ . \end{gathered}$$

Here the $\pi^{j,\pm,k,\alpha}$ are given as sums of residues, of the $p_j=p^{j,k,\alpha}$, in the upper and lower half-plane, respectively. For every j,k,α and τ,ξ^Δ the poles of $p_{j,k,\alpha}(\tau,\xi^\Delta)$ are the roots of $c_N(0,\tau,\kappa,\xi^\Delta)=0$

Looking at (5.11) we first seek an explicit form for g_{lk0} :

$$(6.4)\quad g_{lk0}=\sum_{\mathrm{Im}\kappa_l>0}\mathrm{Res}_{\kappa_l}(-i\kappa)^k\Big(\sum_{|\alpha|\le N^l} b^l_\alpha(t)(\kappa,\xi^\Delta)^\alpha\Big)/c_N(0,\tau,\kappa,\xi^\Delta),$$

simplifying to

$$(6.5)\qquad g_{lk0}(\tau,\xi^\Delta)=\int_{\gamma_+} d\kappa(-i\kappa)^k b_l(\tau,(\kappa,\xi^\Delta))/c_N(0,\tau,\kappa,\xi^\Delta)\ ,$$

with a positively oriented countour $\gamma_+=\gamma_+(\tau,\xi^\Delta)$ in $\mathrm{Im}\,\kappa>0$, surrounding all poles there. Similarly we get

$$(6.6)\qquad g_{lkj}=\int_{\gamma_+}d\kappa\theta(-i\kappa)^k\sum_{|\alpha|\le N^l}b^l_\alpha(\tau)(r_{j,k,\alpha}/c_N^{j+1})(0,\tau,\kappa,\xi^\Delta)\ ,$$

with the polynomials $r_j=r_{j,k,\alpha}$ in $\xi=(\kappa,\xi^\Delta)$ of (5.11). Note, the $r_{j,k,\alpha}$ are of degree $\le Nj-j+|\alpha|+k$. Thus the polynomials in the numerator of (6.6) are of degree $\le Nj-j+k+N^l$. Arguing as for prop.5.5 we find that the g_{lkj} are (finite) sums of homogeneous functions with largest degree $-N-j+k+N^l+1$. After reordering we thus get a homogeneous asymptotic series, and the following result:

<u>Proposition 6.1</u>. The system (5.2) of boundary conditions translates into a system of pseudodifferential equations

$$(6.7)\qquad \sum_{k=0}^{N-1}G_{lk}\psi_k=\varphi_l-B_l\omega\ ,\ l=1,\dots,M\ ,$$

with ψdo's G_{lk} on Γ. After a local coordinate transform near a boundary point onto the coordinates (ν,τ) of sec.3 the symbol $g_{lk}(\tau,\xi^\Delta)$ allows a homogeneous asymptotic expansion

$$(6.8)\qquad g_{lk}(\tau,\xi^\Delta)=\sum_{j=0}^{\infty}g_{lkj}(\tau,\xi^\Delta)\ (\mathrm{mod}\ |\xi^\Delta|)\ \text{at}\ \xi^\Delta=\infty\ ,$$

where g_{lkj} are homogeneous of degree $-N+N^l+1-j+k$ in ξ^Δ, and smooth in τ. Moreover, the highest order terms are given by

$$(6.9)\qquad g_{lk0}(\tau,\xi^\Delta)=\int_{\gamma_+}d\kappa(-i\kappa)^k b_{l,N^l}(\tau,\xi^\Delta+\kappa\vec{n})/c_N(0,\tau,\xi^\Delta+\kappa\vec{n})\ .$$

Here $b_{l,N^l}(\tau,\xi^\Delta)=\sum_{|\alpha|=N^l}b^l_\alpha(\tau)\xi^\alpha$ is the principal part polynomial of $b_l(\tau,\xi)=b_l(\tau,\xi_1,\xi^\Delta)$. In (6.9) we used the notation $\vec{n}=(1,0,..,0)$ $=e_1=$ interior unit normal. $\gamma_+=\gamma_+(\tau,\xi^\Delta)$ is the above countour. For existence of a solution of (5.2') we must find density functions $\psi_0,\dots\psi_{N-1}$ solving the system (6.7). Here we leave the term $B_l\omega$, $\omega\in\ker C$, undetermined, noting just that they are $C^\infty(\Gamma)$-functions, belonging to a <u>finite dimensional</u> space.

In matters of uniqueness we also will translate the system $(5.2)_1$ into a system of ψdo's. A similar procedure may be applied: We now replace B_l, $l=1,\dots,M$, by $D^l_{x_1}$, $l=0,\dots,N-1$, and interior by exterior derivatives. We get the following:

<u>Proposition 6.2</u>. The system $(5.2)_1$ is equivalent to a system of pseudo-differential equations of the form

$$(6.10)\qquad \sum_{k=0}^{N-1}H_{lk}\psi_k=D^l_{x_1}\omega\quad ,\ l=0,\dots,N-1\ ,$$

with ψdo's H_{lk} on Γ . Locally, in the boundary coordinates of sec. 6, the symbol h_{lk} of H_{lk} allows a homogeneous asymptotic expansion

(6.11) $$h_{lk}(\tau,\xi^{\Delta}) = \sum_{j=0}^{\infty} h_{lkj} \quad (\mathrm{mod}\ \xi^{\Delta}) \text{ at } \xi^{\Delta}=\infty ,$$

where h_{lkj} is homogeneous of degree $l+1-N-j+k$. Specifically,

(6.12) $$h_{kl0}(\tau,\xi^{\Delta}) = (-i)^{k}\int_{\gamma_-} d\kappa\, \kappa^{k+l}/c_N(0,\tau,\xi^{\Delta}+\kappa\vec{n}) \ , \ k,l=0,..,N-1.$$

To prove uniqueness - or rather the finite dimension of the null space - we must show that (6.7) and (6.10) together, imposed on a set of density functions ψ_j , imply $\psi_j = 0$ (allow only finitely many linearly independent solutions). Again the right hand side of (6.10) is $C^{\infty}(\Gamma)$ and belongs to a finite dimensional space.

We now are equipped for a result on the boundary problem (5.2)' . For simplicity assume N even, and M=N/2 . We impose

Condition L-S: (The Lopatinski-Shapiro conditions) (i) In the boundary coordinates of sec.3, for every $0\neq\xi^{\Delta}\in \mathbb{R}^{n-1}$, the equation

(6.13) $$c_N(0,\tau,\xi^{\Delta}+\kappa\bar{n}) = 0 \quad , \ \kappa\in \mathbb{C} ,$$

admits exactly k=N/2 roots $\kappa_l=\kappa_l(\tau,\xi^{\Delta})$, $l=1,...,N/2$, in the upper half-plane $\mathrm{Im}\,\kappa >0$, counting multiplicities.

(ii) For every $0\neq\xi^{\Delta}\in \mathbb{R}^{n-1}$ the polynomials

(6.14) $$b_{l,N^l}(\tau,\xi^{\Delta}+\kappa\bar{n}) \ , \ l=1,...N/2 ,$$

are linearly independent mod $c_N^{+}(\kappa)=\Pi_{\mathrm{Im}\kappa_l>0}\ (\kappa-\kappa_l)$.

Theorem 6.3. For an even N the elliptic boundary problem (5.2)' , under Lopatinski-Shapiri conditions, is normally solvable.

Proof. The point is that, under cdn.L-S, (a) we have the matrix $((g_{lk0}(x,\xi^{\Delta})))_{l=1,...N/2,k=0,...,N-1}$ of symbols of rank N/2 , and (b) the matrix of symbols $((h_{lk0}(\tau,\xi^{\Delta})))_{lk=0,...,N-1}$ of rank N/2 as well , and (c) both matrices together of rank N .

More precisely, under cdn.L-S, the polynomial $c_N^{+}(\kappa)$ is of (precise) degree M=N/2. By the Euclidean algorithm we may write

(6.15) $$\beta_l(\kappa)=b_{l,N^l}(\tau,\xi^{\Delta}+\kappa\vec{n}) = \sigma_l(\kappa)c_N^{+}(\kappa)+\rho_l(\kappa) \ , \ \text{degree } \rho_l<M ,$$

with uniquely determined polynomials $\rho_l(\kappa)$ of degree <M . cdn.(ii) of L-S means that the M polynomials $\rho_1,...,\rho_M$ are linearly independent. Looking at (6.9) we find that

$$(6.16)\qquad g_{1k0} = \int_{\gamma_+} d\kappa(-i\kappa)^k \rho_1(\kappa)/c_N \ ,$$

since the integrand involving $\sigma_1(\kappa)c_+(\kappa)$ is regular inside γ_+, so that the integral is 0. Linear independence of the ρ_1 means that

$$(6.17)\quad \rho_1(\kappa)=\sum_{j=0}^{M-1}\rho_{1j}\kappa^j \ , \ P=((\rho_{1j}))_{1=1,..,M,j=1,..,M-1} \text{ invertible.}$$

Focus on the matrix $((g_{1k0}))_{1=1,..,M,k=0,..,M-1}=Z$. To show that Z is invertible we must show that

$$(6.18)\qquad Y=((\int_{\gamma_+} d\kappa\ \kappa^{j+1}/c_N(0,\tau,\kappa,\xi^\Delta)))_{j,1=0,..,M-1}$$

is invertible, since, essentially, Z is the product of Y and P . But, indeed, Y of (6.18) is invertible: Suppose not, then there exists a polynomial $\theta(\kappa)$ of degree < M , not $\equiv 0$, such that

$$\int_{\gamma_+} d\kappa\ \theta(\kappa)\lambda(\kappa)/c_N(\kappa) = 0 \quad \text{for every polynomial } \lambda(\kappa) \text{ of degree } <M.$$

However, $\theta(\kappa)/c_N(\kappa)$ must have poles inside γ_+, since $\frac{1}{c_N}$ has poles of total order M there, while τ is only of degree <M , hence cannot cancel all poles. If κ_0 is a pole of θ/c_N then set $\lambda=\Pi(\kappa-\kappa_1)$, the product running over all other poles of τ/c_N . This choice of $\lambda(\kappa)$ is permitted, but the integral cannot be zero now - it will give the residue of a holomorphic function with a (genuine) simple pole at κ_0 . Thus a contradiction results.

This verifies (a) above, giving a corner of the matrix of rank M . Note that the same argument shows that <u>any</u> M×M-section

$$(6.19)\qquad ((g_{1k0}(\tau,\xi^\Delta)))_{1=1,..,M,k=R,...,R+M-1} \ , \text{ for } R=0,1,..,M,$$

is nonsingular , for all τ,ξ^Δ , $\xi^\Delta\neq 0$. Also, note that (b) follows from a similar argument: just note that $c_N^-(\kappa)=\Pi_{\mathrm{Im}\kappa_1<0}(\kappa-\kappa_1)$ also is of degree M, and that our matrix P above resembles the matrix (6.12). Regarding statement (c): Suppose we have

$$(6.20)\qquad \sum_{k=0}^{N-1} g_{1k0}z_k=0 \ , \ 1=1,..,M \ , \text{ and } \sum_{k=0}^{N-1} h_{1k0}z_k=0 \ , \ 1=0,..,M-1 \ .$$

Repeating the above argument, this means existence of a polynomial $\lambda(\kappa)$ of degree <N (with coefficients $(-i)^k z_k$) such that

$$(6.21)\quad \int_{\gamma_+} d\kappa\frac{\lambda}{c_N}\kappa^j=0 \ , \ j=0,..,M-1 \ , \text{ and } \int_{\gamma_-} d\kappa\ \frac{\lambda}{c_N}\kappa^j=0 \ , \ j=0,..,M-1 \ .$$

Thus it follows that λ/c_N is regular inside both curves $\gamma_\pm$. Since a non-zero polynomial of degree <N cannot cancel all the poles of

$1/c_N$ it follows that $\lambda=0$, i.e., $z_0=\ldots=z_{N-1}=0$. Note we only used half of the matrix h_{1k0} , and have shown that the square matrix

$$(6.22) \qquad (g_{1k0},\ldots,g_{Mk0},h_{0k0},\ldots,h_{M-1,k,0})_{k=0,\ldots,N-1}$$

is nonsingular for every (τ,ξ^{Δ}) , $\xi^{\Delta}\neq 0$.

For the proof of thm 6.3. it now will be necessary to convert the nonsingularity of our matrices (6.19) and (6.22) into ellipticity of corresponding systems of equations. In that respect is should be noticed that the principal symbols g_{1k0} and h_{1k0} are homogeneous in ξ^{Δ} , but not of the same degree.

To remedy this we use the operator $\Lambda=P^{1/2}$, with $P=P_s$ of IV, (5.3), on Γ , with its inverse $\Lambda^{-1}=Q^{1/2}$, both ψdo's in $LC_{\pm 1}$, as discussed in IV,5. Again, we might choose instead $\Lambda^{\pm 1}$ as (inverse) square root of the second order elliptic differential operator

$$(6.23) \qquad L = \sum\omega(1-\Delta)\omega ,$$

with $(1-\Delta)$ in local coordinates, and a subordinated partition$\{\omega\}$ of a finite atlas on Γ . This choice indeed will work, but will require an additional effort we tend to avoid.

In (6.7) and (6.10) introduce

$$(6.24) \qquad \psi_k=\Lambda^{-k}\omega_k , \; k=0,\ldots,N-1 .$$

Also, multiply the 1-th equation (6.7) by Λ^{N-1-N^1} , and the 1-th equation of (6.10) by Λ^{N-1-1}. After this modification we arrive at a new pair of systems of ψde's, now all of order 0, and in the new unknown functions ω_k. The nonsingularity of (6.19) and (6.22) now indeed imply that (α) the modified system (6.7) is elliptic of order 0; (β) the modified system (6.7) and the first M equations of the modified (6.10) again form an elliptic system of N equations in N unknown functions. Green inverses may be constructed.

Conclusion: (1) Equations (6.7) are solvable if $\varphi_1-B_1\omega$ satisfy finitely many linear conditions - i.e. the φ_1 satisfy finitely many conditions, since the $B_1\omega$ are finite dimensional.

(2) Equations (6.7) and (6.10) together , with zero right hand sides, have at most a finite dimensional solution space. Again since the $B_1\omega$ are finite dimensional this means that there can be at most a finite dimensional set ψ_k solving both (6.7) (with $\varphi_k=0$) and (6.10). This completes the proof of thm.6.3.

Problems: 1)Set up a generalized set of Lopatinski-Shapiro conditions working for an elliptic system of R equations in R unknowns, under proper boundary conditions, just as (5.2) . Then generalize thm.6.3 to this case. Hint: Virtually all discussions of the four preceding sections generalize immediately. However, it now may be practical to leave a part of cdn's L-S in a form using matrices of complex integrals. 2) Same as (1) for an even order PDE on a comp-pact C^∞-manifold Ω with boundary. 3) Same as (1) for a PDE mapping between sections of vector bundles over Ω of equal dimension (This is completely formal). 4) Same as (1) for an exterior problem:

(6.25) $\quad c(x,D)u=f$, outside Γ , $B_j u=\varphi_j$ on Γ , $u^{\sim}$, $f^{\sim} \in \mathcal{S}'(\mathbb{R}^n)$,

where again $u^{\sim}, f^{\sim}$ denote the zero-extensions (=0 in Ω, this time). 5) Same as (1), for a Riemann-Hilbert type problem:

(6.26) $\quad c(x,D)u=f$ in $\mathbb{R}^n\setminus\Gamma$, $u,f\in \mathcal{S}'(\mathbb{R}^n)$, $B_j u=\varphi_j$, $j=1,\dots,M$.

where now $B_j = B_{j,+} + B_{j,-}$ with $B_{j\pm}$ containing interior (exterior) derivatives only, resp. 6) Same as (1) for a noncompact domain Ω with "conical boundary Γ ", in the following sense: The homeomorphism $s(x)=\frac{x}{\langle x\rangle}$ of IV,1 maps $\Omega\cup\Gamma$ onto a submanifold of B_1 with closure a submanifold with boundary of B_1^c. (Then the φ_j must satisfy suitable conditions at ∞. For example, in a chart of an infite point, after a coordinate transform onto a piece of a halfspace we might require $\varphi_j \in \mathcal{S}(\mathbb{R}^{n-1})$ - or only $\mathcal{S}'(\mathbb{R}^{n-1})$? . Also, of course, we then must require $u^{\sim}$, $f^{\sim} \in \mathcal{S}'(\mathbb{R}^n)$.

7. Hypo-ellipticity, and the classical parabolic problem.

We now want to focus on some problems with features similar to those of the heat equation - example (4.3) of ch.0. Generally such problems will be of the form of an abstract Cauchy problem

(7.1) $\quad \partial u/\partial t = L(t)u + f(t), \ t>0 \ , \quad u(0) = \varphi ,$

where $L(t):\mathfrak{X} \to \mathfrak{X}$ denotes some family of linear operators, depending on t , and where f(t) is a functions of t taking values in $\mathfrak{X}$. Also, $\varphi\in \mathfrak{X}$, and the solution u(t) takes values in $\mathfrak{X}$.

The theory of abstract equations of this form, in a vector space $\mathfrak{X}$ with topology is well investigated ([Ka_1] , [Fa_1]). One speaks of an evolutionary problem. The abstract differential equation (7.1) often is called an evolution equation.

Some formal comments on evolution equations: First consider the case of a homogeneous equation -i.e. f=0. If a unique solution exists for all $\varphi\in X$ then a family $\{U(t):t\geq 0\}$ of linear operators $U(t):X\to X$ is defined by setting $U(t)\varphi=u(t)$ with the solution u at t for the initial value φ . Actually such a family exists for the initial-value problem at an arbitrary point τ instead of 0:

$$(7.1)_\tau \qquad \partial_t u= L(t)u \ , \ t>\tau \ , \ u(\tau)=\varphi \ .$$

Let $U(\tau,t)$, $t\geq\tau$ be the corresponding family. Then, for general $f(t)$ defined for $t\geq 0$, the solution of (7.1) formally is given by

$$(7.2) \qquad u(t) = U(\tau,t)\varphi + \int_0^t U(\tau,t)f(\tau)d\tau \ .$$

It depends on properties of the family $U(\tau,t)$ to be derived whether the integral in (7.2) exists, and a derivative $\partial_t u(t)$ is meaningful. However, assuming that this is true, and that conventional rules hold, it is confirmed at once by formal differentiation that u(t) of (7.2) satisfies (7.1).

The family $\{U(\tau,t) : t\geq\tau\}$ is called the (family of) <u>evolution operator</u>(s) (or solution operators).

For constant $L(t)=L$ and $f(t)\equiv 0$ the solution formally may be written as $u=\exp(Lt)\varphi$. We get $U(\tau,t)=e^{L(t-\tau)}$. The collection $\{U(t)=U(0,t)=e^{Lt} : t\geq 0\}$, then has the properties of a <u>semi-group</u>:

$$(7.3) \qquad U(0)=1=\text{identity} \ , \ U(t)U(s)=U(s+t) \ , \ s,t\geq 0 \ ,$$

as follows from the fact that then $U(\tau,t)=U(t-\tau)$.

In the cases considered here X will be a space of functions or distributions on some domain or manifold Ω , and $L(t)$ a family of elliptic differential operators on Ω, of the types studied in sec's 1-6: Ω is a "free space", as in sec.1. Or else, $\Omega\subset\subset\mathbb{R}^n$ has smooth boundary -then we define $L(t)$ by a boundary problem (2.1)". That is, $L(t)$ is an unbounded operator $L(t):\text{dom } L = Y \to X=C^\infty(\Omega)$, where (for a domain $\Omega\subset\subset\mathbb{R}^n$ with smooth boundary Γ) , we define

$$(7.4) \qquad \begin{gathered} Y = \text{dom } L = \{u\in X: B_j u=0 \ , \ j=1,\dots,M\} \ , \\ L(t)u = A(t)u \ , \text{ as } u\in Y \ , \end{gathered}$$

with a system $\{A(t_0),B_j\}$, for a fixed t_0 , of PDE's fitting (2.2)"

Generally, $A(t_0)$ is assumed elliptic. But stronger conditions are needed. First we assume that $\partial_t-A(t)$ (as a PDE in (t,x)) is <u>hypo-elliptic</u>. Second, we require conditions insuring (7.1) to have a unique abstract solution in some Hilbert or Banach space X. Such solution $u(t)=u(t,x)$ then will turn out to be smooth, either by hypo-ellipticity or by cor.4.3, because it proves a distribution solution of the PDE $\partial_t u-A(t)u=f$ (under proper boundary cdn's.

There are well developed abstract tools available, for this 2-pivot approach: For example we have a detailed abstract semi-group theory (cf. [HP] , [Yo_1] , [Ka_1], [DS_1]). Also, an abstract theory of evolution equations [Ka_5]. For simplicity we stay with semi-groups - i.e., L(t) of (7.1) is independent of t.

<u>Definition 7.1</u>. A strongly continuous semi-group of a Hilbert or Banach space $\mathcal{X}$ is defined as a 1-parameter family $\{U(t):t\geq 0\}$ of $U(t)\in \mathcal{L}(\mathcal{X})$ which is strongly continuous in t (i.e., $\lim_{t\to t_0} U(t)u= U(t_0)u$ in $\mathcal{X}$, for every $u\in \mathcal{X}$, $t_0\geq 0$), and satisfies (7.4).
The Hille-Yosida-Phillips theorem [DS_1],p.624, states that a semi-group always is linked to a closed linear operator A: dom A $\to\mathcal{X}$, dom A $\subset \mathcal{X}$, called the <u>infinitesimal generator of</u> $\{U(t)\}$, where

$$Au=dU(t)u/dt(0) = \lim_{t\to 0,t>0}((U(t)-U(0))/t) , \tag{7.5}$$

for $u\in$ dom A $=\{$all $u\in\mathcal{X}$ for which the limit exists$\}$.

If $A\in \mathcal{L}(\mathcal{X})$ is a bounded operator, then we have

$$U(t) = \sum_{j=0}^{\infty}(At)^j/j! = \exp(At) . \tag{7.6}$$

Then U(t) is uniformly continuous in t - and even extends to an entire function of t, defined on $\mathbb{C}$. In the general case (7.6) no longer makes sense, although one still might like to write U(t) =exp(At). The theorem describes the class of closed operators occurring as infinitesimal generators of a strongly continuous U(t):

<u>Theorem 7.2</u>. (Hille-Yosida-Phillips) The class of infinitesimal generators of C^0-semigroups coincides with the class of closed operators A such that (i) the resolvent $R(\lambda)=(A-\lambda)^{-1}$ exists for a halfline $\{\lambda>\lambda_0\}$, (ii) for a constant M>0 independent of λ,j we get

$$\|R(\lambda)^j\| \leq M(\lambda-\lambda_0)^{-j} , \lambda>\lambda_0 , j = 1,2,\ldots . \tag{7.7}$$

Then we have

$$U(t) = \lim_{j\to\infty}(1-\tfrac{t}{j}A)^{-j} = \lim_{j\to\infty}\{(-\tfrac{t}{j})^j R(\tfrac{j}{t})^j\} , t>0 , \tag{7.8}$$

and, vice versa,

$$R(\lambda) = - \int_0^{\infty} dt e^{-\lambda t}U(t)dt , \text{Re } \lambda >\lambda_0 . \tag{7.9}$$

For $\varphi\in$ dom A we have $u(t)=U(t)\varphi\in C^1([0,\infty),\mathcal{X})$.

One will recognize well known formulas for the exponential function as basis of (7.8) and (7.9). Note also that, under (7.7), formula (7.8) defines the resolvent in an entire right half-plane.

Looking at (7.1) we will focus on the following two cases:

Case (F) : $L=A=a(x,D)$ with an md-elliptic differential operator on a manifold Ω with conical ends. (or a compact manifold Ω - then A needs only to be elliptic) .

Case (B) : L is an operator of the form (7.4) involving a domain Ω with smooth boundary, an elliptic differential expression $A=a(x,D)$, and Lopatinski-Shapiro-type boundary expressions B_j .

In order to qualify thm.7.2 we will look at (7.1) in the case where $\mathcal{X}$ is a Hilbert or Banach space. Choose $\mathcal{X}=\mathcal{H}=L^2(\Omega)$, in either case. Then (7.4) defines an unbounded operator of $\mathcal{H}$ - in case (B), but this operator is not closed. Its domain is dense in $\mathcal{H}$, and its adjoint L^* exists (its domain contains $C_0^\infty(\Omega)$, dense in $\mathcal{H}$, hence is dense). Thus the closure L^{**} exists. We may look at the problem

$$(7.1') \qquad u(t)\in C^1([0,\infty),\mathcal{H}) \ , \ \partial u/\partial t = L^{**}u \ , \ t\geq 0 \ , \ u(0) = \varphi \in \mathcal{Y} .$$

Similarly, in case (F) , we introduce the operator L by

$$(7.10) \qquad \mathrm{dom}\, L = C_0^\infty(\Omega) \ , \ Lu= Au \ , \ u \in \mathrm{dom}\, L .$$

Again, dom L is dense in $\mathcal{H}$, dom L^* contains $C_0^\infty(\Omega)$, thus is dense; the closure L^{**} exists, and we may look at (7.1') with this L^{**} .

Now, suppose the closed operator L^{**} of $\mathcal{H}$ satisfies the conditions of thm.7.2. Then indeed a solution of (7.1') is given in the form $u(t) = U(t)\varphi$, $U(t) = \exp(L^{**}t)$. Indeed, (7.3) implies

$$(7.11) \qquad \partial u/\partial t(t) = \lim_{h\to 0,h>0}\frac{(U(h)-1)}{h}u(t) = L^{**}u ,$$

in particular, because (7.8) implies $U(t)$: dom $L^{**} \to$ dom L^{**} .

In other words, for $\varphi\in \mathcal{Y}=$dom $L \subset$ dom L^{**}, $u(t)=\exp(L^{**}t)\varphi$ belongs to $C^1([0,\infty),\mathcal{H})$, and solves (7.1'). Moreover, then even u(t) of (7.2) with $\varphi\in$ dom L^{**}, $f\in C([0,\infty),\mathrm{dom}\, L^{**})$ (under graph norm of dom L^{**}) is $C^1([0,\infty),\mathcal{H})$ and solves

$$(7.12) \qquad \partial_t u = L^{**}u+f \ , \ t\geq 0 \ , \ u(0)=\varphi ,$$

by a calculation. Particularly this holds for $\varphi\in \mathcal{Y}$, $f\in C([0,\infty),\mathcal{Y})$. Also one shows uniqueness of such u as a solution of (7.12).

We first focus on case (F). A solution u of (7.12) also is a distribution solution of the PDE $\partial u/\partial t-Au=f$ in $(0,\infty)\times\Omega$. Indeed,

$$(7.13) \ \int u(-\partial\psi/\partial t-A^{\sim}\psi)dxdt= \int(\partial_t u-L^{**}u)\psi dxdt=\langle f,\psi\rangle \ , \ \psi\in C_0^\infty((0,\infty)\times\Omega).$$

Thus, if $f \in C^\infty([0,\infty)\times\Omega)$ and $\partial_t - A$ is hypoelliptic, get $u(t)=u(t,.)$, where $u(t,x) \in C^\infty((0,\infty)\times\Omega)$: u is a classical solution of (7.1).

In fact, we can put to work our boundary hypo-ellipticity theorem (thm.4.4): The solution $u(t)$ of (7.1') is $C^1([0,\infty),\mathcal{H})$. It follows at once that $v=u^\Delta=u$, $t\geq 0$, $=0$, $t<0$, satisfies

$$(7.14) \qquad (\partial_t - A)v = \delta(t)\otimes u(0) + f = \delta(t)\otimes\varphi + f ,$$

in the sense of distributions on $\mathbb{R}\times\Omega$. Clearly the differential expression $C=\partial_t - A$ satisfies (4.5) (with n+1 variables $\nu=t$, $\tau=x$). Using thm.4.4 it follows that v admits an ETE of order 0 , hence it follows that $u \in C^\infty([0,\infty)\times\Omega)$. We have proven thm.7.3, below.

Theorem 7.3. In case (F) assume the differential expression $C=\partial_t - A$ hypo-elliptic. Let the closure L^{**} of the differential operator L defined by A in the domain dom A = $C_0^\infty(\Omega)$ satisfy the assumptions of thm.7.2. Then there exists a solution $u \in C^\infty([0,\infty)\times\Omega)$ of the Cauchy problem (7.1), for general $\varphi \in C_0^\infty(\Omega)$, $f \in C^\infty([0,\infty)\times\Omega)$, supp $f \subset [0,\infty)\times K$, $K \subset\subset \Omega$. The solution is unique if Ω is compact.

Proof. After the above, we only comment on uniqueness: The solution of (7.1') indeed is unique, by thm.7.2. If Ω is compact then $u(t)=u(t,.)$ of thm.7.3 also solves (7.1), hence must be unique.

Remark 7.4. It is well known that such solution needs not to be unique if Ω is non-compact. Of course, the condition at $x=\infty$ that $u(t) \in C^1([0,\infty),\mathcal{H})$, $\mathcal{H}=L^2(\Omega)$, insures uniqueness as well.

Now let us look at case (B). If the semi-group $U(t)=\exp(L^{**}t)$ exists we again get a unique solution of (7.12), for $\varphi \in C^\infty(\Omega\cup\Gamma)$, $f \in C^\infty([0,\infty)\times\Omega\cup\Gamma)$, in the form (7.13). Using a localized version of thm.4.4 (cf. rem.4.8) it then may be concluded, as in the free case, that $u(t,x) \in C^\infty([0,\infty)\times(\Omega\cup\Gamma))$ - assuming that (i) $\partial_t - A$ is hypo-elliptic, and (ii) A is noncharacteristic along Γ :

Theorem 7.5. In case (B), assume that L^{**} is infinitesimal generator of a C^0-semi-group. Assume (i) and (ii) above. Then (1.1) admits a unique solution $u(t)=u(t,x) \in C^\infty([0,\infty)\times(\Omega\cup\Gamma))$, for every $\varphi \in C^\infty(O\cup G)$, $f \in C^\infty([0,\infty)\times(\Omega\cup\Gamma))$, φ and $f(t_0,.)$ satisfying the boundary conditions for all $t_0 \in [0,\infty)$.

Finally we ask: Will our results on elliptic problems, of sec's 1-6, give control on cdn.(7.6) for L^{**} above: Clearly (i), (ii) of thm.7.2 only involve $(L^{**}-\lambda)^{-1} = R(\lambda)$, for L^{**}. Elliptic theory will supply a Fredholm inverse, even a (special) Green inverse for $L-\lambda$: In case (F) we need md-ellipticity of $A-\lambda$, for case (B) we need A elliptic and cdn's L-S for B_j (cf. thm.8.3).

Still, looking at a variety of examples one finds that md-ellipticity of $A-\lambda$ alone in case (F) or ellipticity and cdn. L-S for case (B) though it might be helpful, is not enough.

The insight of the present section is the fact that (i) hypoellipticity of $C=\partial_t-A$ and (ii) the non-characteristic property for Γ with respect to A (or $(0,\infty)\times\Gamma$ with respect to ∂_t-A) reduces the Cauchy problem (for case (F) and case (B)) to a study of the resolvent $(L^{**}-\lambda)^{-1}$, inverse of the closure of L with respect to $\mathcal{H}$.

Commonly one refers to an evolutionary problem (7.1) with hypoelliptic ∂_t-A and well defined semi-group $U(t)=\exp(L^{**}t)$ as a parabolic problem. Often one requires also that U(t) is analytic (cf. Yosida [Yo_1], also Fattorini [Fa_1]). As application we mention:

Example 7.6.

$$\partial u/\partial t= \textstyle\sum_{j,k=1}^{n} \partial/\partial x_j(a_{jk}(x)\partial u/\partial x_k) + f \ , \ t\geq 0 \ , \ u(0,x)=\varphi \ , \tag{7.15}$$

either in free space ($x\in \mathbb{R}^n$) or in a bounded domain Ω with smooth boundary Γ. In the latter case one will need a Dirichlet condition ($u=0$ on Γ) or another self-adjoint boundary condition. Assume the symmetric matrix $((a_{jk}(x)))$ real and >0 for all x (and smooth). In case $\Omega=\mathbb{R}^n$ one assumes that $L-\lambda$ (of (7.10) is md-elliptic of order (2,0), for $\lambda>0$. This requires $a_{jk}\in \psi c_0$, i.e.,

$$a_{jk}^{(\alpha)}(x) = O(\langle x\rangle^{-|\alpha|}) \ , \ x \in \mathbb{R}^n \ , \text{ all } \alpha \ , \tag{7.16}$$

and A uniformly elliptic in $\mathbb{R}^n$.

Under such assumptions the operator $C=\partial_t-A$ indeed satisfies cdn's (i) and (ii) . Moreover, L^{**} then is self-adjoint (admits a spectral decomposition) and semi-bounded below (sec.8,9). Thm.7.2 is applicable for $A=L^{**}$. Hence thm.7.3 (or thm.7.5) apply.

8. Spectral- and semigroup theory for ψdo's.

In this section we shortly consider spectral theory of an unbounded closed linear operator L:dom L→ $\mathcal{H}$ with domain dom L $\subset \mathcal{H}$ of a Hilbert space $\mathcal{H}$. Generally, L will be a realization of a differential expression A in the well known sense, where A=a(x,D), a (locally) of the form (1.1), is given either on $\mathbb{R}^n$, or on a compact manifold Ω, or on a noncompact Ω with conical ends, or on a subdomain $\Omega \subset\subset \mathbb{R}^n$ with smooth boundary as in sec.2.

In each case we set $\mathcal{H}=L^2(\Omega)$ - but methods will apply to properly defined L^2-Sobolev spaces as well. A realization of a PDE A on Ω is defined as a (closed linear) unbounded operator L of $\mathcal{H}$ extending the "minimal operator" L_0 (with $L_0 u=Au$ as $u\in \mathrm{dom}\, L_0 = C_0^\infty(\Omega)$) For $\Omega\subset\subset \mathbb{R}^n$ with smooth Γ we assume L the closure (in $\mathcal{H}$) of an operator of the form (7.4) (that operator is called L' here). Equation $L'u=f$ (for $u\in \mathrm{dom}\, L'$) then corresponds to the boundary value problem (2.2"), while $L=L'^{**}$. In all other cases we set $L=L_0^{**}$, i.e., L is defined as the closure of the minimal operator L_0.

We only consider the elliptic case: For a compact Ω or the case of an $\Omega\subset\subset \mathbb{R}^n$ with boundary assume A elliptic and $A, B_1,\dots B_M$ Lopatinski-Shapiro, respectively. For a noncompact Ω (with conical ends) we assume A md-elliptic (of order $m=(N,m_2)$), $m_2\ge 0$, and the coefficients of A such that $A\in LS_{m,\rho,0}$, with $\rho_1=1$, $\rho_2>0$.

The spectrum $Sp(L)$ is defined as complement of the resolvent set $Res(L) = \{\lambda\in \mathbb{C}: (L-\lambda)^{-1}\in \mathcal{L}(\mathcal{H})\}$. Control of $Sp(L)$ (and of spectral theory of L) depends on criteria for existence of $(L-\lambda)^{-1}$. None of our above criteria of elliptic theory gives existence of a precise inverse of a ψdo, but we have a generous supply of results giving a Green inverse. Let us first omit the case $\Omega\subset\subset \mathbb{R}^n$ with $\Gamma\neq 0$ Note that, for $m_2>0$ in case of conical ends (in general for compact Ω) not only do we have A md-elliptic, but we even have $A-\lambda$ md-elliptic, for every $\lambda\in\mathbb{C}$. Thus a Green inverse G_λ of $A_\lambda=A-\lambda$ exists for every $\lambda\in\mathbb{C}$. Moreover, G_λ may be chosen such that

(8.1) $$(A-\lambda)G_\lambda=1-P_{\ker A_\lambda}\ ,\quad G_\lambda(A-\lambda)=1-P_{\ker A_\lambda^*}$$

with finite dimensional orthogonal projections $P_{\ker X}$ of $\mathcal{H}$.

Theorem 8.1. In case of $\Omega=\mathbb{R}^n$, or Ω compact, or Ω noncompact with conical ends assuming that $m_2>0$, if at least one $\lambda_0\in\mathbb{C}$ exists with

(8.2) $$\ker(L-\lambda_0)=0\ ,\quad \mathrm{im}(L-\lambda_0)\ \text{dense in}\ \mathcal{H}\ ,$$

then L is of compact resolvent. That is, $Sp(L)$ is a discrete set, at most countable, with no finite cluster point. The resolvent $R(\lambda) = (L-\lambda)^{-1}$ is a compact operator of $\mathcal{H}$, for every $\lambda\in Res(L)$.

Proof. First we note that $\ker(L-\lambda_0)=\ker(A-\lambda_0)=\{u\in\mathcal{S}'\cap C^\infty: (A-\lambda_0)u=0\}$. Indeed let $L_\lambda=L-\lambda$. Clearly $L_\lambda u=0$ means existence of $u_j\in C_0^\infty$ with $u_j\to u$, $A_\lambda u_j\to 0$, as $j\to\infty$. Convergence in $\mathcal{H}$ implies convergence in $\mathcal{S}'$. Also $A_\lambda:\mathcal{S}'\to\mathcal{S}'$ is continuous. Thus $A_\lambda u=0$ in $\mathcal{S}'$ follows. Since $0\in C^\infty$ and L_λ is hypo-elliptic we conclude that $u\in C^\infty$. Moreover, from III thm.4.1 we know that $u\in\mathcal{S}$, since A_λ is md-elliptic. Vice versa, every $u\in\mathcal{S}$ with $A_\lambda u=0$ belongs to $\ker L_\lambda$. By our assumption it follows that ker $A_{\lambda_0}=\{0\}$. Hence $P_{\ker A_{\lambda_0}}=0$, and (8.1) yields $G_{\lambda_0}A_{\lambda_0}=1$. Specifically, for $u=A_{\lambda_0}{}^{-1}v$ we get $G_{\lambda_0}v=(G_{\lambda_0}A_{\lambda_0})A_{\lambda_0}{}^{-1}v=A_{\lambda_0}{}^{-1}v$, $v\in \operatorname{im}(\Lambda_0-\lambda_0)$. Since im L_{λ_0} is dense the same holds for $\operatorname{im}(L_0=\lambda_0)$, since L_{λ_0} is the closure of $L_0-\lambda_0$. Therefore $(L-\lambda_0)^{-1}$ is bounded. Its domain is $\mathcal{H}$ (since it is closed). Moreover, $R(\lambda_0)=(L_{\lambda_0})^{-1}=G_{\lambda_0}$ in all of $\mathcal{H}$. Moreover, $G_{\lambda_0}\in O(-m)$ is a compact operator of $\mathcal{H}$, by III,thm.5.1. Thus indeed L is an operator of compact resolvent, and the remaining statements follow from standard functional analysis (cf. Kato[Ka_1], III,6,8 or [DS_1]) . Q.E.D.

<u>Corollary 8.2</u>. Theorem 8.1 holds also for $\Omega\subset\subset \mathbb{R}^n$, $\Gamma\neq 0$: If A is elliptic and $(A,B_1,\dots,B_M)$ satisfy cdn. L-S, then (8.2) for at least one λ_0 implies L to be of compact resolvent.

<u>Proof</u>. This depends on the construction of a 'Green inverse' for the operator L' (i.e., the operator L of (7.4)). The term Green inverse should be redefined for operators acting on a bounded domain of $\mathbb{R}^n$: We will take this in the sense of the well known <u>generalized Greens function</u>, defined as integral kernel of a special Fredholm inverse of an operator of the form (7.4).

<u>Theorem 8.3</u>. Under the general assumptions of sec.2 let A be elliptic. Assume that $(A,B_1,\dots,B_M)$ satisfies cdn.L-S. Then there exists an integral operator $G_\lambda : C^\infty(\Omega\cup\Gamma) \to \operatorname{dom} L'$,

$$(8.3)\qquad G_\lambda u(x) = \int_\Omega g_\lambda(x,y)u(y)dy \ , \ u\in C^\infty(\Omega\cup\Gamma) \ ,$$

such that

$$(8.4)\qquad G_\lambda L'_\lambda u=u-P_\lambda u \ , \ u\in \operatorname{dom} L', \ L'_\lambda G_\lambda v=v-Q_\lambda v \ , \ v\in C^\infty(\Omega\cup\Gamma) \ ,$$

where P_λ, Q_λ are projections of finite rank onto ker L' and a complement of im L' in $C^\infty(\Omega\cup\Gamma)$, and where the integral exists as a Lebesgue (or improper Riemann) integral. We call g_λ a generalized Green's function of L'_λ. We get $g_\lambda(x,y)\in C^\infty(\Omega\times\Omega\setminus\{(x,x)\in \Omega\times\Omega\})$. At $x=y$ g_λ has the same singularity as the distribution kernel of the Green inverse of $C-\lambda$, C as in (2.5), as in sec.3. Calling that kernel $h_\lambda(x,y)$, for a moment, we have $g_\lambda-h_\lambda\in C^\infty(\Omega\times\Omega)$. Moreover, $G_\lambda: C^\infty(\Omega\cup\Gamma)\to \operatorname{dom} L'$ extends to a compact operator $\mathcal{H}\to\mathcal{H}$.

<u>Proof</u>. We may assume $\lambda=0$, since A and $A_\lambda=A-\lambda$ have the same principal part, hence A_λ is elliptic ((A_λ,B_j) is L-S) if and only if A ((A,B_j)) is. Given a $u\in \operatorname{dom} L'$ solving $L'u=f$ - i.e., a solution $u\in C^\infty(\Omega\cup\Gamma)$ of (2.2"). Let C be the md-elliptic expression of sec.3

extending A to $\mathbb{R}^n$, and let G be a Green inverse of C, as in sec.3. With the procedure of sec.3 we convert (2.2") into (3.11). Assume ker C = ker $C^{\sim}=0$ for simplicity. Again we are only interested in achieving equality in (3.11) inside Ω , and it suffices to set $\psi_{N/2}=\dots\psi_{N-1}=0$, then solve the system (with interior derivatives)

(8.5) $$\sum_{j=0}^{N/2-1} B_k G\delta^j_{\psi_j} = -B_k f^{\sim} \ , \ k=1,\dots,M=N/2 \ .$$

With the procedure of section 6 - using the ψdo's Λ defined near (6.23) we set $\psi_j=\Lambda^{-j}\omega_j$, and multiply the k-th equation (8.5) by Λ^{N-1-N_1} . One obtains an elliptic system of order 0 for the ω_j ,

(8.6) $$H\omega = -(\Lambda^{N-1-N_1} B_1 Gf^{\sim})_{1=1,\dots,N/2} \ .$$

With a Green inverse K of H we get $\omega=-K(\Lambda^{N-1-N_1}B_1Gf^{\sim})+P\omega$, P of finite rank. Here K will be a ψdo of order 0. Again assume P=0, just to clarify the idea of proof. In detail, we get

(8.7) $$u = Gf^{\sim} + \sum_{j=0}^{N/2-1} G\delta^j_{\psi_j} \ , \ x\in\Omega \ , \text{ where}$$
$$\psi_j = \sum_{k=1}^{N/2}\Lambda^{-j}K_{jk}\Lambda^{N-1-N_k}[B_kGf^{\sim}]_{\Gamma,\mathrm{int}} \ ,$$

where K_{1k} are certain ψdo's on Γ , of order 0 .

The second term in (8.7) is of the form Zf with an integral operator Z on Ω with $C^\infty(\Omega\times\Omega)$-kernel : Let $x\in\Omega$. For $f\in C_0^\infty(\Omega)$,i.e., f=0 near Γ, get $f^{\sim}\in C^\infty(\mathbb{R}^n)$, $\partial_x^\alpha Gf^{\sim} = \int_\Omega \partial_x^\alpha g(x,y)f(y)dy$, with the distribution kernel g of G. The integral exists as Lebesgue (or improper Riemann) integral. $\partial_x^\alpha g$ is C^∞ except at x=y, where it has a singularity of order $N-n-|\alpha|$. The Λ^j may be written as j-th order differential operators with 0-order ψdo's as coefficients - for example $\Lambda=\Lambda^{-1}P_1$, P_s of IV,(5.3). Hence $K_{k1}\Lambda^{N-1-N_1}B_1Gf^{\sim}=\sum J_{k\alpha}\partial_x^\alpha Gf^{\sim}$, with 0-order ψdo's $J_{k\alpha}$, and a sum over $|\alpha|\le N-1$. For $x\in\Gamma$, $y\in$ supp f the kernels $\partial_x^\alpha g(x,y)$ are $C^\infty(\Gamma)$. Hence we get

(8.8) $$\Lambda^{-k}K_{k1}\Lambda^{N-1-N_1}B_1Gf^{\sim}(x) = \int \upsilon_{k1}(x,y)f(y)dy \ , \ \upsilon_{k1}\in C^\infty(\Omega\times\Omega) \ .$$

Combine this with

(8.9) $$G\delta^j_{\psi_j}(x) = (-1)^j\int_\Gamma \partial^j_{\nu_y} g(x,y)\psi_j(y)dS_y \ ,$$

to conclude that the second term in (8.7) is as described.

An operator $Ku(x)=\int_\Omega k(x,y)u(y)dy$ with $k(x,y)=O(|x-y|^{\varepsilon-n})$ is bounded as a map $L^2(\Omega)\to L^2(\Gamma)$, whenever $\varepsilon>\frac{1}{2}$, by a variant of Schur's lemma (We leave the little exercise to the reader). Conclusion: A bounded map $L^2(\Omega)\to L^2(\Gamma)$ is given by $f\to \Lambda^{-k}K_{kl}\Lambda^{N-1-N_l}B_lGf^{\sim}$, and also by the maps $f\to \psi_j$ defined by the second line of (8.7).

Finally look at the maps $L^2(\Gamma)\to L^2(\Omega)$ induced by $\psi\to G\delta^k_\psi$ defined explicitly by (8.9). We may assume $k\le N/2-1<N-1$, since our sum in (8.7) only extends to these cases. For $|\alpha|\le N/2$ look at $\partial^\alpha_x G\delta^k_\psi= \int_\Gamma dS_y\partial^\alpha_x\partial^k_\nu g(x,y)\psi(y)$ where the kernel also has a singularity of order $N-n-k-|\alpha|\ge 1-n$. These operators are of the form adjoint to those $L^2(\Omega)\to L^2(\Gamma)$ used earlier. Thus they are bounded as well, by the same Schur-type argument. Assuming $N\ge 2$, $f\to G\delta^k_{\psi_k}$ defines a bounded operator of $\mathcal{H}$, for every k of the sum (8.7). The operator even is compact, by III,thm.5.1, since ∇X is bounded, for every first order differential expression ∇ (i.e., we may write X as a finite sum $\sum Q_j(L_jX)$ with $,Q_j\in LC_{-1}$).

The kernel of G has the corresponding properties, hence the two lines of (8.7) define an operator of the form (8.3) with all properties of thm 8.3 (for $\lambda=0$). Let us next admit $P\neq 0$ in (8.6). P may be chosen as an orthogonal projection (in $L^2(\Gamma)$) onto a space $\mathcal{X}\subset C^\infty(\Gamma)$ with $\dim\mathcal{X}<\infty$, by standard elliptic theory applied to H. We still get (8.7) if only the right hand side of (8.6) is $\perp \mathcal{X}$. This amounts to $f\perp \upsilon_j$, $j=1,\dots,R$, with $\upsilon_j\in C^\infty(\Omega\cup\Gamma)$. With the orthogonal projection Q in $\mathcal{H}$ onto $\mathrm{span}\{\upsilon_j\}$ introduce $F=Z(1-Q)$, Z the operator of (8.7). Confirm that $L'F=1-Q$, while F of course still is compact in $\mathcal{H}$ and has a kernel just like Z. In particular F is a right Fredholm inverse of L'. But we saw in sec.7 that $L':\mathcal{Y}\to\mathcal{X}$ is Fredholm. The space ker L' consists of C^∞-functions. We get a left Fredholm inverse W of L' such that $WL'=1-P_{\ker L'}$. Then get $WL'F=W-WQ=F-FP_{\ker L'}$, hence F is an $\mathcal{H}$-bounded Fredholm inverse of L'. The special Fredholm inverse satisfying (8.4) differs from F only by additional $\sum\kappa_j\rangle\langle\lambda_j$, $\kappa_j,\lambda_j\in C^\infty$. Theorem 8.3. is established. (We still assume ker C =ker $C^{\sim}$=0 , but its removal is a technicality.)

<u>Proof of cor.8.2</u>. We get $\|(1-P_\lambda)u\|=\|G_\lambda L'_\lambda(1-P_\lambda)u\|\le c\|L'(1-P_\lambda)u\|$, $u\in \mathrm{dom}\,L'$. Taking closure it follows that

(8.10) $$\|Lu\| \ge c\|u\| \text{ , for all } u\perp \ker L' \text{ , } u\in \mathrm{dom}\,L \text{ ,}$$

This implies ker L = ker L' $\subset C^{\infty}(\Omega\cup\Gamma)$. Clearly we may take closure in (8.4) - i.e. substitute L' by L there. Similarly as in thm.8.1 we conclude $G_{\lambda_0} L_{\lambda_0} u=u$, $u\in$ dom L, $G_{\lambda_0} v=L_{\lambda_0}^{-1}v$, $v\in$ im L_{λ_0}. $L_{\lambda_0}^{-1}$ is closed hence im $L_{\lambda_0}=H$, $R(\lambda)=L_{\lambda}^{-1}$ exists (compact) at $\lambda=\lambda_0$. Q.E.D.

Definition 8.4. The operator L is called hermitian if

(8.11) $(u,Lv) = (Lu,v)$ for all $u,v \in$ dom L .

L is called formally dissipative if

(8.12) $(u,Lu) + (Lu,u) \le 0$ for all $u\in$ dom L .

Both conditions may be verified for L_0 (in case $\Omega=\mathbb{R}^n$, Ω compact, Ω with conical ends) or for L' (in case $\Omega\subset\subset \mathbb{R}^n$) only: (8.11) ((8.12)) holds for dom L if and only if it holds for dom L_0 - or dom L'. Indeed, L is the closure of either L_0 or L', respectively.

In the cases without boundary these conditions may be expressed by direct reference to the differential expression A : 'Hermitian' means that $A=A^*$ (with the Hilbert space adjoint A^* of A). 'Dissipative' amounts to $(u,(A+A^*)u)\ge 0$ for all $u\in C_0^{\infty}(\Omega)$ i.e., the minimal operator of the expression $A+A^*$ is positive. Such expression A will be called self-adjoint (dissipative) , respectively.

For a boundary problem L' we also need the above conditions for the expression A . But, in addition, a postulate on the boundary conditions results: Certain self-adjoint or dissipative boundary conditions (with respect to a given expression A) are to be defined, such that L' is hermitian (dissipative) if and only if A is self-adjoint (dissipative) and the boundary conditions are self-adjoint (dissipative) with respect to A .

For the general case of an L-S-system (A,B_j) the discussion with self-adjoint (dissipative) expression A the discussion of general self-adjoint (dissipative) boundary conditions is quite complicated, and will not be attempted. A simple example will be the case of the Laplace operator $L=\Delta$: Self-adjoint L-S-conditions with real coefficients all will be of the form

(8.13) $$b_0\,\partial_{\nu}u+b_1 u=0 \ , \ x\in \Gamma \ ,$$

with real-valued C^{∞}-functions b_0 , b_1 , $b_0^{\,2} + b_1^{\,2} =1$.
We exclude the case $\Omega\subset\subset \mathbb{R}^n$ for the remainder of sec.8. A result similar to thm.8.5 for boundary problems is found in sec.9.

Theorem 8.5. Assume $\Omega=\mathbb{R}^n$, Ω compact , or Ω noncompact, with coni-

cal ends. Let A be md-elliptic of order (N,m_2) with $N>0$, $m_2\ge 0$.
Assertion: 1) If A is self-adjoint then L is not only hermitian but even a self-adjoint operator of $H=L^2(\Omega)$: It admits a unique orthogonal spectral measure (cf. $[C_2]$,I,thm.3.3) If $m_2>0$ then L is of compact resolvent. It admits a complete orthonormal set of eigenfunctions $\{\varphi_j\}_{j=1,2,\dots}$, with $\varphi_j\in S(\Omega)$.

2) If A is dissipative then L is infinitesimal generator of a semi-group $U_t = e^{tL} \in L(H)$, $t\ge 0$, in the sense of thm.7.2. Moreover, the operators U_t are contractions : we have

$$\|U_t\| = 1 . \tag{8.14}$$

Again, for $m_2>0$ L is of compact resolvent.

Proof. The assertion is equivalent to showing that for $\lambda=\pm i\varepsilon$ (for $\lambda=\varepsilon>0$) the equation $(A-\lambda)u=0$ has no nontrivial distribution solution u with $u\in H$. But we know that for md-elliptic $A+\lambda$ every such u belongs to S. Note that $A-\lambda$, for $\lambda=\pm\varepsilon i$ ($\lambda=\varepsilon$) indeed is md-elliptic for small $\varepsilon>0$ due to $m_2\ge 0$, and since we get $a_N(x,\xi)$ real (or $\operatorname{Re} a_N(x,\xi)\ge 0$), for self-adjoint (dissipative) A.
On the other hand, we clearly get $S\subset \operatorname{dom} L$, hence $(L+\lambda)u=0$. Thus

$$\begin{aligned} 0 &= \operatorname{Im}((L\pm\varepsilon i)u,u) = \varepsilon\|u\|^2 \Rightarrow u = 0 \text{ , for A self-adjoint,} \\ 0 &= \operatorname{Re}((L-\varepsilon)u,u) \le -\varepsilon\|u\|^2 \Rightarrow u = 0 \text{ , for A dissipative.} \end{aligned} \tag{8.15}$$

This argument shows that indeed $\operatorname{im}(L\pm\varepsilon i)$ is dense in H for A self-adjoint (since every $u\in H$, $u \perp \operatorname{im}(L\pm\varepsilon i)$ will be a distribution solution (in $H\subset S'$) of $(A\mp i\varepsilon)u=0 \Rightarrow u=0$). Also $(L\pm\varepsilon i)u=0 \Rightarrow u=0$, by an argument as in (8.15), and, moreover $\|(L\pm i\varepsilon)u\|\ge\varepsilon\|u\|$ implies that $R(\mp\varepsilon i) = (L\pm\varepsilon i)^{-1} \in L(H)$ exists. By a well known result (cf. $[C_2]$,I,cor.2.3, or $[Ka_1]$,V,thm.3.16) this implies self-adjointness of L in H - i.e., L has the desired spectral resolution. If $m_2>0$ one may apply thm.8.1 with $\lambda_0=\pm\varepsilon i$: L is of compact resolvent. This implies discrete spectrum of L, and the orthogonal set $\{\varphi_j\}$.

On the other hand, for dissipative L we similarly get existence of $R(\varepsilon) = (L-\varepsilon)^{-1} \in L(H)$, and, $\|R(\varepsilon)\|\le\varepsilon^{-1}$. This implies (7.7) with M=1, by a simple additional argument. Hence thm.7.2 applies, and we also get (8.14) and the compact resolvent - the latter for $m_2>0$ only - from thm.8.1 again, with $\lambda_0=\varepsilon$. Q.E.D.

9. Self-adjointness for boundary problems.

We discuss an extension of thm.8.5 for boundary problems, for some examples like the cdn's (8.13) for $L=\Delta$, and the self-adjoint case. These matters are very technical. Perhaps we point out the high lights and leave complicated calculations to the reader.

For an A as in sec.2 , $\Omega\subset\subset \mathbb{R}^n$, consider the sesquilinear form

$$(9.1)\qquad Q(u,v) = \int_\Omega dx(\overline{Au}v-\bar{u}Av)dx \ ,\ u,v\in C^\infty(\Omega\cup\Gamma) \ .$$

In coordinates (ν,τ) (sec.2) for $u,v\in C^\infty(\Omega\cup\Gamma)$ let $u^\Delta=(\partial_\nu^j u)_{1=0..N-1}$ $v^\Delta=(\partial_\nu^j v)_{j=0,\dots,N-1}$, for $x\in\Gamma$. By partial integration get

$$(9.2)\qquad Q(u,v) = (Qu^\Delta,v^\Delta)_\Gamma \ ,\ \text{where } (u^\Delta,v^\Delta)_\Gamma = \int_\Gamma dS\, \bar{u}^{\Delta T} v^\Delta \ ,$$

with an $N\times N$-matrix $Q=((q_{jk}(\tau,\partial_\tau)))_{j,k=0...N-1}$ of differential expressions $Q_{jk}=q_{jk}(\tau,\partial_\tau)$ in τ . Here $Q_{jk}=0$ for $j+k\ge N$, and Q_{jk} is of order $N-1-j-k$ for $j+k\le N-1$. Specifically the 'anti-diagonal terms' $Q_{jk}=q_{jk}(\tau)$: $j+k=N-1$, all are <u>functions</u>. They are bounded away from zero if A is elliptic. Note iQ is a formally selfadjoint expression on Γ , we get $\overline{Q(v,u)} = - Q(u,v)$, $u,v \in C^\infty(\Omega\cup\Gamma)$, hence

$$(9.3)\qquad (Qu^\Delta,v^\Delta)_\Gamma = -(u^\Delta,Qv^\Delta)_\Gamma \text{ for all } u^\Delta,v^\Delta \in C^\infty(\Gamma) \ .$$

For simplicity consider the case where all other Q_{jk} are of order 0 as well ($A=\Delta$ is such a case). Also assume that the boundary operators B_j all involve normal derivatives only - i.e., in the coordinates (ν,τ) they may be written as a set of $M=N/2$ linear conditions on the vector u^Δ, without any differentiation for τ, of the form $b^{jT}u^\Delta=0$, $x\in G$, $j=1,\dots,M$. (Again this holds for (8.13).)

Then $Q=((q_{jk}(\tau)))$ is an invertible matrix of functions (as lower right triangular matrix with nonvanishing antidiagonal). The boundary conditions $B_j u=0$, $j=1,..,M$, of (2.2") translate into

$$(2.2)^\Delta\qquad u^\Delta\in Z_x \ ,\ \text{for all } x\in\Gamma \ ,$$

with a linear subspace Z_x of $\mathbb{C}^N$ (all u^Δ with $b_j^T u^\Delta=0$) , at $x\in\Gamma$.

The property of a given $f\in C^\infty(\Omega\cup\Gamma)$ to satisfy the condition

$$(9.4)\qquad (f,Au)=(Af,u) \text{ for all } u\in C^\infty(\Omega\cup\Gamma) \text{ with } B_j u=0,\ j=1,..M=N/2,$$

translates into a condition of the form

$$(9.5)\qquad f^\Delta(x)\in (QZ_x)^\perp = Z_x^* \ ,\ \text{for all } x\in\Gamma \ .$$

Clearly (9.5) may be described by 'adjoint boundary conditions'

$$(9.6)\qquad \beta_j^T f^\Delta = 0 \quad , \ j=1,\dots,N/2$$

again. Then L (defined by (A,B_j)) is hermitian if and only if the boundary conditions (9.6) are equivalent to the conditions $B_j u=0$.

<u>Theorem 9.1</u>. Assume A elliptic and (A,B_j) to satisfy cdn. L-S . Let the differential expression Q of (9.2) be of order 0 (i.e., an invertible N×N-matrix of C^∞-functions on Γ) , and let the boundary operators B_j only contain normal derivatives. Assume L of sec.8 hermitian. Then L is self-adjoint - it has a spectral resolution - and is of compact resolvent as well. There exists a complete orthonormal basis of eigenfunctions.

<u>Proof</u>. We argue similarly as for thm.8.5: Given $f\in \mathcal{H}$ with

$$(9.7)\qquad (f,(L'-z)u)=0 \ , \text{ for fixed } z\in \mathbb{C} \ , \text{ all } u\in \operatorname{dom} L' \ .$$

Self-adjointness of the closure $L=L'^{**}$ follows if we can show that (9.7), for $z=\pm i$, implies $f=0$, ([C₂],I,cor.2.3, or [K_1],V,thm.3.16) Now (9.7) implies that $(A-z)f=0$, $x\in\Omega$, in the sense of distributions. Since A is hypo-elliptic conclude $f\in C^\infty(\Omega)$. Let $f^{\sim}$ and C be as in sec.3. We get $(C-\bar z)f^{\sim} = h$, supp $h\subset\Gamma$. The distribution h must be of the form $h=\sum_{j=0}^R \delta^j_{\psi_j}$, $\psi_j\in \mathcal{D}'(\Gamma)$, $\delta^j_{\psi_j}$ of (3.3).

In fact, we claim that R=N-1. Indeed, let $\nu(x)$ - so far only defined near Γ - be extended into a $C^\infty(\mathbb{R}^n)$-function, also called $\nu=\nu(x)$. For any $u\in C^\infty(\Omega\cup\Gamma)$ conclude that $\nu^N u=v\in \operatorname{dom} L'$, since clearly $v^\Delta=(\partial_\nu^j(\nu^N u))=0$. Thus it may be substituted for u in (9.7), implying $\int_\Omega dx\bar f(A-z)(\nu^N u)=0$ for all $u\in C^\infty(\Omega\cup\Gamma)$. In turn, this yields

$$(9.8)\qquad (f^{\sim},(C-z)(\nu^N\varphi)) = \int_{R^n}\bar f^{\sim}(C-z)(\nu^N\varphi)dx=0 \text{ for all } \varphi\in \mathcal{D}'(\mathbb{R}^n),$$

with C and the "zero-extension" $f^{\sim}$ of f (=0 outside $\Omega\cup\Gamma$) defined as in sec.3. In other words, we get $\nu^N(C-\bar z)f^{\sim}=\nu^N h=0$ in $D'(\mathbb{R}^n)$. A calculation shows that $\langle \nu^N\delta^j_\psi,\varphi\rangle=\langle\psi,\partial_\nu^j(\nu^N\varphi)|_\Gamma\rangle_\Gamma=0$ as $j\le N-1$, but $=j!/N!\ \langle\psi,\partial_\nu^{j-N}\varphi|_\Gamma\rangle_\Gamma$, as $j\ge N$. Thus it follows that $\psi_j=0$, $j\ge N$.

Observe that we obtained a more explicit description of the distributions ψ_j in (3.7), but only in case of $f\in C^\infty(\Omega\cup\Gamma)$. With our matrix Q of (9.2) we get

$$(9.9) \qquad \langle h,\varphi\rangle = \langle \sum_{j=0}^{N-1}\delta_{\psi_j}^{j},\varphi\rangle = \langle Qf^{\Delta},\varphi^{\Delta}\rangle_{\Gamma} \ , \ \varphi \in \mathcal{D}' \ ,$$

as a careful comparison of (3.4) - (3.7) with the above shows.

We were assuming that the matrix Q contains no differentiations but simply is an invertible matrix of smooth functions on Γ. On the other hand, we can meaningfully define the vector $f^{\Delta} = =(\partial_{\nu}^{j}f)_{j=0,\dots,N-1}$ also for our above $f \in \mathcal{H}$: We get

$$(9.10) \qquad (C-\bar{z})f^{\sim} = h = \sum_{j=0}^{N-1}\delta_{\psi_j}^{j} \ ,$$

where C is noncharacteristic on Γ . With the same iterative method as in sec.4, conclude the existence of distributions $f_j \in \mathcal{D}'(\Gamma)$, $j=0,\dots,N-1$ such that, locally, near Γ , in the coordinates (ν,τ),

$$(9.11) \qquad (C-\bar{z})(f^{\sim} - \sum_{j=0}^{N-1} \frac{\nu^j}{j!}\otimes f_j) = \sum_{j=0}^{R}\nu^j\otimes\gamma_j \ , \ \gamma_j\in \mathcal{D}'(\Gamma) \ .$$

Clearly these f_j are uniquely determined. A close examination shows that then $f^{\Delta}=(f_j)_{j=0,\dots,N-1}$, may be substituted in (9.9), above, under our present assumptions. In other words, we get

$$(9.12) \qquad \langle (C-\bar{z})f^{\sim},\varphi\rangle = \langle Qf^{\Delta},\varphi^{\Delta}\rangle_{\Gamma} \ , \ \varphi\in \mathcal{D}'(\mathbb{R}^n) \ .$$

Introducing some complex conjugates we conclude that

$$(9.13) \quad 0 = ((C-\bar{z})f^{\sim},\varphi)=(f,(A-z)\varphi|_{\Omega\cup\Gamma})_{\Omega} = (Qf^{\Delta},\varphi^{\Delta})_{\Gamma} \ , \ \varphi\in \mathcal{D}(\mathbb{R}^n) \ .$$

Or, in other words, equation (9.8) reduces to

$$(9.14) \qquad (Qf^{\Delta},u^{\Delta})=0 \text{ for all } u^{\Delta} \text{ with } b^{jT}u^{\Delta} = 0 \ .$$

Or, recalling self-adjointness of the boundary conditions, one concludes that we must have the boundary conditions $B_jf=0$ satisfied for f in the following generalized form: We have

$$(9.15) \qquad B_jf = b^{jT}f^{\Delta} = 0 \ , \ j=1,\dots,M=N/2 \ .$$

We also have $(A-z)f=0$ and may rederive (3.11) in the present case (with G replaced by G_z). Thus $f^{\sim}$ again is given by a sum of multilayer potentials, only the ψ_j no longer are C^{∞} , but $\psi_j\in \mathcal{D}'(\Gamma)$.

Thus the $\omega_j=\Lambda^{-j}\psi_j$ must satisfy a homogeneous elliptic system of ψdo's. Since ellipticity implies hypo-ellipticity, the ψ_j are $C^{\infty}(\Omega)$ after all. Hence also the f_j are C^{∞} , and $f \in C^{\infty}(\Omega\cup\Gamma)$ follows. In other words we get $f\in \operatorname{dom} L'$. Then, however, we must have

$$(9.16) \qquad 0=\operatorname{Im}(f,(L-z)f) = (\operatorname{Im} z)\|f\|^2 \Rightarrow f=0, \text{ if } \operatorname{Im} z \neq 0. \text{ Q.E.D.}$$

10. C^*-algebras of ψdo's; comparison algebras.

In this section we start from the following observation. From III, thm.3.1 we know that $Op\psi c_0$ is an algebra of bounded operators on $H=L^2(\mathbb{R}^n)$ (in fact, a subalgebra of $L(H_s)$, for every $s=(s_1,s_2)$). Moreover, this algebra contains its (Hilbert space) adjoints, by I,2 and I,6. For each s the norm closure of $Op\psi c_0$ in H_s will define a C^*-subalgebra (called A_s) of $L(H_s)$. Commutators in $Op\psi c_0$ are in $Op\psi c_{-e}$, as we know (cf.I,6). Thus III,thm.5.1 implies commutators to be in $K(H_s)$, $s\in \mathbb{R}^2$. It is known that the compact operators of a Hilbert space X form a norm closed ideal $K(X)$ of $L(X)$. We claim that our algebra A_s contains $K(H_s)$, for every s .

Indeed, $Op\psi c_0 \subset A_s$ clearly contains the operators

(10.1) $\quad s_j(M)$, $s_j(D)$, $j=1,\dots,n$, where $s_j(x)=x_j/\langle x\rangle$,

and $s_j(M)u=s_j(x)u(x)$. The ψdo's (10.1) generate a (unital) subalgebra C_s of A_s, a <u>comparison algebra</u> in the sense of [C_2],V,1. Thus $A_s \supset C_s$ contains $K(H_s)$, by [C_2],V,lemma 1.1. (The proof is simple: An irreducible C^*-algebra containing at least one compact operator $\neq 0$ always contains $K(H)$ (Dixmier [Dx_1], cor.4.1.10). Thus we prove C_s irreducible and get $0\neq C\in K\cap C$ (details cf. [C_1],p.130,thm.1.1.).

Quoting another result on C^*-algebras: The quotient of a C^*-algebra by a closed ideal is a C^*-algebra again. Accordingly, the quotient $A_s^{\vee} = A_s/K(H_s)$ is a commutative C^*-algebra.

Let us focus on $A^{\vee} = A_0^{\vee}$. By the Gelfand-Naimark theorem ([DS_2],IX.3.7 or [C_1],p320,thm.7.7) $A^{\vee}$ is (isometrically isomorphic to) a function algebra $C(\mathbb{M})$ with a compact space $\mathbb{M}$. One will ask about the nature of the space $\mathbb{M}$, and the isomorphism $A^{\vee}\to C(\mathbb{M})$.

These facts are intimately related to normal solvability of the equation Au=f, for $A\in A$. First, by a theorem of Atkinson ([C_1], p.271,thm.4.8) an operator $A\in L(H)$ is Fredholm if and only if its coset $A^{\vee}=\{A+K(H)\}$ is invertible in $L(H)/K(H)$. Second, $A^{\vee}$, as a C^*-subalgebra of $L(H)/K(H)$, contains its $L(H)/K(H)$-inverses: $A^{\vee}\in A^{\vee}$ is invertible in L/K if and only if it is invertible in $A^{\vee}$, by another well known result ([C_1],p.322,l.7.15). Hence Au=f (for $A\in A$) is normally solvable if and only if the continuous function a associated to $A^{\vee}$ by above isometry is invertible in $C(\mathbb{M})$: $a(m)\neq 0$, $m\in \mathbb{M}$

In other words: For an $A\in A$ let us introduce a continuous function $a=\sigma_A\in C(\mathbb{M})$, called the <u>symbol</u> of A (relative to A) using the map $A\to A^{\vee}=A/K(H)\to C(\mathbb{M})$. Then the above chain of arguments gives

<u>Theorem 10.1</u>. An operator $A\in A$ is Fredholm if and only if its

symbol $\sigma_A(m)$, defined on $\mathbb{M}$, does never vanish.

As in [C1], [C2] we call $\mathbb{M}$ the symbol space of the algebra A_0 Asking about the nature of $\mathbb{M}$ and the homomorphism $\Sigma: A \to C(\mathbb{M})$ we suspect a relation between 'algebra symbol' σ_A and 'ψdo-symbol' $a(x,\xi)$ $=$symb(A) for the generators $Op\psi c_0$. These questions were discussed in detail for the algebra *C* with generators (10.1) by E.Herman and author, using methods unrelated to ψdo-theory ([CHe$_1$], thm.36 -the algebra *C* is called $\mathcal{T}_0$ there - (cf. also the algebra called $\mathcal{L}_s$ in [C1],p.135/136,pbm's 1-4). Clearly the ψdo-symbols of (10.1) are

$$(10.2)\qquad a(x,\xi)=s_j(x)\ ,\ \text{for } A=s_j(M)\ ,\ a(x,\xi)=s_j(\xi)\ ,\ \text{for } A=s_j(D).$$

The functions $s_j(x)$ are continuous over $\mathbb{R}^n$. Recalling the directional compactification $\mathbb{B}^n$ of II,(3.1), note that the s_j even extend to functions in $C(\mathbb{B}^n)$ - actually, $\mathbb{B}^n$ is described as the compactification of $\mathbb{R}^n$ allowing continuous extension of $s_1,\ldots,s_n$.

Clearly the ψdo-symbols (10.2) of the generators (10.1) of *C* extend to continuous functions over the compact space

$$(10.3)\qquad \mathbb{B}^n\times\mathbb{B}^n = \{(x,\xi):\ x,\xi\in \mathbb{B}^n\}\ .$$

Moreover, by the Stone-Weierstrass theorem, the $s_j(x), s_j(\xi)$ generate the algebra $C(\mathbb{B}^n\times\mathbb{B}^n)$ - they strongly separate the space (10.3).

The result of [CHe$_1$] regarding *C* may be stated as follows.

<u>Theorem 10.2</u>. The symbol space $\mathbb{M}=\mathbb{M}_C$ of the algebra *C* is (homeomorphic to) the boundary (cf. II,3) of the space (10.3), i.e.,

$$(10.4)\qquad \mathbb{M}^0=\partial(\mathbb{B}^n\times\mathbb{B}^n)=\mathbb{B}^n\times\partial\mathbb{B}^n\cup\partial\mathbb{B}^n\times\mathbb{B}^n = \{(x,\xi)\ :\ |x|+|\xi|=\infty\}$$

Moreover, the symbols of the generators are the restrictions to $\mathbb{M}$ of the continuous extensions to $\mathbb{B}^n\times\mathbb{B}^n$ of their ψdo-symbols:

$$(10.5)\quad \sigma_{s_j(M)}=s_j(x)\ ,\ \sigma_{s_j(D)}=s_j(\xi)\ ,\ \text{as } (x,\xi)\in\mathbb{B}^n\times\mathbb{B}^n\ ,\ |x|+|\xi|=\infty$$

Accordingly an operator $A\in C^0$, with the algebra C^0 finitely generated by the operators (10.1), is a Fredholm operator of $L(H)$ if and only if its ψdo-symbol $a(x,\xi)=$ symb(A) satisfies

$$(10.6)\qquad |a(x,\xi)|\ \geq\ \eta > 0 \text{ for all } |x|+|\xi|\geq 1/\eta\ ,$$

for some η chosen sufficiently small.

The last statement was inserted because it may clarify the relation between thm.10.2 and II,thm.1.6: We may rephrase (10.6) by stating that $A\in C^0\subset Op\psi c_0$ is md-elliptic (of order 0). By II, thm.1.6 this is necessary and sufficient for existence of a K-parametrix (hence also a Green-inverse) in $Op\psi c_0$.

On the other hand, $A\in \mathcal{C}^0$ is a finite sum of finite products of the generators (10.1) hence σ_A is the corresponding sum of products of $s_j(x)$ and $s_j(\xi)$. The $s_j(M)$ and $s_j(D)$ do not all commute, but their commutators are in $Op\psi c_{-e}$, hence their ψdo-symbols vanish at $|x|+|\xi|=\infty$. Since sums and products are finite the statement $\sigma_A\neq 0$ on $\mathbf{M}$ of thm.10.1 indeed is equivalent to (10.6). In other words, for $A\in \mathcal{C}^0$ the abstract discussion leading to thm.10.2 gives a very similar necessary and sufficient criterion on existence of a Fredholm inverse. [In fact, similar considerations for the $\mathcal{C}_s$ guarantee existence of a Green inverse as well if and only if A is md-elliptic of order 0 (cf. [C_1],p.149).]

Let us now state a result for the algebra $\mathcal{A}$. The symbol class ψc_0 forms a sub-algebra of $CB(\mathbb{R}^n\times\mathbb{R}^n)$, the set of all bounded continuous functions on $\mathbb{R}^n\times\mathbb{R}^n$. The closure $(\psi c_0)^c$ (under sup-norm of $CB(\mathbb{R}^n\times\mathbb{R}^n)$ is a (unital) commutative C^*-subalgebra of $CB(\mathbb{R}^n\times\mathbb{R}^n)$. Thus $(\psi c_0)^c$ is a function algebra $C(\mathbf{P})$ with the compact space $\mathbf{P}=\mathbf{P}_{e,0}$ of II,3. The space $\mathbb{R}^n\times\mathbb{R}^n$ is densely imbedded in $\mathbf{P}$, as the space of all maximal ideals $m_{x^0,\xi^0}=\{\varphi(x,\xi):\varphi(x^0,\xi^0)=0\}$, for given $(x^0,\xi^0)\in \mathbb{R}^n\times\mathbb{R}^n$. Or, $\mathbf{P}$ is a compactification of $\mathbb{R}^n\times\mathbb{R}^n$. It is known that a 1-1-corresponce between compactifications of $\mathbb{R}^{2n}$ and C^*-subalgebras of $CB(\mathbb{R}^{2n})$ containing 1 and $C^0(\mathbb{R}^{2n})$ exists: Each compactification defines a subalgebra - the algebra of its continuous functions. Each subalgebra defines a compactification - its maximal ideal space. This relation is order preserving. The largest - called Stone-Cech compactification - belongs to $CB(\mathbb{R}^n)$ itself. The smallest is the 1-point compactification - it belongs to $1+CO(\mathbb{R}^{2n})$.

Our $\mathbf{P}$ will be 'larger than' $\mathbf{B}^n\times\mathbf{B}^n$, since ψc_0 contains the generators $s_j(x)$, $s_j(\xi)$ of $C(\mathbf{B}^n\times\mathbf{B}^n)$. The map $\mathbf{P}\to\mathbf{B}^n\times\mathbf{B}^n$ is given as dual of the injection $C(\mathbf{B}^n\times\mathbf{B}^n)\to (\psi c_0)^c$. (For details cf. Rickart [R_1],III,2; Kelley [Ke_1],ch.5,p149; [C_1],IV,lemma 1.5, and p308f.)

<u>Theorem 10.3</u>. The symbol space of $\mathcal{A}=\mathcal{A}_0$, and $\mathcal{A}_s$, $s\in \mathbb{R}^2$, is given by

$$(10.7)\qquad \mathbf{M} = \mathbf{M}(\mathcal{A}) = \mathbf{M}(\mathcal{A}_s) = \partial\mathbf{P} = \mathbf{P}\setminus(\mathbb{R}^n\times\mathbb{R}^n) = \mathbf{M}_{e,0}\ ,\ (\text{cf.II},3).$$

Moreover, for $A=a(x,D)\in Op\psi c_0$ the symbol σ_A equals the restriction to $\mathbf{M}$ of the continuous extension of $a(x,\xi)$ to $\mathbf{P}$, for all $s\in \mathbb{R}^2$.

<u>Proof</u>. The last sentence of thm.10.3 describes a homomorphism τ: $Op\psi c_0\to C(\partial\mathbf{P})$ since $C=c(x,D)=a(x,D)b(x,D)-(ab)(x,D)\in Op\psi c_{-e}$ has $c=$ symb $C=0$ at $x|+|\xi|=\infty$, hence $\tau_C=0$. We tend to show that τ extends to $\mathcal{A}$, with ker $\tau=\mathcal{K}(\mathcal{H})$, hence the induced $\tau^\vee:\mathcal{A}/\mathcal{K}\to C(\partial\mathbf{P})$ is an isometry. This implies that $\partial\mathbf{P}$ and $\mathbf{M}$ are homeomorphic, proving thm.10.3.

Proposition 10.4. The homomorphism τ satisfies the inequality

$$(10.8)\qquad \inf\{\|A+K\| : K\in \mathcal{K}(\mathcal{H})\} \ge \|\tau_A\| \ ,\ A\in \psi c_0 \ ,$$

with L^2-operator norm $\|A+K\|$ and sup norm $\|\tau_A\|$.

Proof. Note that $|\tau_A|$, the restriction to $\partial\mathbf{P}$ of $|a(x,\xi)|$ extended to $\mathbf{P}$ assumes its maximum $\alpha=\|\tau_A\|$ at some $p\in\mathbf{P}$. There exists a sequence $(x^j,\xi^j)\to p$, $x^j,\xi^j\in\mathbb{R}^n$, where $|x^j|+|\xi^j|\to\infty$. Either we have $|x^j|\to\infty$ or $|\xi^j|\to\infty$ or both. We may assume $|\xi^j|\to\infty$, and that the balls $B_j=\{\xi:|\xi-\xi^j|\le 1\}$ are mutually disjoint. Pick $\varphi\in C_0^\infty(\{|\xi|\le 1\})$ with $\|\varphi\|_{L^2}=1$, and let $\varphi_j(\xi)=\varphi(\xi-\xi^j)$, $u_j=\varphi_j^\vee$, $u=\varphi^\vee$, "$^\vee$"= inverse Fourier transform. Then the $u_j=e^{ix\xi^j}u$ form an orthonormal system of $\mathcal{H}$. Let $b(x)$ be a bounded $C^\infty(\mathbb{B}^n)$-function. With "$\approx$" to be defined write

$$(10.9)\qquad \begin{aligned}\|bAu_j\| &= \|b\int d\!\!\!{}^-\xi e^{ix\xi}a(x,\xi)\varphi_j(\xi)\| \approx \|b\int d\!\!\!{}^-\xi e^{ix\xi}a(x,\xi^j)\varphi_j(\xi)\| \\ &=\|a(x,\xi^j)bu_j(x)\| = \|a(x,\xi^j)bu(x)\| \ .\end{aligned}$$

First consider the case where $|x^j|$ remains bounded. In the compactification $\mathbb{B}^n\times\mathbb{B}^n$ of $\mathbb{R}^{2n}$ (x^j,ξ^j) will converge to (x^0,ξ^0) , where $|\xi^0|=\infty$, $|x^0|<\infty$. We have the estimate

$$\begin{aligned}||a(x,\xi^j)|-\alpha| &\le |a(x,\xi^j)-a(p)| \le |a(x,\xi^j)-a(x^0,\xi^j)| \\ &+ |a(x_0,x^j)-a(x^j,\xi^j)| + |a(x^j,\xi^j)-a(p)| \le c|x-x^0|+c|x^0-x^j|+\varepsilon_j \ ,\end{aligned}$$

where the second and third term at right tend to zero, as $j\to\infty$.

For $\varepsilon>0$ pick a ball $|x-x^0|<\varepsilon/c=\eta$ and conclude that

$$(10.10)\qquad |a(x,\xi^j)|\ge \alpha-\varepsilon-\varepsilon_j,\ \text{as}\ |x-x^0|\le\eta\ ,\ \varepsilon_j\to 0\ \text{as}\ j\to\infty\ .$$

Accordingly, choosing $b\in C_0^\infty(\{|x-x^0|\le\eta\})$, we get $\|a(x,\xi^j)bu\|\ge(\alpha-\varepsilon)\|bu_j\| - \varepsilon_j$. Next, to control "$\approx$" , we get

$$(10.11)\qquad \|b\int e^{ix\xi}(a(x,\xi)-a(x,\xi^j)\varphi(\xi-\xi_j)d\!\!\!{}^-\xi\| \le c/\langle\xi^j\rangle \to 0\ ,\ j\to\infty\ .$$

Furthermore, $bAu_j=Abu_j+Cu_j$, with $C\in\mathcal{K}(\mathcal{H})$. Since $\{u^j\}$ is orthonormal it weakly converges to 0, hence $Cu_j\to 0$ strongly. All in all we conclude that, for any $K\in\mathcal{K}$, we have

$$(10.12)\qquad \|(A+K)bu_j\| \ge (\alpha-\varepsilon)\|bu_j\| - \varepsilon_j\ ,\ \varepsilon_j\to 0\ ,\ \text{as}\ j\to\infty\ .$$

Here $\|bu_j\|=\|bu\|\ne 0$. As $j\to\infty$ get $\|A+K\|\ge\alpha-\varepsilon$. This for all $\varepsilon>0$, hence for $\varepsilon=0$. With "inf" over $\mathcal{K}$ get (10.8) so far only if x_j is bounded

In the leftover case $|x^j|\to\infty$ we follow the same argument, but now require a more precise evaluation of the error terms. For simplicity of calculations we discuss only the case n=1. Choose

$\varphi(\xi)=1$, as $|\xi|\le 1$, = 0 elsewhere. We get $u(x)=2\frac{\sin x}{x}$, $u_j=e^{ix\xi^j}u$. Now we have

$$||a(x,\xi^j)| - \alpha| \le |a(x,\xi^j)-a(x^j,\xi^j)| + |a(x^j,\xi^j)-a(p)|$$
$$\le c/\langle x^j\rangle\, |x-x^j| + \varepsilon_j ,$$

hence (10.10) now holds with $|x-x^0|\le\eta$ replaced by $|x-x^j|\le\eta\langle x^j\rangle$. We now will pick $b=b_j$ depending on j , according to $b_j=1$ in

$|x-x^j|\le\frac{\eta}{2}\langle x^j\rangle$, $b_j\in C_0^\infty(\{|x-x^j|<\eta\langle x^j\rangle\})$, $|b_j|\le 1$, insuring that

$$(10.13)\qquad c\langle x^j\rangle^{-1/2} \le \|b_j u_j\| \le C\langle x^j\rangle^{-1/2} , \ j=1,2,\dots .$$

A choice is $b_j(x)= b((x-x^j)/\langle x^j\rangle)$, $b\in C_0^\infty(|x|<\eta)$, $b=1$, $|x|\le\eta/2$.

It is fairly evident that now it suffices to show that

$$(10.14)\qquad \|b_j(x)\int đ\xi e^{ix\xi}(a(x,\xi)-a(x,\xi^j))\varphi(\xi-\xi^j)\| = O(\varepsilon_j\langle x^j\rangle^{-1/2}) ,$$

and that

$$(10.15)\qquad \|[b_j,A]u_j\| = O(\varepsilon_j\langle x^j\rangle^{-1/2}) , \ \varepsilon_j = o(1) , \text{ as } j\to\infty .$$

Indeed, we then may introduce $v_j= b_j u_j\langle x^j\rangle^{1/2}$ with $C>\|v_j\|\ge c>0$. One easily shows that $v_j\to 0$ weakly, so that $\|Kv_j\|\to 0$ for compact K . Repeating the arguments of (10.12) one gets $\|Av_j\|\ge(\alpha-\varepsilon)\|v_j\|-\varepsilon_j$, hence $\|(A+K)v_j\|\ge (\alpha-\varepsilon)\|v_j\|-\varepsilon_j$, again proving (10.8).

Let us offer some details: For (10.14) look at

$$\int d\eta\varphi(\kappa)(a(x,\xi^j+\kappa)-a(x,\xi^j))e^{ix\kappa} = \int_{-1}^{1} e^{ix\kappa}(a(x,\xi^j+\kappa)-a(x,\xi^j))d\kappa$$

$$=\frac{e^{ix}}{x}(a(x,\xi^j+1)-a(x,\xi^j))-\frac{e^{-ix}}{x}(a(x,\xi^j-1)-a(x,\xi^j)$$

$$-\frac{1}{ix}\int_{-1}^{1} e^{ix\kappa}a_{|\xi}(x,\xi^j+\kappa)d\kappa = o(\tfrac{1}{x}) \text{ as } |x-x^j|\le\eta\langle x^j\rangle , \ j\to\infty .$$

implying (10.14). For (10.15): We get $[b_j,A] \approx \langle b_j,a\rangle(x,D)$ with the Poisson bracket $\langle b_j,a\rangle=b_{j|\xi}a_{|x}-a_{|\xi}b_{j|x} = -b_{j|x}a_{|\xi} = \beta(x,\xi)$. Clearly $\|\beta(x,D)u_j\|=\|b_{j|x}\int_{-1}^{1} đ\eta a_{|\xi}(x,\xi^j+\eta)\|=O(\varepsilon_j\langle\xi^j\rangle)$ as desired.

For the error term expressed by "$\approx$" we involve I,(5.7) for N=1. From I,(5.7) and I,thm.6.1 we get the error as $\|\gamma_j(x,D)u_j\|$ with

$$\gamma_j(x,\xi)=-2\int_0^1 \tau d\tau(1-\tau)\int đyđ\eta e^{-iy\eta}a_{|\xi\xi}(x,\xi-\tau\eta)b_{|xx}(x-y) ,$$

where $\gamma_{j|\xi}(x,\xi) = O(\langle\xi^j\rangle^{-2})$. Thus the error term obeys the same estimate, and we get (10.15). For the weak convergence of v_j observe that, for $\varphi \in C_0^\infty(\mathbb{R})$, we have $(v_j,\varphi) = 0$ for large j, since $\mathrm{supp}\, v_j \subset \{|x-x^j| \le \eta\langle x^j\rangle\}$ where $\eta<1$ may be assumed so that $|x| \ge |x^j|-\eta\langle x^j\rangle \to \infty$, as $j\to\infty$. Since $\|v_j\| \le C$ and C_0^∞ is dense in $\mathcal{H}$ we get $|(v_j,\varphi)| \le |(v_j,\psi)| + C\|\varphi-\psi\| < \varepsilon$ with suitable $\psi \in C_0^\infty$. Q.E.D.

To continue with the proof of thm.10.3 observe that (10.8) implies the map $\tau^\vee : Op\psi c_0{}^\vee \to C(\partial\mathbb{P})$ to have a continuous extension to $\mathcal{A}^\vee = \mathcal{A}/\mathcal{K}$, since the left hand side of (10.8) is just the norm $\|A^\vee\|$. Then $\tau^\vee$ induces a (continuous) algebra homomorphism $\mathcal{A}^\vee \to C(\partial\mathbb{P})$. Likewise τ itself extends to a continuous homomorphism $\tau:\mathcal{A}\to C(\partial\mathbb{P})$, and $\tau=\tau^\vee\circ\kappa$ with the canonical injection $\kappa:\mathcal{A}\to\mathcal{A}+\mathcal{K}(\mathcal{H})$. We claim that $\tau^\vee$ defines an isomorphism between $\mathcal{A}^\vee$ and $C(\partial\mathbb{P})$. This indeed will establish thm.10.3, since then the dual $\tau^{\vee\prime}$ will give a homeomorphism $\mathbb{M} \leftrightarrow \partial\mathbb{P}$, the latter being the maximal ideal space of $C(\partial\mathbb{P})$.

Proposition 10.5. An operator $A\in Op\psi c_0$ is Fredholm in $\mathcal{L}(\mathcal{H})$ if and only if $\tau_A \ne 0$ in all of $\partial\mathbb{P}$.

Proof. Clearly $\tau_A\neq 0$ on all of $\partial\mathbb{P}$ means that A is md-elliptic (of order 0). Such an operator will have a Green inverse $B\in\mathcal{O}(0)$ by II,thm.1.6 and III,thm.4.2. Clearly, B is a Fredholm inverse in $\mathcal{L}(\mathcal{H})$ as well, so that B must be Fredholm. Vice versa, let $A\in Op\psi c_0$ be Fredholm in $\mathcal{L}(\mathcal{H})$. Then we have $\Pi_s^{-1}A\Pi_s = A + \Pi_s^{-1}[A,\Pi_s]$ Fredholm too, for all $s\in\mathbb{R}^2$, since $\Pi_s^{-1}[A,\Pi_s] \in \mathcal{O}(-1)\subset\mathcal{K}(\mathcal{H})$. Here $\Pi_s = \pi_s(x,D) = \langle x\rangle^{s_2}\langle D\rangle^{s_1}$, as in III,(3.1). Recall that $\Pi_s:\mathcal{H}_s\to\mathcal{H}$ is an isometry between $\mathcal{H}_s$ and $\mathcal{H}$. The above means that not only is $A:\mathcal{H}\to\mathcal{H}$ Fredholm but also $A:\mathcal{H}_s\to\mathcal{H}_s$. For clarity let us denote the latter operator by A_s. Then $A_s:\mathcal{H}_s\to\mathcal{H}_s$ is Fredholm for all s and we have $A_0=A$. All operators A_s have the same index, $\nu(A_s) = \dim\ker A_s - \dim\ker A_s^*$. However, as s_1 increases, for fixed s_2, one finds that $\dim\ker A_s$ decreases while $\dim\ker A_s^*$ increases. Since the difference $\psi(A_s)$ is constant, they both must be constant (Here A_s^* denotes the adjoint of A_s with respect to the pairing

$$(10.16)\qquad (u,v) = (\pi_{-s}(x,D)u,\pi_s(x,D)v)_{L^2}\ ,\ u\in\mathcal{H}_s\ ,\ v\in\mathcal{H}_{-s}\ ,$$

not with respect to the usual Hilbert space pairing $(u,v)_s$. Thus we have $A_s^*:\mathcal{H}_{-s}\to\mathcal{H}_{-s}$, in fact, $A_s^*=\bar{a}(M_r,D)$ with a symbol independent of s. This explains why $\dim\ker A_s^*$ increases with increasing s_j.) The fact that $\ker A_s\subset\mathcal{H}_s$, $\dim\ker A_s$ =const then implies that $\ker A_s\subset\mathcal{S}$, and $\ker A_s^*\subset\mathcal{S}$ as well. Both spaces are independent of s . For a more detailed discussion cf. [C_1], ch.III .

It is clear then, that a special Fredholm inverse B of A, constructed as in III,4 - i.e., $AB=1-P_{\ker A^*}$, $BA=1-P_{\ker A}$, also defines a Green inverse of A. Thus, if $A:\mathcal{H}\to \mathcal{H}$ is Fredholm, it follows that $a=\mathrm{symb}(A)\in \psi c_0$ is md-elliptic, hence $\tau_A\neq 0$ on $\partial\mathbb{P}$. Q.E.D.

Finally, in order to show that $\tau^\vee$ is an isomorphism, assume that $A\in \mathbf{A}$ and that $\tau_A(p^0)\neq 0$ for some $p^0\in \partial\mathbb{P}$. Then $\tau_{A^*A}(p)\geq c>0$ in N_{p^0} , a neighbourhood of p^0 in $\partial\mathbb{P}$. (Note that clearly τ is a *-homomorphism.)

Observe that τ maps onto a dense subset of $C(\partial\mathbb{P})$ (containing the set $\psi c_0|\partial\mathbb{P}$) . But a *-homomorphism $\upsilon:\mathcal{X}\to\mathcal{Y}$ between C^*-algebras always defines an isometry $\upsilon^\Delta:(\mathcal{X}/\ker \upsilon)\to\mathcal{Y}$ (cf.[C_1],p.323,thm.7.17) Therefore its image is necessarily a closed subalgebra of $\mathcal{Y}$. In our case it follows that $\mathrm{im}\,\tau = C(\partial\mathbb{P})$, since $\mathrm{im}\,\tau$ is dense in $C(\partial\mathbb{P})$. There exists a function $\chi\in C(\partial\mathbb{P})$ such that $\chi=1$ near p^0 , $\mathrm{supp}\,\chi \subset N(p^0)$, while $\chi\geq 0$. Conclude the existence of $X\in \mathbf{A}$ with $\tau_X=\chi$. Then the operator $XA^*A+(1-X) = Z$ has $\tau_Z=\chi\tau_{A^*A}+(1-\chi)\neq 0$ on all of $\partial\mathbb{P}$. There exists a sequence $Z_j\in \mathrm{Op}\psi c_0$, $Z_j\to Z$. Then $\tau_{Z_j}\to\tau_Z$. For large j we get $\tau_{Z_j}\geq\eta>0$. By prop.10.5 Z_j is Fredholm. There exists a Fredholm inverse Y_j with $\tau_{Y_j}=1/\tau_{Z_j}$ $\Rightarrow \|\tau_{Y_j}\|\leq 1/\eta$. Using (10.8) we get an inverse mod $\mathcal{K}(\mathcal{H})$, called W_j with $\|W_j\|\leq \frac{2}{\eta}$, and a Fredholm inverse V_j with $\|V_j\|\leq \frac{3}{\eta}$. This implies that Z is Fredholm as well (cf. [C_1], corollary on p.267).

But if A were compact, so is XA^*A compact. Hence 1-X then is a Fredholm operator. The latter is impossible: Then we get a Fredholm inverse $B\in \mathbf{A}$ and $\tau_B\tau_{(1-X)}=1$ on all of $\partial\mathbb{P}\Rightarrow 0=\tau_{B(p^0)}(1-\chi(p_0))$ $=1$, a contradiction. Thus $\tau_A=0$ implies A compact, and $\tau^\vee$ must be an isomorphism. Theorem 10.3 is established for s=0. For general s observe that $\mathbf{A}_s=\Pi_s^{-1}\mathbf{A}\Pi_s$, using the isometry $\Pi_s:\mathcal{H}_s\to \mathcal{H}$. Also, $\Pi_s^{-1}A\Pi_s=A + K$, $K\in \mathcal{K}(\mathcal{H}_t)$, for all t, showing that $A\in \mathrm{Op}\psi c_0$ has the same symbol for all s. Q.E.D.

Chapter 6. HYPERBOLIC FIRST ORDER SYSTEMS

0. Introduction.

We now focus on hyperbolic theory. It the present chapter 6 we look at a first order system of pseudodifferential equations on $\mathbb{R}^n$ of ν equations in ν unknown functions, of the form

$$(0.0) \qquad \partial u/\partial t = iKu \ , \ K=k(t,x,D) \quad ,$$

with a $\nu\times\nu$-matrix of symbols $k=((k_{jl}))$, $k_{jl}\in\psi c_e$, for fixed t. In essence 'hyperbolic' means that either the matrix $k(t,x,\xi)$ is self-adjoint, or at least has real eigenvalues, both mod ψc_0 .

In sec.1 we discuss a symmetric hyperbolic first order system of PDE's, using a method of mollifiers, after K.O.Friedrichs. The case of a ψdo K(t) may be treated by a similar technique (sec. 2). Here we use the weighted L^2-Sobolev spaces of ch.III.

In sec.3 we will look at properties of the evolution operator $U=U(\tau,t)$. We find that $U(\tau,t)$ is of order 0, while $\partial_t^p\partial_\tau^q U$ is of order (p+q)e. In sec.4 we discuss strictly hyperbolic systems, no longer symmetric hyperbolic, by the method of symmetrizer.
In sec.5 we discuss a global Egorov type result for a single hyperbolic equation, proving existence of a characteristic flow. Actually our flows are of more general type, called particle flow, using a generalized principal symbol of K , no longer homogeneous in ξ .

This flow is related to the family $A \leftrightarrow U^{-1}AU$ of automorphisms of the C^*-subalgebra $\mathcal{A}$ of $L(H)$ generated by Opψc_0 (V,10), where $U=U(\tau,t)$ is the evolution operator: The restriction of the flow to $|x|+|\xi|=\infty$ is the dual of the induced automorphism of $\mathcal{A}^\vee=\mathcal{A}/K(H)$. In sec.6 we discuss the action of these flows on symbols. Sec.7 deals with propagation of singularities and maximal ideals of $\mathcal{A}$.

General theory of hyperbolic N-th order (systems or) equations will be reduced to the case of 1-st order systems, as already prepared in sec.4. This will be discussed in ch.7.

1. First order symmetric hyperbolic systems of PDE.

We enter the theory of hyperbolic (pseudo-)differential equations from our discussion in 0,6. There it was seen that a single linear first order PDE of the form $\partial u/\partial t=Lu$, $L=\sum b_j\partial_{x_j}+p$, with real-valued $b_j(x)$, essentially is equivalent to an ODE along a set of curves in (x,t)-space. The curves were called the characteristic curves. They were given as solution curves of the system of ODE's $\frac{dx}{dt}=-b(x)=-(b_1(x),\ldots,b_n(x))$.

If $u(x)$ is no longer a scalar, but a vector function, and the equation a system of equations, then the simple approach of 0,6 no longer works - although the essence of it can be saved if there is only one x-variable - cf.[CH],II, ch.V, for example. We regard it as a principal merit of ψdo-calculus that it will restore the relationship between the PDE and its 'characteristic flow', i.e. the system of characteristic curves. First we study abstract existence and uniqueness results.

For a first result, for PDE's only, look at the problem

$$(1.1)\qquad \partial_t u + iKu = f \ , \ t\in\mathbb{R} \ , \ u(0)=\varphi \ ,$$

where

$$(1.2)\qquad L = iK = \sum_{j=1}^n a^j\partial_{x_j} + a^0 \ , \ \bar{a}^{jT}(x) = a^j(x) \ , j=1,\ldots,n \ ,$$

is a first order $\nu\times\nu$-matrix of differential expressions in x with real symmetric principal part, and coefficients independent of t. Let L be defined on $\mathbb{R}^n$, and let $a_j\in C^\infty(\mathbb{R}^n)$, a_j , $a_{j|x_l}$ bounded.

The adjoint expression L^* of L is given by

$$(1.3)\qquad L^* = -iK^* = -L+\beta(x) \ , \ \beta(x) = \bar{a}^{0T}+a^0-\sum a^j_{|x_j} \ .$$

Clearly it also has bounded coefficients.

We regard (1.1) as an ODE in t with dependent variable u taking values in the Hilbert space $H=L^2(\mathbb{R}^n,\mathbb{C}^\nu)$, assuming that $f(t)\in H$, for every t. It then is natural to interpret L as an unbounded operator of H, defined in a suitable domain $\mathrm{dom}\, L\subset H$.

In this setting problem (1.1) suggests that we look at the family $\{e^{iKt}\}$ possibly generated by the <u>infinitesimal generator</u> $L=iK$. Formally $L+L^*=\beta(x)$ is a bounded matrix function. Thus $L+\gamma$ is formally <u>accretive</u> (<u>dissipative</u>) whenever the constant $\gamma\in\mathbb{R}$ is large (small). In other words, we have, correspondingly,

$$(1.4)\qquad \begin{aligned} 2\mathrm{Re}(u,(L+\lambda)u) &= (u,Lu)+(Lu,u)+2\lambda(u,u) = (u,(L+L^*+2\lambda)u) \\ &= (u,(\beta(x)+2\lambda)u) \geq 0 \ (\text{or } \leq 0) \text{ for all } u\in C_0^\infty(\mathbb{R}^n) \ . \end{aligned}$$

By V,thm.7.2 the operator L will generate a strongly continuous group $\{E(t) = e^{Lt} = e^{iKt} : t \in \mathbb{R}\}$,such that

$$(1.5) \qquad \mathrm{dom}\, L = \{u \in \mathcal{H} : d/dt(E(t)u)|_{t=0} \text{ exists in } \mathcal{H}\} ,$$

and that $u(t) = E(t)u$, for $u \in$ dom L , satisfies

$$(1.6) \qquad u(t) \in C^1(\mathbb{R}) , \ du/dt - Lu = 0 \ , \ t \in \mathbb{R} , \ u(0) = u ,$$

provided that we can verify the assumptions for all real $|\lambda| \ge \gamma$.

The t-derivative in (1.6) exists in strong convergence of $\mathcal{H}$. We get $E(t) \in L(\mathcal{H})$, and, with operator norm $\|.\|$ of $\mathcal{H}$, M of V,(7.7),

$$(1.7) \qquad \|E(t)\| \le Me^{\gamma|t|} \ , \ t \in \mathbb{R} .$$

Then the inhomogeneous Cauchy problem (1.1) may be written as

$$(1.8) \qquad \frac{d}{dt}(E(t)u(t)) = E(t)f(t) \ , \ t \in \mathbb{R} \ , \ (Eu)(0) = \varphi .$$

An integration then solves (1.1) : For $f(t) \in C(\mathbb{R},\mathcal{H})$, $\varphi \in \mathcal{H}$, we get

$$(1.9) \qquad u(t) = E(-t)\varphi + \int_0^t E(\tau - t)f(\tau)d\tau .$$

Note that (1.1) and (1.9) are equivalent only if $\varphi \in$ dom L .

<u>Theorem 1.1.</u> Let $L = L_0^{**}$ be the closure of the <u>minimal</u> <u>operator</u> L_0 (with domain dom $L_0 = C_0^\infty(\mathbb{R}^n)$) induced by the expression L . Then, under our above formal assumptions on L and its coefficients, L and K=-iL satisfy all of the above properties, so that the group E(t) is well defined, and the Cauchy problem (1.1) admits the solution (1.9) for every $f \in C(\mathbb{R},\mathcal{H})$, $\varphi \in$ dom L .

For the proof we need the lemma, below.

<u>Lemma 1.2</u>. Under the assumptions of theorem 1.1 the closures of the minimal operators of L and its adjoint expression L^* are mutually adjoint unbounded closed operators of $\mathcal{H}$.

(This often is stated by saying that the concepts of <u>weak</u> and <u>strong</u> L^2<u>-solution</u> coincide, under the assumptions given.)

Assuming Lemma 1.2 correct we next notice that

$$(1.10) \qquad |\mathrm{Re}\,(L_0u,u)| = \tfrac{1}{2}|(Lu,u)+(u,Lu)| \le \gamma(u,u) \ , \ u \in C_0^\infty(\mathbb{R}^n) ,$$

hence for real λ we have

$$(1.11) \qquad (\lambda-\gamma)(u,u) \le \mathrm{Re}((L_0+\lambda)u,u) \ , \ u \in \mathrm{dom}\, L_0 .$$

Taking closure we get

$$(1.12) \qquad (\lambda-\gamma)\ \|u\|^2 \le \|u\|\,\|(L+\lambda)u\| \quad , \; u \in \text{dom } L \; ,$$

which implies that $(L+\lambda)$ is one to one and has closed range ,and

$$(1.13) \qquad \|(L+\lambda)u\| \ge |\lambda-\gamma|\,\|u\| \quad , \; u \in \text{dom } L \; , \text{ for all } \lambda>\gamma \; .$$

However, in (1.10) we may replace L_0 by the minimal operator of the adjoint expression. Since its closure is the adjoint of L, we get exactly the same properties , and the estimate (1.13) also for the adjoint operator L^* of L . It follows that also $(L+\lambda)^*$ is is one to one, and has closed range. Accordingly $L+\lambda$ is 1-1 and onto, and has an inverse $R(\lambda)=(L+\lambda)^{-1} \in \mathcal{L}(\mathcal{H})$, for all $\lambda>\gamma$.

Similarly we get existence of $R(\lambda)=(L+\lambda)^{-1}$,for all $\lambda<-\gamma$. For real λ satisfying $|\lambda|>\gamma$ we get $\|(L+\lambda)^{-1}\|\le\frac{1}{|\lambda|-\gamma}$. This implies the assumptions of the Hille-Yosida theorem, and thm.1.1 is proven.

Proof of Lemma 1.2. Let L and M denote the closures of L_0 and L_0^* , respectively. Clearly L and M are in adjoint relation. That is,

$$(1.15) \qquad (Lu,v) = (u,Mv) \; , \text{ for all } u \in \text{dom } L \; , \; v \in \text{dom } M \; ,$$

as follows for $u,v \in C_0^\infty(\Omega)$ by definition, and then for general u,v by taking closure. In other words, we have $L \subset M^*$, and are left with showing that $M^* \subset L$. Let $f \in \text{dom } M^*$, i.e.,

$$(1.16) \qquad f\in \mathcal{H} \; , \; (f,Mu) = (g,u) \; , \text{ for all } u \in C_0^\infty(\mathbb{R}^n) \quad ,$$

for some $g \in \mathcal{H}$ (defined as $g = M^*f$) .Then we must show that f $\in$ dom L and that $g = Lf$ .In other words a sequence $f_k\in C_0^\infty(\Omega)$ must be constructed such that $f_k \to f$, $Lf_k \to g$, (in $\mathcal{H}$) , as $k \to \infty$.

Introduce the convolution operator J_ε , $\varepsilon>0$, by setting

$$(1.17) \qquad J_\varepsilon u(x) = \int\varphi_\varepsilon(x-y)u(y)dy \; , \; \varphi_\varepsilon(x) = \varepsilon^{-n}\varphi((x-y)/\varepsilon) \; ,$$

with some $\varphi \in C_0^\infty(\mathbb{R}^n)$, supp $\varphi \subset \{|x|\le 1\}$, $\varphi\ge0$, $\int\varphi dx = 1$,φ even. It is clear that $f_\varepsilon=J_\varepsilon f\to f$ (in $\mathcal{H}$) as $\varepsilon\to0$, for every $f\in \mathcal{H}$. Indeed, for continuous f with compact support one finds

$$|f_\varepsilon-f|(x)=|\int\varphi_\varepsilon(x-y)(f(x)-f(y))dy|\le\text{Max}\{|f(x)-f(y)|:|x-y|\le\varepsilon\}\to0,\ \varepsilon\to0,$$

while supp $(f_\varepsilon-f)$ stays within a fixed compact set, implying that $f_\varepsilon\to f$ in $\mathcal{H}$. On the other hand, $J_\varepsilon=V_\varepsilon^{-1}J_1V_\varepsilon$ with the unitary (dilation) operator $V_\varepsilon u(x)=\varepsilon^{n/2}u(\varepsilon x)$, shows that $\|J_\varepsilon\|=c$ is independent of ε . Also $C_0^\infty(\mathbb{R}^n)$ is dense in $\mathcal{H}$.

For $f\in \mathcal{H}$ satisfying (1.16) let $h_\varepsilon=\psi_\varepsilon f_\varepsilon=\psi_\varepsilon J_\varepsilon f$, with $\psi_\varepsilon=\varphi_\varepsilon{}^\wedge = \varphi^\wedge(\varepsilon x)$, as in (1.17). Then $h_\varepsilon\to f$ in $\mathcal{H}$ holds as well. Also, $h_\varepsilon\in C_0^\infty$

for every $\varepsilon>0$. To prove lemma 2.1 we must show that also $Lh_\varepsilon \to g$. But we get $Lh_\varepsilon = \psi_\varepsilon Lf_\varepsilon + [L,\psi_\varepsilon]f_\varepsilon$, where the commutator $[L,\psi_\varepsilon]$ is a multiplication by the matrix $\varepsilon\sum a^j(x)\varphi^\wedge_{|x_j}(\varepsilon x)$. The latter tends to zero uniformly on $\mathbb{R}^n$. Hence $[L,\psi_\varepsilon]f_\varepsilon \to 0$, and we must show $Lf_\varepsilon \to 0$. Let us write $a^j(x) = ((a^j_{kl}(x)))$,and introduce the vector $\varphi^k_{\varepsilon,x}(y) = (\varphi^{k,j}_{\varepsilon,x}(y))_{j=1,\dots,m} = (\delta_{kj}\varphi_\varepsilon(y-x))$ .For the k-th component of Lf_ε we get (summing over all double indices) ,

$$(Lf_\varepsilon)_k(x) = \int (a^j_{kl}(x)\partial_{x^j} + a^0_{kl}(x))\varphi_\varepsilon(y-x)f_l(y)dy$$

$$(1.18)\qquad = \int ((-\partial_{y^j} a^j_{ql}(x) + a^0_{ql}(x))\varphi^{k,q}_{\varepsilon,x}(y))f_l(y)\,dy$$

$$= (L^*\varphi^k_{\varepsilon,x}, f) + T_k f ,$$

with

$$T_k f = \int (\partial_{y_j}(a^j_{ql}(y) - a^j_{ql}(x)) + (a^0_{ql}(x) - a^0_{ql}(y))\varphi^{k,q}_{\varepsilon,x}(y)f_l(y)dy.$$

But (1.16) implies

$$(1.19)\qquad (L^*\varphi^k_{\varepsilon,x}, f) = (\varphi^k_{\varepsilon,k}, g) = (J_\varepsilon v)_k .$$

As $\varepsilon \to 0$,this converges to g (in H) .Therefore the lemma is established if we can show that $Tf \to 0$ in H , as $\varepsilon \to 0$.

For the latter we first note that $\|T\|$ is bounded for all $\varepsilon > 0$. Indeed this follows because the first derivatives of $\varphi_\varepsilon(y)$ are $O(\varepsilon^{-n-1})$, while the support is in a ball of radius ε ,and the factors $a(y)-a(x)$ are $O(\varepsilon)$ .It follows that the integral kernel $t_\varepsilon(x,y)$ satisfies Schur's condition $\int |t_\varepsilon(x,y)| \begin{cases} dx \\ dy \end{cases} \le c$ for all x,y. On the other hand ,if $f \in C_0^\infty(\mathbb{R}^n)$,then a partial integration, removing the derivative from $\varphi_\varepsilon(x-y)$ implies that $Tf \to 0$ in H , as $\varepsilon \to 0$. Combining the two facts we conclude that $Tf \to 0$ for all $f \in H$, q.e.d.

2. First order symmetric hyperbolic systems of ψde's on $\mathbb{R}^n$.

In this section we shall work on the Cauchy problem for a system of pseudo-differential equations over $\mathbb{R}^n$, of the general form

$$(2.1)\qquad \partial u/\partial t + iK(t)u = f(t) \quad , t_1 \le t \le t_2 , \quad u(0) = \varphi .$$

For convenience we assume a compact interval $0 \in I = [t_1,t_2]$.

We assume that K(t) are matrices of ψdo's in the sense of ch.I :

$$(2.2)\qquad K(t) = k(t,M_1,D) = ((k_{jl}(t,M_1,D)))_{j,l=1,\dots,\nu}\ ,$$

where the functions $k_{jl}(t,x,\xi)$ are symbols, for every fixed t.

Again we regard (2.1) as a first order ODE for u taking values in a space $\mathcal{H}_s\otimes\mathbb{C}^\nu$ of all $u=(u_1,\dots,u_\nu)$, $u_j\in\mathcal{H}_s$ (of III,(3.1)) For simplicity we write $\mathcal{H}_s$ instead of $\mathcal{H}_s\otimes\mathbb{C}^\nu$. We assume $K(t)\in Op\psi c^\nu_e$, e=(1,1), introducing the class ψc^ν_m of ν×ν-matrix-valued symbols with entries in ψc_m. By III,thm.3.1 get $K(t)\in L(\mathcal{H}_t,\mathcal{H}_{t-e})$, for all $t\in\mathbb{R}^2$. In general we do not get $K\in L(\mathcal{H}_s)$ but may interprete K as an unbounded operator of $\mathcal{H}_s$ with domain $\mathcal{H}_{s+e}$, e = (1,1).

We consider initial values $\varphi\in\mathcal{H}_s$, and require $f\in C(I,\mathcal{H}^s)$. The following precise assumptions on the k(t,x,ξ) are imposed:

$$(2.3)\quad \langle x\rangle^{-1+|\beta|}\langle\xi\rangle^{-1+|\alpha|}k^{(\alpha)}_{(\beta)}(t,x,\xi)\in C(I,CB(\mathbb{R}^{2n},L^\nu))\ ,\ \alpha,\beta\in\mathbb{N}^n\ .$$

and

$$(2.4)\quad \langle x\rangle^{|\beta|}\langle\xi\rangle^{|\alpha|}(k-k^*)^{(\alpha)}_{(\beta)}(t,x,\xi)\in C(I,CB(\mathbb{R}^{2n},L^\nu))\ ,\ \alpha,\beta\in\mathbb{N}^n\ .$$

Here L^ν denotes the algebra of complex ν×ν-matrices,and we write CB(X,Y) for the space of all bounded continuous functions from X to Y. Also $\mathbb{N}^n$ is the class of all n-multi-indices $\alpha=(\alpha_1,\dots,\alpha_n)$, $\alpha_j\in\mathbb{N}$, $\mathbb{N}=\{0,1,2,3,\dots\}$,and we have written $K^*=k^*(t,M_1,D)$ for the Hilbert space adjoint $K^*(t)$, a ψdo with symbol given by I,(5.9). Alternately we may express (2.3) and (2.4) by writing $k\in C(I,\psi c^\nu_e)$ and $k-k^*\in C(I,\psi c^\nu_0)$,using the Frechet topology of ψc_m .

Clearly (2.4) implies that the skew-symmetric part $1/2(K(t)-K^*(t))$ of K(t) is of order zero hence bounded in every $\mathcal{H}_s$. This motivates the notation "symmetric hyperbolic system" of ψde's. As a special case we may choose K as a differential operator with the properties of K in (1.1), (1.2), but K may depend on t here.

In the functional analytical sense the solution of (2.1) no longer involves a group e^{iKt} ,since now K(t) depends on t. While abstract existence results (cf. Kato [Ka_5]) could be used we find it practical to use the technique of lemma 1.2 again (cf. [Tl_1]).

Theorem 2.1. Under the assumptions (2.3), (2.4) the Cauchy problem (2.1) admits a unique solution $u\in C(I,\mathcal{H}_s)\cap C^1(I,\mathcal{H}_{s-e})$, for each $\varphi\in\mathcal{H}_s$, and $f\in C(I,\mathcal{H}_s)$, where $s\in\mathbb{R}^2$ is arbitrary.

Remark. As a consequence of uniqueness the solution u of (2.1) is independent of s, if $\varphi\in\mathcal{S}=\cap\mathcal{H}_s$, and $f\in C(I,\mathcal{H}_s)$ for all s. Then u also takes values in $\mathcal{S}$, and u , $\partial_t u$ are continuous in every $\mathcal{H}_s$.

Proof. The key of the proof is the so-called energy estimate (prop.2.3) .We first shall solve (2.1) for a "mollified" operator $K_\varepsilon(t)$ instead of $K(t)$,where $K_\varepsilon(t)$ are bounded operators of H_s . Under such conditions Picard's method of succesive approximation may be used to solve the Cauchy problem. A limit $\varepsilon \to 0$, using the Arzela-Ascoli theorem, will provide the desired solution of (2.1).

As in section 1 we choose a regularizer (or mollifier)

$$(2.5) \qquad J_\varepsilon = (\psi_\varepsilon *) = \psi^\wedge(\varepsilon D) \quad , \quad \varepsilon > 0 ,$$

with a fixed $C_0^\infty(\mathbb{R}^n)$-function ψ, supp $\psi \subset \{|x|< 1\}$, $\psi \geq 0$, $\int \psi dx = 1$, and $\psi_\varepsilon(x) = \varepsilon^{-n}\psi(x/\varepsilon)$. Observe that $\psi^\wedge \in S$. We introduce

$$(2.6) \quad K_\varepsilon(t) = \psi^\wedge(\varepsilon M)K(t)\psi^\wedge(\varepsilon D) = k_\varepsilon(t,M_1,D),\ k_\varepsilon = k(t,x,\xi)\psi^\wedge(\varepsilon x)\psi^\wedge(\varepsilon\xi).$$

Since $\psi^\wedge \in S$, it is clear that the matrix k_ε has all its entries in ψt_0, using (2.3). Therefore $K_\varepsilon(t)$ are bounded operators of every H_s, for $\varepsilon>0$, $t\in I$. In fact, we get $K_\varepsilon(t) \in C(I,H_s)$, for $\varepsilon>0$. Thus the Cauchy problem $(2.1)_\varepsilon$ (with K_ε instead of K in (2.1)) admits a unique solution $u=u_\varepsilon \in C^1(I,H_s)$, for each $\varepsilon > 0$ ([Dd_1]) .

Letting $\varepsilon \to 0$ we attempt to construct a uniformly convergent subsequence of $u_\varepsilon(t)$ with limit function $u(t)$ solving (2.1). We shall need more properties of J_ε and $K_\varepsilon(t)$, in that respect.

Proposition 2.2. a) The family $J_\varepsilon = \psi^\wedge(\varepsilon D)$, $0 \leq \varepsilon \leq 1$ defines a bounded function $[0,1] \to L(H_s)$, for each $s \in \mathbb{R}^2$, which is norm continuous in $(0,1]$, and strongly continuous at 0. We have $J_0=1$. If ψ is choosen as an even function , then also all J_ε are hermitian symmetric in the Hilbert space $H = H_0$.

b) The family $K_\varepsilon(t)$ defines a bounded function $[0,1] \times I \to L(H_s,H_{s-e})$ which is norm continuous in $(0,1] \times I$, and strongly continuous in $[0,1] \times I$. We have $K_0(t) = K(t)$, $t \in I$.

Proposition 2.3. (Energy estimate). Let $u(t)\in C(I,H_s) \cap C^1(I,H_{s-e})$ be a solution of $(2.1)_\varepsilon$,for given ε , s , φ , f(t). There exists a constant c independent of ε,s,φ and f such that

$$(2.7) \qquad \|u(t)\|_s^2 \leq \|\varphi\|_s^2 e^{c|t|} + e^{c|t|}\int_0^{|t|} e^{-c\tau}\|f(\tau \operatorname{sgn} t)\|_s^2\, d\tau ,$$

with the norm $\|.\|_s$ of H_s .

A proof, based on the symmetry of K(t) expressed in (2.4), and some commutator relations of I,7 is postponed to the end of sec.2.

Note that the energy estimate (2.7) implies that

$$(2.8) \qquad \|u_\varepsilon(t)\|_s \leq c , \quad , 0 \leq \varepsilon \leq 1 \quad , t \in I ,$$

with c independent of ε and t . Let us first assume that φ and f satisfy the assumptions of thm.2.1 for all s . The uniqueness of the solution u_ε contained in Picard's theorem then implies that u_ε is independent of s , and that $u_\varepsilon \in C^1(I,H_s)$, for all s. Using (2.8) and the differential equation $(2.1)_\varepsilon$ it follows that

$$(2.9)\quad \|u^{\cdot}_\varepsilon(t)\|_s \le \|K_\varepsilon(t)u_\varepsilon(t)\|_s + \|f(t)\|_s \le c\|u_\varepsilon(t)\|_{s-e} + \|f(t)\|_s \le c.$$

We also were using prop.2.2. Similarly,

$$(2.10)\qquad \|\langle M\rangle\langle D\rangle u_\varepsilon(t)\|_s \le c_s\|u_\varepsilon(t)\|_{s+e} \le c ,$$

all with constants independent of t and ε . In (2.10) we used that the norm $\|\langle M\rangle\langle D\rangle u\|_s$ is equivalent to the norm $\|u\|_{s+e}$ (cf. III,3).

The operator $C=(\langle M\rangle\langle D\rangle)^{-1}$ is compact in H_s, for all s, by III thm.5.1. Thus each set $C_c=\{u\in H_s : \|C^{-1}u\|\le c\}$ is conditionally compact in H_s: Each sequence $u_j\in C_c$ has a convergent subsequence.

The family $\{u_\varepsilon : 0\le\varepsilon\le 1\}$ satisfies the assumptions of the Arzela-Ascoli theorem. There exists $\varepsilon_j\to 0$, $\varepsilon_j>0$, such that $u_{\varepsilon_j}(t) \to u(t)\in C(I,H_s)$, uniformly in I. Under the present restrictions on φ and f this holds for all s. Using another Cantor diagonal scheme ε_j may be chosen such that this convergence holds in every H_s, s $\in \mathbb{Z}^2$, and then for all H_s, $s\in\mathbb{R}^2$. In particular, we get $u\in C(I,S)$.

Next one confirms that

$$u(0) = \lim_j u_{\varepsilon_j}(0) = \varphi\ ,\quad \lim_j u^{\cdot}_{\varepsilon_j}(t) = f(t) - i \lim K_{\varepsilon_j}(t)u_{\varepsilon_j}(t)$$

$$(2.11)\qquad = f(t) - iK(t)u(t)\ ,\text{in } S \text{ and every } H_s\ ,\text{uniformly over I .}$$

Since u_{ε_j} and its derivative $u^{\cdot}_{\varepsilon_j}$ converge uniformly the limit function u(t) is in $C^1(I,H_s)$,for every s ,and in $C^1(I,S)$ .Also , $u^{\cdot}_{\varepsilon_j} \to u^{\cdot}$,and u solves the Cauchy problem $(2.1)_0 = (2.1)$.

Now consider the general case of $\varphi\in H_s$, $f\in C(I,H_s)$, for fixed s. We may construct sequences $\varphi_j\in S$, $f_j\in C(I,S)$ such that $\varphi_j\to\varphi$, $f_j(t)\to f(t)$ in H_s, uniformly in I. For example one might choose

$$(2.12)\qquad \varphi_j = \psi^{\sim}(M/j)\psi^{\sim}(D/j)\varphi\ ,\quad f_j(t) = \psi^{\sim}(M/j)\psi^{\sim}(D/j)f(t)\ ,$$

with ψ as in (2.5), using prop.2.2. Then, letting u_j be the solution of (2.1) with φ_j and f_j, it follows that u_j-u_l solves (2.1) for $\varphi_j-\varphi_l$ and f_j-f_l. Thus the energy inequality (2.7) implies

$$(2.13)\quad \|u_j - u_l\|_s \le c\{\|\varphi_j - \varphi_l\|_s + \mathrm{Max}_{t\in I}\|f_j - f_l\|_s\} \to 0\ ,\ j,l\to\infty.$$

Thus $\{u_j\}$ is a Cauchy sequence in $C(I,H_s)$, and $u = \lim u_j \in C(I,H_s)$. Similarly $u \in C^1(I,H_{s-e})$, and u satisfies (2.1), proving existence. Uniqueness of the solution is a consequence (2.7). Q.E.D.

We now are left with the discussion of prop.s 2.2 and 2.3. We first discuss a few commutator properties of the ψdo's involved, referring to the machinery prepared in I,7.

1) The families $\{L_\varepsilon = \psi^\wedge(\varepsilon M) : 0\le\varepsilon\le 1\}$, and $\{J_\varepsilon = \bar{F}L_\varepsilon F\}$, are bounded in $L(H_0)$. Both have strong limit 1, as $\varepsilon\to 0$, in $L(H_0)$. (Indeed the family of functions $\{\psi^\wedge(\varepsilon x) : 0\le\varepsilon\le 1\}$ is bounded, and $\lim_{\varepsilon\to 0}\psi^\wedge(\varepsilon x) = \psi^\wedge(0) = \int đx\, \psi(x) = 1$, pointwise, for all x.)

2) The family $\{K(t) : t\in I\}$ is bounded in $Op\psi c^\nu_e$. (This follows directly from (2.3).)

3) The family $\{K(t) - K^*(t) : t\in I\}$ is bounded in $Op\psi c^\nu_0$. (Again this is a consequence of (2.4).)

4) The families $\{[\Pi_s, J_\varepsilon]/\varepsilon\}$, and $\{[\Pi_s, L_\varepsilon]/\varepsilon\}$, where $0<\varepsilon\le 1$, and $\Pi_s = \langle M\rangle^{s_2}\langle D\rangle^{s_1}$, with fixed $s\in\mathbb{R}^2$, are bounded in $Op\psi c_{s-e}$. Indeed we may apply I, lemma 7.2 onto

$[\Pi_s, J_\varepsilon]\Lambda_s^{-1} = [\langle M\rangle^{s_2}\langle D\rangle^{s_1}, \psi^\wedge(\varepsilon D)]\langle D\rangle^{-s_1}\langle M\rangle^{-s_2} = [\langle M\rangle^{s_2}, \psi^\wedge(\varepsilon D)]\langle M\rangle^{-s_2}$, noting that $(\langle x\rangle^{s_2})^{(\theta)} \in \psi c_{0,s_2-1}$, while $\{(\psi^\wedge(\varepsilon x))^{(\theta)}/\varepsilon : 0\le\varepsilon\le 1\}$ is bounded in $\psi c_{s_1-1,0}$, as $|\theta|=1$. For the latter one confirms that

$$(2.14)\quad \partial_\xi^\theta \omega(\varepsilon\xi) = \varepsilon^l \omega^{(\theta)}(\varepsilon\xi) = O(\varepsilon^l\langle\varepsilon\xi\rangle^{-l}) \le c\varepsilon^l(1+\varepsilon|\xi|)^{-l} = O(\langle\xi\rangle^{-|\theta|})$$

with $l = |\theta|$, for any $\omega\in S$. Similarly for L_ε.

5) The family $\{[\Lambda_s, K(t)] : t\in I\}$ is bounded in $OP\psi c^\nu_s$. This follows from (2.3) and I, lemma 7.2 again.)

6) The families $\{[K(t), J_\varepsilon]/\varepsilon\}$, and $\{[K(t), L_\varepsilon]/\varepsilon\}$, where $0<\varepsilon\le 1$, and $t\in I$, are bounded in $Op\psi c^\nu_0$. Indeed we again may apply I, lemma 7.2, using that $k^{(\theta)}(t,x,\xi)\in\psi c^\nu_{e^2}$, $k_{(\theta)}(t,x,\xi) \in \psi c^\nu_{e^1}$, $\{(\psi^\wedge(\varepsilon\xi))^{(\theta)}/\varepsilon : 0<\varepsilon\le 1\}$ is bounded in ψc_{-e^1}, for $|\theta|=1$.

7) The family $\{[J_\varepsilon, L_\varepsilon] : 0<\varepsilon\le 1\}$ is bounded in $Op\psi c_{-e}$. (Proof similar to the above.)

<u>Proposition 2.4</u>. We have, with the inner product $(u,v)_s$ of H_s

$$(2.15)\quad |(u,K_\varepsilon(t)u)_s - (K_\varepsilon(t)u,u)_s| \le c_s\|u\|_s^2,\ 0\le\varepsilon\le 1,\ t\in I,\ s\in\mathbb{R}^2,\ u\in S,$$

where the constant c_s is independent of ε, t, and u.

<u>Proof.</u> Let $v = \Pi_s u = \langle M\rangle^{s_2}\langle D\rangle^{s_1}u$. Abbreviate $K(t)=K$, $J_\varepsilon = J$, $L_\varepsilon = L$, $\Pi_s = \Pi$:

$$(K_\varepsilon u,u)_s - (u,K_\varepsilon u)_s = ((K-K^*)v, LJv) + (Kv,[J,L]v) + ([K,J]v,Lv)$$

$$(2.16)\qquad + (v,[K,L]Jv) + ([\Pi,LKJ]u,v) + (v,[LKJ,\Pi]u,\ \text{where } [\Pi,LKJ] =$$

$$= [\Pi,L]KJ + L[\Pi,K]J + LK[\Pi,J]\quad ,$$

$(.,.)$=inner product of $\mathcal{H}$. We assumed $J_\varepsilon^*=J_\varepsilon$ (ψ even) and used that

$$(2.17)\qquad \|u\|_s^2 = \|\Pi_s^{-1}v\|_s^2 = \|v\|^2 = \|v\|_0^2 .$$

For the first 4 terms at right we use, in this order, (1) and (3), (2) and (7), (1) and (6), (1) and (6). For the last two terms, the commutator split as stated, use (1), (2), (4), and (1), (5), q.e.d.

Proof of proposition 2.3, for $u \in C^1(I,\mathcal{S})$. First let the solution u of $(2.1)_\varepsilon$ be in $C^1(I,\mathcal{S})$. Then $\mu(t)=\|u(t)\|_s$ is $C^1(I,\mathbb{R})$. We get

$$(2.18)\qquad \begin{aligned} d\mu(t)/dt &= \mu^\cdot(t) = (u^\cdot(t),u(t))_s + (u(t),u^\cdot(t))_s \\ &= -i((K_\varepsilon u,u)_s - (u,K_\varepsilon u)_s) + 2\,\mathrm{Re}(f,u)_s \quad . \end{aligned}$$

Thus (2.15) implies

$$(2.19)\qquad \begin{aligned} |\mu^\cdot(t)| &\le c_s\|u(t)\|_s^2 + \|f(t)\|_2^2 + \|u(t)\|_s^2 \\ &=(c_s+1)\mu(t) + \|f(t)\|_s^2 \quad , \quad \mu(0) = \|\varphi\|_s^2 . \end{aligned}$$

The Cauchy problem (2.19), for an ordinary differential inequality may be integrated in the usual way. It follows that

$$(2.20)\qquad \mu(t)e^{-ct} - \|\varphi\|_s^2 \le \int_0^t e^{-c\tau}\|f(\tau)\|_s^2 d\tau \quad ,$$

proving (2.7) for $t \ge 0$. Similarly for $t \le 0$.

In the general case $u\in C(I,\mathcal{H}_s) \cap C^1(I,\mathcal{H}_{s-e})$ it no longer is evident that $\mu(t)$ is differentiable since the right hand side of (2.19) no longer is meaningful. The discussion below shows that $\mu^\cdot \in C(I)$ with (2.19) still exists. Then (2.20), (2.7) follow again.

Instead of μ we first form $\mu_\delta(t)=(u(t),J_\delta L_\delta u(t))_s$, $t\in I$, $\delta>0$. Abbreviating as in (2.16) again we note that $JLu \in C^1(I,\mathcal{S})$. The inner product of $\mathcal{H}_s$ is trivially extended as $(g,h)_s=\langle \Lambda_s g,\Lambda_s \bar{h}\rangle$, $g\in \mathcal{S}'$, $h\in \mathcal{S}$. Hence we may carry out a differentiation as above:

$$(2.21)\qquad \begin{aligned} \mu_\delta^\cdot(t) &= (u^\cdot,JLu)_s+ (u,JLu^\cdot)_s=-i((K_\varepsilon u,JLu)_s-(u,JLK_\varepsilon u)_s) \\ &\quad + (f,JLu)_s + (u,JLf)_s . \end{aligned}$$

We get restricted to the case $\varepsilon = 0$ since for $\varepsilon > 0$ we may keep $\delta = 0$ and carry out (2.16), using the extended inner product.

For $\varepsilon = 0$ consider the term, with $M=M_\delta=JL=J_\delta L_\delta$, $v=\Pi_s u=\Pi u$,

$$(2.22)\qquad \begin{aligned} (Ku,J_\delta L_\delta u)_s- (u,J_\delta L_\delta Ku)_s &= ((K-K^*)v,Mv) + (Kv,[\Pi,M]u) \\ &+([\Pi,K]u,Mv) +(v,[K,M]v)+(v,[MK,\Pi]u) + ([\Pi,K]u,[\Pi,M]u) \ , \end{aligned}$$

by a calculation. As $\delta\to 0$ the right hand side of (2.22) converges to

$$(2.23)\qquad ((K-K^*)v,v) + ([\Pi,K]u,v) + (v,[K,\Pi]u) = p(t) ,$$

uniformly for $t \in I$. Indeed, the second, fourth and sixth term, at right, tend to zero, using (2) and (4) ,(6) , (5) and (4) ,respectively. The remaining three terms of (2.22) converge to to the terms of (2.23) ,in the order of listing, using (1) and (3) ,(1) and (5), and (2)and (4) ,for $[MK,\Pi] = M[K,\Pi] + [M,\Pi]K \to [K,\Pi]$ in $L(H_s,H_0)$, respectively. Each convergence is uniform, for $t \in I$.

Using this in (2.21) ,as $\delta \to 0$,we find that

$$(2.24)\qquad \lim_{\delta\to 0}\mu_\delta^{\cdot}(t) = p(t) + 2\,\mathrm{Re}\,(f,u)_s \text{ uniformly for } t \in I .$$

Since μ_δ and $\mu_\delta^{\cdot}$ both converge uniformly it follows that the limit $\mu(t)$ is $C^1(I)$,and $\lim \mu_\delta^{\cdot} = \mu^{\cdot}$ .Hence we indeed get (2.19) estimating the terms (2.23) with (3),(5). This completes the proof.

3. The evolution operator and its properties.

As a consequence of the existence and uniqueness theorem of sec.2 we observe that, for every fixed $t \in I$, the assignment $\varphi \to u(t)$ defines a linear operator $U(t): H_s \to H_s$, where $u(t)$ denotes the solution of (2.1) for $f = 0$. This operator $U(t)$ will be called the evolution operator (or the solution operator) of the Cauchy problem . As an immediate consequence of (2.7) we find that

$$(3.1)\qquad \|U(t)\varphi\|_s \le \|\varphi\|_s \quad , \varphi \in H_s , t \in I .$$

Moreover from thm.2.1 it follows that the operator family $\{ U(t) : t \in I \}$ is strongly continuous in $L(H_s)$, while the derivative $U^{\cdot}(t) = dU(t)/dt$ exists in strong operator convergence of $L(H_s,H_{s-e})$,and is $C(I,L(H_s,H_{s-e}))$,under the assumptions stated. Similarly, under the assumptions of thm.1.1, the evolution operator of (1.1) will define a strongly continuous group in $L(H)$.

With thm.2.1 $U(t)={}_sU(t)$ exists for every $s\in \mathbb{R}^2$, and we get ${}_sU(t)\varphi={}_{s'}U(t)\varphi$ whenever $\varphi\in H_s \cap H_{s'}$. One obtains a well defined continuous map $S'\to S'$, of order 0 in the sense of III,3. We denote it by $U(t)\in O(0)$. Clearly $U^{\cdot}$ exists in S' , and $U^{\cdot}\in O(e)$. Also,

$$(3.2)\qquad U^{\cdot}(t) + iK(t)U(t) = 0 , t \in I , U(0) = 1 = \text{identity}.$$

Somewhat more generally we may consider the Cauchy problem

$$(3.3)\qquad du/dt + iK(t)u = f \quad , t \in I \quad , \quad u(t_0) = \varphi ,$$

with given $\tau_0 \in I$. Then I no longer must contain 0. This problem reduces to (2.1) if the new variable $t+t_0$ is introduced. Consider

$$(3.4)\quad \dot{v} + iK(t+t_0)v = f(t+t_0)\ ,t \in I-t_0 = [t_1-t_0,t_2-t_0],\ v(0)=\varphi.$$

Clearly v solves (3.4) if and only if $u(t)=v(t+t_0)$ solves (3.3). Thm.2.1 applies to (3.4), which is of the general form (2.1) with all assumptions satisfied also for the coefficient $K(t+t_0)$ and $f(t+t_0)$. Hence for each $f \in C(I,H_s)$ there exists a unique solution $u \in C(I,H_s) \cap C^1(I,H_{s-e})$, solving (3.3). Thus get a more general evolution operator $U(\tau,t)$: $S'\to S'$, mapping $H_s \to H_s$, $s\in \mathbb{R}^2$, satisfying

$$(3.5)\quad \partial_t U(t_0,t) + iK(t)U(t_0,t) = 0\quad ,\ t \in I\quad ,\ U(t_0,t_0) = 1\ .$$

If K(t) is independent of t we get $U(t_0,t) = U(t-t_0) = e^{-iK(t-t_0)}$. so that $U(t_0,t)$ is determined by the group U(t) , similar as in thm.1.1. For reasons similar as above we have

$$(3.6)\qquad \|U(t_0,t)\|_s \le c_s\quad ,\ t,t_0 \in I\ .$$

Also U and $\partial U/\partial t$ are strongly continuous for each fixed t_0. Regarding dependence of $U(t_0,t)$ on t_0 we consider the dependence of v in (3.4) on the parameter t_0. We strengthen our assumptions (2.3)-(2.4) on k=symb(K(t)) by requiring that also $k\in C^1(I,\psi c_e)$, i.e.,

$$(3.7)\qquad \langle x\rangle^{-1+|\alpha|}\langle\xi\rangle^{-1+|\beta|}\partial_t k_{(\beta)}^{(\alpha)}(t,x,\xi)\ \in\ C(I,CB(\mathbb{R}^{2n},L^\nu))\quad .$$

Let us return to the mollified equation $(3.4)_\varepsilon$, i.e., (3.4) with K replaced by K_ε of (2.6) . The coefficient $K_\varepsilon(t+t_0)$ is $C^1(I,L(H_s))$, with the set $I = \{(t,t_0) \in \mathbb{R}^2 : t_0\ ,\ t + t_0 \in I\}$. Classical theory of linear ODE's with bounded coefficients implies that $v_\varepsilon(t,t_0)$ solving $(3.4)_\varepsilon$ for f=0 is $C^1(I,L(H_s))$. Moreover, $\partial_{t_0} v_\varepsilon = w_\varepsilon$ solves the formally differentiated Cauchy problem

$$(3.8)\qquad \dot{w}_\varepsilon + iK_\varepsilon(t+t_0)w_\varepsilon + i\dot{K}_\varepsilon(t+t_0)v_\varepsilon = 0\ ,\ w_\varepsilon(0) = 0\ ,$$

where v_ε is given as solution of (3.4). Assuming first again that $\varphi \in S$, the technique of sec.2 may be repeated. One finds that $d/dt(\|w_\varepsilon\|_s^2) = O(\|w_\varepsilon\|_s^2 + \|v_\varepsilon\|_{s+e})$, using (3.7). This implies

$$(3.9)\qquad \|w_\varepsilon\|_s^2 \le e^{c|t|}\|v_\varepsilon\|_{s+e}^2\ \le\ e^{2c|t|}\|\varphi\|_{s+e}^2\quad ,$$

with c independent of ε,t,t_0 , and φ . The family $\{v_\varepsilon(t_0,t):0<\varepsilon\le 1\}$ is equi-continuous; it has bounded first partials for t and t_0 . It is bounded, and maps into a conditionally compact set. Thus ε_k, k=1,2,..., may be found, $\varepsilon_k\to 0$, v_{ε_k} converging to $v(t_0,t)\in C(I,S)$, in every H_s. Also $\partial_t v_{\varepsilon_k} = -iK_{\varepsilon_k}(t)v_{\varepsilon_k}(t)$ converges in S, hence $\partial_t v \in C(I,S)$. Moreover, the function $z_\delta(t)=(v(t_0+\delta,t)-v(t_0,t))/\delta$ solves

$$(3.10) \qquad \dot{z_\delta}(t)+iK(t+t_0)z_\delta(t)=-i(\Delta K/\delta)v^\delta \ , \ z_\delta(0)=0 \ ,$$

with $v^\delta(t)=v(t_0+\delta,t)$, $\Delta K=K(t+t_0+\delta)-K(t+t_0)$. For the difference $p=z_\delta-z_{\delta'}$, δ , $\delta'>0$, with $\Delta' K=\Delta K$, for $\delta=\delta'$, we get

$$(3.11) \quad \dot{p}+iK(t+t_0)p=-i(\Delta K/\delta\ (v^\delta-v^{\delta'})+(\Delta K/\delta-\Delta' K/\delta')v^{\delta'} \ , \ p(0)=0 \ ,$$

where $\Delta K/\delta=\delta^{-1}\int_{t+t_0}^{t+t_0+\delta}\dot{K}(\kappa)d\kappa$ is a bounded family in $L(H_s,H_{s-e})$ since $K^\cdot(t)$, $t\in I$, is bounded in $\mathrm{Op}\psi c_e^\nu$. Moreover, $\Delta K/\delta-K^\cdot(t+t_0)\to 0$ in $L(H_s,H_{s-e})$, by (3.7) and III, thm.1.1. Also, $v^\delta-v^{\delta'}\to 0$ in each H_s, since $v\in C(I,S)$. The right hand side of (3.11) tends to 0 in each H_s, as $\delta,\delta'\to 0$, uniformly in I. Using (2.7) on (3.11) get $p\to 0$ as $\delta,\delta'\to 0$, in each H_s, uniformly for $(t_0,t)\in I$, hence $\partial_{t_0}v\in C(I,S)$ exists. Taking limits in (3.10) conclude that $w=\partial_{t_0}v$ satisfies

$$(3.12) \qquad \dot{w}+iK(t+t_0)w+i\dot{K}(t+t_0)v=0 \quad , \quad w(0)=0 \ .$$

Clearly (3.12) implies that $\dot{w}$ is continuous, so that $v\in C^1(I,S)$.

Finally, if $\varphi\in H_s$ we again construct $\varphi_j=LJ\varphi\in S$, $\varphi_j\to\varphi$ in H_s and then will get $\|v-v_j\|_s\to 0$, $\|\dot{v}-\dot{v}_j\|_{s-e}\to 0$, as in the proof of thm.2.1, with v_j the solution for φ_j . Using (3.12) we also get $\|w-w_j\|_{s-e}\to 0$, for $w_j=\partial_{t_0}v_j$. Consequently,

$$(3.13) \qquad v(t_0,t)\in C(I,H_s)\cap C^1(I,H_{s-e}) \ .$$

Similarly $u(t_0,t)=v(t_0,t_0+t)$, the solution of(3.3), satisfies

$$(3.14) \qquad u(t_0,t)\in C(I\times I,H_s)\cap C^1(I\times I,H_{s-e}) \ .$$

Thm.3.1, and cor.3.2, below, now can be left to the reader:

<u>Theorem 3.1.</u> Under the assumptions of (2.3), (2.4), and (3.7) on the symbol of K(t) the evolution operator $U(\tau,t)$, $\tau,t\in I$, of the Cauchy problem (2.1) has the following properties.

(i) $U(\tau,t) : S'\to S'$ is an operator of order 0 ; it continuously maps $S\to S$, and $H_s\to H_s$, $s\in\mathbb{R}^2$. Its first partials $\partial_\tau U$, $\partial_t U$ exist in strong operator convergence of $L(H_s,H_{s-e})$, $s\in\mathbb{R}^2$. They are $\in O(e)$.

(ii) $U(\tau,t)$ is bounded and strongly continuous over $I\times I$, in $L(H_s)$, $s\in\mathbb{R}^2$. $\partial_\tau U$ and $\partial_t U$ both are bounded and strongly continuous in $L(H_s,H_{s-e})$, $s\in\mathbb{R}^2$, $t_0,t\in I$.

(iii) For $t,\tau,\kappa \in I$ we have

(3.15) $$U(\tau,\tau) = 1 \ , \ U(t,\kappa)U(\tau,t) = U(\tau,\kappa) \ , \ U(t,\tau)U(\tau,t) = 1 \ .$$

(iv) $U(\tau,t)$ satisfies the two differential equations

(3.16) $$\partial_t U(\tau,t)+iK(t)U(\tau,t) = 0 \ , \ \partial_\tau U(\tau,t)-iU(\tau,t)K(\tau) = 0, \ t,\tau \in I,$$

(v) The operator $U(\tau,t)$ is uniquely determined by properties (i), (ii), (iii), and one of the differential equations (iv).

Corollary 3.2. The evolution operator $U(\tau,t)$ is invertible in $\mathcal{S}$, $\mathcal{S}'$ and in every $\mathcal{H}_s$, for $\tau,t \in I$. Its inverse is given by $U^{-1}(\tau,t) = U(t,\tau)$. Moreover, the family

(3.17) $$V(\tau,t) = U^{-1*}(\tau,t) = U^*(t,\tau) \quad , \ t,\tau \in I \ ,$$

is the evolution operator of the adjoint equation's Cauchy problem

(3.18) $$v^{\cdot} + iK^*(t)v = g \quad , \ t \in I \quad , \ v(\tau) = \psi \ ,$$

where "*" means the $L^2(\mathbb{R}^n)$-adjoint.
For $f \in C(I,\mathcal{H}_s)$ the unique solution of (3.3) is given by

(3.19) $$u(t) = U(\tau,t)\varphi + \int_\tau^t U(\kappa,t)f(\kappa)d\kappa \quad .$$

Next we notice that higher τ,t-derivatives of U will exist if higher derivatives of K(t) are assumed. For example, let

(3.20) $$\text{symb}(K) = k \in C^\infty(I,\psi c_e^\nu) \quad .$$

Then it is clear that $K \in C^\infty(\mathcal{H}_s,\mathcal{H}_{s-e})$,for all s. Accordingly it follows from the differential equations (3.16) that $\partial_t U$ and $\partial_\tau U$ have first order partials for t and τ in $L(\mathcal{H}_s,\mathcal{H}_{s-2e})$,in strong operator convergence. Thus the second order partials of U for t,τ exist and are strongly continuous in $L(\mathcal{H}_s,\mathcal{H}_{s-2e})$. Using this in (3.16) again we find that the third order partials of U exist in $L(\mathcal{H}_s,\mathcal{H}_{s-3e})$, etc. By iteration one finds the corollary, below.

Corollary 3.3. Under the assumption of (2.3), (2.4), (3.7), and (3.20) $U(\tau,t)$ has partials of all orders existing in strong operator convergence of $\mathcal{S}$ and $\mathcal{S}'$. We have $\partial_t^j\partial_\tau^l U \in O((j+l)e)$. That derivative is strongly continuous in every $L(\mathcal{H}_s,\mathcal{H}_{s-(j+l)e})$, $s \in \mathbb{R}^2$.

Remark 3.4. It should be noticed that diffentiability in thm.3.1 and cor.3.3 may be strengthened if we replace (3.7) or (3.20) by

(3.21) $$k \in C^\infty(I,\psi c_{e^\Delta}^\nu) \ ,$$

(or, resp., $k\in C^1(I,\psi c^{\vee}_{e^\Delta}))$, where $e^\Delta=e^1=(1,0)$ or $e^\Delta=e^2=(0,1)$. With the stronger assumption get $\partial_\tau U,\ \partial_t U\in O(-e^\Delta)$, existence, continuity of $\partial_t U,\partial_\tau U$ in strong convergence of $L(H_s,H_{s-e^\Delta})$, (for C^1), and $\partial_t^j\partial_\tau^l U\in O(j+l)e^\Delta)$, corresponding existence, continuity (for C^∞). Proof by reinspection of (3.16), under the new assumptions.

4. N-th order strictly hyperbolic systems and symmetrizers.

The hyperbolic equations encountered in ch.0 were of order 2. Correspondingly we plan an existence and uniqueness theorem for an N-th order hyperbolic system (4.1), below, with a detailed theory to be worked out in ch.VII. With $I=[t_1,t_2]$, $t_0\in I$, let

$$(4.1)\quad Lu=\sum_{j=0}^N(-i)^j A_{N-j}(t)(\tfrac{d}{dt})^j u=f(t),\ t\in I,\ d^l u/dt^l=\varphi_l,\ t=t_0,\ 0\le l<N$$

where $A_k(t)=a_k(t,x,D_x)$ are k-th order differential operators in x on $\mathbb{R}^n$, with smooth coefficients, and $A_0(t)=1$. Again (4.1) will be regarded as an ODE for $u=u(t)$, taking values in $L^2(\mathbb{R}^n)$ or $H_s(\mathbb{R}^n)$.

Using a standard method of ODE convert (4.1) to a first order system. In correspondence with rem.3.4 use a <u>comparison</u> <u>operator</u>

$$(4.2)\quad \Lambda^{-1}=\Lambda_{e^\Delta}{}^{-1}=\Pi_{e^\Delta}=\pi_{e^\Delta}(x,D)=\langle x\rangle\langle D\rangle,\ e^\Delta=e,\ =\langle D\rangle,\ ^\Delta=^1,\ =\langle x\rangle,\ ^\Delta=^2.$$

Note that $\Lambda_{e^\Delta}\in Op\psi c_{-e^\Delta}$. With the new dependent variables

$$(4.3)\qquad u_j=\Lambda^j D_t^j u,\ j=0,1,\dots,N-1,$$

the Cauchy problem (4.1) (i.e., (5.7)) is equivalently written as

$$(4.4)\qquad v^{\cdot}+iK(t)v=g(t),\ t\in I,\ v(t_0)=\psi,$$

where we have introduced the (column-) vectors

$$(4.5)\quad v=(u_0,\dots,u_{N-1})^T,\ g=(0,\dots,0,i\Lambda^{N-1}f)^T,\ \psi=(\varphi_0,\dots,(-i\Lambda)^{N-1}\varphi_{N-1})^T,$$

and the square matrix of ψdo's

$$(4.6)\qquad K=\Lambda^{-1}\begin{pmatrix} 0 & -1 & 0 & \dots & 0\\ 0 & 0 & -1 & \dots & 0\\ \cdot & \cdot & \cdot & \cdot & \cdot\\ \cdot & \cdot & \cdot & \cdot & \cdot\\ 0 & 0 & \dots & 0 & -1\\ \Lambda^N A_N, & \Lambda^N A_{N-1}\Lambda^{-1}, & & \dots & ,\Lambda^N A_1\Lambda^{-N+1}\end{pmatrix}.$$

Clearly $u=u_0$ solves (4.1) if and only if v solves (4.4). K of (4.6) is a matrix of ψdo's. If (4.1) is a single equation for one unknown function then (4.4) is an $N\times N$-system. In general, if (4.1) is an $\nu\times\nu$-system, then (4.4) will be an $N\nu\times N\nu$-system. We desire to match the system (4.4) with the assumptions of thm.2.1, or better, thm.3.1 and cor.3.3 (or rem.3.4), but find it generally impossible to achieve the symmetry condition (2.4). All other conditions translate into natural conditions for equation (4.1):

We assume that $\mathrm{symb}(K(t))_{jl}\in C^\infty(I,\psi c_{e^\Delta})$, with the 3 choices of e^Δ corresponding to (4.2). This means that

$$(4.7)\qquad a_j(t,x,\xi) = \sum_{|\alpha|\le j} a_{\alpha,N-j}\xi^\alpha \in C^\infty(I,\psi c^\nu_{je^\Delta})\ ,j=1,\dots,N\ .$$

Or, as a possibly stronger condition ,we require

$$(4.8)\qquad a_{\alpha,j}(t,x)\in C^\infty(I,\psi c^\nu_{(0,(N-j)f_2)})\ ,\ f=(f_1,f_2)=e^\Delta\ .$$

Let us be guided by the case of (4.1), for scalar $u(t,x)$ (i., e., $\nu=1$), being constant coefficient hyperbolic with respect to t (ch.0,sec.8, ex'le (e), or VII,1). We then think of $e^\Delta=e^1$, $\Lambda=\lambda(D)$, $\lambda(\xi)=\langle\xi\rangle^{-1}$. The symbol σ of the operator matrix $K_0=\Lambda K$ then has the property that,up to lower order terms, $\det(\sigma+z)=\sum_{j=0}^N z^{N-j}a_j(\xi)\lambda^j(\xi)$ $=P_N(z,\xi/|\xi|)+O(\langle\xi\rangle^{-1})$, where $P_N(\tau,\xi)$ is the principal part polynomial of $L=P(D_t,D_x)=\sum a_{N-j}(D_x)D_t^j$. If L is hyperbolic then the roots of $P_N(\lambda,\xi)$ are real, as $\xi\ne0$. L is strictly hyperbolic if and only if they are real and distinct. For $|\xi|=\infty$ this happens if and only if $\mathrm{symb}(K_0)$ has real and distinct eigenvalues.

Accordingly, from now on, we focus on a system (2.1), where

$$(4.9)\qquad K_0(t) = \Lambda_f K(t)=k_0(t,x,D) + \kappa(t,x,D)\quad ,$$

with $k_0\in C^\infty(I,\psi c^\nu_0)$, $\kappa\in C^\infty(I,\psi c^\nu_{-e})$, the $\nu\times\nu$-matrix $k_0(t,x,\xi)$ having ν real and distinct eigenvalues , for all $(t,x,\xi)\in I\times\mathbf{M}_{e,0}$, with $f_2|x|+f_1|\xi|\ge\eta_0$, with the compactification $\mathbf{M}_{e,0}$ of II,3, $e^\Delta=(f_1,f_2)$ We then will speak of a <u>strictly hyperbolic system of type</u> $f=e^\Delta$.

<u>An N-th order</u> $\nu\times\nu$-<u>system (4.1) will be called strictly hyper bolic of type</u> e^Δ <u>if it leads to a strictly hyperbolic system (4.4)</u>

We will rederive all results of sec,s 2.3.4 when the symmetry condition is replaced by a weaker condition, using "symmetrizers". A (global) <u>symmetrizer</u> is defined as a one-parameter family $\{R(t)=r(t,M_w,D):t\in I\}$ of zero-order ψdo's, $r(t,.,.)\in C^\infty(I,\psi c^\nu_0)$, such that, in the Hilbert space $\mathcal{H}$, we have $R^*(t)=R(t)\ge c>0$, where R(t) is md-elliptic, uniformly over I, and that

(4.10) $(RK)^*(t)-(RK)(t)=K^*(t)R(t)-R(t)K(t) \in C^\infty(I,Op\psi c_0)$.

Here "*" denotes the $\mathcal{H}$-adjoint. The operator R(t), in other words, is a bounded hermitian and positive definite operator $\mathcal{H} \to \mathcal{H}$. Since we use the Weyl representation $R(t)=r(t,M_w,D)$ of I,3, it is no loss of generality to assume $r(t,x,\xi)$ hermitian and $\geq c>0$.

Symmetrizers were first studied by Friedrichs and Lax [FL]. For an application similar to ours cf. Taylor [T_1]. We achieve a simplification of the method of [T_1], using prop.4.1, below.

Proposition 4.1. Given a symmetrizer $R(t) = r(t,M_w,D)$ (satisfying above assumptions). Then the (unique) positive square root S(t) of R(t) , defined for every bounded positive self-adjoint Operator $\mathcal{H}\to\mathcal{H}$, is a ψdo of order 0 again, together with its inverse $S^{-1}(t)$. Moreover, S, S^{-1} are md-elliptic, and we get S , $S^{-1}\in C^\infty(I,Op\psi c_0)$.

Proof. Write $S(t)=\frac{i}{2\pi}\int_\Gamma d\lambda\sqrt{\lambda}(R(t)-\lambda)^{-1}$ as a complex curve integral with a suitable branch of $\sqrt{\lambda}$ and a curve Γ in the upper halfplane surrounding the spectrum of S(t). The resolvent $(R(t)-\lambda)^{-1}$ is a Green inverse coinciding with the K-parametrix of II,thm.1.6, mod $O(-\infty)$. In particular, Γ may be laid such that $R(t)-\lambda$ is md-elliptic for $t\in I$, $\lambda\in\Gamma$. The terms of the resulting asymptotic expansion of the symbol of the integrand are analytic in λ and smooth in t. The 0-order term yields $s_0(t,x,\xi)=\frac{i}{2\pi}\int_\Gamma \frac{\sqrt{\lambda}d\lambda}{r(t,x,\xi)-\lambda}$, using a harmless interchange of integrals. Lower order terms contain higher powers of $r-\lambda$ in the integrand, but are well defined symbols of proper order. One gets an asymptotic sum $\sum s_j$ defining a symbol of S(t) (mod $O(-\infty)$). Then all other properties follow.

Notice that prop.4.1 legalizes the transformation

(4.11) $S(t)u(t) = v(t) \ , \ t \in I$,

for the evolution problem (2.1). That problem is taken to

(4.12) $\partial_t v+iK^\sim(t)v = g(t)=S(t)f(t) \ , \ t\in I \ , \ v(0) = \psi=S(0)\varphi$,

with

(4.13) $K^\sim(t) = S(t)K(t)S^{-1}(t) +iS^\cdot(t)S^{-1}(t)$.

Clearly $K^\sim\in C^\infty(I,Op\psi c_f)$, by prop.4.1. Moreover, we confirm that

(4.14) $K^{\sim *}(t)-K^{\sim}(t)=S^{-1}(K^{*}R-RK)S^{-1}-i(S^{-1}S^{\cdot}+S^{\cdot}S^{-1})\in C^{\infty}(I,Op\psi c_{0})$

using calculus of ψdo's. For $e^{\Delta}=e$ all assumptions of thm.2.1, thm. 3.1, cor.3.2, cor.3.3 now hold for $K^{\sim}$. We have S, $S^{-1}\in O(0)$, they map $H_s\to H_s$, and $S\to S$, and $S'\to S'$. For $e^{\Delta}=e$ we even have the stronger conditions of rem.3.4. The result is summarized in thm.4.2, below.

<u>Theorem 4.2</u>. Let the system (2.1) of ψdo's satisfy $k\in C^{\infty}(I,\psi c^{\nu}_{e^{\Delta}})$, and assume that there exists a symmetrizer R(t), satisfying above conditions. Then the statements of thm.2.1, thm.3.1, cor.3.2, cor. 3.3 still hold. Moreover, for $e^{\Delta}\neq e$, the solution u of (2.1) satisfies $\partial_t^j u\in H_{s-je^{\Delta}}$ (not only $\in H_{s-je}$), as f(t), $\varphi\in H_s$. Also, for the solution operator $U(\tau,t)$ we get $\partial_\tau^j\partial_t^l U\in O((j+l)e^{\Delta})$.

Now we will show that a strictly hyperbolic system (2.1) (or (4.1)) of type $e^{\Delta}=e$, e^1, or e^2, in the above sense, always has a global symmetrizer, so that thm.4.2 applies. Indeed, we have (4.9), where the eigenvalues of $k_0(t,x,\xi)$ are real and distinct uniformly over $f_2|x|+f_1|\xi|\geq\eta_0$ in the compactification $\mathbb{M}_{e,0}$. It follows that $k_0(t,x,\xi)$ is diagonalizable for $f_2|x|+f_1|\xi|\geq\eta_0$, and that its eigenvalues $\mu_j(t,x,\xi)$ may be arranged for

(4.15) $$\mu_1(t,x,\xi) < \mu_2(t,x,\xi) < \dots < \mu_\nu(t,x,\xi),$$

Moreover, we must have

(4.16) $$\mu_{j+1}(t,x,\xi)-\mu_j(t,x,\xi)\geq\eta_1>0,\ j=1,\dots,\nu-1,\ f_2|x|+f_1|\xi|\geq\eta_0.$$

with a constant $\eta_1>0$, independent of t,x,ξ. By well known perturbation arguments for matrices it then follows that the functions μ_j are C^{∞} for $t\in I$, $f_2|x|+f_1|\xi|\geq\eta_0$.

The point is that the spectral projections $p_j(t,x,\xi)$ of the symbol $k_0(t,x,\xi)$ defined for $f_2|x|+f_1|\xi|\geq\eta_0$, may be extended to symbols $p_j\in C^{\infty}(I,\psi^{\nu}_{0})$, and that then $R(t)=r(t,M_w,D)$ with the symbol

(4.17) $$r(t,x,\xi)=\sum\bar{p}_j^{T}(t,x,\xi)p_j(t,x,\xi)+r_{-e}(t,x,\xi)=r_0+r_{-e},$$

for a suitable term r_{-e}, will define a symmetrizer. Note r(t,x,ξ) of (4.17) is symmetric modulo lower order. Its first term r_0 is in

vertible, since $r_0u=0$ implies $p_ju=0$, $j=1,\dots\nu$, hence $u=\sum p_ju=0$. We

have $k_0=\sum\lambda_lp_l$, hence $r_0k_0=\lambda_j\bar{p}_j^{T}p_j$, using that $p_jp_l=p_j\delta_{jl}$. Also,

with $\bar{p}_j^{T}\bar{p}_l^{T}=\bar{p}_j^{T}\delta_{jl}$, and $\bar{k}_0^{T}=\sum\lambda_l\bar{p}_l^{T}$, we get $\bar{k}_0^{T}r_0=\lambda_j\bar{p}_j^{T}p_j=r_0k_0$, for

$t \in I$, $f_2|x|+f_1|\xi|\geq\eta_0$. If we can show that $p_j(.,x,\xi)\in\psi c_0$ then we can hope to use calculus of ψdo's , and a proper choice of r_{-e} to get the desired properties for $R(t)$.

First the eigenvalues $\mu_j(t,x,\xi)$ are solutions of the characteristic equation $\Theta(\lambda,t,x,\xi)=\det(k_0(t,x,\xi)-\lambda) =0$, with Θ, a polynomial in λ with highest coefficient $(-1)^\nu$, and other coefficients in ψc_0. Its roots are distinct, by assumption, and are bounded by the matrix norm of $k_0(t,x,\xi)$ which is bounded above. For the first derivatives one finds

$$\mu_{j(\beta)}^{(\alpha)}=-(\Theta_{(\beta)}^{(\alpha)}/\Theta_{|\lambda})(\mu_j(t,x,\xi),t,x,\xi) \tag{4.18}$$

$$= \Theta_{(\beta)}^{(\alpha)}(\mu_j(t,x,\xi),t,x,\xi)/(\Pi_{j=1,j\neq l}^{\nu}(\mu_j(t,x,\xi)-\mu_l(t,x,\xi))$$

which is found to be $O(\langle\xi\rangle^{-|\alpha|}\langle x\rangle^{-|\beta|})$, using condition (4.16), and the boundedness of μ_j . Similarly, for the higher derivatives, including t-derivatives, one finds by successively differentiating (4.18) recursively that $\partial_t^j\mu_{j(\beta)}^{(\alpha)} = O(\pi_{-|\alpha|e^1-|\beta|e^2}(x,\xi))$, so that indeed $\mu_j\in C^\infty(I,\psi c_0)$. In all the above we were always assuming $f_2|x|+f_1|\xi|\geq\eta_0$, and all estimates are derived only for those x,ξ . For $e^\Delta=e$ it is clear that any extension of the restriction of μ_j to $|x|+|\xi|\geq2\eta_0$ to a $C^\infty(I\times\mathbb{R}^{2n})$-function will satisfy the same estimates, thus give symbols in $C^\infty(I,\psi c_0)$ equal to the μ_j, for large $|x|+|\xi|$. For $e^\Delta=e^1$ use a cut-off $\chi(\xi)$, $=1$ in $|\xi|\geq2\eta$, $=0$ near $\{|\xi|\leq\eta\}$, and use $\mu_j\chi$ instead of μ_j . Similarly for $e^\Delta=e^2$.

For the p_j we have a complex integral representation

$$p_j(t,x,\xi) = i/2\pi\int_{|\lambda-\mu_j|=\eta_1/2}(k_0-\lambda)^{-1}d\lambda \ , \tag{4.19}$$

with the constant η_1 of (4.16). Here (4.19) may be differentiated under the integral sign, where the integration path may be kept constant during differentiation. It is clear at once that we get $p_j\in C^\infty(I,\psi c_0^\nu)$, after modification outside $\{f_2|x|+f_1|\xi|\geq2\eta\} = \Xi_\eta$. For $e^\Delta\neq e$ again choose χp_j as p_j, with χ as for μ_j. We summarize:

<u>Proposition 4.3</u>. For a strictly hyperbolic system of type e^Δ, with $e^\Delta= e$, e^1 , e^2 , and $k_0\in C^\infty(I,\psi c_0^\nu)$, the eigenvalues μ_j of k_0 are symbols in $C^\infty(I,\psi c_0)$, and the spectral projections p_j are symbols in $C^\infty(I,\psi c_0^\nu)$, both after suitable modification outside some $I\times\Xi_\eta$.

<u>Theorem 4.4</u>. Let equation (2.1) be strictly hyperbolic of type e^Δ and assume that $k_0\in C^\infty(I,\psi c_0^\nu)$. Then there exists a symmetrizer, and thm.4.2 is applicable.

<u>Proof</u>. For $e^{\Delta}=e$ let r be given by (4.17) with r_{-e} to be determined For $e^{\Delta}\neq e$ add $(1-\chi)$ with $\chi(\xi)$ or $\chi(x)$ to r_0 at right of (4.17), respectively. Observe that $R_0=\frac{1}{2}(r_0(t,x,D)+r_0(t,x,D)^*)$ has the proper form, by calculus of ψdo's , while R_0 is a bounded self-adjoint operator of $\mathcal{H}$,by III, thm.1.1. Evidently R_0 is md-elliptic of order 0 , and so is $R_0-\gamma$, $0\le\gamma\le 1/n$, because, for $z\in\mathbb{C}^n$, we get

$$(4.20)\qquad \bar{z}^T r_0 z = \sum|p_j z|^2 \ge n^{-1}|z|^2 .$$

Thus we conclude that the self-adjoint operator $R_0(t)-\gamma$ is Fredholm, for $\gamma<1/n$, hence can only have discrete spectrum there, by II,4. Moreover, the eigenfunctions of this operator are in $\mathcal{S}$, hence its spectral projections are operators in $O(-\infty)$. Thus we may define $R(t)=R_0(t)(1-P_\gamma(t))$, for example, where $P_\gamma(t)$ is the orthogonal projection in $\mathcal{H}$ onto the span of all eigenvectors to eigenvalues $\le\gamma<1/n$. Here γ must be selected locally, for $t\in\Delta\subset I$, such that none of the eigenvalues ever equals γ . Then $P(t)\in C^\infty(\Delta,\psi c_{-\infty})$, and we may define a global R(t), for all $t\in I$, as a weighted mean, using a suitable partition of unity $1=\sum\chi_j(t)$. This completes the proof of thm.4.4.

<u>5. The particle flow of a single hyperbolic ψde</u>.

In this section we again consider the Cauchy problem (2.1) but assume $\nu=1$, i.e., a scalar problem

$$(5.1)\qquad \dot{u}+ik(t,M_1,D)u = f\ ,\ t\in I\ ,\ \ u(\tau)=\varphi\ ,$$

with a compact interval I , containing τ , and a <u>complex-valued</u> symbol $k\in C^\infty(I,\psi c_e)$. The conditions (2.3), (2.4) then amount to

$$(5.2)\quad k(t,x,\xi)=k_1(t,x,\xi)+k_0(t,x,\xi),\ k_1 \text{ real-valued, } k_0\in C(I,\psi c_0)$$

Under these assumptions thm.2.1, and thm.3.1 apply. For j=0,1 we assume $k_j\in C^\infty(I,\psi c_{je})$ and then have cor.3.2 and cor.3.3 as well. Assume $e^{\Delta}=e$; the cases $e^{\Delta}\neq e$ mean stronger assumptions, to be studied later on. We will get back to systems in ch.9.

In this section we focus on the result below, regarded as one of the focal points. Thm.5.1 will be called <u>Egorov</u>'s theorem, although what we present will be an amended global version of the result of Egorov [Eg_1].

<u>Theorem 5.1.</u> Conjugation $A \to A_{t,\tau} = U(\tau,t)^{-1}AU(\tau,t)$ with the evolution operator $U(\tau,t)$ of (5.1) constitutes an order preserving automorphism of the algebra $Op\psi c$ of pseudodifferential operators with symbol in $\psi c = \psi c_\infty$, as defined in I,7, for all $t,\tau \in I$. In other words, for $A = a(M_1,D) \in Op\psi c_r$, $r \in \mathbb{R}^2$, we have

$$(5.3) \qquad A_{\tau,t} = U(t,\tau)AU(\tau,t) = a_{\tau,t}(M_1,D) ,$$

with a certain symbol $a_{\tau,t} \in \psi c_r$, for each $\tau,t \in I$.

The proof of thm.5.1 will be prepared by discussing a few consequences of the theorem providing hints towards the proper approach of proof. First consider the case r=0. In V,10 we have discussed the C^*-subalgebra $\mathcal{A}_s$ of $\mathcal{L}(\mathcal{H}_s)$ obtained as norm closure of $Op\psi c_0$ in $\mathcal{L}(\mathcal{H}_s)$. It was seen that the <u>symbol space</u> of $\mathcal{A}_s$ -i.e., the maximal ideal space of the commutative C^*-algebra $\mathcal{A}_s/\mathcal{K}_s$ - is given as boundary $\mathbb{M}_{e,0}$ of $\mathbb{R}^{2n}$ in the compactification $\mathbb{P}_{e,0}$ allowing continuous extension of symbols $a \in \psi c_0$. We trivially have

$$(5.4) \qquad U(t,\tau)\mathcal{K}_s U(\tau,t) = \mathcal{K}_s \quad , \ \mathcal{K}_s = \mathcal{K}(\mathcal{H}_s) .$$

From thm.5.1 we conclude that

$$(5.5) \qquad U(t,\tau)\mathcal{A}_s U(\tau,t) = \mathcal{A}_s .$$

Accordingly there is an induced automorphism of the commutative C^*-algebra $\mathcal{A}_s/\mathcal{K}_s \cong C(\mathbb{M}_{e,0})$. It is well known that every such automorphism of an algebra $C(\mathbb{M}_{e,0})$ of continuous functions must have the form $\varphi(x) \to \varphi(\nu(x))$, with a homeomorphism $\nu\colon \mathbb{M}_{e,0} \leftrightarrow \mathbb{M}_{e,0}$. The map ν is defined as dual of the automorphism. Thus we have

$$(5.6) \qquad \nu' = \nu'_{\tau,t} \ , \ \nu'\colon \mathbb{M}_{e,0} \leftrightarrow \mathbb{M}_{e,0} \ , \ \tau,t \in I ,$$

a family of homeomorphisms induced by the automorphisms (5.5) with

$$(5.7) \qquad a_{\tau,t} = a \circ \nu_{\tau,t} \ , \text{ as } |x|+|\xi| = \infty , \text{ for all } a \in \psi c_0 .$$

In other words, the map $a \to a_{\tau,t}$ of the symbol of $A = a(x,D)$ under the automorphism $A \to U^{-1}AU$ of thm.5.1 is given by a certain family of homeomorphisms, as far as the values of a at ∞ are concerned.

Perhaps this justifies the idea to attempt construction of $a_{\tau,t}$ over all of $\mathbb{R}^{2n}$ from a family of diffeomorphisms

$$(5.9) \qquad \nu_{\tau,t} : \mathbb{R}^{2n} \to \mathbb{R}^{2n} \quad , \ \tau,t \in I ,$$

extending continuously onto the compactification $\mathbb{P}_{e,0}$, where

(5.10) $$\nu'_{\tau,t} = \nu_{\tau,t}|\mathbb{M}_{e,0} ,$$

while we expect $U^{-1}AU \approx (a \circ \nu_{\tau,t})(x,D)$. Let us assume that

(5.11) $$\nu_{\tau,\tau} = \text{identity} = \text{id} \ , \ \nu_{\tau,\kappa} = \nu_{t,\kappa} \circ \nu_{\tau,t} \quad \tau,\kappa,t \in I ,$$

because the corresponding relations for ν' follow from (3.15). Actually (5.11) points to the fact that $\nu_{\tau,t}$ is a <u>flow</u> generated by a system of differential equations (details later on).

Our condition that $\nu_{\tau,t}$ extends continuously to the compactification $\mathbb{P}_{e,0}$ amounts to the statement that

(5.12) $$a \in \psi c_0 \quad \Rightarrow \quad a \circ \nu_{\tau,t} \ \in \psi c_0 \ .$$

(We shall see that even $a \in \psi c_r$ implies $a \circ \nu \in \psi c_r$, for all $r \in \mathbb{R}^2$.)

An explicit construction of $\nu_{\tau,t}$ and a proof of thm.5.1 results if we assume that the symbol $a_{\tau,t}$ of (5.3) is of the form

(5.13) $$a_{\tau,t} = a \circ \nu_{\tau,t} + r_{\tau,t} \qquad r_{\tau,t} \in \psi c_{-e} .$$

For simplicity of notation assume $\tau=0$, and write $U(t) = U(0,t)$, $A_t = A_{0,t}$, $a_t = a_{0,t}$, $\nu_t = \nu_{0,t}$, $\nu'_t = \nu'_{0,t}$. From thm.3.1 we know that $U(t)$,hence A_t admits a strong t-derivative in $L(H_s,H_{s-e})$ .Using (3.16) we find that

(5.14) $$A_t^{\cdot} = U^{\cdot}(t)AU(t) + U(t)AU^{\cdot}(t) = iU^{-1}(t)[K(t),A]U(t) = i[K(t),A]_t .$$

Notice that the commutator $[K(t),A]$ of two ψdo's is readily evaluated in terms of ψdo-calculus. From I,(5.7) we conclude that

(5.15) $$i[K(t),A] = \langle k,a\rangle(M_1,D) + R_t \ , \ R_t \in \mathrm{Op}\psi c_{-e} \ ,$$

with the <u>Poisson bracket</u>

(5.16) $$\langle k,a\rangle = \sum_{|\theta|=1} (k^{(\theta)}a_{(\theta)} - a^{(\theta)}k_{(\theta)}) \ .$$

Assuming (5.13), and in addition that $r_t^{\cdot}$ exists and $r_t^{\cdot} \in \psi c_{-e}$,

(5.17) $$a_t^{\cdot}(x,\xi) = \partial_t(a \circ \nu_t)(x,\xi) = ((\langle k(t),a\rangle) \circ \nu_t)(x,\xi) \quad ,$$

modulo a symbol of order $-e$. It is practical to replace $k(t)$ by the real-valued $k_1(t)$ of (5.2), effecting an error term $\langle k_0,a\rangle$ of order $-e$ again. Applying the chain rule to the left hand side and (5.14) to the right hand side of (5.17), with

(5.18) $$\nu_t(x,\xi) = (x_t(x,\xi),\xi_t(x,\xi)) \qquad , \ x,\xi \in \mathbb{R}^n ,$$

a comparison gives the relations

$$(5.19)\quad x_t^{\cdot}=k_{1|\xi}(t,x_t,\xi_t)\ ,\ \xi_t^{\cdot}=-k_{1|x}(t,x_t,\xi_t)\ ,\ t\in I,\ x_0=x\ ,\ \xi_0=\xi$$

modulo terms of order -e again.

We try to satisfy (5.19) precisely, not only modulo -e, observing that (5.19) gives a Cauchy problem for a nonlinear system of 2n ODE's in the 2n variables x_t,ξ_t . This is a <u>hamiltonian</u> <u>system</u> with real-valued C^∞-coefficients $k_{1|\xi}$, $k_{1|x}$. Locally, for a small (t,x,ξ)-set, the solution of (5.19) exists uniquely, and depends smoothly on the initial parameters x,ξ, by basic results on ODE's. The local solution may be continued into all of $\mathbb{R}^{2n}\times I$ provided that we can derive apriori estimates insuring boundedness of (x_t,ξ_t) over I, for any given finite x,ξ. Such estimates will be obtained in sec.6. Hence for $\alpha,\beta\in \mathbb{Z}^n$, $\partial_x^\alpha\partial_\xi^\beta x_t(x,\xi)$, $\partial_x^\alpha\partial_\xi^\beta \xi_t(x,\xi)\in C^\infty(I\times\mathbb{R}^{2n})$, by standard results (cf. [CdLv], for example).

Note that the same argument for $U(\tau,t)$ instead of $U(t)=U(0,t)$ will lead to the same system (5.19) of differential equations, but with initial conditions at τ, not at 0. Thus we get (5.11) as an consequence. Particularly, $\nu_{t,\tau}\circ\nu_{\tau,t}$ = identity, so that $\nu_{\tau,t}$ is indeed a family of homeomorphisms (even diffeomorphisms) of $\mathbb{R}^{2n}$ onto itself. This family will be called a <u>particle</u> <u>flow</u> of equation (5.1) .Clearly the particle flow is <u>not</u> uniquely determined by (5.1), since the symbol k_1 is only fixed up to an additional term of order -e. However, it is unique at $|x|+|\xi|=\infty$, as will be seen.

<u>Theorem 5.2</u>. For any given real-valued $k_1\in \psi c_e$ the flow $\nu_{\tau,t}$ defined above (following the solution of (5.19) through (x°,ξ°) (at $t=\tau$) to $t=t$) induces a family $a\to a_{\tau,t}=a\circ\nu_{\tau,t}$ of automorphisms $\psi c_m\leftrightarrow\psi c_m$, for all m. Moreover, we have $a_{\tau,t}\in C^\infty(I\times I,\psi c_m)$, $m\in\mathbb{R}^2$.

The proof will be discussed in sec.6.

It is evident, after thm.5.2, that the map $a\to a_{\tau,t}=a\circ\nu_{\tau,t}$ also provides an automorphism of the sup-norm closure of the function algebra ψc_0 , the maximal ideal space of which defines the compactification $\mathbf{P}_{e,0}$. The associate dual map of this automorphism will be a homeomorphism $\mathbf{P}_{e,0}\leftrightarrow\mathbf{P}_{e,0}$, defined as the continuous extension of $\nu_{\tau,t}$ onto $\mathbf{P}_{e,0}$. The restriction $\nu'_{\tau,t}$ to $M_{e,0}=\partial\mathbf{P}_{e,0}$ will be uniquely determined by the system (5.1). This map coincides with the homeomorphism $\nu'_{\tau,t}:\mathbf{M}_{e,0}\to\mathbf{M}_{e,0}$ previously discussed (cf.(5.6)).

<u>Proof of thm.5.1</u>. From (5.11) we get $\partial_t(\nu_{t,\tau}\circ\nu_{\tau,t})=0$. From the ODE's (5.19), using the Poisson bracket of (5.16), we get

$$(5.20)\qquad \partial_t x_{t,\tau}+\langle k_1,x_{t,\tau}\rangle=0\ ,\quad \partial_t\xi_{t,\tau}+\langle k_1,\xi_{t,\tau}\rangle=0\ ,$$

with $\langle\, , \rangle$ of $x_{t,\tau}(x,\xi)$ taken componentwise.

For an arbitrary smooth $a(x,\xi)$ and $a_{t,\tau}=a\circ\nu_{t,\tau}$ we get

$$\partial_t a_{t,\tau}(x,\xi) + \langle k_1, a_{t,\tau}\rangle = 0 \quad , \tag{5.21}$$

as follows by using (5.20) in

$$\begin{aligned}\partial_t a_{t,\tau}(x,\xi) &= \partial_t a(x_{t,\tau}(x,\xi),\xi_{t,\tau}(x,\xi)) \\ &= a_{|x}(x_{t,\tau},\xi_{t,\tau})\partial_t x_{t,\tau} + a_{|\xi}(x_{t,\tau},\xi_{t,\tau})\partial_t\xi_{t,\tau} \quad .\end{aligned} \tag{5.22}$$

For $a \in \psi c_m$ let us use the operator

$$P_{\tau,t}= U(t,\tau)A_{t,\tau}U(\tau,t) \ , \ A_{t,\tau}=a_{t,\tau}(x,D) \ , \ a_{t,\tau}=a\circ\nu_{t,\tau} . \tag{5.23}$$

Clearly $A_{t,\tau}\in C^\infty(I\times I, L(H_s,H_{s-m}))$, by III,3, and thm.5.2. Since $\partial_t U(\tau,t)\in C(I\times I, L(H_s,H_{s-e}))$ in strong operator topology, by thm. 3.1, we get $P_{\tau,t} \in C^1(I\times I, L(H_s,H_{s-m-e}))$ in strong convergence and

$$\partial_t P_{\tau,t} = U(t,\tau)\{(\partial_t a_{t,\tau})(x,D) + i[K(t),A_{t,\tau}]\}U(\tau,t) = V_t \ , \tag{5.24}$$

by (5.14). Using (5.15) and (5.21), and calculus of ψdo's, and $U\in O(0)$ (by thm.3.1) we find that $V_t\in O(m-e)$. Also V_t is strongly continuous in t as a map $H_s\to H_{s-m+e}$. Hence Vt is integrable, and

$$P_{\tau,t}-P_{\tau,\tau} = U(t,\tau)A_{t,\tau}U(\tau,t)-A = \int_\tau^t d\sigma V_\sigma \in L(H_s,H_{s-m+e}) \ , \tag{5.25}$$

This confirms that at least (5.3) holds modulo an additive term in $O(r-e)$. To get more precision note that $V_t=U(t,\tau)W_tU(\tau,t)$, with $W_t \in Op\psi c_{m-e}$, hence (5.25) assumes the form

$$U(\tau,t)AU(t,\tau) = a_{t,\tau}(M_1,D) - \int_t^t d\kappa\ U(\kappa,t)W_\kappa U(t,\kappa) \ . \tag{5.26}$$

Clearly (5.25') may be iterated. Using (5.26) on W_k expresses $U(\kappa,t)W_\kappa U(t,\kappa)$ as a sum $T_\kappa+R_\kappa$, with $T_\kappa\in Op\psi c_{r-e}$, remainder R_κ in $O(r-2e)$, as in the integral of (5.26), etc. Thus, for all N,

$$U(\tau,t)AU(t,\tau) = a_{t,\tau}(M_1,D) + \textstyle\sum_{j=1}^N T^N_{t\,\tau} + X^N \quad , \tag{5.27}$$

with $T^N_{t,\tau} \in Op\psi c_{r-Ne}$, X_N of order $r-(N+1)e$. Then we get

$$U(\tau,t)AU(t,\tau) = a_{t,\tau}(M_1,D) + \textstyle\sum_{j=1}^\infty T^N_{t,\tau} + X^\infty \ , \tag{5.28}$$

with $X^\infty\in Op\psi c_{-\infty}$, and an asymptotic sum at right, using I,6. Q.E.D.

A generalization of thm.5.1 to hyperbolic $\nu\times\nu$-systems was discussed in [CE]. We will discuss this in ch.9 for a larger and more natural algebra of ψdo's (called $Op\psi s$), rather than for $Op\psi c$.

6. The action of the particle flow on symbols.

In this section we again look at the particle flow $\{v_{\tau t}\}$ In particular we prove thm.5.2. Assumptions are as in thm.5.1.

In order to simplify the calculations let us write

$$(6.1)\quad f=f(x,\xi)=f(x,\xi,t,\tau)=x_{\tau t}(x,\xi)\ ,\ \varphi=\varphi(x,\xi)=\varphi(x,\xi,t,\tau)=\xi_{\tau t}(x,\xi).$$

Writing $k_1=k$, for simplicity, we then have (5.19) in the form

$$(6.2)\qquad f^{\cdot}=k_{|\xi}(f,\varphi)\ ,\ \varphi^{\cdot}=-k_{|x}(f,\varphi)\ ,\ f=x\ ,\ \varphi=\xi \text{ at } t=\tau\ .$$

Proposition 6.1. The functions f , φ satisfy the apriori-estimates

$$(6.3)\qquad 0<c\le\langle f\rangle/\langle x\rangle\le C\ ,\ 0<c\le\langle\varphi\rangle/\langle\xi\rangle\le C\ ,$$

for $t,\tau\in I$, $x,\xi\in\mathbb{R}^n$, with constants c,C independent of x,ξ,t,τ.

Proof. One derives the following inequalities for $\langle f\rangle^{\cdot}$, $\langle\varphi\rangle^{\cdot}$:

$$(6.4)\qquad \begin{aligned}\langle f\rangle^{\cdot} &= f.f^{\cdot}/\langle f\rangle = k_{|\xi}.f/\langle f\rangle = O(\langle f\rangle)\ ,\\ \langle\varphi\rangle^{\cdot} &= \varphi.\varphi^{\cdot}/\langle\varphi\rangle = -k_{|x}.\varphi/\langle\varphi\rangle = O(\langle\varphi\rangle)\ .\end{aligned}$$

Integrating (6.4), under the initial conditions $\langle f\rangle=\langle x\rangle$, $\langle\varphi\rangle=\langle\xi\rangle$ at $t=\tau$ we get

$$(6.5)\qquad \log(\langle f\rangle/\langle x\rangle) = O(1)\ ,\ \log(\langle\varphi\rangle/\langle\xi\rangle) = O(1)\ .$$

Since $\langle f\rangle$, $\langle\varphi\rangle\ge 1$ this indeed give (6.3), q.e.d.

Note that prop.6.1 gives the apriori estimates required for continuation of the local solution of (6.2) into all of I , for any choice of initial values x,ξ . As noted before, we then also get continuity of derivatives $\partial_\tau^j\partial_t^l f_{(\beta)}^{(\alpha)}$, $\partial_\tau^j\partial_t^l\varphi_{(\beta)}^{(\alpha)}$ of all orders j,l,α,β , as $t,\tau\in I$, $x,\xi\in\mathbb{R}^n$.

Proposition 6.2. For $j,l=0,1,\ldots.,$ and $\alpha,\beta\in\mathbb{N}^n$ we have

$$(6.6)\quad \partial_\tau^j\partial_t^l f_{(\beta)}^{(\alpha)}=O(\langle x\rangle^{1-|\beta|}\langle\xi\rangle^{-|\alpha|})\ ,\ \partial_\tau^j\partial_t^l\varphi_{(\beta)}^{(\alpha)}=O(\langle x\rangle^{-|\beta|}\langle\xi\rangle^{1-|\alpha|})\ .$$

Proposition 6.3. For a C^∞-function $b(x,\xi)$ consider the composition

$$(6.7)\qquad c(x,\xi) = b(f(x,\xi),\varphi(x,\xi))\ .$$

The derivative $c_{(\beta)}^{(\alpha)}$ is in the span of the terms

$$(6.8)\qquad (\nabla_x^r \nabla_\xi^s b)(f,\varphi)\Pi_{j=1}^r f^{(\alpha^j)}_{(\beta^j)}\ \Pi_{l=1}^s\ \varphi^{(\alpha'^l)}_{(\beta'^l)}$$

where $|\alpha^j|+|\beta^j|\ge 1$, $|\alpha'^l|+|\beta'^l|\ge 1$, $\sum\alpha^j+\sum\alpha'^l=\alpha$, $\sum\beta^j+\sum\beta'^l=\beta$,

and where the tensors $\nabla_x^r\nabla_\xi^s b$ and vectors $f^{(\alpha^j)}_{(\beta^j)}$, $\varphi^{(\alpha'^l)}_{(\beta'^l)}$ are to be contracted in arbitrary order of indices, f with ∇_x , φ with ∇_ξ .

This proposition follows by induction.

For the proof of prop.9.2 we also use induction and first assume j=l=0 and $\alpha+\beta=1$. Note that the case $j=l=\alpha=\beta=0$ is a matter of (6.3). Differentiating (6.2) we obtain

$$(6.9)\qquad \begin{aligned} f^{(\alpha)\cdot}_{(\beta)} &= k_{|\xi x}f^{(\alpha)}_{(\beta)} + k_{|\xi\xi}\varphi^{(\alpha)}_{(\beta)}\ ,\ f^{(\alpha)}_{(\beta)}=x^{(\alpha)}_{(\beta)} \\ \varphi^{(\alpha)\cdot}_{(\beta)} &= -k_{|xx}f^{(\alpha)}_{(\beta)} - k_{|x\xi}\varphi^{(\alpha)}_{(\beta)}\ ,\ \varphi^{(\alpha)}_{(\beta)}=\xi^{(\alpha)}_{(\beta)} \end{aligned}\quad ,\ \text{as } t=\tau\ .$$

Here we have written $k_{|x\xi}f=\sum k_{|x\xi_j}f_j$, but $k_{|\xi x}f=\sum k_{|\xi x_j}f_j$, etc.

Multiplying the first equation (6.9) by $\langle\varphi\rangle/\langle f\rangle$ we get, as matrix,

$$(6.10)\qquad \begin{pmatrix} \langle\varphi\rangle/\langle f\rangle f^{(\alpha)\cdot}_{(\beta)} \\ \\ \varphi^{(\alpha)\cdot}_{(\beta)} \end{pmatrix} = \begin{pmatrix} k_{|\xi x} & k_{|\xi\xi}\langle\varphi\rangle/\langle f\rangle \\ \\ -k_{|xx}\langle f\rangle/\langle\varphi\rangle & -k_{|x\xi} \end{pmatrix} \begin{pmatrix} \langle\varphi\rangle/\langle f\rangle f^{(\alpha)}_{(\beta)} \\ \\ \varphi^{(\alpha)}_{(\beta)} \end{pmatrix} .$$

The 2n×2n-matrix in (6.10), called P_1 , has bounded coefficients, since $k\in\psi c_e$. Let the vector at right in (6.10) be called $p=\binom{p^1}{p^2}$. The left hand side in (6.10) is not $p^\cdot$, but we get

$$(6.11)\qquad p^{1\cdot} = \langle\varphi\rangle/\langle f\rangle f^{(\alpha)\cdot}_{(\beta)} - ((\varphi k_{|x})/\langle\varphi\rangle^2 + (fk_{|\xi})/\langle f\rangle^2)p^1\ ,$$

hence (6.10) yields the linear ODE $p^\cdot=Pp$ with a matrix P differing from P_1 only in the upper left n×n-corner. There we find the additional term $-((f.k_{|\xi})/\langle f\rangle^2+(\varphi.k_{|x})/\langle\varphi\rangle^2)$, also bounded in x,ξ , so that P also is a bounded matrix. We get

$$(6.12)\qquad |p|^{2\cdot}=2p.p^\cdot = 2p.Pp = O(|p|^2)\ .$$

This may be integrated for $\log(|p|/|p(\tau)|)=O(|t-\tau|)=O(1)$. Calculating $|p(\tau)|$ from (6.9) we get (6.6) for j=l=0, $|\alpha|+|\beta|=1$. Proceeding in our induction proof, assume (6.6) true for j=l=0 and $|\alpha|+|\beta|\le r-1$, and consider a pair α,β with $|\alpha|+|\beta|=r\ge 2$. We claim that the vector p at right of (6.10) now will satisfy

$$(6.13)\qquad p^\cdot = Pp + q\ \ ,\ p(\tau) = 0\ ,\ q = \binom{q^1}{q^2}\ ,$$

with the above matrix P , and with q^1 , q^2 in the span of

$$(6.14)\qquad (\langle\varphi\rangle/\langle f\rangle\nabla_x^s\nabla_\xi^{s'+1}k(f,\varphi))\Pi_{j=1}^s f_{(\beta^j)}^{(\alpha^j)}\ \Pi_{l=1}^{s'}\ \varphi_{(\beta'^l)}^{(\alpha'^l)}\quad ,$$

and of

$$(6.15)\qquad (\nabla_x^{s+1}\nabla_\xi^{s'}k(f,\varphi))\Pi_{j=1}^s\ f_{(\beta^j)}^{(\alpha^j)}\ \Pi_{l=1}^{s'}\ \varphi_{(\beta'^l)}^{(\alpha'^l)}\quad .$$

respectively. In (6.14) and (6.15) we assume that $s+s'\le r$, and $1\le|\alpha_j|+|\beta_j|<r$, $1\le|\alpha'^l|+|\beta'^l|<r$, $\sum\alpha^j+\sum\alpha'^l=\alpha$, $\sum\beta^j+\sum\beta'^l=\beta$. Again contraction of $\nabla_x^r\nabla_\xi^s k$ with the products of vectors is in arbitrary order of indices. Relations (6.13), (6.14), (6.15) follow by repeated differentiation of (6.9), using (6.8).

The induction hypothesis implies that $q=O(\langle x\rangle^{-|\beta|}\langle\xi\rangle^{1-|\alpha|})$, by direct examination of the terms (6.13), (6.14). Then (6.13) implies $|p|^{2\cdot} = O(|p|^2+|q|^2)$, $|p(\tau)|^2=0$. This may be integrated for $|p|=O(|q|)=O(\langle x\rangle^{-|\beta|}\langle\xi\rangle^{1-|\alpha|})$, completing the proof for j=l=0.

Next we look at t and τ-derivatives of f,φ . First it is evident for the vector of (6.13) that also $p^\cdot=O(\langle x\rangle^{-|\beta|}\langle\xi\rangle^{1-|\alpha|})$, just looking at (6.13). This gives (6.6) for j=0, l=1. In fact, we may differentiate (6.13) for t . The right hand side then contains only terms involving the 0-th and 1-th t-derivatives of $f_{(\beta)}^{(\alpha)}$ and $\varphi_{(\beta)}^{(\alpha)}$, already known to satisfy (6.6) . Using this, one finds that also $p^{\cdot\cdot}=O(\langle x\rangle^{-|\beta|}\langle\xi\rangle^{1-|\alpha|})$. In fact, the procedure may be iterated to get the same for all $\partial_t^l p$. This implies (6.6) for j=0 and all l=0,1,2,... .

Finally let us look for τ-derivatives. In that respect it is convenient to notice that $(f,\varphi)=(x_{\tau\tau},\xi_{\tau t})$ satisfies the system

$$(6.16)\qquad f'=-k_{|\xi}(f,\varphi)\ ,\ \varphi'=k_{|x}(f,\varphi)\ ,\ "'" = "\partial/\partial\tau"\quad .$$

Indeed, relations (5.20), (5.21) indicate that, for fixed τ,x^0,ξ^0, the curves $(x,\xi) = \nu_{t,\tau}(x^0,\xi^0)$ are the characteristic curves of the single first order PDE (5.21) in the unknown function $a(t,x,\xi)=a_{t\tau}(x,\xi)$. This equation differs from (5.17) only by a sign, hence it follows that $\nu_{t,\tau}$ must be the 'reverse flow' satisfying (6.16), just as $\nu_{\tau,t}$ satisfies (5.19) . This clearly implies (6.16). However, (6.16) implies an ODE like (6.13) with "$\cdot$" replaced by "'" , for the same vector p with components $\langle\varphi\rangle/\langle f\rangle f_{(\beta)}^{(\alpha)}=p^1$, $\varphi_{(\beta)}^{(\alpha)}=p^2$. It follows at once that (6.6) holds for l=0, j=0,1,2,..., for reason of symmetry. But mixed t,τ-derivatives may also be estimated in this way: For example, writing $p'=Qp+r$, we get $p'^{\cdot}=Qp^\cdot+$

$Q^{\cdot}p+r^{\cdot}=QPp+Qq+Q^{\cdot}p+r^{\cdot}$, where the right hand side consists of terms already estimated. Hence $p^{\cdot\cdot}=\partial_\tau\partial_t p$ may be estimated. Similarly for all mixed τ,t-derivatives, completing the proof of prop.6.2.

Note that now the <u>proof of thm.5.2</u> is a matter of induction. First of all it follows from prop.6.3 that $a_{\tau,t}=a\circ\nu_{\tau,t}=a(f,\varphi)\in\psi c_m$ for all $t,\tau\in I$, whenever $a\in\psi c_m$: All terms (6.8) obey their proper estimate, using (6.6) . Moreover, also the derivatives $\partial_\tau^j\partial_t^l a_{\tau,t}$ exist in the Frechet topology of ψc_m, so that $a_{\tau,t}\in C^\infty(I\times I,\psi c_m)$. For example,

$$(6.17)\qquad \partial_t a_{\tau,t} = a_{|x}(f,\varphi)f^{\cdot} + a_{|\xi}\varphi^{\cdot} = O(\pi_m(x,\xi)) ,$$

using (6.6), prop.6.1, and the estimates for the symbol $a\in\psi c_m$. Similarly for all higher derivatives, by induction.

<u>7. Propagation of maximal ideals and propagation of singularities</u>.

We have noted before that the particle flow induces a class of homeomorphisms $\nu'_{\tau t}:\mathbb{M}\to\mathbb{M}$, of the space $\mathbb{M}=\mathbb{M}_{e,0}$, i.e., the maximal ideal space of the C^*-algebra $A=A_0$ of V,10. In particular $\nu'_{\tau t}$ is defined as restriction to $\mathbb{M}=\partial\mathbb{P}$ of the continuous extension to $\mathbb{P}$ of $\nu_{\tau t}$ defined by (5.18),(5.19). The algebra symbol $\sigma_{A_{\tau t}}$ of $A_{\tau t}=U(t,\tau)AU(\tau,t)$ is given as

$$(7.1)\qquad \sigma_{A_{\tau t}} = \sigma_A\circ\nu'_{\tau t} ,$$

for every $A\in A$. Similar for $A\in A_s$, the H_s-closure of $Op\psi c_0$.

It is interesting to note that the above facts immediately translate into a result concerning propagation of wave front sets - and, correspondingly, propagation of singularities - under hyperbolic evolution. The key is the following result (cf. [Tl_2]).

<u>Theorem 7.1</u>. For any distribution $u\in S'(\mathbb{R}^n)$ the wave front set WF(u) (as defined in II,5) is given as

$$(7.2)\qquad WF(u) = \cap\{char(A,C): A\in C^\circ , Au\in S\} ,$$

with the <u>characteristic set of</u> A relative to C of V,10 defined by

$$(7.3)\qquad char(A,C) = \{m\in\mathbb{W} : \sigma_A(m)=0\} .$$

<u>Proof</u>. Suppose $a(x^\circ,\xi^\circ)\neq 0$ for some $(x^\circ,\xi^\circ)\in\mathbb{W}$, $A=a(x,D)\in C^\circ$, with $Au\in C^\infty(\mathbb{R}^n)$. Then it follows that, with respect to some cut-offs $\varphi_{x^\circ}(x)\,\psi_{\xi^\circ}(\xi)$, as in II,5 , we must have A md-elliptic with

respect to $\varphi_{x^0}(x)\psi_{\xi^0}(\xi) = \varphi(x,\xi)$, in the sense of II,3. By II,thm. 3.3 we then get a local left parametrix E_1 such that $E_1 A=\varphi(M_r,D)+K_1$, $K_1 \in O(-\infty)$, $E_1 \in Op\psi c_0$. It follows that $(\psi_{\xi^0}(\varphi_{x^0}u)^\wedge)^\vee = E_1 Au-K_1 u\in S$, hence $(x^0,\xi^0)\in W\backslash WF(u)$, by II,prop.5.1. In other words, we have "$\subset$" in (7.2). Vice versa, if $(x^0,\xi^0)\in W\backslash WF(u)$, then there exist cut-offs as above with $\psi(D)\varphi(x)u\in S$, by II, prop.5.2. It is clear then that $B=\psi(D)\varphi(x)\in C^\infty$, while (x^0,ξ^0) does not belong to $\mathrm{char}_C B$. Hence we also get "$\supset$", q.e.d.

Thm.7.1 first prompts us to define an *A*-wave front set

$$WF(u,A) = \cap\{\mathrm{char}(A,A): A\in Op\psi c_0\ ,\ Au\in S\}\ , \tag{7.4}$$

where

$$\mathrm{char}(A,A) = \{m\in W(A): \sigma_A(m)=0\}\ , \tag{7.5}$$

with the wave front space $W(A)=\iota^{-1}W = \iota^{-1}(\mathbb{R}^n\times\partial\mathbb{B}^n)$ of A, using the surjective map $\iota: M_{e,0} \to M$ of II,3. Clearly we get

$$\iota WF(u,A) \subset WF(u)\ . \tag{7.6}$$

We also may define the larger sets

$$\begin{aligned} ZF(u,A) &= \cap\{\mathrm{char}_{md}(A,A) : A\in Op\psi c_0 : Au\in S\}\ , \text{ with} \\ \mathrm{char}_{md}(A,A) &= \{m\in M(A) : \sigma_A(m)=0\}\ . \end{aligned} \tag{7.7}$$

Note that $ZF(u,A)$ (and $ZF(u,C) = ZF(u)$ defined similarly, with reference to C instead of A) also addresses singularities of $u\in S'$ at $x=\infty$. In fact,

$$ZF(u)\supset WF(u) \cup WF(u^\wedge)\ . \tag{7.8}$$

<u>Theorem 7.2</u>. Consider the solution $u(t)=U(\tau,t)\varphi$ of the Cauchy problem (3.3) with $f(t)=0$. We have

$$WF(u(t),A)= \nu'_{\tau t}(WF(\varphi,A))\ ,\ ZF(u(t),A)=\nu'_{\tau t}(ZF(\varphi,A))\ . \tag{7.9}$$

Similarly, for $f(t)\neq 0$, the sets at left in (7.9) are contained in the unions of the sets at right with the sets $\left.{WF \atop ZF}\right\}(U(\kappa,t)f(\kappa),A) =\nu'_{\kappa t}(\left\{{WF \atop ZF}\right\}(f(\kappa),A))$, κ between τ and t.

<u>Proof</u>. Observe that $ZF(u(t),A)=\cap\{\mathrm{char}_{md}(A,A):U(t,\tau)AU(\tau,t)\varphi\in S\}$, where $\mathrm{char}_{md}(A,A)=\{m:\sigma_A=0\}=\{m:\sigma_{A_{\tau t}}(\nu'_{t\tau}(m))=0\}=\nu'_{\tau t}(\mathrm{char}_{md}(A_{\tau t},A))$

Therefore, $ZF(u(t),A)=\nu'_{\tau t}\{\mathrm{char}_{md}(A_{\tau t},A):A_{\tau t}u\in S\}=\nu'_{\tau t}(ZF(\varphi,A))$.

Similarly for $WF(u(t),A)$, and for $\left.{WF \atop ZF}\right\}(U(\kappa,t)f(\kappa),A)$. Q.E.D.

Corollary 7.3. Let $k_0(t,x,\xi)=\mathrm{symb}(K_0(t))$ of (4.9) allow an asymptotic expansion $k_0=\sum k_{0j}$ (mod $|\xi|$) at $\xi=\infty$, as in V,(5.10). Assume $k_j \in C^\infty$. for $t\in I$, $|\xi|=1$, $x\in K\subset\subset \mathbb{R}^n$, all K. Then we also get formulas (7.9) for WF(u) instead of WF(u,$\mathcal{A}$) .

For a proof we only must show that (i) C^0 in thm.7.2 may be replaced by the algebra $\mathcal{B}$ of all $A\in \mathrm{Op}\psi c_0$ with symbols having asymptotic expansions, and (ii) that conjugation with $U(\tau,t)$ leaves $\mathcal{B}$ invariant - assuming the asymptotic expansion on k_0 , while (iii) the particle flow leaves $\mathbf{M}$ invariant. Details of these (rather formal) discussions are left to the reader.

Remark 7.4. A propagation law for the generalized singularity sets ZF(u) may be derived as well, under proper assumptions on k_0 .

Problems. (1) Do the algebras *VGS* and *ΨGL* (cf.VIII,5) have commutators in $\mathcal{K}(L^2(\mathbb{R}^n))$? (2) For the $\mathcal{L}(L^2(\mathbb{R}^n))$-norm closure $\mathcal{A}$ of an algebra of (1), give a useful definition of WF(A,$\mathcal{A}$) and ZF(A,$\mathcal{A}$).

Chapter 7. HYPERBOLIC DIFFERENTIAL EQUATIONS

0. Introduction.

A polynomial $P(x) = \sum_{|\alpha|\le N} a_\alpha x^\alpha$, $a_\alpha \in \mathbb{C}$ (and the PDE P(D)) is called hyperbolic with respect to a vector $h \in \mathbb{R}^n$ if (i) $P_N(h) \neq 0$ (ii) for some real τ_0 we have $P(\xi+i\tau h) \neq 0$ as $\xi \in \mathbb{R}^n$, $\tau \le \tau_0$.

The above definition was given by Garding [Ga_1]. Its importance for the Cauchy problem of P(D)u=f in the half space $x.h \ge 0$, with data given at the hyperplane x.h=0 becomes evident if we try to apply the Fourier-Laplace method of ch.0,4: Let h=(1,0,...,0). Taking the Fourier transform in $(x_2,\dots,x_n)$ and the Laplace transform in x_1 will convert P(D)u=f into $P(\xi+i\tau h)u^\wedge = f^\wedge$, where the imaginary part τ of the ξ_1-variable must be $\le \tau_0$, for some τ_0, in accordance with the rules of the Laplace transform of ch.0,(3.14). Thus (ii) insures that we can write $u^\wedge = f^\wedge / P(\xi+i\tau h)$, as $\tau \le \tau_0$. Specifically, for $f = \delta/\kappa_n$, the inverse transform $\left(\frac{1}{P(\xi+i\tau h)}\right)^\vee = e$ defines a fundamental solution of P(D). An analysis shows that (i),(ii) insure existence of e as a complex integral. Moreover, e=0 outside a cone $|x| \le cx_1$. This leads to Garding's theorem:

Theorem 0.1. A differential polynomial P(D) admits a fundamental solution $e \in \mathcal{D}'(\mathbb{R}_n)$ with support in a strictly convex cone $0 \in \boldsymbol{h} \subset \{0\} \cup \{x \cdot h > 0\}$ if and only if P is hyperbolic with respect to h.

For a distribution e with the properties of thm.0.1 and a general $f \in \mathcal{D}'$ with support in the half-space $x.h \ge 0$ the convolution product $u = e * f \in \mathcal{D}'$ is always defined, and supp u $\le \{x.h \ge 0\}$ ([C_1],I, thm.8.1,for example), and P(D)u=f. In fact, the noncharacteristic Cauchy problem of the half space $x.h \ge 0$ is well posed if and only if P is hyperbolic with respect to h (written as "$P \in$ hyp(h)"). For a detailed discussion of such results cf.[Hr_1].

For PDE's with variable coefficients we do not know results of similar precision. Existence theorens assume strict hyperbolicity (cf. VI,4) which is not a necessary condition. Still we discuss some useful algebraic properties of hyperbolic polynomials in sec. 1 below, focusing on the cone $\boldsymbol{h}$ of time-like vectors. In sec.2 we consider <u>strictly hyperbolic</u> (differential) polynomials. The class s-hyp(h) of strictly hyperbolic polynomials consists precisely of all $P\in$ hyp(h) with only simple real characteristic surfaces (i.e., the $P\in$ hyp(h) of principal type). "s-hyp(h)" is a property of the principal part of P, insensitive against lower order perturbations

A PDE a(x,D) with <u>variable coefficients</u> is called s-hyp(h) at $x=x^0$ if $a(x^0,D)\in$ s-hyp(h). For an a(x,D), strictly hyperbolic over a domain $\Omega\subset R^n$ the vector may depend on x. If is can be chosen constant, we speak of a <u>normally strictly hyperbolic</u> a(x,D). It is convenient then to assume $h=(1,0,\dots,0)$ and distinguish the time direction h. This leads to sec.3, considering an N-th order $\nu\times\nu$-system, looking like an ODE in t with coefficients PDE's in x. We prove existence and uniqueness for a <u>relativistic</u> and a <u>nonrelativistic</u> normally strictly hyperbolic Cauchy problem.

In sec.4 we simplify the assumptions, essentially by linking strictly hyperbolic differential operators with the strictly hyperbolic ψdo's of ch.VI. In sec.5 we discuss the 'hyperbolic features' - region of dependence and influence, and finite propagation speed. In sec.6 we derive a local existence result, and apply it to hyperbolic problems on a manifold.

<u>1. Algebra of hyperbolic polynomials</u>.

We follow $[Hr_1]$, discussing some algebraic properties of hyperbolic polynomials. No differential operators will get involved.

<u>Proposition 1.1</u>. A homogeneous polynomial $P = P_N$ is hyp(h) if and only if (i) $P(h) \neq 0$, and (ii) for each fixed $\xi \in \mathbb{R}^n$ the equation $P(\xi + \tau h) = 0$ has only real roots.

<u>Proof.</u> Since $P = P_N$ the equivalence of conditions (i) is evident. For a homogeneous $P \in$ hyp(h) let $P(\xi + \tau h) = 0$. Using that P is homogeneous we also get $P(\sigma\xi+\sigma\tau h) = P((\xi+\sigma(\mathrm{Re}\,\tau)h)+i(\mathrm{Im}\,\tau)\sigma) = 0$, for all real σ. Since P is hyperbolic, we get $\sigma\,\mathrm{Im}\,\tau \geq \tau_0$, for real σ, implying that $\mathrm{Im}\,\tau = 0$. Hence all roots of $P(\xi+\tau h)=0$ are real. Vice versa if all roots are real, we get $P(\xi+i\tau h)\neq 0$ for all real $\tau\neq 0$, implying (ii) with $\tau_0=0$. Hence P is hyperbolic.

<u>Corollary 1.2</u>. A hyperbolic homogeneous polynomial has real coefficients, except perhaps for a common complex factor $\neq 0$.

<u>Proof.</u> We have $P(\xi+\tau h)=P(h)\tau^N+\dots+P(\xi)$. Thus $P(\xi)/P(h)$ must equal the product of all the (real) roots of $P(\xi+\tau h)=0$, hence must be real, for all real ξ . This implies that $P(\xi)/P(h)$ has real coef-

ficients, since all its values, for real ξ, are real.

Proposition 1.3. If P is hyp(h) , then P is hyp(-h).
Proof. We have $P_N(-h)=(-1)^N P_N(h)\neq 0$, i.e., condition (i) for -h. Consider the polynomial of the single variable τ

$$(1.1)\qquad p(\tau) = P(\xi+i\tau h) = P_N(ih)\{\tau^N + p_{N-1}(\xi)\tau^{N-1} + \dots\}$$

of exact degree N. We must have $p_1(\xi)=\sum_{j=1}^N \tau_j(\xi)$, with the N roots $\tau_j(\xi)$ of $p(\tau)=0$. Also, $p_1(\xi)$ must be a first degree polynomial in ξ, i.e., a linear function of ξ. We have Re $\tau_j \geq \tau_0$ for all ξ, thus Re $p_1(\xi) \geq N\tau_0$. For a linear function this implies that Re $p_1(\xi)$ = const. Accordingly, Re τ_k = Re $p_1 - \sum_{j\neq k}$ Re $\tau_j \leq$ Re $p_1 -(N-1)\tau_0$ is also bounded from above. Accordinglg we also have $p(\tau)\neq 0$ for all $\xi\in\mathbb{R}^n$, and $\tau\leq\tau_1$, q.e.d.

Proposition 1.4. The principal part P_N of a $P\in$ hyp(h) is hyp(h).
Proof. We note that $P_N(\xi) = \lim_{|\sigma|\to\infty} Q_\sigma(\xi)$, with $Q_\sigma(\xi)=\sigma^{-N}P(\sigma\xi)$. Let $q_\sigma(\tau) = Q_\sigma(\xi+i\tau h)$, then also $\lim q_\sigma(\tau)=q(\tau)=P_N(\xi+i\tau h)$, for all τ . Suppose $q(\tau)=0$, for some real τ . There must be some complex root τ_σ of $q_\sigma(\tau) = 0$, $|\sigma| > \sigma_0$, with $\lim \tau_\sigma = \tau$. But this implies $\sigma^N q_\sigma(\tau_\sigma) = P(\sigma\xi+i\tau_\sigma\sigma h) = 0$, so that σ Re $\tau_\sigma \geq \tau_0$. Since Re $\tau_\sigma\to\tau$ this is possible only if $\tau=0$. Accordingly, $P_N(\xi+i\tau h)\neq 0$ as τ is real and $\neq 0$. Therefore P_N is hyperbolic. (Note that the highest coefficient of $q_\sigma(\tau)$ is $P_N(h)(i)^N$ again, hence is independent of σ . Therefore q_σ is close to q on some sufficiently large circle $|\tau|=\sigma_0$, so that Rouché's theorem supplies the root τ_σ .)

The converse of prop.1.4 is false: A polynomial P with $P_N\in$ hyp(h) needs not be hyp(h). For discussion see the proof of prop. 2.6. Given a hyperbolic homogeneous P_N , the polynomial $P=P_N+R$, deg R <N , will be hyperbolic, if (and only if), with constant c ,

$$(1.2)\qquad \sum_\alpha |R^{(\alpha)}(\xi)|^2 \leq c \sum_\alpha |P_N{}^{(\alpha)}(\xi)|^2 , \ \xi\in\mathbb{R}^n .$$

We will discuss this in sec.2.

A polynomial P (or a differential expression P(D)) will be called strictly hyperbolic (or Petrovsky-hyperbolic) (with respect to a real vector $h\neq 0$) if (i) $P_N(h)\neq 0$, and (ii) for each $\xi\in\mathbb{R}^n$, linearly independent of h , the equation $P_N(\xi+\tau h)=0$ has real and distinct roots. We use the notation s-hyp(h) instead of hyp(h) , for strictly hyperbolic polynomials or expressions. Note that "s-hyp(h)" is a property of the principal part P_N. The lower order terms do not enter the definition. We will show in sec.2 that "s-hyp(h)" implies "hyp(h)". Thus $P_N \in$ s-hyp(h) has the property that all $P=P_N+R$, deg $R<N$, are hyp(h) .

For a polynomial $P\in$ hyp(h) , $0\neq h\in \mathbb{R}^n$, we introduce $\mathbf{h}=\mathbf{h}(P,h)$ as the set of all $\eta\in \mathbb{R}^n$ such that the polynomial $p(\lambda)=P_N(\eta+\lambda h)$ has only negative real roots. Prop.1.1 and prop.1.4 imply that the roots are real. The highest coefficient $P_N(h)$ is constant, hence λ changes continuously with η . Thus $\mathbf{h}\subset \mathbb{R}^n$ is an open set. For $\eta=h$ we get $p(\lambda)=P_N(h)(1+\lambda)^N$, with all roots $\lambda=-1$. Hence $h\in \mathbf{h}$, and $\mathbf{h}$ is non-void. Also, $\mathbf{h}$ is a cone: If $\lambda\in \mathbf{h}$, and $\rho>0$, then $P_N(\rho\eta+\lambda h)=\rho^N P_N(\eta+h\lambda/\rho)=0$ implies that $\lambda=\rho(\lambda/\rho)<0$. In terms of special relativity, for the special hyperbolic expression $D_2^2+\ldots+D_n^2-D_1^2$, this cone $\mathbf{h}$ plays the role of the cone of time-like vectors.

Theorem 1.5. For $P \in$ hyp(h) the set $\mathbf{h} = \mathbf{h}(P,h)$ is an open convex cone. Moreover we have $P \in$ hyp(η) for all $\eta \in \mathbf{h}$.

The proof requires some preparations.

Proposition 1.6. The set $\mathbf{h}$ equals the connected component of h in the open set $\{\eta\in \mathbb{R}^n: P_N(\eta)\neq 0\}=\mathbf{m}$. Moreover, $\mathbf{h}$ is starshaped with with respect to h : For $\eta\in \mathbf{h}$ the straight segment $\overline{\eta h}$ also is in $\mathbf{h}$.

Proof. For $\eta \in \mathbf{h}$ we get $P_N(\eta+\tau h)=P_N(h)\ \Pi_{j=1}^N(\tau-\tau_j)$, and all τ are <0 . Hence $P_N(\eta) = P_N(h)\ \Pi(-\tau_j) \neq 0$. We already know that $\mathbf{h}$ is open. For an η on the boundary $\partial\mathbf{h}$ all roots τ_j still are ≤ 0 , but at least one must be zero, so that $P_N(\eta)=0$. Hence $\partial\mathbf{h}$ is a subset of $\partial\mathbf{m}$. To complete the proof it suffices to show that $\mathbf{h}$ is starshaped, since then it must be connected, and cannot be a proper subset of the component of $\mathbf{m}$ containing h; otherwise some boundary point of $\mathbf{h}$ would be in the interior of $\mathbf{m}$. Let $\eta\in \mathbf{h}$, and $\zeta=\lambda\eta+\mu h$, $\lambda,\mu > 0$, $\lambda+\mu=1$. Write $P_N(\zeta+\tau h)=P_N(\eta\lambda+(\tau+\mu)h)=\lambda^N P_N(\eta+h(\tau+\mu)/\lambda)$. This shows that for a root τ of $P_N(\zeta+\tau h) = 0$ the expression $\tau' = (\tau+\mu)/\lambda$ must be a root of $P_N(\eta+\tau h)=0$, hence is negative. Hence also $\tau=\lambda\tau'-\mu< 0$. It follows that $\zeta\in \mathbf{h}$, so that $\mathbf{h}$ is starshaped.

Proposition 1.7. For $\eta \in \mathbf{h}$ we have $P(\xi+i\tau h+i\sigma\eta)\neq 0$, as $\mathrm{Re}\ \tau < \tau_0$, $\mathrm{Re}\ \sigma \leq 0$, $\xi \in \mathbb{R}^n$, with τ_0 for h as in the definition of hyp(h).
Proof. This evidently is true for $\mathrm{Re}\ \sigma=0$, by definition of hyp(h). Consider the roots of the polynomial $q(\sigma)=P(\xi+i\tau h+i\sigma\eta)=i^N P_N(\eta)\sigma^N+ + \ldots$, as $\mathrm{Re}\ \tau < \tau_0$. As τ varies these roots must vary continuously, since the highest coefficient of the polynomial is independent of τ . By our above remark none of the roots may cross the

imaginary σ-axis so that the number of roots of $q(\sigma) = 0$ within $\mathrm{Re}\,\sigma < 0$ stays constant, as τ varies within the half plane $\mathrm{Re}\,\tau < \tau_0$. Therefore it suffices to show that q has no roots with $\mathrm{Re}\,\sigma < 0$ whenever τ is real and $|\tau|$ is large.

Assume $\tau \neq 0$, and write $\sigma = \tau\mu$, so that $q = 0$ takes the form $P(\xi+i\tau(h+\mu\eta)) = 0$. We know that all roots of $P_N(h+\mu\eta) = 0$ are negative real, since $\eta \in \boldsymbol{h}$. Consider the family of polynomials $r_\tau(\mu) = (i\tau)^{-N}P(\xi+i\tau(h+\mu\eta)) = P_N(\eta)\mu^N + \ldots$. Clearly we get $\lim_{\tau\to\infty} r_\tau(\mu) = P_N(h+\mu\eta)$. Accordingly $r_\tau(\mu) = 0$ implies $\mathrm{Re}\,\mu < 0$, since μ must be close to one of the roots of $P_N(h+\mu\eta) = 0$, as $|\tau|$ is large. In other words, for $\tau<0$, with large $|\tau|$, all roots $\sigma = \mu\tau$ of $q(\sigma) = 0$ must have positive real part, q.e.d.

Proof of Thm.1.5. Let us apply prop.1.8 for real σ,τ , where $\tau=\varepsilon\sigma$, $\varepsilon > 0$, assuming $\tau_0 < 0$. We get

$$(1.3) \qquad P(\xi+i\sigma(\eta+\varepsilon h)) \neq 0 \text{ , as } \sigma < \tau_0/\varepsilon \text{ , } \xi \in \mathbb{R}^n .$$

We know that $P_N(\eta+\varepsilon h)\neq 0$, due to $\eta+\varepsilon h \in \boldsymbol{h}$, since $\boldsymbol{h}$ is starshaped and a cone. It follows that $P \in \mathrm{hyp}(\eta+\varepsilon h)$, $\varepsilon > 0$. But $\boldsymbol{h}$ is open, hence contains $\eta_\varepsilon = \eta-\varepsilon h$, for small $\varepsilon>0$. Applying the above for η_ε instead of η one concludes that $P \in \mathrm{hyp}(\eta)$, for every $\eta \in \boldsymbol{h}$.

Finally, to prove convexity of $\boldsymbol{h}$, let $\eta^1, \eta^2 \in \boldsymbol{h}$. It follows that $P \in \mathrm{hyp}(\eta^j)$, $j=1,2$. Moreover, we get $\boldsymbol{h} = \boldsymbol{h}(P,\eta^j)$, $j=1,2$, since h and the η^j, all are in the same connected component of the set $\boldsymbol{m}$. Since $\boldsymbol{h}$ then is star-shaped with respect to η^1, the segment from η^1 to η^2 must be in $\boldsymbol{h}$, so that $\boldsymbol{h}$ indeed is convex, q.e.d.

As an example consider the 'wave operator' $\partial^2_{x_1}-\partial^2_{x_2}-\ldots-\partial^2_{x_n}$. This operator is hyperbolic and strictly hyperbolic with respect to the "time-like" direction $h=(1,0,\ldots,0)$. Indeed, we get $P(\xi)=|\xi^\sim|^2-\xi_1^2$, $\xi^\sim=(\xi_2,\ldots,\xi_n)$, hence $P(\xi+\tau h)=|\xi^\sim|^2-(\xi_1-\tau)^2=0$ amounts $\tau=\xi_1\pm|\xi^\sim|$. The two roots are real and distinct except if $\xi^\sim=0$. We have $P=P_N$ homogeneous, and $P_N(h)=1\neq 0$. The cone $\boldsymbol{h}$ of time-like vectors is given by $\boldsymbol{h}=\{\xi: |\xi^\sim|<\xi_1\}$, as follows from prop.1.6.

2. Hyperbolic polynomials and characteristic surfaces.

In this section we shall investigate the relation between "hyp(h)" and "s-hyp". It turns out that the set s-hyp(h) consists precisely of all $P\in$ hyp(h) with simple characteristics. Here the concept of characteristic (surface) must be understood in the sense of ch.0,8. There we already defined the concepts of simple and multiple characteristic. Presently we look at these definitions under the aspect of constant coefficient expressions.

A differential polynomial P(D) is said to be of principal type if $P_{N|\xi}(\xi)\neq 0$ for all $0\neq\xi\in\mathbb{R}^n$. For a principal type expression all (real) characteristic surfaces are simple. Consider a surface $\varphi(x)=\varphi(x^0)$, near x^0, where $P_N(\varphi_{|x}(x^0))=0$ (cf. ch.0,(8.3)). If $h\in\mathbb{R}^n$ is any vector and $\xi=\varphi_{|x}(x^0)$, then the polynomial $p(\tau) = P_N(\xi+\tau h)$ has a root at $\tau = 0$. If h can be found such that 0 is a simple root of $p(\tau)$ then the surface is called simply characteristic at x^0. We have given the same definition for characteristic surfaces in ch.0,8. The multiplicity at x^0 is defined as the lowest multiplicity of the root 0 of $p(\tau)$ atteinable for various h .

For a real root ξ of $P_N(\xi)=0$ let $\varphi=\xi\cdot x$, so that $\varphi_{|x}=\xi$, and $P_N(\varphi_{|x}) = 0$. It follows that, for a differential polynomial P(D), every real root ξ of the principal part polynomial P_N gives raise to a family $x.\xi=c$ of real characteristic hyperplanes, with normal vector ξ . Vice versa, for an arbitrary real characteristic surface Γ of P(D) the tangent planes, at all points $x^0\in\Gamma$, and their parallels, all are real characteristic (hyper-)surfaces.

Now let P(D) be of principal type. Clearly Taylor's formula yields $p(\tau)=P_N(\xi+\tau h)=P_N(\xi)+\tau P_{N|\xi}(\xi)\cdot h+\ldots$. Hence if ξ is a real root of $P_N=0$, then we may set $h=P_{N|\xi}(\xi)\neq 0$, and get $p(\tau)=\tau P_{N|\xi}(\xi)^2+ +\ldots$, which clearly has a simple root at $\tau=0$. Vice versa, for an expression P(D), let $P_{N|\xi}(\xi)=0$, at some $\xi\neq 0$. Using Euler's formula for the homogeneous function P_N we find that $NP_N(\xi)=\xi.P_{N|\xi}(\xi) =0$. Thus we get $p(\tau)=O(\tau^2)$ for any choice of h. Or, the surfaces with normal ξ are multiple characteristics. We have proven:

Proposition 2.1. A constant coefficient operator P(D) is of principal type if and only if all of its real characteristic surfaces are simply characteristic.

Elliptic operators (in the sense of ch.0,sec.8(d)) have no real characteristics, hence cannot have multiple real characteristics. Thus an elliptic operator always is of principal type.

Proposition 2.2. An operator $P(D)\in$ s-hyp(h) is of principal type.

Proof. Let $P\in$ s-hyp(h), and let $P_N(\xi)=0$, for $0\neq\xi\in\mathbb{R}^n$. We know that $P_N(h)\neq 0$, so that $\xi=\lambda h$ is impossible. Hence ξ is linearly independent of h, and $P\in$ s-hyp(h) implies that $P_N(\xi+\tau h)=0$ has a simple root at $\tau=0$. Hence the planes normal to ξ are simple characteristics. It follows that all characteristics are simple, q.e.d.

Theorem 2.3. We have hyp(h)$\subset$ s-hyp(h). For a polynomial $P\in$ hyp(h) we have $P\in$ s-hyp(h) if and only if all real characteristics of $P(D)$ are simple. In other words, s-hyp(h) consists precisely of all $P\in$ hyp(h) which are of principal type.

We prove thm.2.3 in several steps.

<u>Proposition 2.4</u>. Let $P=P_N\in$ hyp(h) be homogeneous, and let the polynomial R of degree <N satisfy (1.2). Then also $P_N+R\in$ hyp(h). Moreover, if $P=P_N\in$ s-hyp(h) then $P_N+R\in$ hyp(h) for every R, deg R < N , regardless of (1.2). In particular, s-hyp(h)$\subset$ hyp(h).

Proof. Observe that $P_N+R\in$ s-hyp(h) implies $P_N\in$ s-hyp(h), so that the second statement implies $P_N+R\in$ hyp(h), reducing the third statement to the second. To reduce the second statement to the first, observe that $P_N\in$ s-hyp(h) must be of principal type, by prop.2.2. Conclude that $P_{N|\xi}(\xi)$ is homogeneous of degree N-1, while $|P_{N|\xi}(\xi)|\neq 0$ for all $|\xi|=1$. It follows that, with some $c_0>0$,

$$(2.1)\qquad \sum_\alpha |P_N^{(\alpha)}(\xi)|^2 \ge |P_{N|\xi}(\xi)|^2 \ge c_0\,|\xi|^{2N-2}\ ,\ \text{for all } \xi\in\mathbb{R}^n\ .$$

However (2.1) implies that (1.2) holds for all R, deg R <N .

To prove the first statement it is convenient to write (1.2) as $R\prec\!\!\prec P_N$, with the order relation "$\prec\!\!\prec$" of [Hr_1]. Or, $F\prec\!\!\prec G$ means

$$(2.2)\qquad \sum_\alpha |F^{(\alpha)}(\xi)|^2 \le c\sum_\alpha |G^{(\alpha)}(\xi)|^2\ ,$$

with some constant c, for all $\xi\in\mathbb{R}^n$. One confirms easily that this defines a partial ordering compatible with the algebraic structure of the ring of polynomials. Using Taylor's formula, we get

$$(2.3)\qquad F^\sim(\xi) = \{\sum|F^{(\alpha)}(\xi)|^2\}^{1/2} \le c(\eta)\,F^\sim(\xi+\eta)\ ,$$

for all $\xi\in\mathbb{R}^n$, $\eta\in\mathbb{C}^n$, with a constant $c(\eta)$ independent of ξ. Accordingly, $F\prec\!\!\prec G$ implies $F_\eta\prec\!\!\prec G_\eta$, for the (real or complex) translations $F_\eta(\xi)=F(\xi-\eta),\ldots$, . Evidently $F\prec\!\!\prec G$ implies that the order of F cannot exceed the order of G . Moreover, let $G=G_N$ be homogeneous and $F=\sum_{j=1}^N F_j$, with homogeneous parts F_j of degree j, then $F\prec\!\!\prec G$ implies that $F_j\prec\!\!\prec G$ for all j. Indeed it follows at once that also $F(t\xi)\prec\!\!\prec G(t\xi)=t^N G(\xi)$, for all t>0, hence that $F(t\xi)$ $\prec\!\!\prec G(\xi)$. For a collection $t_0,\ldots,t_N$ of distinct positive numbers the van der Monde determinant is different from zero, and we get

$$(2.4)\qquad \sum_{k=0}^N t_j^k F_k(\xi) = F(t_j\xi) \prec\!\!\prec G\ ,\ j=0,\ldots,N\ .$$

Thus indeed, we get $F_j\prec\!\!\prec G$, $j=0,\ldots,N$.

Let $P_N \in \mathrm{hyp}(h)$. From thm.1.5 we conclude that $h+\mathrm{Re}\,\zeta$ is time-like for sufficiently small $|\zeta|$. Accordingly, $P_N(\xi+i(h+\zeta))\neq 0$ as $|\zeta|\le 2\sqrt{n}\delta>0$, $\xi\in\mathbb{R}^n$. Using Cauchys estimate $|\alpha|$-times, on the analytic function $f(\zeta)=P_N(\xi+i(h+\zeta))$ of n complex variables ζ, get

$$(2.5)\quad |P_N^{(\alpha)}(\xi+ih)/P_N(\xi+ih)| \le (c/\delta)^{|\alpha|} \operatorname{Max}_{|\zeta_j|\le\delta} |P_N(\xi+i(h+\zeta))/P_N(\xi+ih)|$$

with c independent of ξ. For every fixed ζ, the polynomial $\varphi(t) = f(t\zeta)$ cannot vanish, as $|t|\le 2$, assuming $|\zeta_j|\le\delta$, i.e., $|z|\le\sqrt{n}\delta$. This gives $\varphi(1)=\varphi(0)\Pi_{j=1}^{N}((\tau_j-1)/\tau_j)$, with the roots t_j, $|t_j|\ge 2$, of φ. Hence the Max at the right hand side of (2.5) is estimated by $|\varphi(1)/\varphi(0)|\le(1/2)^N$, as $|\zeta|\le\sqrt{n}\delta$. We have proven the following.

<u>Proposition 2.5</u>. If $P=P_N\in\mathrm{hyp}(h)$, then for all α and $\xi\in\mathbb{R}^n$ we have

$$(2.6)\qquad |P_N^{(\alpha)}(\xi+ih)/P_N(\xi+ih)|\le c\ ,\ c \text{ independent of } \xi \text{ and } \alpha\ .$$

If $P\prec P_N$ then we have seen that $P_j\prec P_N$, hence $P_j(\xi+ih) \prec P_N(\xi+ih)$, by translation invariance. This and (2.6) implies

$$|P_j(\xi+ih)|^2\le c\sum|P_N^{(\alpha)}(\xi+ih)|^2\le c|P_N(\xi+ih)|^2,\ \xi\in\mathbb{R}^n\ ,\ j=0,\dots,N-1\ ,$$

where again c is independent of ξ. Using homogeneity we get $|P_j(\xi+i\tau h)|=|\tau|^j|P_j(\xi/\tau+ih)|\le c|\tau|^j|P_N(\xi/\tau+ih)|=c|\tau|^{j-N}|P_N(\xi+i\tau h)|$. Let $R = P_0+P_1+\dots+P_{N-1}$, then we get $|R(\xi+i\tau h)|\le c|\tau|^{-1}|P_N(\xi+i\tau h)|$, Therefore $P = P_N + R$ satisfies the estimate

$$(2.7)\qquad |P(\xi+i\tau h)| \ge (1-c|\tau|^{-1})|P_N(\xi+i\tau h)| \neq 0\ ,\ \text{as } \tau < -2/c\ .$$

This establishes prop.2.4.

<u>Proposition 2.6</u>. Suppose all polynomials P with principal part P_N are hyp(h). Then P_N is of principal type, and P_N is s-hyp(h).
<u>Proof.</u> It is clear that P_N must be hyp(h), so that $P_N(\xi+\tau h) = 0$ has only real roots, and $P_N(h)\neq 0$. If P_N is not strictly hyperbolic then, for some ξ, linearly independent of h, there must be a multiple root τ_0 of $P_N(\xi+\tau h)=0$. To prove prop.2.6 we will show that in this case not every $P=P_N+Q$, $\deg Q\le N-1$, can be hyperbolic. We may assume $\tau_0=0$. Then the polynomial $p(\tau)=P_N(\tau\xi+h)=\tau^N P_N(\xi+h/\tau)$ has degree $N-k\le N-2$.

Now we will show: If, under above assumptions, $P=P_N+R$ is hyp(h) then the polynomial $p^{\sim}(\tau)=P(\tau\xi+h)$ also must have degree $\le N-2$. For a general R of degree N-1 or less, this can not be true.

Therefore not every P_N+R can be hyp(h), and prop.2.6 follows.

Consider the equation $P(\tau(\xi+\sigma h))=0$, as a polynomial equation in 2 variables σ,τ .With $\kappa=\frac{1}{\tau}$ one may write it in the form

$$(2.8)\qquad P_N(\xi+\sigma h) + \kappa r(\kappa,\sigma) = 0 \ , \ r(\kappa,\sigma) = \kappa^{N-1}R(\tfrac{\xi+\sigma h}{\kappa}) \ ,$$

with a polynomial $r(\kappa,\sigma)$ in κ, deg r $\le N-1$. At $\kappa=0$ we get the equation $P_N(\xi+\sigma h)=0$, which, as we know, has the k-fold root $\sigma=0$, $k\ge 2$. From Rouché's theorem we derive the existence of N roots $\sigma_j(\kappa)$, $j=1,\dots,N$, of (2.8), near the roots of $P_N(\xi+\tau h)=0$, for all $|\kappa|\le\delta$. These roots may be organized into "Puiseux series" -i.e. power series in $\kappa^{1/l}$, with suitable integers $l>0$. In particular each of the k roots approaching 0, as $\kappa\to 0$, has an expansion $\sum_{m\ge s} a_m\kappa^{m/l}$. Let s be choosen such that a_s is the smallest coefficient $\neq 0$. Clearly $s\ge 0$, since the series vanishes for $\kappa=0$. If $0<s<l$, then $\tau\sigma(1/\tau) \approx a_s\tau^{1-s/l} =|a_s||\tau|^{\gamma}\exp\{i(\arg a_s+\gamma \arg\tau)\}$, where $0<\gamma=1-s/l<1$. For real τ we may set $\arg\tau=\nu\pi$, with an integer ν . Here ν may be choosen such that $\arg a_s+\nu\gamma\pi$ is not a multiple of π, since l must be ≥ 2. Then we get $|\tau \operatorname{Im}\sigma_j(1/\tau)| \to\infty$, as $|\tau|\to\infty$. However, we have $0=P(\tau\xi+h(\operatorname{Re}\tau\sigma_j(1/\tau))+ih\operatorname{Im}\tau\sigma_j(1/\tau))$, so that $\operatorname{Im}\tau\sigma_j(1/\tau)$ remain bounded by definition of hyp(h). Thus we get a contradiction unless $r\ge l$. It follows that $\sigma_j(\kappa)=O(|\kappa|)$, as $\kappa\to 0$, for at least $k\ge 2$ of the roots σ_j , while all others are O(1) .

Finally observe that $\lambda_j(\tau)=\tau\sigma_j(1/\tau)-1$ are the N roots of $f(\lambda)=P(\lambda h+(\tau\xi+h))=0$, for sufficiently large $|\tau|$. It follows that that $f(0) = P_N(h)\sum_{j=1}^N(-\lambda_j(\tau)) = P(\tau\xi+h) = O(|\tau|^{N-k})$. Hence indeed $p^{\sim}(\tau) = P(\tau\xi+h)$ must have degree $\le N-2$, q.e.d.

<u>Corollary 2.7</u>. If $P\in$ s-hyp(h) , then also $P\in$ s-hyp(η) , for every time-like vector $\eta\in \boldsymbol{h}$.

<u>Proof.</u> The cone $\boldsymbol{h}$ of time-like vectors clearly is determined by the principal part P_N only. Also we know that $P_N+R=P$ is hyp(η) for every R of degree N-1 or less, because each such P is hyp(h) , hence also hyp(η), using thm.1.5. Therefore $P\in$ s-hyp(η) follows from prop.2.6, q.e.d.

Clearly thm.2.3 is a consequence of prop.2.4 and prop.2.6. The fact that, for $P_N\in$ hyp(h) the condition $P\prec\!\!\!\!\bigcirc P_N$ is implied by $P\in$ hyp(h), was proven by Chaillou [Chu_1] and Svensson [Sv_1] .

3. The hyperbolic Cauchy problem for variable coefficients.

A linear differential expression L=a(x,D), as in ch.0,(8.1), will be called strictly hyperbolic with respect to h, $0\neq h\in \mathbb{R}^n$, at x^0, if the polynomial $a(x^0,\xi)$ in ξ is s-hyp(h). If this holds for all $x^0\in \Omega\subset \mathbb{R}^n$, where the vector h is allowed to change with x^0, then L is called strictly hyperbolic in Ω. We shall mainly focus on the case of h being independent of x^0 in some domain Ω. Then we call L strictly and normally hyperbolic with respect to h, in Ω. We will write the latter property as L∈ s-hyp(h)=s-hyp(h,Ω). For a general set M we say L∈ s-hyp(h,M) if L∈ s-hyp(h,Ω) for open $\Omega \supset M$.

In this section we will assume h to be one of the coordinate vectors. This is no restriction, in view of the fact that "s-hyp" is a property of the principal part L_N of L, while the principal part transforms like a contravariant tensor of N variables when new independent variables are introduced. A linear transform $x=my$ with constant matrix m takes $a(x,D_x)$ into $a^{\sim}(y,D_y)$ with principal part polynomial $a^{\sim}_N(y,\eta)=a_N(my,m^{-T}\eta)$, where $m^{-T}=(m^{-1})^T$. Choosing $h^0=m^Th=(1,0,\dots,0)$ will make the transformed expression $a^{\sim}(y,D_y)$ s-hyp(h^0). In fact, for a more general transform $x=m(y)$ the constant matrix m may be replaced by the Jacobian matrix $J=\partial x/\partial y$. Then $h^{\sim}=J^Th$ will depend on y. One may use this - under proper conditions - to transform an expression L=a(x,D), strictly hyperbolic with variable h_x, onto one, $L^{\sim}$, which also is normally hyperbolic.

Of special interest will be the case of a second order strictly hyperbolic expression L written in the form

$$(3.1)\qquad L = \sum a_{jk}D_jD_k + \text{lower order terms}\ ,\ a_{jk} = a_{kj}\ .$$

The principal part polynomial is given by $a_2(\xi)=\sum a_{jk}\xi_j\xi_k$. We get

$$(3.2)\quad a_2(\xi+\tau h)=\tau^2a_2(h)+2\tau a_2(h,\xi)+a_2(\xi)\ ,\ \text{with } a_2(h,\xi)=\sum a_{jk}h_i\xi_k.$$

We know that the a_{jk} are real-valued (up to a common complex factor). Cdn. "s-hyp(h)" then means that
tor). "s-hyp(h)" then means (i) $a_2(h)\neq 0$, (ii) $a_2(h,\xi)^2-a_2(h)a_2(\xi)>0$, as $\xi\neq\gamma h$. Here (ii) yields $a_2(h)a_2(\xi)<0$ for $\xi\in N=\{h\}^{\perp}$. Assuming $a_2(h)>0$ we get the form a_2 negative definite over N. Hence, up to a nonvanishing complex factor all second order s-hyp-operators are of the 'classical' form: The principal symbol matrix $((a_{jk}))$ has one positive and n-1 negative eigenvalues.

We now assume $h=h^0=(1,0)$, $0\in\mathbb{R}^n$, in Euclidean space $\mathbb{R}^{n+1}=\{(t,x): t\in\mathbb{R},\ x\in\mathbb{R}^n\}$, and an expression $L=a(t,x,D)$ of the form

$$(3.3)\quad L=\sum_{|\alpha|+k\le N} a_{\alpha,k}(t,x)D_x^{\alpha}D_t^k=\sum_{k=0}^N A_{N-k}(t)D_t^k,\quad A_{N-k}=\sum_{|\alpha|\le N-k} a_{\alpha,k}D_x^{\alpha},$$

with principal part

$$(3.4)\quad L_N=\sum_{|\alpha|+k=N} a_{\alpha,k}(t,x)D_x^{\alpha}D_t^k=\sum_{k=0}^N B_{N-k}(t)D_t^k,\quad B_{N-k}=\sum_{|\alpha|=N-k} a_{\alpha,k}D_x^{\alpha}.$$

Let $\Omega=I\times\mathbb{R}^n$, $I=[t_1,t_2]$, and $a_{\alpha,k}\in C^{\infty}(\Omega)$, for all α,k. For complex-valued $a_{\alpha,k}$ the expression (3.3) is s-hyp(h^0) if and only if

$$(3.5)\quad a_{0,0}(t,x)\neq 0,\quad \sum_{|\alpha|+k=N} a_{\alpha,k}(t,x)\xi^{\alpha}\tau^k=0 \text{ has real distinct roots,}$$

as polynomial in τ, for all $(t,x)\in\Omega$, and all $0\neq\xi\in\mathbb{R}^n$, by a simple calculation. It is convenient to replace the first cdn.(3.5) by $a_{00}(t,x)=1(=A_N(t))$, as can be achieved by a normalizing factor.

So far, in this chapter, we assumed complex-valued coefficients. However, the results, below, are valid for the case of a $\nu\times\nu$-system as well. From now on we assume that $a_{\alpha,k}(t,x)\in\mathbb{L}^{\nu}$, with the algebra $\mathbb{L}^{\nu}$ of all complex $\nu\times\nu$-matrices. A $\nu\times\nu$-matrix-valued expression is called s-hyp(h^0) if the <u>determinant</u> of its symbol is s-hyp(h^0). For L of (3.5) this means that $a_{0,0}(t,x)$=identity, and

$$(3.6)\quad p(\tau)=\det\Big(\sum_{|\alpha|+k=N} a_{\alpha,k}(t,x)\xi^{\alpha}\tau^k\Big) \text{ has } N\nu \text{ real distinct roots },$$

for all $(t,x)\in\mathbb{R}^{n+1}$, $0\neq\xi\in\mathbb{R}^n$.

This condition ties up with the conditions on L in VI,(4.1). There we were transforming to a problem of the form VI,(4.4) with intent to apply thm.4.4, requiring "strictly hyperbolic of type e or e'". The latter is identical for $x\in K\subset\subset\mathbb{R}^n$ with our present "s-hyp(h^0)" - while at $x=\infty$ the conditions do not match.

We focus on the <u>local</u> "<u>non-relativistic</u>" Cauchy problem

$$(3.7)\quad Lu=f,\ (t,x)\in\Omega,\quad \partial_t^j u=\varphi_j,\ \text{as } t=t_0,\ j=0,\dots,N-1,$$

for given f (defined in Ω) and φ_j (defined for $t=t_0\in I$), assumed C_0^{∞}-functions. The function (distribution) u is to be found.

The term '<u>non-relativistic</u>' expresses that the initial conditions $\partial_t^j u=\varphi_j$ are imposed at a given constant time t_0. It will be practical to also consider "<u>relativistic</u>" problems with initial conditions on a more general surface $\Sigma\subset\Omega$. We find that the study of relativistic Cauchy problems is only a matter of a 'space-time-transform' -i.e., a transformation of independent variable. Recall the cone $\boldsymbol{h}=\boldsymbol{h}_{t,x}$ of time-like vectors. $\boldsymbol{h}$ is convex, $h^0\in\boldsymbol{H}$, and L of (3.3) is s-hyp(η) at (t,x), for every $\eta \in H_{t,x}$ (cor.2.7).

A smooth hypersurface $\Sigma \subset \Omega$ will be called space-like if its surface normal is a time-like vector, at each $(t,x) \in S$. Since h is convex and $h^{\circ} \in h$, a space-like hypersurface has a unique projection to t=0. Thus it is given by an equation of the form $t=\Theta(x)$. Assume such space like $\Sigma \subset \Omega$ given, $\Theta(x) \in C^{\infty}(\mathbb{R}^n)$ and $\Theta(x)=t_0 \in I$ for large $|x|$. The relativistic Cauchy problem (for Σ) is defined by

$$(3.8) \qquad Lu=f \ , \ (t,x) \in \Omega \ , \ \partial_{\nu}^{j} u=\varphi_j \ , \ (t,x) \in \Sigma \ , \ j=0,\ldots,N-1 .$$

Here ∂_{ν} denotes the normal derivative to S, ∂_{ν}^{k} defined as in V,3. We choose the normal pointing to increasing t. Assume f, $\varphi_j \in C_0^{\infty}$ given, defined on Ω and Σ, respectively.

The relativistic Cauchy problem may be reduced to the nonrelativistic problem by a coordinate transform. For simplicity we only discuss the case where the cones $h_{t,x}=h_x$ are independent of t (for example, if L or its principal part are independent of t).

Let us consider a transformation of the form

$$(3.9) \qquad t' = t'(t,x) \ , \ x' = x \ , \ (t,x) \in \Omega=I\times\mathbb{R}^n ,$$

inverted by

$$(3.10) \qquad t = t'(t,x') \ , \ x = x' \ , \ (t',x') \in \Omega' ,$$

defining a diffeomorphism $\Omega \leftrightarrow \Omega'$, which takes the surface Σ onto the plane $t'=0$ (i.e., require that $t'(\Theta(x),x)=0$). One then verifies that the cdn's $\partial_{\nu}^{j}u=\varphi_j$ of (3.8) transform to cdn's of the form in (3.7), with modified data. On the other hand, the DE Lu=f transforms to $L'u=f'$, $(t',x') \in \Omega'$ of the general form (3.3). We get

$$(3.11) \qquad a_N'(x^{\Delta}{}',\xi^{\Delta}{}') = a_N(x^{\Delta}(x^{\Delta}{}'),J\xi^{\Delta}{}') \ , \ J = ((\partial x^{\Delta}{}'_j/\partial x^{\Delta}{}_l)) ,$$

where, for a moment, we are introducing the notation $x^{\Delta}=(t,x)$, $x^{\Delta}{}' = (t',x')$, and with $a_N(x^{\Delta},\xi^{\Delta}) = a_N(t,x,\tau,\xi)$. In the Jacobian matrix J the row index is j, and the column index is l. One calculates that $J\xi^{\Delta}{}' = (t'_{|t}\tau',\xi'+t'_{|x}\tau') = (0,\xi') + \tau' h(t',x')$, with

$$(3.12) \qquad h(t',x') = (\partial t'/\partial t,\partial t'/\partial x_1,\ldots,\partial t'/\partial x_n)(x^{\Delta}(x^{\Delta}{}')) .$$

Thus $0=\det a_N'(x^{\Delta}{}',\xi^{\Delta}{}') = \det a_N(x^{\Delta}(x^{\Delta}{}'),(0,\xi'+\tau' h^{\Delta x^{\Delta}}))$ has real distinct roots τ' if L is s-hyp(h) for h of (3.12), all (t',x').

Let $t'(t,x)=t-\Theta(x)$, so that $t(t',x')=t'+\Theta(x')$, and get $h=(1,\nabla\Theta(x'))$ independent of t'. Clearly h is normal to Σ, as $(t,x) \in \Sigma$. Thus indeed there we get $h \in h_{t,x}$ since Σ was assumed space like. Now, if $h_{t,x}=h_x$ is independent of t then $h(t'x')\in h_{t,x'}$ for all t,x , and the transformed problem is s-hyp(h_0), and nonrelativistic. Clearly the set Ω' contains a smaller $\Omega''=I''\times\mathbb{R}^n$.

Note that the same transformation still works if h is time dependent, if it is assumed that S is not only space-like, but, afortiori, has its normal at x in the intersection $\cap\{h_{t,x}:t\in I\}=$ $=h_x$. Clearly h_x is a convex cone with nonvoid interior, for all x.

Theorem 3.1. Assume L of (3.3) in s-hyp(h^0), and, moreover, assume that the transformed problem VI,(4.4), with the matrix of VI,(4.6) is strictly hyperbolic of type e or e'. Then there exists a unique solution $u(t,x)\in C^\infty(I,H_s)$, for all $s\in\mathbb{R}^2$, of the nonrelativistic Cauchy problem (3.7), for every f and $\varphi_j\in C_0^\infty$.

The proof is an immediate consequence of VI,thm.4.4.

We observed that the condition "strictly hyperbolic of type e^Δ" coincides with (i) "s-hyp(h^0) on all compact sets $K\subset\Omega$" and (ii) an additional condition obtainable by modifying L outside an arbitrarily given compact set K_0. If S is a surface as above, then it follows easily that a transformation $x'=x$, $t'=t-\Theta(x)$ will not influence cdn.(ii), since we were setting $\Theta(x)=t_0$ for large $|x|$. As a consequence we have the result, below.

Theorem 3.2. Let L satisfy the assumptions of thm.3.1, and let $\Sigma\subset$ Ω^{int} be a space-like surface with unique projection onto the plane

t=0, and parallel to t=0, for large $|x|$. Let $\tau_1=\text{Min }\Theta$, $\tau_2=\text{Max }\Theta$, taken over $\mathbb{R}^n$. Assume that Σ is not only space-like, but even that its normal is contained in the cone $h_x=\cap\{h_{t,x}: (t,x)\in M\}$, with the "slab" $M=\{\tau_1-t_1\le t-\Theta(x)\le t_2-\tau_2\}$. Then there exists a unique solution u of the relativistic Cauchy problem (3.8), defined for $(t,x)\in M$, and each f, $\varphi_j\in C_0^\infty$, such that the function $u(t-\Theta(x),x)=u^\sim(t,x)$ satisfies $u^\sim\in C^\infty([\tau_1-t_1,t_2-\tau_2],H_s)$ for all $s\in\mathbb{R}^2$.

The two above results, thm.3.1 and thm.3.2 will be improved in sec.5, below, after discussing 'finite propagation speed'.

4. The cone h for a strictly hyperbolic expression of type e^Δ.

In sec.3 we already used the fact, that an expression L of the form (3.3) is s-hyp(h^0,K), $h^0=(1,0)$, for every $K\subset\subset I\times\mathbb{R}^n$, whenever the first order system VI,(4.4) is strictly hyperbolic of type e or e'. Discussing this in detail, note the symbol of the operator $K_0=\Lambda K$, K of VI,(4.6), is of the form $k_0+\kappa_0$, where the $\kappa_0\in$

$C^\infty(I,\psi c_{f-e})$, while the matrix $k_0(t,x,\xi)$ coincides with the matrix in VI,(4.4) except in the last row. All elements in rows 1 to N-1 equal 0 or 1, as in VI,(4.6), but the elements of the last row are

$$(4.1)\qquad (\pi_{-e^\Delta}(x,\xi))^{N-j}a_{N-j}(t,x,\xi)\ ,\ j=0,\dots,N-1$$

with $$a_{N-j}=\sum_{|\alpha|\le N-j} a_{\alpha,j}(t,x)\xi^\alpha\ ,$$

in the order listed.

We may choose k_0 and κ_0 above for the decomposition VI,(4.9). Then the condition "strictly hyperbolic of type e^Δ " implies that $k_0(t,x,\xi)$ has real and distinct eigenvalues $\mu_1<\mu_2<\dots<\mu_N$ for all $(t,x,\xi)\in I\times\mathbb{M}_{e,0}$ with $|\xi|\ge\eta_0$. It was pointed out before that the functions $\mu_j(t,x,\xi)$ then must be continuous in this compact set. Under the present special conditions, the matrix k_0 and its eigenvalues even extend to continuous functions over $\{I\times\mathbb{M}^\Delta\times\mathbb{B}^n: |\xi|\ge\eta_0\}$, with $\mathbb{B}^n$ of II,3, and the compactification $\mathbb{M}^\Delta$ allowing continuous extension of all functions $a(x)\in\psi c_0$. Here we assume that

$$(4.2)\qquad a_{\alpha,j}(t,x)\in\psi c_{(N-j)(e^\Delta-e^1)}\ \text{ as } |\alpha|\le N-j\ .$$

Accordingly, for every $|\xi^\circ|=1$, the limits $\lim_{\rho\to\infty}\mu_j(t,x,\rho\xi^\circ)=\mu_j(t,x,\infty\xi^\circ)$ exist, and define $N\nu$ real-valued continuous functions over $I\times\mathbb{M}^\Delta\times\{|\xi^\circ|=1\}$, assuming mutually distinct values at each point of this compact space. A calculation shows that $\tau_j=-\mu_j(t,x,\infty\xi^\circ)$ are the roots of the equation

$$(4.3)\qquad \det\Big(\sum_{j+l=N}\upsilon^{-l}\tau^j b_l(t,x,\xi^\circ)\Big)=0\ ,$$

where $b_{N-j}(t,x,\xi)=\sum_{|\alpha|=N-j}a_{\alpha,j}(t,x)\xi^\alpha$ are the principal parts of the a_{N-j} , and we have assumed $b_0=1$ again. Also, we have set $\upsilon=\langle x\rangle$ for $e^\Delta=e$, and $\upsilon=1$ for $e^\Delta=e^1$.

Note that (4.3) may be rewritten as

$$(4.4)\qquad p(\tau\upsilon)=\det\sum_{j+l=N}b_l(t,x,\xi^\circ)(\tau\upsilon)^j=0\ ,$$

with the polynomial $p(\tau)$ of (3.6) , taken at t,x,ξ° . In other words, the roots of (3.6) are given as the numbers $-\upsilon\mu_j$.

We have proven the following result.

Proposition 4.1. a) Assume that the coefficients of L of (3.3) satisfy (4.2), and that the equivalent first order system VI,(4.4) is strictly hyperbolic of type e^Δ , where $e^\Delta=e$ or $e^\Delta=e^1$. Then the differential expression L is s-hyp(h^0,K) for every $K\subset\subset I\times\mathbb{R}^n$. Moreover, we even have $L \in$ s-hyp(h^0,$I\times\mathbb{R}^n$), uniformly, insofar as the minimum difference between two roots of (3.6) is bounded below by $c_0>0$ for all $|\xi^0|=1$, as $e^\Delta=e^1$, and by $c_0\langle x\rangle$, as $e^\Delta=e$. Furthermore, the roots of (3.6) are bounded by a constant C_0 as $e^\Delta=e^1$, and by $C_0\langle x\rangle$ as $e^\Delta=e$, for all $t\in I$, $x\in\mathbb{R}^n$, $|\xi^0|=1$.

Remark. Note that a different choice of k_0 and κ_0 in VI(4.9) does not influence the above derivation, since the limit $\lim_{\rho\to\infty}$ will give the same polynomial as in (4.3).

Note that our knowledge about the roots of (3.6) gives the following information about the cone $h=h_{t,x}$ of time-like vectors.

Proposition 4.2. Under the assumptions of prop.4.1 the cone $h_{t,x}$, for L of (3.3), at (t,x), is given by

$$(4.5)\qquad h=\{(\tau,\xi):\xi\in\mathbb{R}^n\ ,\ \tau>-\upsilon(x)|\xi|\mu_j(t,x,\infty,\tfrac{\xi}{|\xi|})\ ,\ j=1,\dots,N\nu\}\ ,$$

where $\upsilon(x)=\pi_{e^\Delta-e^1}(x,\xi)=1$ as $e^\Delta=e^1$, $=\langle x\rangle$ as $e^\Delta=e$.

Corollary 4.3. If $e^\Delta=e^1$, the cone $h_{t,x}$ contains a cone $\{\tau\ge C_0|\xi|\}=h_0$ for all $(t,x)\in I\times\mathbb{R}^n$. If $e^\Delta=e$ then, at least, $h_{t,x}$ contains the cone $h_{|x|}=\{\tau\ge C^0\langle x\rangle|\xi|\}$, for every $(t,x)\in I\times\mathbb{R}^n$. Here C_0 is some positive constant independent of t and x.

All above statements are evident, after our remarks.

Now assume that an expression L as in (3.3) is given, with $a_{\alpha,k}$ satisfying (4.2). For all $(t,x)\in I\times\mathbb{R}^n$, $|\xi^0|=1$, let equation (3.6) have real and distinct roots, bounded by C_0, with differences bounded below by c_0, just as in the statement of prop.4.1.

Proposition 4.4. Under above assumptions the system VI,(4.4) is strictly hyperbolic of type e^1.

Indeed, we will just make the same decomposition into k_0 and κ_0, as above. The negative eigenvalues $\tau_j=-\mu_j$ then again satisfy

$$(4.6)\qquad \sum_{k+l=N}\langle\xi\rangle^{-l}a_l(t,x,\xi)\tau^k=0\ ,$$

with a_{N-j} of (4.1) or (3.3). We may wite this as

$$(4.7)\qquad p(\tau;t,x,\tfrac{\xi}{\langle\xi\rangle})+\sum_{k+l=N}\tau^k\sum_{|\alpha|<l}\langle\xi\rangle^{l-|\alpha|}a_{\alpha,N-l}(t,x)(\tfrac{\xi}{\langle\xi\rangle})^\alpha=0\ ,$$

where $p(t)=p(\tau,t,x,\xi)$ denotes the polynomial of (3.6). For $\xi=\infty\xi^0$ equation (4.7) assumes the form $p(\tau;t,x,\xi^0)=0$, and the above assumptions then state that its roots are real and distinct. On the other hand, the polynomial in τ of (4.7) has its highest coefficient equal to 1, independent of t,x,ξ. Thus the roots may locally be arranged into continuous functions. They are distinct at the compact set $I\times\mathbb{M}^{\Delta}\times\partial\mathbb{B}^n$, hence they will stay real and distinct in some neighbourhood $I\times\mathbb{M}^{\Delta}\times\{\xi\in\mathbb{B}^n,\ |\xi|\ge\eta_0\}$. But this is precisely the statement of prop 4.4. Q.E.D.

5. Regions of dependence and influence; finite propagation speed.

Let us again focus on the Cauchy problems (3.7), (3.8). We can significantly improve on the existence results thm.3.1 and thm 3.2, insofar as it may be shown that "uniform s-hyp(h^0)" is sufficient - it will imply the condition on the system VI,(4.4). Also the solution u(t) is $C_0^\infty(\mathbb{R}^n)$ for each fixed t, not only in $\mathcal{S}=\cap \mathcal{H}_s$.

To simplify the assumptions let us assume that the conditions of thm.3.1 are satisfied for <u>all</u> compact intervals $I=[t_1,t_2]$.

Consider the adjoint L^* of our expression (3.3), defined as

$$(5.1)\qquad L^* = \sum_{|\alpha|+k\le N} D_x^\alpha D_t^k \,\bar{a}^T_{\alpha,k}(t,x) \quad .$$

<u>Proposition 5.1.</u> We have $L\in$ s-hyp(h^0) if and only if L^* is s-hyp(h^0). Moreover, the cones $h_{t,x}$ of time-like directions for L and L^* coincide, at each (t,x), and a surface Σ is space-like for L if and only if it is space-like for L^*.

For L with complex-valued coefficients we know that "s-hyp", the cones $h_{t,x}$, and "space-like" are completely determined by the principal part of L. Cor.1.3 implies that the principal parts of L and L^* differ only by a nonvanishing complex factor $\gamma(t,x)$: We get $a_N(t,x,\tau,\xi)/g(t,x)=b$ real-valued, for some $g(t,x)$, hence the principal part of L^* evaluates as $a_N^*=\bar{g}b=(\bar{g}/g)a_N$. Since "s-hyp" implies "hyp" this proves the proposition for complex-valued expressions L. For general matrix-valued expressions we still have $a_N^*=(\bar{a}_N)^T$, and det $a_N^* = (\bar{g}/g)\det((\bar{a}_N)^T)$ follows because det a_N is a hyperbolic polynomial. Thus again the proposition follows, q.e.d.

In the following let us assume that L, L^* both satisfy the assumptions of thm.3.1. Consider a region $\Omega_0\subset\mathbb{R}^{n+1}$ as in fig.5.1. In detail, the boundary $\partial\Omega_0$ is piece-wise smooth, and consists of two parts, Σ_1 and Σ_2, both space like surfaces, with equations $t=\Theta_j(x)$, $j=1,2$.

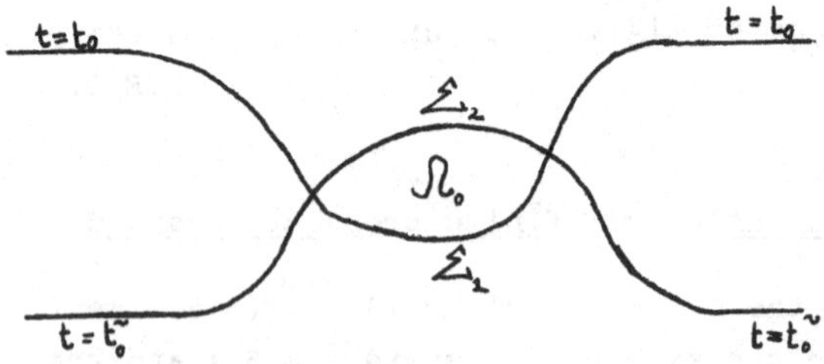

Fig.5.1

The surfaces Σ_j intersect at positive angle in some smooth compact n-1-dimensional surface. Moreover, we assume Θ_j extended over all of $\mathbb{R}^n$, and that the extended surfaces $t=\Theta_j(x)$ satisfy the assumptions of thm.3.2. Assume the compact region Ω_0 defined by

(5.2) $$\Omega_0 = \{(t,x) : x\in \mathbb{R}^n,\ \Theta_1(x)\le t\le\Theta_2(x)\} .$$

<u>Lemma 5.2</u>. Suppose a solution u of the relativistic or nonrelativistic Cauchy problem ((3.8) or (3.7)) with f=0 is defined in the compact region Ω_0 , and assume that $u=u_{|t}=\ldots=\partial_t^{N-1}u=0$ on $\Sigma_1 \cap \partial\Omega_0$. Then u vanishes identically in all of Ω_0 .

<u>Proof</u>. Note that the relativistic Cauchy problem (3.8), for L^* rather than L , and for the surface Σ_2 , admits a solution v . Here we assume that $\varphi_j=0$, but will allow an arbitrary C_0^∞-function f . Existence of v then is a matter of thm.3.2. Note that u and v both are $C^\infty(\Omega_0)$. By a partial integration we get

(5.3) $$\int_{\Omega_0} \bar{u}^T f dtdx = \int \bar{u}^T L^* v dtdx = 0 .$$

Indeed, we get Lu=f=0, by assumption. No boundary terms may appear because the Cauchy data of u and v vanish on $\Sigma_1 \cap\partial\Omega_0$ and on $\Sigma_2 \cap\partial\Omega_0$, respectively. Thus the partial integration gives 0 . Since this is true for arbitrary smooth f with compact support, we conclude that u=0 in Ω_0 , q.e.d.

<u>Proposition 5.3</u>. Assume that L, L^* both satisfy the assumptions of thm.3.1. Let $\Sigma: t=\Theta(x)$ be space-like surface satisfying the assumptions of thm.3.2, and let u be the solution of the relativistic Cauchy problem of thm.3.2, for the surface Σ , and for data f,φ_j where $f\in C_0^\infty(I\times\mathbb{R}^n)$, for every I , and φ_j of compact support on Σ . Then, if Σ^Δ is any space-like surface $t=\Theta^\Delta(x)$, $x\in \mathbb{R}^n$, with the property that supp f, and supp φ_j all lie above Σ^Δ , in (t,x)-space then u vanishes identically in the set $\Theta(x)\le t\le\Theta^\Delta(x)$ (see fig.5.2).

<u>Proof</u>. This is a matter of constructing "lenses" onto which lemma

5.2 may be applied. First one notices that, for each fixed t the function u(t,.) has compact support: For example, one may work with a surface Σ^Δ carrying a spherical bulge, such as

$t=t_0+\mathrm{Max}\{\sqrt{|x|^2-2a.x}\ ,b\}=\Theta^\Delta(x)$, with $0<b<|a|^2$, smoothened at the rim, where b must be chosen sufficiently close to $|a|^2$. Using cor.4.3 one finds that such Θ^Δ satisfies the assumptions of

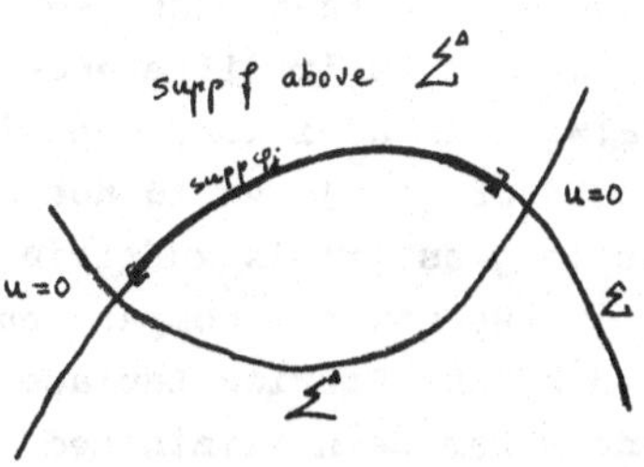

Fig.5.2

the lemma, provided that the maximum slope is sufficiently small. This does not limit the height such a bulge can 'climb'. The latter works if $e^\Delta=e^1$, while for $e^\Delta=e$ a somewhat more complicated lense must be chosen. Once we have shown that u(t,x) has compact support for all t, we may examine the solution in successive slabs $t_{j-1}\le t\le t_j$, where $t_j=\varepsilon j$, for small $\varepsilon>0$. The surface $t=t_j$ outside Σ^Δ, $t=t_{j-1}$ inside Σ^Δ $t=\Theta^\Delta$ otherwise then, after smoothening, may be seen to satisfy the assumptions of thm.3.2 for existence of the solution v for the adjoint problem, as in lemma 4.2. Then an argument as in the lemma will give u=0 below that surface. Q.E.D.

Theorem 5.4. Let L of (3.3) have its coefficients satisfying (4.2) Assume that L is s-hyp(h^0, $\mathbb{R}^n$) uniformly, in the sense that the roots τ_j of equation (3.6) are real and distinct for all $(t,x)\in\mathbb{R}^{n+1}$, $0\ne\xi\in\mathbb{R}^n$, while, on $|\xi|=1$, $t_1\le t\le t_2$, we have $|\tau_j|\le C_0$, $|\tau_j-\tau_l|>c_0$, with $0<c_0$, C_0 independent of t,x,ξ. Assume Σ to be a space-like surface satisfying the assumptions of thm.3.2. Then the relativistic Cauchy problem, for the surface Σ, and arbitrary smooth data f,φ_j, where φ_j are compactly supported while $(\mathrm{supp}\ f)\cap I\times\mathbb{R}^n\subset\subset\mathbb{R}^{n+1}$ for all $I\subset\subset\mathbb{R}$, admits a unique solution $u\in C^\infty(\mathbb{R}^{n+1})$. Moreover, we have $(\mathrm{supp}\ u)\cap(I\times\mathbb{R}^n)$ compact for all compact I.

Proof. Let us set $e^\Delta=e^1$. Then confirm that L as well as L^* satisfy the assumptions of prop.4.4. Accordingly, the corresponding first order systems VI,(4.4) are strictly hyperbolic of type $e^1=e^\Delta$, and we have the assumptions of prop.5.3 satisfied. Thus our assertion follows. Q.E.D.

Remark 5.5. It might be interesting to note that our above discussion will still work in the case where we require that (4.2) holds for $e^\Delta=e$ – i.e., we have

$$a_{\alpha,j}(t,x)\in\psi c_{(N-j)e^2}\quad ,\text{as } |\alpha|\le N-j\ . \tag{5.4}$$

However, we then must require the roots of (3.6) to be bounded by $C_0\langle x\rangle$ and their differences bounded below by $c_0\langle x\rangle$, $c_0>0$, as $|\xi|=1$. In addition, one will require the same conditions on the roots of (4.7), valid not only for large $|\xi|$ but also for all $|\xi|$, as long as $|x|$ is sufficiently large. We will not discuss details.

We now may compare our results of thm.5.4 with those achieved by the Fourier-Laplace method in ch.0,4. Note that use of ψdo's has been eliminated, except in the proofs. For a Cauchy problem as in thm.5.4 we have finite propagation speed: The solution $u(t,x)$ stays $\equiv 0$ for sufficiently large $|x|$, as long as t stays within a compact interval. There is a "region of influence" : For a change of data - within a compact set K - the "influence" on the solution $u(t,x)$ will be felt only inside a conically shaped region, expanding from K towards increasing times. There is a "domain of dependence": The values of u at or near (t^0,x^0) depend only on the values of the data within a conically shaped region with tip at (t^0,x^0), expanding backward in t.

For a point (t^0,x^0) one can define a precise region of dependence $RD(t^0,x^0)$ and region of influence $RI(t^0,x^0)$ as the backward part and the forward part of the envelope of all space-like surfaces through (t^0,x^0). This envelope will have the normal cone of the cone h_{t^0,x^0} as a tangent cone, at (t^0,x^0) . Hence, at that point (t^0,x^0) , it may be decomposed into a forward and backward part. The regions RD and RI also may be described by the bicharacteristics of L, in the sense of ch.0,8, at least for $\nu=1$. Or, it may be linked to the "particle flow" of VI,5,6, generalized to systems of the form VI,(4.4) - cf. ch.IX .

6. The local Cauchy problem; hyperbolic problems on manifolds.

Notice that thm.5.4 implies an existence result for a local hyperbolic Cauchy problem. Assume first that, in a set $\Omega=I\times B_3$, with $B_\gamma=B_\gamma(0)=\{x\in\mathbb{R}^n:|x|<\gamma\}$, $I=[t_1,t_2]$, we have a differential expression of the form (3.3) defined. Assume that $L\in$ s-hyp(h^0,Ω) , and let the coefficients $a_{\alpha,j}(t,x)$ be $C^\infty(\Omega)$. Assume a space-like surface $\Sigma\subset\Omega^{\text{int}}$. Then ask for existence of a function $u(t,x)$, defined near Σ , satisfying $Lu=f$ (near Σ) and $\partial_t^j u=\varphi_j$ on Σ, $j=0,..N-1$, where f and φ_j have support within $I\times B_1$.

To solve this problem we will modify and globalize the expression L as follows. Letting $L=a(t,x,D_t,D_x)$, with the polynomial $a(t,x,\tau,\xi)=\sum_{|\alpha|+k\le N} a_{\alpha,k}(t,x)\xi^\alpha\tau^k$, and with a function $\theta(\rho)\in$

$C^{\infty}((0,\infty))$ satisfying $\tau(\rho)=\rho$ as $\rho\in[0,1]$, $0\le\theta(\rho)<2$ in $(0,\infty)$, $\theta(\rho)=0$ as $\rho\ge 2$. Then define $L^{\Delta} = a^{\Delta}(t,\theta(|x|)\frac{x}{|x|},D_t,D_x)$, $t\in I$, $x\in \mathbb{R}^n$. Observe that $L=L^{\Delta}$ within $I\times B_1^c$. Also the coefficients of L^{Δ} are independent of x as $|x|\ge 2$ - we have $L^{\Delta} = a(t,0,D_t,D_x)$ there. In particular, L^{Δ} is uniformly s-hyp(h^0,$I\times\mathbb{R}^n$), in the sense of prop. 4.4. We may modify the surface Σ outside $I\times B_1^c$ to obtain Σ^{Δ} satisfying the assumptions of thm.3.2, at least close to Σ. Since $\Sigma=\Sigma^{\Delta}$ within $I\times B_1^c$ we may extend the initial conditions imposed there by setting $\varphi_j=0$, $f=0$ outside $I\times B_1^c$. In this way we obtain a Cauchy problem satisfying all assumptions of thm.3.2. The solution exists and its restriction to $I\times B_1^c$ will supply a solution of the local problem.

Now, for a general set $\Omega=I\times\Omega^0$, where $\Omega^0 \subset\subset \mathbb{R}^n$ we may solve the corresponding problem, using a partition of unity. Assume L , of the general form (3.3), defined in $I\times\Omega^{\sim}$ with an open set $\Omega^{\sim}\supset\Omega^c$, and let it be s-hyp(h^0,$I\times\Omega^{\sim}$) . Construct a finite covering of Ω by balls $B_{\varepsilon}(x^k)$ with the property that $B_{3\varepsilon}(x^k)\subset\Omega^{\sim}$, and a partition of unity $\sum\omega_k=1$, supp $\omega_k \subset B_{\varepsilon}(x^k)$. The solution of

(6.1) $Lu=f$, near Σ , $\partial_t^j u=\varphi_j$ at Σ , $j=0,\dots,N-1$,

where $\Sigma\subset I\times\Omega$ is a space-like surface, and where supp $\varphi\subset I\times\Omega$, supp $\varphi_j\subset\Omega$, may be obtained as a superposition of the solutions of (6.1) for $f \leftrightarrow \omega_k f$, $\varphi_j \leftrightarrow \omega_k\varphi_j$, to be regarded first as a problem within $I\times B_{3\varepsilon}(x^k)$. The solution of the latter problem may be extended zero along Σ , for a small time interval, because it also will vanish within $I\times B_{\varepsilon}(x^k)$ outside the region of influence. The problem within $I\times B_{3\varepsilon}(x^k)$ may be translated and dilated into a problem in $I\times B_3$. We have proven the following result.

<u>Theorem 6.1</u>. For a differential expression L of the general form (3.3) with smooth coefficients defined in $I\times\Omega^{\sim}$, and s-hyp(h^0,$I\times\Omega^{\sim}$) there, and a space-like surface $\Sigma\subset\subset I\times\Omega$, where $\Omega\subset\subset\Omega^{\sim}$, the Cauchy problem (6.1) admits a unique local solution, defined in a sufficiently small neighbourhood of Σ .

<u>Remark</u>: The local uniqueness simply is a matter of an argument like that in the proof of lemma.5.2.

Next we notice that the concepts "s-hyp(h)", "s-hyp(h,Ω^0)" are meaningful for differential expressions defined on a differentiable manifold Ω. Since we have coordinate invariance we may interpret h as a vector of the cotangent space $T^*\Omega_{x^0}$, at a point x^0 , and define L to be s-hyp(h) if this is so in any set of coordinates. If we are given a field of cotangent vectors $h=h(x)\in T^*\Omega_x$, for $x \in \Omega^0$, an open set $\subset\Omega$, then we shall define $L\in$ s-hyp(h) = s-hyp(h,Ω^0) if $L\in$ s-hyp(h(x)) , at each $x\in\Omega^0$. We then will have a cone $\mathit{h}_x\subset T^*\Omega_x$ of cotangent vectors ar x - it is the cone h of the differential polynomial of L at x, in any set of local coordinates; note that the polynomial is now defined on $T^*\Omega_x$, and that its principal part is defined as well. It determines h_x as a subset of $T^*\Omega$. A space-like surface $\Sigma\subset\Omega$ will be defined as an n-1-dimensional submanifold with normal at $x\in\Sigma$ contained in h_x. Here we will assume that L is s-hyp(h,Ω^-) with some domain $\Omega^-\subset\Omega$, and that $\Sigma\subset\Omega^0\subset\subset\Omega^-$. Then again we may formulate a <u>hyperbolic</u> <u>Cauchy problem</u>: For compactly supported $f\in C_0^\infty(\Omega)$ and $\varphi_j\in C_0^\infty(\Sigma)$, j=0,..., N-1 (where L is of order N) find a solution $u\in C^\infty(\Omega')$ of

$$(6.2) \qquad Lu=f \ , \ x\in\Omega' \ , \ \partial^j_{h(x)}u=\varphi_j \ , \ x\in\Sigma \ , \ j=0,1,\dots,N-1 \ ,$$

where $\Omega'\subset\Omega^0$ is some (sufficiently small) open neighbourhood of Σ, and with the j-fold derivative $\partial^j_{h(x)}u$ (for τ, at $\tau=0$) of the function $u(x+\tau h(x))$, in some fixed set of local coordinates.

<u>Theorem 6.2</u>. The solution u(x) of problem (6.2) exists and is unique, up to the choice of Ω' .

The proof is by localization, exactly as for thm.6.1.

Notice that we have admitted $\nu\times\nu$-systems, in the above . Correspondingly we may allow an expression L mapping between ν-dimensional complex vector bundles on Ω , in thm.6.2, and then specify data f , φ_j which are sections in vector bundles.

Clearly the notions of finite propagation speed, region of influence and dependence *RI* and *RD* extend literally as well.

The case of a second order scalar operator again merits special attention: This will be related to relativity theory. We have seen that, locally, up to a nonvanishing complex factor, a second order expression L is of the form (3.1) , with a quadratic form $a_2(\xi)$ of rank n and signature n-2 .

Chapter 8. PSEUDO-DIFFERENTIAL OPERATORS

AS SMOOTH OPERATORS OF $L(H)$.

0. Introduction.

In this chapter we look at a different way to introduce pseudodifferential operators. From their introduction in ch.I - using I,(1.5)- they appear as a technical device inspired by the Fourier-Laplace method. It may be surprising that there is a natural definition of ψdo's at least for some special symbol classes: Certain algebras of ψdo's appear as "smooth" subalgebras of $L(H)$. For example, the algebra $\mathrm{OpCB}^{\infty}(\mathbb{R}^{2n})$, with $\mathrm{CB}^{\infty}(\mathbb{R}^{2n})=\psi t_0$ the class of $a(x,\xi) \in C^{\infty}(\mathbb{R}^{2n})$, bounded with all derivatives, coincides with the class of operators $A\in L(H)$ which are both "translation smooth" and "gauge smooth". That is, with the family $\{T_z:z\in \mathbb{R}^n\}$ of translation operators $T_z u(x)=u(x-z)$, and the family $\{M_\zeta:\zeta\in \mathbb{R}^n\}$ of "gauge multiplications" $M_\zeta u(x)=e^{i\zeta x}u(x)$, we get $T_{-z}AT_z\in C^{\infty}(\mathbb{R}^n,L(H))$ and $M_{-\zeta}AM_\zeta\in C^{\infty}(\mathbb{R}^n,L(H))$. Notice that both T_z and M_ζ are groups of unitary operators on H, representations of the Lie-group $\mathbb{R}^n$ in $U(H)$.

In terms of ch.I, $\mathrm{Op}\psi t_0=\mathrm{OpCB}^{\infty}(\mathbb{R}^{2n})$, a subalgebra of $L(H)$ by III,1, is the class of translation and gauge smooth operators.

There are other natural actions of Lie groups on $H=L^2(\mathbb{R}^n)$, such as the dilations $u(x)\to u(\gamma x)$, with $\gamma\neq 0$, or the rotations $u(x)\to u(ox)$, with a rotation $o:\mathbb{R}^n\to\mathbb{R}^n$, or the linear group $u(x)\to u(mx)$ with a nonsingular real matrix m. If we require dilation and rotation smoothness, in addition to translation and gauge smoothness we obtain a more practical algebra (called $\mathrm{Op}\psi s_0$) with a calculus of ψdo's and symbol decay similar to $\mathrm{Op}\psi c_0$. With the linear group we get a still smaller algebra, called $\mathrm{Op}\psi l_0$.

Similar symbols were studied by Weinstein and Zelditch [WZ].

We will give a completely independent presentation of these facts, introducing ψdo's again, without reference to ch.I. In particular, we will work with symbol classes ψt_m, ψs_m, and ψl_m exclusively, in ch.9 and ch.10 below, derived from the <u>natural classes</u> ψt_0, ψs_0, and ψl_0. These classes are similar, but not identical to ψc_m. They allow a (slightly more complicated) calculus of ψdo's.

In sec's 1,2,3 we are concerned with the 'ΨDO-theorem' describing the new characterization of $\mathrm{Op}\psi t_0$. In sec.4 and 5 we discuss the classes $\mathrm{Op}\psi s_0$ and $\mathrm{Op}\psi l_0$. Sec's 6 and 7 are concerned with symbols in ψx_m for general m, with calculus of ψdo',s, and decay properties of the new symbols. Sec.8 deals with corresponding Lie-algebras and Beals-Dunau-criteria.

An earlier version of our theory was announced in [C4] and distributed as lecture notes of a seminar at Berkeley in 1984, but remained unpublished.

1. ψdo's as smooth operators of $L(H)$.

Before giving a natural introduction to the concept of pseudo differential operator we make some observations concerning smoothness of continuous linear operators on the Hilbert space $H=L^2(\mathbb{R}^n)$.

$\mathbb{R}^n$ is an abelian group under vector addition. The characters of this group are the functions $e^{ix\xi}$, $\xi\in\mathbb{R}^n$, the character group is $\mathbb{R}^n$ again. For $z\in\mathbb{R}^n$ define the translation operator $T_z\in L(H)$ by

(1.1) $$T_z u(x) = u(x-z) \ , \ u \in H \ .$$

For each character $e^{i\zeta x}$ we define the gauge operator $M_\zeta=e^{i\zeta M}$ by

(1.2) $$M_\zeta u(x) = e^{i\zeta x}u(x) \ .$$

Note that T_z and M_ζ are unitary operators of H. Moreover,

(1.3) $$T_{z+z'} = T_z T_{z'} \ , \ M_{\zeta+\zeta'} = M_\zeta M_{\zeta'} \ , \ \text{for all } z,z',\zeta,\zeta' \ ,$$

which expresses that $z \to T_z$ and $\zeta \to M_\zeta$ each define a unitary representation of the group $\mathbb{R}^n$ on the Hilbert space H.

We want to emphasize: The differentiable manifold $\mathbb{R}^n$ with above group structure, is a Lie group, but the representations (1.1) and (1.2) are imperfect insofar as $z\to T_z$ and $\zeta\to M_\zeta$ are not smooth as maps $\mathbb{R}^n\to L(H)$. For fixed $u\in H$ the functions $T_z u=f_z$, $M_\zeta u=g_\zeta$ are continuous but need not be differentiable ($u(x)=\mathrm{sgn}(x)e^{-|x|}$, for example). For $u\in C_0^\infty$ we get f_z, $g_\zeta\in C^\infty(\mathbb{R}^n,H)$ but not in general.

In fact, it is known that the operator families T_z and M_ζ are strongly continuous in their parameter z or ζ , but are not norm continuous: Choosing $u \in H$ with support in a ball $|x|\le\varepsilon$ (or $|x-x^0|\le\varepsilon$ with $x^0(\zeta-\zeta')=\pi$) with small ε one confirms that

(1.4) $$\|T_z-T_{z'}\| \ge \sqrt{2} \ , \ \|M_\zeta-M_{\zeta'}\| \ge \sqrt{2} \ , \text{ as } z\ne z' \ , \ \zeta\ne\zeta' \ .$$

Similarly, for a given operator $A \in L(H)$ the families

(1.5) $$A_{z,0} = T_z^{-1}AT_z \ , \ A_{0,\zeta} = M_\zeta^{-1}AM_\zeta \ , \ z,\zeta \in \mathbb{R}^n \ ,$$

are strongly continuous, but in general not uniformly continuous.

The pseudodifferential operators within the algebra $L(H)$ will be defined as those operators $A \in L(H)$ for which both families $A_{z,0}$ and $A_{0,\zeta}$ are not only norm continuous, but even are $C^{\infty}(\mathbb{R}^n, L(H))$, $L(H)$ being equipped with the norm topology.

Let us denote the class of all such operators by ΨGT. That is,

$$(1.6) \qquad \Psi GT = \{A \in L(H): A_{z,0}\ ,\ A_{0,\zeta} \in C^{\infty}(\mathbb{R}^n, L(H))\} \ .$$

We will show in the following that every operator $A \in \Psi GT \subset L(H)$ has a unique representation in the form

$$(1.7) \qquad Au(x) = (2\pi)^{-n}\int d\xi \int dy e^{i\xi(x-y)} a(x,\xi)u(y) \ ,$$

where $a(x,\xi)$ is a complex-valued function in

$$(1.8) CB^{\infty}(\mathbb{R}^{2n}) = \{b(x,\xi) \in C^{\infty}: \text{partials } b^{(\alpha)}_{(\beta)} \text{ of all orders are bounded}\}$$

called the symbol of A . (1.7) holds for $u \in C_0^{\infty}(\mathbb{R}^n)$, but determines the operator A since C_0^{∞} is dense in H. Even for $u \in C_0^{\infty}$ the existence of integrals in (1.7) will be a point of discussion.

Interestingly, the existence of a representation (1.7) with symbol function $a \in CB^{\infty}(\mathbb{R}^n)$ characterizes the operator class ΨGT : For any $a \in CB^{\infty}(\mathbb{R}^{2n})$ the integrals in (1.7) exist for all $u \in C_0^{\infty}(\mathbb{R}^n)$ and $Au(x)$ of (1.7) belongs to H. The operator $A: C_0^{\infty}(\mathbb{R}^n) \to H$ thus defined extends to a map $\in L(H)$. Moreover, A belongs to ΨGT .

In the remainder of this section we make a few observations, preparing the detailed discussion of the result mentioned above.

First, observe the 'commutator relation'

$$T_{-z}M_{\zeta}T_z u(x) = e^{i\zeta(x+z)}u(x) = e^{iz\zeta}M_{\zeta}u(x) \ .$$

Or,

$$(1.9) \qquad T_{-z}M_{\zeta}T_zM_{-\zeta} = e^{iz\zeta} \ .$$

Thus the subgroup of the unitary group $U(H)$ of H generated by all T_z and M_{ζ} , $z,\zeta \in \mathbb{R}^n$, consists precisely of the operators

$$(1.10) \qquad E_{z,\zeta,\varphi} = e^{i(\varphi+x\zeta)}T_z \ ,\ \varphi \in \mathbb{R} \ ,\ z,\zeta \in \mathbb{R}^n \ .$$

This group is called the Heisenberg group. We denote it by GH. Get

$$(1.11) \qquad E_{z,\zeta,\varphi}E_{z',\zeta',\varphi'} = E_{z+z',\zeta+\zeta',\varphi+\varphi'-\zeta'z} \ .$$

Thus GH is isomorphic to the group obtained by equipping the space $gh = \mathbb{R}^n \times \mathbb{R}^n \times \mathbb{S}^1 = \{(z,\zeta,\varphi): z,\zeta \in \mathbb{R}^n, \varphi \in \mathbb{S}^1 = \mathbb{R}/2\pi\mathbb{Z}\}$ with the group operation

$$(1.12) \quad (z,\zeta,\varphi)\circ(z',\zeta',\varphi') = (z+z',\zeta+\zeta',\varphi+\varphi'-\zeta'z) \ ,\ \text{unit} = (0,0,0) \ .$$

Clearly $\mathcal{GH}$ with the manifold structure of gh is a Lie group. Our above condition describing $\Psi\mathcal{GT}$ is equivalent to the following:

<u>For</u> $A \in L(H)$ <u>let</u> $A_{z,\zeta} = E_{z,\zeta,\varphi}^{-1} A E_{z,\zeta,\varphi}$. (Note that $A_{z,\zeta}$ is independent of φ.) <u>Then we have</u>

$$A_{z,\zeta} \in C^\infty(gh, L(H)) . \tag{1.13}$$

Indeed, (1.13) trivially implies the condition of (1.6). Vice versa, note that (1.9) implies

$$A_{z,\zeta} = T_{-z} A_{0,\zeta} T_z = M_{-\zeta} A_{z,0} M_\zeta . \tag{1.14}$$

If (1.6) holds then use that
$\|A_{z,\zeta} - A_{z,\zeta'}\| = \|A_{0,\zeta} - A_{0,\zeta'}\|$, $\|A_{z,\zeta} - A_{z',\zeta}\| = \|A_{z,0} - A_{z',0}\|$, showing that $A_{z,\zeta}$ is continuous in z for fixed ζ and continuous in ζ for fixed z , and either continuity is uniform in the other variable. This clearly implies norm continuity of $A_{z,\zeta}$ in both variables. Similarly (1.6) and (1.14) imply existence of

$$\partial_\zeta^\alpha A_{z,\zeta} = T_{-z} \partial_\zeta^\alpha A_{0,\zeta} T_z \ , \ \partial_z^\beta A_{z,\zeta} = M_{-\zeta} \partial_z^\beta A_{z,0} M_\zeta \ , \tag{1.15}$$

in operator norm topology, uniformly in the other variable. Since $\partial_\zeta^\alpha A_{0,\zeta}$ and $\partial_z^\beta A_{z,0}$ are still $C^\infty(\mathbb{R}^n, L(H))$, we may repeat the above conclusion and find that both functions (1.15) are in $CB(gh)$.

By a standard argument we then conclude that all mixed derivatives $\partial_z^\alpha \partial_\zeta^\beta A_{z,\zeta}$ exist (and are $CB(\mathbb{R}^{2n})$) as well. Indeed, this needs to be shown at $z=\zeta=0$ only, in view of the group property. Thus look at $\chi A_{z,\zeta}$ with a cut-off function $\chi \in C_0^\infty$, $\chi(z,\zeta)=1$ near 0. For $u,v \in H$ consider $p=(u,\chi A_{z,\zeta} v)$. Let $p_{(\beta)}^{(\alpha)} = \partial_z^\alpha \partial_\zeta^\beta p$. We get $p^{(\alpha)}, p_{(\beta)} \in H_\gamma = L^2(Q_\gamma)$, $Q_\gamma = \{(z,\zeta): |z_j| \le \gamma,\ |\zeta_j| \le \gamma,\ j=1,\dots,n\}$, for large γ, and all α,β. Parsevals relation for Fourier series implies that the (multi-)sequence $\{q_{\theta,\upsilon}: \theta,\upsilon \in \mathbb{Z}^n\}$ of Fourier coefficients of p satisfies $\theta^\alpha q_{\theta,\upsilon}$, $\upsilon^\beta q_{\theta,\upsilon} \in l^2 = L^2(\mathbb{Z}^{2n})$, for all α,β . Thus also $\theta^\alpha \upsilon^\beta (1+|\theta|^2+|\upsilon|^2)^k \in l^2$ for $k=0,1,\dots$, and all α,β. It follows that $p_{(\beta)}^{(\alpha)} \in H_k$, for all α,β, with the Sobolev space $H_k = H_{k,0}$ of III,3. By Sobolev's lemma we thus get $p_{(\beta)}^{(\alpha)} \in C^k(\mathbb{R}^{2n})$, for all k. Looking at p in dependence of u,v we get $\|p^{(\alpha)}\|$, $\|p_{(\beta)}\| \le c\|u\|\|v\|$, with norms in H_γ and H, resp., and with c undependent of u,v . Following this estimate through the above argument - Parseval, Sobolev, etc.,..., we find that also $\|p_{(\beta)}^{(\alpha)}\|_{C^k} \le c\|u\|\|v\|$, for each α,β . This implies $(\chi A_{z,\zeta})_{(\beta)}^{(\alpha)} \in L(H)$, and even $\|(\chi A_{z,\zeta})_{(\beta)}^{(\alpha)}\| \le c$. Then one concludes that $\partial_z^\alpha\partial_\zeta^\beta A_{z,\zeta}=(A_{z,\zeta})_{(\beta)}^{(\alpha)}$ exist in norm convergence, for all α,β . Thus indeed (1.6) and (1.13) are equivalent.

Secondly, we want to examine existence of the integrals in (1.7). Note that the inner integral $\int dy$ exist trivially, as a proper Riemann or Lebesgue integral, and equals $(2\pi)^{n/2}e^{ix\xi}u^\wedge(\xi)$, where $u^\wedge(\xi)$ denotes the Fourier transform of $u \in C_0^\infty(\mathbb{R}^n)$. Moreover, it is well known that $u^\wedge$ is a rapidly decreasing function. That is $u^\wedge(\xi) = O((1+|\xi|)^{-k})$,for every k=0,1,2, ... , and the same is true for the derivatives $u^{\wedge(\alpha)}$. Since $a(x,\xi)\in CB^\infty(\mathbb{R}^{2n})$ is bounded, and $|e^{i\xi(x-y)}|=1$, it then is clear that also the outer integral $\int d\xi$ exists as improper Riemann or Lebesgue integral and that that there is a bound for the integral independent of x .

In order to show that the right hand side of (1.7) defines a function in $\mathcal{H}$ we prove that $x^\alpha Au(x)$ also is bounded, for every α . Indeed $x^\alpha=((x-y)+y)^\alpha$ is a finite linear combination of terms $(x-y)^\beta y^\gamma$, $\beta+\gamma=\alpha$, while $\int d\xi a(x,\xi)\int dye^{i\xi(x-y)}(x-y)^\beta y^\gamma u(y) =$

$$= (-i)^{|\beta|}\int d\xi a(x,\xi)\partial_\xi^\beta\int dye^{i\xi(x-y)}(y^\gamma u(y)) .$$

Again the inner integral equals $(2\pi)^{n/2}e^{ix\xi}(x^\gamma u)^\wedge$, hence is a function in $\mathcal{S}$, while $a(x,\xi)$ has all derivatives bounded. It follows that a partial integration may be carried out, leaving no boundary terms at infinity, giving

$$= (2\pi)^{n/2}(+i)^{|\beta|}\int d\xi a^{(\beta)}(x,\xi)e^{ix\xi}(x^\gamma u)^\wedge(\xi) = O(1) .$$

Thus indeed $x^\alpha Au(x)$ is bounded for all α , hence $(1+|x|^2)^n Au$ is bounded, and $Au \in L^2$.

In fact, integrals of the form (1.7) already have been investigated in I,1, under weaker conditions on the symbol $a(x,\xi)$.

2. The ΨDO-theorem.

As prepared in sec.1 we will now prove a theorem linking the concept of pseudodifferential operator, introduced by formula (1.7) to the class ΨGT of continuous operators on the Hilbert space $\mathcal{H} = L^2(\mathbb{R}^n)$ which are 'smooth' with respect to the translation operators, and the 'gauge transforms' $u \to e^{i\zeta x}u$ (or, equivalently, *gh*-smooth). Here we were using the standard representation

GH of the Heisenberg group gh, i.e. $GH=\{E_{z,\zeta,\varphi}:(z,\zeta,\varphi)\in gh\}$, with $gh = \mathbb{R}^n\times\mathbb{R}^n\times\mathbb{S}^1$, $\mathbb{S}^1 = \mathbb{R}/(2\pi\mathbb{Z})$, where

$$(2.1) \qquad E_{z,\zeta,\varphi}:H \to H \quad , \quad (E_{z,\zeta,\varphi}u)(x)=e^{i(\varphi+\zeta x)}u(x-z) \ .$$

Especially,all translations $T_z:u(x) \to u(x-z)$,for a constant vector z, and all "gauge transforms" $u(x) \to e^{i(\varphi+\zeta x)}u(x)$,for constants $\varphi\in \mathbb{R}$, $\zeta \in \mathbb{R}^n$,belong to this class.

For $A \in L(H)$, a continuous operator on H , we are interested in the function $A_{z,\zeta}$: $GT \to L(H)$ obtained by conjugating A with the operators $E_{z,\zeta,\varphi}$:

$$(2.2) \qquad A_{z,\zeta} = E^{-1}_{z,\zeta,\varphi}AE_{z,\zeta,\varphi} \ .$$

Note that $A_{z,\zeta}$ does not really depend on φ ,since the constants $e^{i\varphi}$ commute with all of $L(H)$. Hence $A_{z,\zeta}$ may be regarded as a function $GT'\to L(H)$,where GT' denotes the quotient of GT modulo its normal subgroup $\{e^{i\varphi} :\varphi \in \mathbb{R}\}$ .Clearly GT' is isomorphic to $\mathbb{R}^{2n}$,as the composition law (1.12) shows. We may write

$$(2.3) \qquad A_{z,\zeta} = E^{-1}_{z,\zeta}AE_{z,\zeta} \ , \ E_{z,\zeta} = E_{z,\zeta,0} \ ,$$

and henceforth regard $A_{z,\zeta}$ as a function $\mathbb{R}^{2n} \to L(H)$.

We restate the essence of our discussion in sec.1 as a theorem, below. Here, for a general class X of functions over $\mathbb{R}^{2n}$, and class Y of ψdo's we write

$$(2.4) \qquad \mathrm{Op}\ X = \{a(x,D) : a \in X\} \ , \ \mathrm{Symb}\ Y = \{a : a(x,D) \in Y\} \ .$$

Also, for a single function $a(x,\xi)$ we will write $\mathrm{Op}(a)=a(x,D)$, $\mathrm{Symb}\ A = a(x,\xi)$, where $A= \mathrm{Op}(a) = a(x,D)$ denotes the formal pseudodifferential operator (1.7) belonging to the symbol $a \in X$.

<u>Theorem 2.1</u>. Let $\psi t_0 = CB^\infty(\mathbb{R}^{2n})$ denote the class of all complex-valued $a(x,\xi)$ with partial derivatives $a^{(\alpha)}_{(\beta)}(x,\xi) = \partial^\beta_x\partial^\alpha_\xi a(x,\xi)$ of all orders bounded over $\mathbb{R}^{2n}$. Then we have

$$(2.5) \qquad \mathrm{Op}\psi t_0 = \{A \in L(H) : A_{z,\zeta} \in C^\infty(\mathbb{R}^{2n},L(H))\} = \Psi GT \ .$$

Here the partial derivatives of $A_{z,\zeta}$ are assumed to exist in the convergence of the operator norm $\|.\|$ of $L(H)$,defined by

$$(2.6) \qquad \|A\| = \sup\ \{\|Au\|/\|u\| : u \in H \ , \ u \neq 0 \} \ , \ A \in L(H) \ .$$

The ψdo A of (1.7) is defined for $u \in C^\infty_0(\mathbb{R}^n)$ by integrating in the order stated, and for general $u \in H$ by taking continuous extension from the dense subspace C_0^∞.

With pseudodifferential operator notation we have

$$(2.7) \qquad E_{z,\zeta} = e^{i\zeta x} e^{-izD} ,$$

using that

$$(e^{izD}u)(x) = \int d\!\!/\xi e^{i(x+z)\xi} u^\wedge(\xi) = u(x+z) .$$

With (2.7) we may differentiate (2.2) formally, for

$$(2.8) \qquad \partial_{z_j} A_{z,\zeta}|_{z=\zeta=0} = i[D_j, A] \ , \ \partial_{\zeta_j} A_{z,\zeta}|_{z=\zeta=0} = -i[x_j, A] \ ,$$

with formal commutators between the operator $A \in L(H)$,and the unbounded differentiation or multiplication operators D_j ,and x_j . Similarly the higher derivatives are related to iterated commutators between A , D_j ,and x_l .This provides a link of thm.2.1 to a slightly different characterization of more general classes of ψdo's as operators $S \to S$ with well defined iterated commutators with differentiations and multiplications (cf. R.Beals, [B_2]) .

Looking for a proof of thm.2.1 we first will discuss the fact that, given a symbol $a \in CB^\infty(\mathbb{R}^{2n})$, the formal expression $A = a(x,D)$ of (1.7) defines an operator in $L(H)$. Actually, this is already done (cf. III, thm.1.1), but we offer another proof here, with an approach first published in [CC]. (It may be a bit more complicated, but it fits the present general approach.)

<u>Theorem 2.2</u>. An operator $A = a(x,D)$ with symbol $a \in \psi t_0 = CB^\infty(\mathbb{R}^{2n})$ is bounded in $H = L^2(\mathbb{R}^n)$. More precisely: The operator $A: C_0^\infty \to H$ of (1.7) extends to a continuous operator $H \to H$.

<u>Proof</u>. It turns out that we need only boundedness of 'a few' derivatives of the symbol, for this proof. Consider the case of n=1, for simplicity. The case of general n may be handled analogously.

For m=1,2,... introduce the function

$$(2.9) \qquad \gamma_m(t) = e^{-t}\, t^{m-1}/(m-1)! \ , \text{ as } t\ge 0 \ , = 0 \ , \text{ as } t<0 \ .$$

One checks easily on the following facts:

<u>Proposition 2.3</u>. We have $\gamma_m \in C^{m-2}(\mathbb{R}) \cap C^\infty(\mathbb{R}\setminus\{0\})$. The derivative $\gamma_m^{(m-1)}$ still is piecewise continuous with a single jump of magnitude 1 at t=0 . We have $\gamma_m^{(j)}(t) = O(e^{-|t|/2})$, as $t \ne 0$. Moreover,

$$(2.10) \qquad (\partial_t + 1)^m \gamma_m = \delta \ ,$$

with the Dirac distribution δ . (In other words, γ_m is a fundamental solution of the differential operator $(\partial_t+1)^m$).

For k=0,1,2,... let $CB^k(\mathbb{R}^n)$ be the space of $f(x,\xi)\in C^k(\mathbb{R}^n)$ with bounded derivatives up to order k including. Also, define

$$q(x,\xi) = \gamma_2(x)\gamma_2(\xi) \ , \ x,\xi \in \mathbb{R} . \tag{2.11}$$

Corollary 2.4. Every function $a(x,\xi)$ over $\mathbb{R}^2$ with $(\partial_x+1)^2(\partial_\xi+1)^2 a = b \in CB(\mathbb{R}^2)$ may be written in the form

$$a(x,\xi) = \int dy d\eta q(x-y,\xi-\eta)b(y,\eta) \ , \ x,\xi \in \mathbb{R} . \tag{2.12}$$

Then $a\in CB^2(\mathbb{R}^2)$, with derivatives given by differentiating (2.12) under the integral sign.

We also introduce the 'pseudo-differential operator' $Q = q(x,D)$, where of course q is not in ψt_0 , but still the right hand side of (1.7) with a=q defines a function Qu(x) for every $u \in C_0^\infty(\mathbb{R})$. In fact, $Qu(x) = \gamma_2(x) F^{-1}\gamma_2 Fu(x)$, so that

$$Q = \gamma_2(x)\gamma_2(D)= \kappa_1\gamma_2(x)(\frac{1}{(1+ix)^2} *) . \tag{2.13}$$

It is clear from (2.13) that $Q \in L(H)$. In fact one finds at once that Q is an integral operator with kernel $\kappa_1^2\gamma_2(x)/(1+i(x-y))^2 \in L^2(\mathbb{R}^2)$ - i.e., Q is of Schmidt class. Moreover, we may write

$$Q=V^*U \ , \ V=(1-ix)^{-1}(1-\partial_x)^{-1} \ , \ U=(1+ix)(1+\partial_x)Q \ , \tag{2.14}$$

where both operators U and V are of Schmidt class. Indeed, V and U have integral kernels in $L^2(\mathbb{R}^2)$, given by

$$\nu_x(y)=\frac{1}{1+ix}\gamma_1(x-y), \ \mu_x(y)=\kappa_1^2(1+ix)(1+\partial_x)(\frac{1}{(1+i(x-y))^2}\gamma_2(x)). \tag{2.15}$$

Now thm.2.2 will follow from prop.2.5, below.

Proposition 2.5. For a with $b=(1+\partial_x)^2(1+\partial_\xi)^2 a\in CB(\mathbb{R}^2)$ we have

$$\|Au\|_{L^2} \le c\|u\|_{L^2} \ , \ c= 2\pi[[[U]]][[[V]]]\|b\|_{L^\infty(R^2)} , \tag{2.16}$$

with the Schmidt-norms $[[[U]]] = \|\mu_x(y)\|_{L^2}$, $[[[V]]]= \|\nu_x(y)\|_{L^2}$.

Proof. Use (2.12) to write

$$\begin{aligned} Au(x)&=\int d\!\!\!/\xi u^\wedge(\xi)e^{ix\xi}\int dz d\zeta q(x-z,\xi-\zeta)b(z,\zeta) \\ &= \int dz d\zeta b(z,\zeta)\int d\!\!\!/\xi q(x-z,\xi-\zeta)e^{ix\xi}u^\wedge(\xi) \ , \end{aligned} \tag{2.17}$$

the integral interchange being legitimate, for $u \in C_0^\infty$, hence $u^\wedge \in \mathcal{S}$. The inner integral may be written as $q(x-z,D-\zeta)$, using (1.7).

For an operator $A=a(x,D)$, $a \in \psi t_0$ (general n) one finds that

$$A_{z,0}u(x) = \int d\!\!{}^-\xi a(x+z,\xi)e^{i(x+z)\xi}u^\wedge(\xi)e^{-iz\xi} = a(x+z,D)u(x), \tag{2.18}$$

where $a(x+z,D)$ denotes the ψdo with symbol $a(x+z,\xi)$, of course, with the constant parameter $z \in \mathbb{R}^n$. Similarly,

$$A_{0,\zeta}u(x) = \int d\!\!{}^-\xi a(x,\xi)e^{i(\xi-\zeta)x}\int d\!\!{}^-y\, e^{-iy(\xi-\zeta)}u(y) = \tag{2.19}$$

$$\int d\!\!{}^-\xi a(x,\xi)e^{i(\xi-\zeta)x}\, u^\wedge(\xi-\zeta) = \int d\!\!{}^-\xi a(x,\xi+\zeta)e^{i\xi x}\, u^\wedge(\xi) = a(x,D+\zeta)u(x) .$$

Combining (2.18) and (2.19) we get

$$A_{z,\zeta} = a(x+z,D+\zeta) \quad . \tag{2.20}$$

Clearly (2.20) holds for $a=q$ as well. Thus (2.17) may be written

$$Au(x) = \int dz d\zeta b(z,\zeta)Q_{-z,-\zeta}u(x) . \tag{2.21}$$

This leads into the following estimate for the inner product (.,.)

$$|(Au,v)| = |\int dz d\zeta \bar{b}(z,\zeta)(Q_{-z,-\zeta}u,v)| \le \|b\|_{L^\infty}\int|(Q_{-z,-\zeta}u,v)|dz d\zeta,$$

where

$$|(Q_{z,\zeta}u,v)| = |(UM_\zeta T_z u, VM_\zeta T_z v)| \le \tfrac{1}{2}\{\|UM_\zeta T_z u\|^2 + \|VM_\zeta T_z v\|^2\}.$$

Let $w=T_z u$, keeping z fixed. Using Parseval's relation we get

$$\int d\zeta\|UM_\zeta w\|^2 = \int d\zeta\int dx|(\mu_x,M_\zeta w)|^2 = \int dx\int d\zeta|\int dy \mu_x(y)w(y)e^{iy\zeta}|^2$$

$$= 2\pi\int dx dy|\mu_x(y)|^2|w(y)|^2 .$$

Recalling now the meaning of w :

$$\int dz d\zeta\|UM_\zeta T_z u\|^2 = 2\pi\int dx dy dz|\mu_x(y)|^2|u(y-z)|^2 = 2\pi\|u\|^2\int dx dy|\mu_x(y)|^2 .$$

Similarly,

$$\int dz d\zeta\|VM_\zeta T_z v\|^2 = 2\pi\|v\|^2\int dx dy|\nu_x(y)|^2$$

Summarizing, we have derived an estimate of the form

$$|(Au,v)| \le \tfrac{1}{2}\{S^2\|u\|^2 + T^2\|v\|^2\} \ , \ u,v \in C_0^\infty, \tag{2.22}$$

with constants S,T. Here we let $v=v_j$ run through a sequence in C_0^∞ with L^2-limit Au/S^2, concluding that $\|Au\| \le ST\|u\|$, $u \in C_0^\infty$. Noting the precise values of the constants S,T we get (2.16). Q.E.D.

The case of general n = 2,3,... follows analogously: Define $q(x,\xi)=\Pi_{j=1}^{n}(\gamma_2(x_j)\gamma_2(\xi_j))$, and the corresponding operator $Q=q(x,D)$ We get (2.12) with $b(x,\xi)=\Pi_{j=1}^{n}\{(1+\partial_{x_j})^2(1+\partial_{\xi_j})^2\}a(x,\xi)$. Also, the U and V are tensor products of the 1-dimensional operators, their Schmidt norms are products of 1-dimensional Schmidt norms.

Corollary 2.6. In the general case of n dimensions, we have

$$\|A\|_{L^2(\mathbb{R}^n)} \le c_1^n \sup\{|\Pi_{j=1}^{n}\{(1+\partial_{x_j})^2(1+\partial_{\xi_j})^2\}a(x,\xi)| : x,\xi\in \mathbb{R}^n\}, \tag{2.23}$$

where

$$c_1=\{\iint dxdy|\frac{\gamma_1(x-y)}{1+ix}|^2\iint dxdy|(1+ix)(1+\partial_x)\{\frac{\gamma_2(x)}{(1+i(x-y))^2}\}|^2\}^{1/2} \tag{2.24}$$

Next we prove that $Op\psi t_0 \subset \Psi GT$. From thm.2.2 it follows that, for $A\in Op\psi t_0$, all derivatives of $A_{z,\zeta}$ exist in norm convergence of $L(H)$. For example, the difference quotient

$$\nabla_{\eta,j}A_{z,\zeta} = (a(x+z+\eta e^j,D+\zeta) - a(x+z,D+\zeta))/\eta \ , \ \eta \ne 0 \ , \tag{2.25}$$

is a ψdo with symbol

$$\nabla_{\eta,j}a(x,\xi) = \eta^{-1}(a(x+\eta e^j,\xi) - a(x,\xi)) = \int_0^1 dt\, a_{|x_j}(x+t\eta e^j,\xi) \ ,$$

(using (2.20) where $z=\zeta=0$, without loss of generality). Therefore,

$$(\nabla_{\eta,j}a)_{(\beta)}^{(\alpha)}(x,\xi) = \int_0^1 dt\, a_{(\beta+e_j)}^{(\alpha)}(x+t\eta e^j,\xi) \ . \tag{2.26}$$

But the uniform boundedness of all derivatives of $a(x,\xi)$ implies that each derivative $a_{(\beta)}^{(\alpha)}(x,\xi)$ is uniformly continuous over $\mathbb{R}^{2n}$:

$$a(x+h,\xi) - a(x,\xi) = h\int_0^1 \nabla_x a(x+th,\xi) = |h|O(1) \ , \tag{2.27}$$

for example, where $O(1)$ is bounded for all $x,h,\xi \in \mathbb{R}^n$ .Similarly,

$$a_{(\beta)}^{(\alpha)}(x+h,\xi+\eta) - a_{(\beta)}^{(\alpha)}(x,\xi) = O(|h|+|\eta|) \ . \tag{2.28}$$

Thus the integral at right of (2.26) equals

$$a_{(\beta+e_j)}^{(\alpha)}(x,\xi) + O(|\eta|) \tag{2.29}$$

with $O(.)$ independent of x,ξ. Thus (2.23) implies

$$\lim_{\eta\to 0}\|\nabla_{\eta,j}A_{z,\zeta} - a_{|z_j}(x+z,D+\zeta)\| = 0 \ . \tag{2.30}$$

Similarly for all partial derivatives, by induction.

3. The other half of the ΨDO- theorem.

Note that we have proven one half of thm.1.2 :Every ψdo in $\mathrm{Op}\psi t_0$ is contained in the class ΨGT. To attack the other half we must answer the following question: Given $A \in L(H)$ with $A_{z,\zeta} \in C^\infty$, how to define a symbol $a \in \psi t_0$ such that $A=a(x,D)$? We will give a formula for $a(x,\xi)$ in terms of A, first in the case $A=\mathrm{Op}(A)=a(x,D) \in \mathrm{Op}\psi t_0 \subset \Psi GT$, constructing a left inverse for the map $a \to \mathrm{Op}(a)$. Again let n=1 first. Departing from (2.12), given $a \in \psi t_0$, we get

$$a(z,\zeta) = \int dx d\xi \{\gamma_2(-x)\gamma_2(-\xi)e^{-ix\xi}\}\{b(z+x,\zeta+\xi)e^{ix\xi}\} , \tag{3.1}$$

with $\quad b(x,\xi) = (1+\partial_x)^2(1+\partial_\xi)^2 a(x,\xi)$,

writing $\rho(x,\xi)=\gamma_2(-x)\gamma_2(-\xi)e^{-ix\xi}$ as product of 2 Schmidt kernels:

$$\rho(x,\xi)=\int d\eta \frac{\gamma_1(\xi-\eta)}{(1+i\xi)^2}\{(1+\partial_\eta)(1+i\eta)^2\rho(x,\eta)\}=\int d\eta v(\xi,\eta)\bar{u}(x,\eta) . \tag{3.2}$$

In fact, both kernels

$$v(\eta,\xi) = \gamma_1(\xi-\eta)/(1+i\xi)^2 , \tag{3.3}$$

and

$$u(x,\eta) = (1+\partial_\eta)\{(1-i\eta)^2\gamma_2(x)\gamma_2(\eta)e^{ix\eta}\} \tag{3.4}$$

are $L^2(\mathbb{R}^2) \cap L^1(\mathbb{R}^2)$, as easily checked. Using (3.2) write (3.1) as

$$a(z,\zeta) = \int dx d\eta \bar{u}(x,\eta)\int d\xi b(x+z,\xi+\zeta)e^{ix\xi}v(\xi,\eta) , \tag{3.5}$$

using a Fubini-type interchange of integrals. We introduce

$$w(x,\eta) = \int d\xi e^{ix\xi}v(\xi,\eta) , \tag{3.6}$$

and the integral operators U , V , W with kernels $u(x,y)$, $v(x,y)$ and $w(x,y)$, respectively. We get $W=\sqrt{2\pi}F^*V$, with the Fourier transform F. Note that $w \in L^2(\mathbb{R}^2)$, so that W is Schmidt. For two Schmidt operators U,V consider the inner product of their kernels

$$\mathrm{trace}(VU^*) = \int dx dy \bar{u}(x,y)v(x,y) . \tag{3.7}$$

It is known that (3.7) may serve as a generalization of the matrix trace, introducing a "trace" for products of Schmidt operators, hence the notation. With this notation (3.5) assumes the form

$$a(z,\zeta) = \mathrm{trace}((B_{z,\zeta}W)U^*) , \tag{3.8}$$

where we used (2.20) for the ψdo B=b(x,D) with symbol b of (3.1).

Next we notice that (2.20) implies

$$B = (1+\partial_z)^2(1+\partial_\zeta)^2 A_{z,\zeta}|_{z=\zeta=0} \;. \tag{3.9}$$

This shows that a(z,ζ) of (3.8) is well defined for a general operator $A \in \Psi GT$. In fact, for general $A \in \Psi GT$ we have

$$B_{z,\zeta}= (1+\partial_z)^2(1+\partial_\zeta)^2 A_{z,\zeta} \in C^\infty(\mathbb{R}^2, L(H)) \;. \tag{3.10}$$

The product map (A,B)→ AB for $A\in L(H)$, $B\in S(H)$, is continuous in A,B, in the norms of $L(H)$ and $S(H)$= Schmidt class: We have

$$[[[AB]]] \le \|A\|\cdot[[[B]]] \;,\; [[[BA]]]\le \|A\|\cdot[[[B]]] \;. \tag{3.11}$$

Limits may be taken inside the norm, hence a of (3.8) $\in CB^\infty(\mathbb{R}^2)$.

In other words, our left inverse S:A → a (with a defined by (3.8)) of the map O: $\psi t_0 \to \Psi GT$ (with Oa = Op(a)) is a well defined map $\Psi GT \to \psi t_0$. To show it to be an inverse of O we prove that SA=0 implies A=0. Indeed, then S is 1-1, and $S(OS-1)=SOS-S=(SO-1)S= 0$, hence $OS=1$ as well, so that S and O are inverses; S maps ΨGT onto ψt_0 and we have the full equivalence of the conditions of thm.2.2.

At this time we return to n dimensions: Define the constant coefficients PDE $P = p(\partial_z,\partial_\zeta)$ by

$$P = \Pi_{j=1}^n (1+\partial_{z_j})^2(1+\partial_{\zeta_j})^2 = p(\partial_z,\partial_\zeta) \tag{3.12}$$

For $A \in L(H)$ with $A_{z,\zeta} \in C^\infty(\mathbb{R}^{2n}, L(H))$ define $B_{z,\zeta}\in L(H)$ by

$$B_{z,\zeta} = PA_{z,\zeta} = p(\partial_z,\partial_\zeta)A_{z,\zeta} \tag{3.13}$$

With $B=B_{0,0}$ one finds that $B_{z,\zeta}=E^{-1}_{z,\zeta}BE_{z,\zeta}$,(i.e.(2.2) holds for B) We still have $B_{z,\zeta}\in C^\infty(\mathbb{R}^{2n}, L(H))$, by (3.13), so that $B\in \Psi GT$. Let

$$b(z,\zeta) = Pa(z,\zeta) = p(\partial_z,\partial_\zeta)a(z,\zeta) \;, \tag{3.14}$$

and note that we get (3.1) again in the form

$$a(z,\zeta) = \int dx d\xi q(-x,-\xi)e^{-ix\xi}\{b(z+x,\zeta+\xi)e^{ix\xi}\} \;, \tag{3.15}$$

$$\text{with} \qquad q(x,\xi)= \Pi_{j=1}^n\{\gamma_2(x_j)\gamma_2(\xi_j)\} \;.$$

All earlier arguments of sec.3 may be repeated: With the 1-dimensional (3.3),(3.4) define U_n, V_n on $L^2(\mathbb{R}^n)$ with kernels

$$u_n(x,\eta)= \Pi_{j=1}^n u(x_j,\eta_j) \;,\; v_n(\eta,\xi)= \Pi_{j=1}^n v(\eta_j,\xi_j) \in L^2(\mathbb{R}^{2n}) \;, \tag{3.16}$$

and an operator W_n with kernel

(3.17) $$w_n(x,\eta) = \Pi_{j=1}^n w(x_j,\eta_j) \in L^2(\mathbb{R}^{2n}) .$$

Again U_n, V_n, W_n are of Schmidt class. With an analogue of (3.7) as definition of trace we then get

(3.18) $$a(z,\zeta) = \mathrm{trace}((B_{z,\zeta}W_n)U_n^*) ,$$

valid for all $a \in \psi t_0$, and $A=\mathrm{Op}(a)$, with $B_{z,\zeta}$ of (3.13). The right hand side defines a map S: $\Psi GT \to \psi t_0$, a left inverse of O.

Now we prove that S is 1-1. Let $\omega(x,\xi) \in D(\mathbb{R}^{2n})$. Then we claim that a function $\varphi(x,\xi) \in D(\mathbb{R}^{2n})$ can be found such that

(3.19) $$\int dzd\zeta\, \varphi(z,\zeta)q(z-x,\zeta-\xi)e^{-ix\xi} = \omega(x,\xi) .$$

Indeed, using prop.2.3 or cor.2.4 we obtain φ in the form

(3.20) $$\varphi(x,\xi) = p(-\partial_x,-\partial_\xi)(\omega(x,\xi)e^{ix\xi}) .$$

Let the right hand side of (3.18) vanish identically, for some A. Using (3.7) we get (writing U for U_n again, etc.)

(3.21) $$\mathrm{trace}((B_{z,\zeta}W)U^*) = \frac{1}{\kappa_n}\int d\eta(u_\eta,B_{z,\zeta}F^*v_\eta) = 0 \text{ for all } z,\zeta ,$$

where we write the kernels $u(x,\eta)=u_\eta$, $v(x,\eta)=v_\eta$ as families of vec tors u_η, $v_\eta \in H$, depending on the parameter $\eta \in \mathbb{R}^n$, using the inner product $(.,.)$ of H. The n-dimensional (3.2) assumes the form

(3.22) $$e^{-ix\xi}q(-x,-\xi) = \int d\eta v_\eta(x)\bar{u}_\eta(\xi) ,$$

while (3.21) may be written as

(3.23) $$e^{iz\zeta}\int d\eta(E_{z,\zeta}u_\eta,BF^*E_{\zeta,-z}v_\eta) = 0 ,$$

using that

(3.24) $$E_{z,\zeta}F^* = e^{iz\zeta}F^*E_{\zeta,-z} .$$

Multiply (3.23) by φ of (3.20) and integrate: $\int d\eta$ may be pulled out; its integrand is $L^1(\mathbb{R}^{2n})$. Thinking of BF^* as an integral operator, our attention will be directed to the kernel

(3.25) $$k(x,\xi) = \int dzd\zeta d\eta\varphi(z,\zeta)e^{iz\zeta}(E_{z,\zeta}\bar{u}_\eta)(x)(E_{\zeta,-z}v_\eta)(\xi) .$$

Calculate k : Use the definition of $E_{z,\zeta}$, (3.22) and (3.19) for

$$k(x,\xi) = \int dzd\zeta\varphi(z,\zeta)e^{iz\zeta}e^{-i\zeta x}e^{-iz\xi}\int d\eta\bar{u}_\eta(x-z)v_\eta(\xi-\zeta)$$

$$(3.26) \qquad = \int dzd\zeta\varphi(z,\zeta)e^{iz\zeta}e^{-i\zeta x}e^{-iz\xi}e^{-i(x-z)(\xi-\zeta)}q(z-x,\zeta-\xi)$$

$$= \int dzd\zeta\varphi(z,\zeta)e^{-ix\xi}q(z-x,\zeta-\xi) = \omega(x,\xi) .$$

To use (3.26) we first assume B to be an operator of finite rank. Let $BF^* = \sum b^j\rangle\langle c^j$, b^j , $c^j \in H$. Also let $\omega(x,\xi)=\chi(x)\psi(\xi)$, with (real-valued) functions χ, $\psi \in D(\mathbb{R}^n)$. Then, by a calculation,

$$\int dzd\zeta d\eta\varphi(z,\zeta)(E_{z,\zeta}u_\eta, BF^*E_{\zeta,-z}v_\eta) = \int dxd\xi\omega(x,\xi)b_j(x)c^j(\xi)$$

$$(3.27) \qquad = (\chi, BF^*\psi) .$$

For general $B \in \Psi GT$ with (3.21) (or (3.23)) focus on $B_j=P_jBP_j$, P_j denoting the orthogonal projection onto span$\{\varphi_1,\ldots\varphi_j\}$, with some orthonormal base $\{\varphi_j: j=1,2,\ldots\}$ of H . Such B_j is of finite rank, so that (3.27) holds. The left hand side of (3.27), for B_j, converges to that with B, hence to 0, (by (3.23)): We get $B-B_j= Q_jB + P_jBQ_j$, $Q_j= (1 - P_j)$. Denote the factor of $\varphi(z,\zeta)$ at left in (3.27) by $X(B)$, for a moment. By Schwarz' inequality we get

$$(3.28) \qquad |X(P_jBQ_j)| \le [[[U]]]\cdot[[[(1-R_j)V]]] , \; R_j=E_{\zeta,-z}{}^*FP_jF^*E_{\zeta,-z} .$$

This expression goes to zero, as $j\to\infty$, for fixed z,ζ, and it is bounded in z,ζ. The first is true because, letting $\kappa_j(y)$ be the expansion coefficient of the kernel $v(x,y)$ in the orthogonal expan sion with respect to the base $\psi_j=E_{\zeta,-z}{}^*F\varphi_j$, we have $[[[V]]]^2 = = \sum_{j=1}^\infty\int|\kappa_j(y)|^2dy<\infty$, so that $[[[(1-R_j)V]]]^2 = \sum_{j=n}^\infty\int|\kappa_j(y)|^2dy \to 0$.

For the boundedness we use that $|X(P_jBQ_j)| \le [[[U]]]\cdot[[[V]]]$. Similarly for $X(Q_jB)$. By Lebesgues theorem the integral at left of (3.27) (with $B=B_j$) goes to zero. Hence

$$(3.29) \qquad (\chi,P_jBF^*P_j\psi) \to 0 \; , \; j \to \infty \; , \text{ for all } \chi,\psi \in D(\mathbb{R}^n) .$$

It follows that B=0. This gives A=0, in view of the result below.

<u>Proposition 3.1</u>. For an $X \in \Psi GT$, assume that either $(1+\partial_{z_j})X_{z,\zeta}=0$, or $(1+\partial_{\zeta_j})X_{z,\zeta}=0$, for some j, and $z=\zeta=0$. Then it follows that X=0.

<u>Proof</u>. These relations are true for all z,ζ, if they are true for

$z=\zeta=0$, by (2.2). Using that $X_{z,\zeta}\in C(\mathbb{R}^{2n},L(H))$ conclude that

$$(3.30)\qquad 0=\int_{-\infty}^{0}(1+\partial_t)X_{te^j,0}e^t dt = X+\int_{-\infty}^{0}X_{te^j,0}(1-\partial_t)e^t dt = X\ ,$$

by partial integration, from $(1+\partial_{z_j})X_{0,0}=0$. Similarly the other.

This proves prop.3.1, and completes the proof of thm.2.1.

4. Smooth operators; the ψ^*-algebra property; ψdo-calculus.

To summarize our accomplishment in sec's 1 to 3: Let us call an operator $A\in L(H)$ translation smooth if the family $A_{z,0}=T_z^{-1}AT_z$ of (1.5) is $C^\infty(\mathbb{R}_n,L(H))$, and gauge smooth if the other family $A_{0,\zeta}=M_\zeta^{-1}AM_\zeta$ of (1.5) is $C^\infty(\mathbb{R}^n,L(H))$. Then the essence of thm.1.1 may be reexpressed as follows.

Theorem 4.1. An bounded operator A on $H=L^2(\mathbb{R}^n)$ is both translation smooth and gauge smooth if and only if it is a pseudo-differential operator with symbol $a(x,\xi)\in CB^\infty(R^n)$, with $Au(x)$ expressible by (1.7) for u in the dense subspace $D(\mathbb{R}^n)$ of H. Here the symbol a of A is given by formula (3.8), for n=1 (or by (3.17) for general n) $W=(2\pi)^{n/2}FV$, U and V denoting the integral operators with kernel (3.4) and (3.3) (or (3.15)). The trace is defined by (3.7), and the (Green's) function $\gamma_j(t)$ by (2.9) .

Several comments are in order, to illuminate this result. First, it is trivial, that the class ΨGT of operators which are both, translation- and gauge- smooth, forms a subalgebra of $L(H)$, just using the common rules of differentiation. In fact, the algebra ΨGT is *-invariant - it trivially contains its adjoints, since $A^*_{z,0}=A_{z,0}{}^*$, $A^*_{0,\zeta}=A_{0,\zeta}{}^*$, and $\|B_j\|\to 0$ implies $\|B_j^*\|=\|B_j\|\to 0$.

The *-algebra ΨGT no longer is a C^*-subalgebra of $L(H)$, but it carries a Frechet topology induced by the natural norms of CB^∞:

$$(4.1)\quad \|A\|_k=\|A_{z,\zeta}\|_{C^k}=\sup\{\|\partial_z^\alpha\partial_\zeta^\beta A_{z,\zeta}\|:z,\zeta\in\mathbb{R}^n,|\alpha+\beta|\le k\},\ k=0,1,2,..$$

The norms (4.1) establish ΨGT as a Frechet space. The limit A of a sequence $A_j\in\Psi GT$ has $A_{z,\zeta}=\lim(A_j)_{z,\zeta}$ in the norms (4.1), implying $A_{z,\zeta}\in C^\infty(\mathbb{R}^{2n},L)$ so that $A\in\Psi GT$. Involution and algebra operations are continuous, for similar reason, looking at $C^\infty(\mathbb{R}^{2n},L)$. Therefore ΨGT , with this topology, is a Frechet-*-algebra.

Note that the same topology may be just as well described in terms of the symbols: For $a\in\psi t_0$ we introduce the norms

$$(4.2)\ \|a\|_k=\|a\|_{CB^k(\mathbb{R}^{2n})}=\sup\{|a^{(\alpha)}_{(\beta)}(x,\xi)|:x,\xi\in\mathbb{R}^{2n},|\alpha+\beta|\le k\},\ k=0,1,..$$

If $A = Op(a)$, then

(4.3) $\partial_z^\alpha\partial_\zeta^\beta A_{z,\zeta}=Op(a_{(\alpha)}^{(\beta)})_{z,\zeta}$, $a_{(\alpha)}^{(\beta)}(z,\zeta)=\mathrm{trace}(p(\partial_z,\partial_\zeta)\partial_z^\alpha\partial_\zeta^\beta A_{z,\zeta}WU^*)$,

by (2.20) and (3.8) . By (2.23) and the formulas of sec.3 we get:

<u>Proposition 4.2</u>. Let $A=Op(a)$. There exists a constant c depending on n only (expressible in a form similar to (2.24)) such that

(4.4) $\|a\|_k \le c\|PA_{z,\zeta}|_{z=\zeta=0}\|_k$, $\|A\|_k \le c\|Pa\|_k$, $P=\Pi_j((1+\partial_{z_j})(1+\partial_{\zeta_j}))^2$.

The topologies on $\Psi GT=Op(\psi t_0)$ of (4.1) and (4.2) are equivalent.

As another remarkable fact: If $A \in \Psi GT$ possesses an inverse in $L(H)$, (i.e. if there exists an $L^2(\mathbb{R}^n)$-bounded operator $B=A^{-1}$ with $ABu=BAu=u$ for all $u\in H$), then we have $A^{-1}\in \Psi GT$ as well: We get $A^{-1}{}_{z,0}=A_{z,0}{}^{-1}\in C^\infty$, and may calculate the derivatives in the usual way. For example, $\partial_{z_j}(A^{-1}{}_{z,0})=-A_{z,0}{}^{-1}(\partial_{z_j}A_{z,0})A_{z,0}{}^{-1}$, ...

We summarize:

<u>Theorem 4.3</u>. The class $\Psi GT = Op(\psi t_0)$ is a Frechet-*-algebra under either of the set of norms (4.1) or (4.2). Moreover, $\Psi GT \subset L(H)$ contains all its inverses (with respect to $L(H)$).

Algebras with the properties of thm.4.3 were investigated by B. Gramsch, who uses the notation 'ψ^*-algebra' for a *-subalgebra of $L(H)$ containing all its $L(H)$-inverses. In many respects, notably regarding the Fredholm property of its operators, a ψ^*-algebra behaves like a C^*-algebra.

We notice that the resolvent $R(\lambda)= (A-\lambda)^{-1}$ (in the sense of $H=L^2(\mathbb{R}^n)$) is a well defined operator of ΨGT as well, for every point λ of the resolvent set Res(A) , an open subset of $\mathbb{C}$, assuming that $A \in \Psi GT$. In fact, the complex derivative $dR(\lambda)/d\lambda = -R(\lambda)^2$ belongs to ΨGT as well. Moreover, for the difference quotient $Q = \frac{1}{h}(R(\lambda+h)-R(\lambda))$ we get $Q_{z,\zeta} = \frac{1}{h}\{(A_{z,\zeta}-\lambda-h)^{-1}-(A_{z,\zeta}-\lambda)^{-1}\}$. One finds easily that $\lim_{h\to 0}\partial_z^\alpha\partial_\zeta^\beta Q_{z,\zeta}$ exists, for each z,ζ, uniformly as $z,\zeta \in \mathbb{R}^n$. Accordingly, the resolvent is complex differentiable within the Frechet-*-algebra ΨGT, defined on Res(A).

Commonly, for a (complex-valued) holomorphic function $\varphi(\lambda)$, holomorphic in a connected neighbourhood N of the spectrum Sp(A) of A, and a simple closed contour Γ containing Sp(A) in its interior, one defines $\varphi(A)$ by the '<u>Dunford</u> <u>integral</u>'

(4.5) $$\varphi(A) = \frac{i}{2\pi}\int_\Gamma (A-\lambda)^{-1}d\lambda ,$$

existing in ΨGT since its integrand is continuous. We have proven:

Proposition 4.4. For $A \in \Psi GT$ and a function φ with above properties we have $\varphi(A) \in \Psi GT$. The symbol $r_\lambda(x,\xi)$ of the resolvent $R(\lambda) \in \Psi GT$ is holomorphic in Res(A), as a function of λ taking values in ψt_0 with topology of (4.2). The symbol of $\varphi(A)$ is

$$\text{symb}(\varphi(A))(x,\xi) = \frac{i}{2\pi} \int_\Gamma r_\lambda(x,\xi) d\lambda \ . \tag{4.6}$$

Extensions of prop.4.4 are possible, applying to the case where the contour Γ touches the spectrum in a single point, or Γ may run through ∞ on the extended plane, or that $A \in \text{Op}\psi t_m$, $m \neq 0$. Also, the same questions will arise for other symbols.

The symbol a of an operator $A \in \Psi GT$ plays a role similar to an integral kernel. In fact, we may write (1.7) as

$$Au(x) = \kappa_n \int d\xi a(x,\xi) e^{ix\xi} u^\wedge(\xi) \ , \ u^\wedge = Fu \ , \tag{4.7}$$

showing that AF^* has integral kernel $a(x,\xi)e^{ix\xi} = e^{ix\xi}\text{symb}(A)$.

The question arises for a composition formula, linking the symbols a,b of A, $B \in \Psi GT$ with the symbol c of C=AB, similar as well known for integral kernels. Such a formula, involving a 'very singular' integral, called finite part, was discussed in I,4:

$$c(x,\xi) = \kappa_n^2 \int a(x,\xi-\eta) b(x-y,\xi) e^{-iy\eta} dy d\eta \ . \tag{4.8}$$

The integrand clearly is not $L^1(\mathbb{R}^{2n})$, and the integral was defined by a special proceedure (cf.I,(4.19)). Similarly, a formula linking a=symb(A) with $a^\sim$=symb(A^*): We have (cf.I,(4.6))

$$a^\sim(x,\xi) = \kappa_n^2 \int \bar{a}(x-y,\xi-\eta) e^{-iy\eta} dy d\eta \ , \tag{4.9}$$

where again the integral is a finite part.

Different formulas are known for differential operators

$$A = a(x,D) = \sum_{|\alpha| \le N} a_\alpha(x) D_x^\alpha \ , \ a_\alpha \in C^\infty(\mathbb{R}^n) \ . \tag{4.10}$$

We know that A of (4.1) may be written in the form (1.7) with

$$a(x,\xi) = \sum_{|\alpha| \le N} a_\alpha(x) \xi^\alpha \ , \tag{4.11}$$

a polynomial in ξ with coefficients depending on x . Indeed,

$$(D_{x_j} u)^\wedge(\xi) = \xi_j u^\wedge(\xi) \ , \ (D_x^\alpha u)^\wedge(\xi) = \xi^\alpha u^\wedge(\xi) \ , \tag{4.12}$$

whence

$$Au(x) = a(x,D)u(x) = \kappa_n^2 \int d\xi \int dy e^{i\xi(x-y)} a(x,\xi) u(y) \ . \tag{4.13}$$

A differential operator A of (4.10) never belongs to ΨGT, except, perhaps, if $a_\alpha=0$ for $\alpha\neq 0$. For differential operators $A=a(x,D)$, $B=b(x,D)$ with symbols $a(x,\xi)$, $b(x,\xi)$ we have composition and adjoint formulas given by the Leibniz' formula of differentiation, expressed as follows : $AB= C= c(x,D)$ and $A^*=a^\sim(x,D)$ again are differential operators with symbols given by

$$(4.14)\quad c = \sum_\gamma \frac{(-i)^{|\gamma|}}{\gamma!} a^{(\gamma)}(x,\xi)b_{(\gamma)}(x,\xi) \ , \ a^\sim = \sum_\gamma \frac{(-i)^{|\gamma|}}{\gamma!} a^{(\gamma)}_{(\gamma)}(x,\xi) \ ,$$

where the sums at right are finite, since a is a polynomial in ξ.

We have seen in I,5, I,6 that (4.14) generalizes to ψdo's of rather general symbol classes, giving Taylor-Leibniz-type composition formulas, involving as sum like (4.14) but also a remainder term. This was seen in I,5 for operators with symbol in ψt, and even in ST_1. However, a Taylor-Leibniz formula proved to be useful only if the remainders - given in form of finite part integrals - become small, and an asymptotic expansion results. In that respect the symbol class ψt proved to be impractical: we always were working with $\psi h_{m,\rho,\delta}$, where both $\rho_j>0$, and $0\le\delta<\rho_1$.

A <u>calculus of ψdo's</u> will be possible again for operators $A\in L(H)$ which also are <u>rotation smooth</u> and <u>dilation smooth</u>, in addition to the translation and gauge smoothness already discussed.

Here we refer to the groups O_n of rotations $o:\mathbb{R}^n\to\mathbb{R}^n$ and $\mathbb{R}^+$ of dilations $\sigma:\mathbb{R}^n\to\mathbb{R}^n$, with an orthogonal n×n-matrix o (of determinant 1) and a $\sigma>0$, $\sigma\in\mathbb{R}$. Define the groups of unitary operators

$$(4.15)\qquad O_o u(x) = u(ox) \ , \ o\in O_n \ , \ S_\sigma u(x) = \sqrt{\sigma}u(\sigma x) \ , \ \sigma\in\mathbb{R}^+ \ .$$

An operator $A\in L(H)$ is called <u>rotation smooth</u> (<u>dilation smooth</u>) if

$$(4.16)\qquad O_o^* A O_o \in C^\infty(O_n, L(H)) \ , \ S_\sigma^* A S_\sigma \in C^\infty(\mathbb{R}^+, L(H)) \ ,$$

respectively. We denote the class of operators $A\in L(H)$ with all 4 above smoothness properties by ΨGS , and will investigate this class in sec.5 below. Clearly $\Psi GS \subset \Psi GT= Op(\psi t_0)$, so that ΨGS consists of pseudodifferential operators with symbol having all derivatives bounded. We will see that the symbols of ΨGS have derivatives <u>decaying at ∞ , and stronger so, as the order of differentiation increases</u>.

Instead of requiring dilation and rotation smoothness, we even might require smoothness of $A\in L(H)$ with respect to the class of all maps $u(x)\to u(gx)$, where $g=((g_{jk}))$ is a real invertible n×n-matrix (in addition to translation and gauge smoothness). This defines yet another class $\Psi GL\subset \Psi GT\subset \Psi GT$ of ψdo's, studied in sec.5.

5. The operator classes ΨGS and ΨGL , and their symbols.

As already observed, the classes $\Psi GS \supset \Psi GL$ of sec.4 are contained in ΨGT, thus consist of ψdo's (1.7) with symbol in $CB^{\infty}(\mathbb{R}^{2n})$ The additional smoothness of $A_{z,\zeta}$ will lead to stronger conditions for symb(a), to be worked out next.

The maps $u(x)\rightarrow u(gx)$, for a general matrix $g\in GL(\mathbb{R}^n)$, together with the translations $u(x)\rightarrow u(x-z)$ and gauge transforms $u(x)\rightarrow e^{iz\zeta}$ of sec.1 generate a larger Lie-subgroup of $U(H)$, we will write as

$$(5.1) \qquad GL = \{T_{g,z,\zeta,\varphi} : (g,z,\zeta,\varphi) \in gl\}$$

where

$$(5.2) \qquad T_{g,z,\zeta,\varphi}u(x) = (\det g)^{1/2}\, e^{i(\zeta x+\varphi)}u(gx+z) \ , \ u \in H \ ,$$

$$gl = \{(g,z,\zeta,\varphi): g \in GL(\mathbb{R}^n) \ , \ z,\zeta\in \mathbb{R}^n \ , \ \varphi\in \mathbb{S}^1\} \ ,$$

The group operation in GL (or gl) is best decribed by introducing the linear maps g: $\mathbb{R}^n\rightarrow\mathbb{R}^n$ and λ: $\mathbb{R}^n\rightarrow\mathbb{R}$, for a given $(g,z,\zeta,\varphi)\in gl \subset \mathbb{R}^{n^2+2n+1}$ by setting $g(x)=gx+z$, $\lambda(x)=\zeta x+\varphi$. Write

$$(5.3) \qquad T_{g,z,\zeta,\varphi}u(x) = T(g,\lambda)u(x) = (\det g)^{1/2}e^{i\lambda(x)}(u\circ g)(x) \ .$$

We get

$$(5.4) \quad T(g,\lambda)T(g',\lambda') = T((g,\lambda)\circ(g',\lambda')) \ , \ T((g,1)^{-1}) = T(g,\lambda)^{-1} \ ,$$

where $(g,\lambda)=(g,z,\zeta,\varphi)\in gl$, $(g',\lambda')=(g',z',\zeta',\varphi')\in gl$, and

$$(5.5) \quad (g,\lambda)\circ(g',\lambda') = (g'\circ g,\lambda+\lambda'\circ g) \ , \ (g,\lambda)^{-1} = (g^{-1},-\lambda\circ g^{-1}) \ .$$

For a pair of maps $g(x)=gx+z$, $\lambda(x)=\zeta x+\varphi$ intoduce the matrix

$$(5.6) \qquad M = M_{g,\lambda} = \begin{pmatrix} 1, \zeta^T, \varphi \\ 0, g, z \\ 0, 0, 1 \end{pmatrix} \ ,$$

The group operations (5.5) correspond to matrix multiplication and inversion -i.e., $(g,\lambda)\circ(g',\lambda') \leftrightarrow M'M$, $(g,\lambda)^{-1} \leftrightarrow M^{-1}$, by a calculation. Thus $(g,\lambda) \leftrightarrow M$ defines an isomorphism between GL and the Lie-subgroup of the linear group $GL(\mathbb{R}^{n+2})$ consisting of all matrices (5.6). Actually we must calculate mod 2π in the variable φ , i.e., work with the quotient of the matrix group modulo the subgroup of all $(g,z,\zeta,\varphi)=(1,0,0,2k\pi)$, to obtain an isomorphism.

Speaking in terms of Lie-groups we find that (5.6) describes a representation of the Lie-group ***gl*** in $GL(\mathbb{R}^{n+2})$.

We introduce groups ***GS*** and ***gs*** from the unitary maps used for definition of ΨGS : $GS\subset GT$ consists of all $T_{\sigma o,z,\zeta,\varphi}$, with a rotation o and $0<\sigma\in\mathbb{R}$. Define $gs=\{(\sigma,\omega,z,\zeta,\varphi):(\sigma o,z,\zeta,\varphi)\in gl\}$, where $\sigma>0$, $o^T o=1$, det o =1 . A matrix representation of ***gs*** is given by (5.6) again, since ***gs*** is identified with a subgroup of ***gl*** .

We thus have introduced a chain $GT=GH\subset GS\subset GL\subset U(H)$ of subgroups of the unitary group $U(H)$, where ***GX*** , ***X=T,H,S,L***, all are unitary representations of Lie-groups ***gt=gh, gs, gl*** . Note that ***gx, x=h,t,s,l***, are connected. For each ***gx*** we also have the finite matrix representation (5.6). Clearly, the relation between ***gx*** and its corresponding Lie-algebra (called ***ax***) is described by the discussion of ch.0,9, using the representation (5.6). It is suggested that the representation ***GX*** of ***gx*** on the infinite dimensional space ***H*** is governed by similar principles.

Such principles stand behind the discussion of ΨGL and ΨGS, below. - This is why we decided to present sec.9 of ch.0. However, this central part of Lie-group theory will not be needed, below.

To characterize the symbol classes ψl_0 and ψs_0 belonging to ΨGL and ΨGS we assume that $A = a(x,D)$ is given with $a \in CB^\infty(\mathbb{R}^{2n})$, and ask for the symbol (if any) of the operator

$$(5.7)\qquad Z(g) = Z(g,A) = R_g^{-1}AR_g \ ,\ R_g u(x) = u(gx) \ .$$

Note that the function $u_g(x) = u(gx) \in \mathcal{S}$ has the Fourier transform

$$(5.8)\qquad u_g^\wedge(x) = (\det g^-)u(g^{-t}) = (\det g^-)u_{g^{-t}}(x) \ ,$$

using the abbreviation $g^- = g^{-1}$, $g^{-t} = (g^{-1})^t$. Accordingly,

$$(5.9)\qquad Au_g(x) = \kappa_n(\det g^-)\int d\xi e^{ix\xi}a(x,\xi)u^\wedge(g^{-t}\xi) \ .$$

Another transformation of integration variable yields

$$(5.10)\qquad Z(g)u(x) = Au_g(g^- x) = (\det g^-)\kappa_n\int e^{ixg^{-t}\xi}a(g^- x,\xi)u^\wedge(g^{-t}\xi)d\xi$$

With another substitution of integration variable we get:

<u>Proposition 5.1</u>. For $A = a(x,D) \in \Psi GT$ the operator $Z(g,A)=R_g^{-1}AR_g$ is in ΨGT again. We have

$$(5.11)\qquad \mathrm{symb}(Z(g,A))(x,\xi) = a(g^{-1}x,g^t\xi) \ .$$

Note that the symbol (5.11) indeed belongs to $\psi t_0=CB^\infty$.

Now we can evaluate <u>linear</u> <u>group</u> <u>smoothness</u>; it amounts to

$$(5.12)\qquad Z(g)\in C^{\infty}(GL(\mathbb{R}^n),L(H)) ,$$

with derivatives existing in norm convergence. Clearly (5.12) implies the existence of the partial derivatives

$$(5.13)\qquad (\partial_{g_{11}})^{\alpha_{11}}\ldots(\partial_{g_{nn}})^{\alpha_{nn}} Z(g_{11},\ldots g_{nn}) \text{ at } g_{jl}=\delta_{jl} ,$$

for all integers $\alpha_{jl}\geq 0$.

Formally we may execute the differentiations of (5.12) by just differentiating the symbol. For example, using that

$$(5.14)\qquad \partial_{g_{pq}}(g^{-1}) = -g^{-1}h^{pq}g^{-1} ,\ h^{pq}= ((\delta_{pi}\delta_{qj}))_{i,j=1,\ldots,n} ,$$

hence

$$(5.15)\qquad \begin{aligned} \partial_{g_{jl}}(g^{-1}x)_m|_{g=1} &= -(h_{jl}x)_m = -\delta_{jm}x_l \\ \partial_{g_{jl}}(g^t\xi)_m|_{g=1} &= (h_{lj}\xi)_m = \delta_{lm}\xi_j \end{aligned} ,$$

we get

$$(5.16)\qquad \begin{aligned} &\partial_{g_{jl}}a(g^{-1}x,g^t\xi)|_{g=1} = \\ &= \Sigma_{m=1}^n(-a_{|x_m}(x,\xi)\delta_{jm}x_l + a_{|\xi_m}(x,\xi)\delta_{lm}\xi_j) \\ &=(\xi_j a_{|\xi_l} - x_l a_{|x_j})(x,\xi) . \end{aligned}$$

Our attention is directed to the folpde's

$$(5.17)\qquad \varepsilon_{jl} = \xi_j\partial_{\xi_l} - x_l\partial_{x_j} ,\quad j,l=1,\ldots,n ,$$

since (5.16) formally implies

$$(5.18)\qquad \partial_{g_{jl}}Z(1) = (\varepsilon_{jl}a)(x,D) .$$

<u>Proposition 5.2</u>. For $A=a(x,D)\in\Psi GT$, $u\in S$ and $x\in\mathbb{R}^n$ we have

$$(5.19)\qquad \partial_{g_{jk}}Z(g,A)u(x) = \Sigma_{p=1}^n g^-_{pj}(\varepsilon_{pk}a)(g^{-1}x,g^t\xi)u(x)$$

where we again abbreviate $g^{-1}=((g^-_{pq}))$.

Note that (1.7) may be differentiated under the integral sign for whatever parameter the symbol a might depend on as long as $u\in S$ (hence $u^\wedge\in S$) and the derivatives of the symbol a (for the parameter) are of polynomial growth in (x,ξ), uniformly in the parameter. We will omit the function $u\in S$ in (5.19), derive a formal relation for the operators, to be checked for polynomial growth.

We have $R_{hg}u(x) = R_g u(hx) = R_g R_h u(x)$, so that

$$(5.20)\qquad Z(hg,A) = R_{h^{-1}g^{-1}} A R_g R_h = Z(h,B) \ , \text{ where } B = Z(g,A) \ .$$

With D_k=derivative in the direction of (the n×n-matrix) k we get

$$(5.21)\qquad \begin{aligned} D_{hg}Z(g,A) &= \{\tfrac{1}{\varepsilon}\{Z(g+\varepsilon hg)-Z(g)\}\}|_{\varepsilon\to 0} \\ &= \{\tfrac{1}{\varepsilon}\{Z((1+\varepsilon h)g,A)-Z(g,A)\}\}|_{\varepsilon\to 0} = D_h Z(1,B) \ . \end{aligned}$$

Setting $h=h^{jk}=((\delta_{jp}\delta_{1q}))_{p,q=1,\dots,n}$ get $D_h Z(1,B)=\partial_{g_{j1}} Z(1,B) =$ $(\varepsilon_{j1}b)(x,D)$, $b(x,\xi)=a(g^{-1}x,g^t\xi)$, by (5.18). The left hand side is

$$\{\tfrac{1}{\varepsilon}\{Z(g+\varepsilon t)-Z(g)\}\}|_{\varepsilon\to 0} = t_{pq}\partial_{g_{pq}} Z(g,A) \ , \ t_{pq}=(hg)_{pq}=\delta_{jp}g_{1q} \ .$$

Together we have (summing over indices occurring twice)

$$(5.22)\qquad \delta_{jp}g_{1q}\partial_{g_{pq}} Z(g,A) = g_{1q}\partial_{g_{jq}} Z(g,A) = (\varepsilon_{j1}b)(x,D)$$

where

$$\begin{aligned} (\varepsilon_{j1}b)(x,\xi) &= (\xi_j\partial_{\xi_1} - x_1\partial_{x_j})a(g^-_{pq}x_q, g_{qp}\xi_q) \\ &= g^-_{rj}(g_{sr}\xi_s)a_{|\xi_p}(g^{-1}x,g^t\xi) - g_{1r}(g^-_{rs}x_s)a_{|x_p}(g^{-1}x,g^t\xi)g^-_{pj} \\ &= \{g^-_{rj}g_{1p}\eta_r\partial_{\eta_p}a(y,\eta) - g_{1r}g^-_{pj}y_r\partial_{y_p}a(y,\eta)\}|_{y=g^{-1}x,\eta=g^t\xi} \end{aligned}$$

Summarizing we get

$$(5.23)\qquad (\varepsilon_{j1}b)(x,\xi) = g^-_{rj}g_{1p}(\varepsilon_{rp}a)(g^{-1}x,g^t\xi) \ .$$

Substituting (5.23) into (5.22) the formal relation (5.19) follows

Going again through the arguments, starting at (5.21), find the polynomial growth condition satisfied. Thus the formal argument without u(x) leads to a rigorous proof of (5.19).

Notice that (5.19) may be written in the form

$$(5.24)\qquad \partial_{g_{jk}} Z(g,A) = \textstyle\sum_{p=1}^n g^-_{pj} Z(g,(\varepsilon_{pk}a)(x,D)) \ ,$$

as a consequence of (5.11). This shows that (5.19) (or (5.24)) may be iterated to obtain all partial derivatives (5.13) of arbitrary order. These all exist as linear operators $\mathcal{S}\to\mathcal{S}$, as $A\in \Psi GT$. Moreover, derivatives (5.13) are linear combinations of b(x,D) with

$$(5.25)\qquad b = (\varepsilon_{j_1 l_1}\varepsilon_{j_2 l_2}\cdots)a \quad ,$$

finite application of the ε_{jl} to a, with coefficients rational in the coefficients of g, with no further assumptions.

<u>Theorem 5.3</u>. Let the symbol class ψl_0 be defined as the set of all $a \in CB^\infty(\mathbb{R}^{2n})$ such that all finite applications (5.25) also are contained in $CB^\infty(\mathbb{R}^{2n})$. Then we have $\Psi GL = Op\psi l_0$.

For ΨGS we state a similar result, using the folpde's

$$(5.26)\qquad \eta_{jl}=\varepsilon_{jl}-\varepsilon_{lj}=(\xi_j\partial_{\xi_l}-\xi_l\partial_{\xi_j})+(x_j\partial_{x_l}-x_l\partial_{x_j}),\ j,l=1,\dots,n,$$

$$\eta_{00}=\sum_{p=1}^n\varepsilon_{pp}=\sum_{p=1}^n\xi_p\partial_{\xi_p}-\sum_{p=1}^n x_p\partial_{x_p}.$$

<u>Theorem 5.4</u>. Let the symbol class ψs_0 be defined as the set of all $a \in CB^\infty(\mathbb{R}^{2n})$ such that all finite applications of η_{jl} to a are in $CB^\infty(\mathbb{R}^{2n})$ again. Then we have $\Psi GS = Op\psi s_0$.

Before discussing the proofs we note:

<u>Proposition 5.5</u>. Introduce (in addition to (5.17),(5.26))

$$(5.27)\qquad \varepsilon_{p0}=\eta_{p0}=\partial_{x_p},\ \varepsilon_{0p}=\eta_{0p}=\partial_{\xi_p},\ p=1,\dots,n.$$

Then ψl_0 and ψs_0 consist precisely of all $a\in C^\infty(\mathbb{R}^{2n})$ allowing arbitrary finite application of ε_{jl} (for ψl_0) and η_{jl} (for ψs_0), as bounded functions on $\mathbb{R}^{2n}$, for $j,l=0,\dots,n$ with ε_{jl} or η_{jl} defined.

For the proof of prop.5.5 we note the commutator relations

$$(5.28)\qquad [\varepsilon_{p0},\varepsilon_{jl}]=-\delta_{pl}\varepsilon_{j0},\ [\varepsilon_{0p},\varepsilon_{jl}]=\delta_{pj}\varepsilon_{0l},\ p,j,l=1,\dots,n,$$

verified by a calculation. (5.28) shows that, a product (5.25) may be written as a combination of products with ε_{0p} and ε_{p0} pulled out to the right. Thus the condition of thm.5.3 implies that of prop.5.5, while the reverse is evident. Similarly for thm.5.4.

<u>Proof of thm.5.3</u>. It is clear from remarks around (5.24) that all derivatives (5.13) exist in convergence of $\mathcal{S}$ if only $A=a(x,D)\in \Psi GT$ (i.e., $a\in\psi t_0$). Moreover, the derivatives are linear combinations of b(x,D) with b of the form (5.25), as noted. Thus they belong to $L(H)$ - even to ΨGT - if $a\in\psi s_0$, with a uniform bound on compact regions of $GL(\mathbb{R}^n)$. This implies that $A=a(x,D)$, $a\in\psi s_0$, must be $GL(\mathbb{R}^n)$-smooth. For example we get

$$(5.29)\qquad Z(g(t))u-Z(g(0))u=\int_0^t d\tau\left(\frac{d}{d\tau}Z(g(\tau))\right)u,\ u\in\mathcal{S},$$

for a smooth curve g(t). Here the integrand $I(\tau)$ satisfies $\|I(t)\|\le c\|u\|$, with L^2-norms, and c independent of u and t. First conclude from (5.29) that Z(g) is continuous in g. Similarly all derivatives (5.13) are continuous as maps $GL(\mathbb{R}^n)\to L(H)$. Knowing this we omit $u\in S$ in (5.29) and divide by t. As $t\to 0$ get existence of the partials in norm convergence. Iterating we confirm $GL(\mathbb{R}^n)$-smoothness.

Vice versa, if $A \in L(H)$ is GL-smooth, in addition to translation and gauge smoothness, we get $A=a(x,D)$ with $a\in \psi t_0$. As in sec.1 one shows that $T(\gamma,\lambda)^{-1}AT(g,\lambda)=A(g,\lambda)$ has continuous mixed partials for z,ζ,g. All derivatives (5.13) are translation and gauge smooth, hence belong to $Op\psi t_0$. Hence $a=\text{symb}\, A \in \psi s_0$, q.e.d.

The proof of thm.5.4 is analogous to the above. For dilation and rotation smoothness focus on the derivatives along the curves

$$(5.30)\qquad g(t) = te^s \ ,\ 0 < t < \infty\ ,\ s=((s_{jk})) = -((s_{kj}))\ ,$$

with the matrix exponential function $e^s=1+s+s^2/2!+\ldots$. It is known that a rotation o may be written in the form $o=e^s$, with skew-symmetric s . Thus, the derivatives (5.13) must be replaced by

$$(5.31)\qquad ((\partial_t)^{\alpha_{00}}(\partial_{s_{12}})^{\alpha_{12}}\ldots)Z(te^s)\ ,\ \text{at } t=1\ ,\ s=0\ ,\ s^T=-s\ ,$$

where only derivatives for s_{jl} with $j<l$ occur. For $g=te^s$ we get

$$(5.32)\qquad \begin{aligned} t\partial_t Z(te^s) &= \Sigma_{jl} g_{jl}\partial_{g_{jl}} Z(g) = \Sigma_{jlp} g_{jl} g^-_{pj}(\varepsilon_{pl}a)(g^-x,g^tD) \\ &= \Sigma_p(\varepsilon_{pp}a)(g^-x,g^tD) = (\eta_{00}a)(g^-x,g^tD) = Z(te^s,(\eta_{00}a)(x,D))\ , \end{aligned}$$

using (5.19). Similarly, for $j < l$,

$$(5.33)\qquad \partial_{s_{jl}} Z(e^s) = \Sigma_{pq}(\partial_{s_{jl}}e^s)_{pq}\partial_{g_{pq}}(Z(g))\ .$$

For s=0 we get

$$(5.34)\qquad \partial_{s_{jl}} e^s|_{s=0} = h^{jl}- h^{lj}\ ,\ h^{pq} = ((\delta_{\mu p}\delta_{\nu q}))_{\mu,\nu=1,\ldots,n}\ .$$

Accordingly, using (5.18),

$$(5.35)\qquad \partial_{s_{jl}} Z(e^s)|_{s=0} = (\varepsilon_{jl}-\varepsilon_{lj})a(x,D) = (\eta_{jl}a)(x,D)\ .$$

For an analogue to (5.19) we repeat the discussion leading to (5.23) as follows: A modification of (5.21) yields

$$(5.36)\qquad \partial_{s_{jl}} Z(e^s o,A)|_{s=0} = \partial_{s_{jl}} Z(e^s,B)|_{s=0} = (\eta_{jl}b)(x,D)\ ,$$

with $B = b(x,D) = a(o^tx,o^tD)$. Using (5.23) we get

$$(5.37)\qquad (\eta_{jl}b)(x,\xi) = (o_{jr}o_{lp}(\varepsilon_{rp}a) - o_{lr}o_{jp}(\varepsilon_{rp}a))(o^tx,o^t\xi)$$

$$= o_{jr} o_{lp} (\eta_{rp} a)(o^t x, o^t \xi) .$$

Accordingly, we have the following substitute for (5.24):

$$(5.38) \quad \partial_{s_{jl}} Z(e^s o, A)|_{s=0} = \Sigma_{r,p=1}^n o_{jr} o_{lp} Z(o, A_{rp}) , \; A_{rp} = \eta_{rp} a(x,D) .$$

We need the Campbell-Hausdorff formula of ch.0,9, quickly re-derived as follows here: For $s \in N(s_0)$, a neighbourhood of s_0 write

$$(5.39) \quad e^s = e^{\lambda(s,s_0)} e^{s_0} , \; \lambda = \log(e^s e^{-s_0}) = -\Sigma_{m=1}^{\infty} \tfrac{1}{m}(1-e^s e^{-s_0})^m .$$

Clearly $\lambda(s,s_0)$, is well defined (and holomorphic in the coefficients s_{jl} , $s_{0\,jl}$), as long as $|1-e^s e^{s_0}| < 1$. Since $e^{\lambda} = e^s e^{s_0}$ is orthogonal we get $e^{\lambda^t} = e^{-\lambda}$, hence $\lambda^t = \log(e^{-\lambda}) = -\lambda$, so that λ is skew symmetric (and real-valued, for real-valued s, s_0) . For fixed s_0 the function $\lambda(s,s_0)$ has a local inverse at s_0 . We get

$$(5.40) \quad \partial_{s_{jl}} Z(e^s)|_{s=s_0} = \Sigma_{\mu<\nu} (\partial\lambda_{\mu\nu}/\partial s_{jl})(s_0, s_0) \partial_{\lambda_{\mu\nu}} Z(e^{\lambda} e^{s_0})|_{\lambda=0} .$$

Or, with holomorphic functions $\kappa_{jl\mu\nu}(s)$, $j<l$, $\mu<\nu$,

$$(5.41) \quad \partial_{s_{jl}} Z(e^s) = \Sigma_{\mu<\nu} \kappa_{jl\mu\nu}(s) \Sigma_{r,p} o_{\mu r} o_{\nu p} Z(e^s, (\eta_{pq} a)(x,D)) .$$

Note that (5.41) is the desired substitute for (5.24); It may be iterated, showing that derivatives of arbitrary order

$$(5.42) \quad ((\partial_{s_{12}})^{\alpha_{12}} \ldots .) Z(e^s)$$

may be expressed as linear combinations of

$$(5.43) \quad Z(e^s, (\eta_{j_1 l_1} \eta_{j_2 l_2} \ldots a)(x,D))$$

with coefficients holomorphic in s_{jl}. The proof of thm.5.3. now may be repeated with η_{jl} instead of ε_{jl}, for a proof of thm.5.5 (Note that (5.32) way be iterated as well). Q.E.D.

6. The Frechet algebras ψx_0, and the Weinstein-Zelditch class.

We already stated that the symbol classes ψs_0 and ψl_0 consist of locally classical symbols. More precisely we have the following

Proposition 6.1. For any open set $Q \subset\subset \mathbb{R}^n$ the sets $\psi x_0 | Q \times \mathbb{R}^n$, $x = s, l$, of restrictions of symbols $a \in \psi x_0$ to $Q \times \mathbb{R}^n = \{x \in Q, \xi \in \mathbb{R}^n\}$ are subsets of the Hoermander class $S_{0,1,0}(Q)$, (cf. [Hr₂]). That is, we have

$$(6.1) \quad a_{(\beta)}^{(\alpha)} = O(\langle\xi\rangle^{-|\alpha|}) , \text{ as } x \in Q^{\sim} \subset\subset Q .$$

Proof. It suffices to look at ψs_0, since $\psi l_0 \subset \psi s_0$. Then we get

$$(6.2)\quad \sum_j \xi_j \partial_{\xi_j} a = \eta_{00} a + O(1),\ (\xi_j \partial_{\xi_l} - \xi_l \partial_{\xi_j}) a = \eta_{jl} a + O(1),\ x \in Q,\ a \in \psi s_0 .$$

Since $\eta_{jl} a = O(1)$ as well, we get

$$(6.3)\quad \lambda_{00} a = \sum_j \xi_j \partial_{\xi_j} a = O(1)\ ,\ \lambda_{jl} a = (\xi_j \partial_{\xi_l} - \xi_l \partial_{\xi_j}) a = O(1)\ .$$

Multiplying (6.2) by ξ_l and ξ_j and adding we get

$$(6.4)\quad |\xi|^2 a_{|\xi_j} = \xi_j \lambda_{00} a - \sum_l \xi_l \lambda_{jl} a = O(|\xi|)\ ,$$

or,

$$(6.5)\quad (1+|\xi|) a_{|\xi_j} = O(1)\ ,\ \text{i.e.},\ a_{|\xi_j} = O(\langle\xi\rangle^{-1})\ .$$

This chain of arguments may be iterated, to prove

$$(6.6)\quad a^{(\alpha)}(x,\xi) = O(\langle\xi\rangle^{-|\alpha|})\ ,\ x \in Q\ .$$

Starting with $a_{(\beta)}$ instead of a we get (6.1), q.e.d.

Let us work out a global result, with the same argument.

Proposition 6.2. For $a \in C^\infty(\mathbb{R}^{2n})$ let

$$(6.7)\quad a_{00} = \eta_{00} a\ ,\ a_{p0} = \eta_{p0} a = a_{|x_p}\ ,\ a_{0q} = \eta_{0q} a = a_{|\xi_q}\ ,$$

$$a_{pq} = \eta_{pq} a\ ,\ p,q = 1,\dots,n\ .$$

Then there exist symbols $\gamma^{1j}_{pq} \in \psi c_{e^1-e^2}$, $\gamma^{2j}_{pq} \in \psi c_{e^2-e^1}$ such that

$$(6.8)\quad a_{|x_j} = \sum_{p,q=0}^n \gamma^{1j}_{pq} a_{pq}\ ,\ a_{|\xi_j} = \sum_{p,q=0}^n \gamma^{2j}_{pq} a_{pq}\ .$$

Here we recall the symbol class ψc_m , $m = (m_1, m_2)$, defined by

$$(6.9)\quad \psi c_m = \{a \in C^\infty(\mathbb{R}^{2n}) : a^{(\alpha)}_{(\beta)}(x,\xi) = O(\langle\xi\rangle^{m_1-|\alpha|}\langle x\rangle^{m_2-|\beta|})\ \}$$

For symmetry reason it suffices to discuss only one of the identities (6.8). For the second one write (6.2), more explicitly, as

$$(6.10)\quad \sum_p \xi_p \partial_{x_p} a = a_{00} + \sum_p x_p a_{p0}\ ,\ (\xi_p \eta_{\xi_q} - \xi_q \eta_{\xi_p}) a = a_{pq} + (x_p a_{q0} - x_q a_{p0})\ .$$

Then (6.3) may be replaced by

$$(6.11)\quad (1+|\xi|^2) a_{|\xi_j} = a_{0j} + \xi_j (a_{00} + \sum_p x_p a_{p0}) + \sum_p \xi_p (a_{pj} + x_p a_{j0} - x_j a_{p0})\ ,$$

by the same derivation. Dividing (6.11) by $\langle\xi\rangle^2 = 1+|\xi|^2$, we obtain a formula of type (6.8) for $a_{|\xi_j}$, noting that the quotients $\langle\xi\rangle^{-2}$,

$x_p/\langle\xi\rangle^2$, $\xi_q/\langle\xi\rangle^2$, $x_p\xi_q/\langle\xi\rangle^2$ all are in $\psi c_{e^2-e^1}$. Q.E.D.

As a consequence of prop.6.2 we get the result, below.

Theorem 6.3. For $a \in \psi s_0$ we have the estimates

$$(6.12)\quad a^{(\alpha)}_{(\beta)}(x,\xi) = O((\langle x\rangle/\langle\xi\rangle)^j) \ , \ -|\beta| \le j \le |\alpha| \ , \text{ for all } \alpha,\beta \ .$$

with O(.)-constant independent of x,ξ but not necessarily of α,β .

Note that (6.12), for $|\alpha|+|\beta|=1$, is a consequence of (6.8), since $a \in \psi s_0$ has all a_{pq} bounded. For general α,β we use that $a_{pq} \in \psi s_0$, again, so that we may iterate (6.8). Q.E.D.

The class Z_n of all symbols a satisfying (6.12) was introduced by Weinstein [We_1] and Zelditch [Ze_1]. Thm.6.3 amounts to

$$(6.13)\qquad \psi l_0 \subset \psi s_0 \subset Z_n \ .$$

Let us observe that all three classes ψs_0 , ψl_0 , and Z_n are symmetric in x and ξ. Therefore in prop.6.1 x and ξ may be interchanged. We also get an x-decay of derivatives on compact ξ-sets.

The inclusions of (6.13) are proper. For example, let n=1 and

$$(6.14)\qquad a(x,\xi) = \chi(x^2+\xi^2)\omega(x/\xi)\cos x \ ,$$

where $\chi(t), \omega(t) \in C^\infty(\mathbb{R})$, $\chi=1$ for $t\ge1$, $=0$ near 0, $\omega=1$ near $t=1$, $=0$ in $|t-1| \ge 1/2$. Clearly all derivatives of a , for x and ξ , are bounded. Moreover, a and all its derivatives vanish, as $\frac{x}{\xi}$ is outside the interval (1/2,2) -i.e., unless we have $x\le 2\xi$, and $\xi\le 2x$ both. In other words, we have $\langle x\rangle \le 2\langle\xi\rangle$, and $\langle\xi\rangle \le 2\langle x\rangle$ in supp a .

Accordingly all (positive and negative) powers of $\langle x\rangle/\langle\xi\rangle$ are bounded in supp a . Hence we also have all $(\langle x\rangle/\langle\xi\rangle)^j a^{(\alpha)}_{(\beta)}(x,\xi)$ bounded. In particular, it follows that $a\in Z_1$. On the other hand it is clear that $\varepsilon_{11}a(x,\xi) = \xi a_{|\xi}-xa_{|x} = -x\sin x$, as $x=\xi$, $|x| \ge 1$. Thus $\varepsilon_{11}a$ is unbounded ,and $a \in \psi l_0$ follows.

Next we focus on topological and algebra properties of ψx_0.

Proposition 6.4. The classes ψx_0, $x=t,l,s$, are algebras under the 'composition product' $a^\Delta b$ defined by $(a^\Delta b)(x,D)=a(x,D)b(x,D)$, as well as the 'pointwise product' $(ab)(x,\xi)=a(x,\xi)b(x,\xi)$.

Proof. By definition ψx_0 are algebras under "Δ". Pointwise we get

$$(6.15)\quad \varepsilon_{jl}(ab) = (\varepsilon_{jl}a)b + a(\varepsilon_{jl}b) \ , \ \eta_{jl}(ab) = (\eta_{jl}a)b + a(\eta_{jl}b).$$

By (5.19) or (5.35) $a,b\in \psi x_0$ implies $\varepsilon_{jl}a$, $\varepsilon_{jl}b$, $\eta_{jl}a$, $\eta_{jl}b\in \psi x_0$, hence $\varepsilon_{jl}(ab)$ (or $\eta_{jl}(ab)\in \psi t_0$. Similarly for iterated ε_{jl}. Q.E.D.

It is natural to introduce on ψl_0 the Frechet topology of the countable class of semi-norms

(6.16) $\|\varepsilon_{p_N q_N}\cdot\ldots\cdot\varepsilon_{p_1 q_1} a\|_{L^\infty} = \|a\|_{\pi\kappa}$, $\pi=(p_1,\ldots,p_N)$, $\kappa=(q_1,\ldots,q_N)$,

with ε_{pq} of (6.1) , for $p,q \ge 1$, and

(6.17) $\varepsilon_{p0} = \eta_{p0} = \partial_{x_p}$, $\varepsilon_{0q} = \eta_{0q} = \partial_{\xi_q}$, (ε_{00} undefined) .

Similarly, for ψs_0, using the η_{jl} instead of the ε_{jl}. Or, for ψt_0, using only the expressions (6.17).

On the other hand, for $\mathrm{Op}\psi x_0 = \Psi GX$, one may use the semi-norms

(6.18) $\|\nabla^\alpha a(x,D)\| = \|\|a\|\|_\alpha$,

with the $L(H)$-operator norm $\|\cdot\|$, and $\nabla^\alpha A$ denoting the partial derivatives of $A_{g,z,\zeta}$, or $A_{\sigma,o,z,\zeta}$, or $A_{z,\zeta}$, at the unit of the group. Here α will be a multi-index of the dimension of the group.

<u>Proposition 6.5</u>. The topologies (6.16),and (6.18) are equivalent.
<u>Proof</u>. Consider the case of ψl_0 . We know that

(6.19) $\nabla^\alpha(a(x,D)) = a_\alpha(x,D)$, $a_\alpha = \Pi_{j=1}^{|\alpha|}\varepsilon_{p_j q_j} a$,

with some selection $\varepsilon_{p_j q_j}$, corresponding to the multi-index α .

From cor.2.6 we conclude that

(6.20) $\|\|a\|\|_\alpha = \|a_\alpha(x,D)\| \le c \sum_{|\beta|,|\gamma|\le n} \|a_{\alpha(\gamma)}^{(\beta)}\|_{L^\infty}$,

where the terms of the sum at right are all norms of type (6.16), since the ε_{jl} include the expressions (6.15).

On the other hand, for $A\in \Psi GS$ we have a=symb(A) given by the trace formula (3.18). For $b=\Pi_{j=1}^N \varepsilon_{p_j q_j} a$ we get $B=b(x,D)=\nabla^\alpha A$ with the corresponding combination ∇^α of $\partial_{g_{pq}}$, ∂_{z_p} and ∂_{ζ_q} on $A_{g,z,\zeta}$ at $(g,z,\zeta)=(1,0,0)$, as follows from (2.20) and (5.19). Use (3.18) for

(6.21) $\|b\|_{L^\infty} \le c \sum_{|\beta|,|\gamma|\le 2n} \|(\nabla^\alpha a)_{(\beta)}^{(\alpha)}(x,D)\|$.

Again, the terms of the sum, at right, are semi-norms occuring in (6.18) . This shows indeed that, in case of ψl_0, the topologies are equivalent. Similarly for ψs_0. Q.E.D.

<u>Theorem 6.6</u>. The classes ΨGX, $X=T,S,L$, are ψ^*-subalgebras of $L(H)$, under the operator product and the topology (6.16) or (6.18).

<u>Proof</u>. Prop.6.4 and prop.6.5 make ψx_0 (or ΨGX) Frechet algebras.

The adjoint invariance and inverse-closed-property follow from

$$(6.22)\qquad A^*_{g,z,\zeta} = (A_{g,z,\zeta})^* \ , \ (A^{-1})_{g,z,\zeta} = (A_{g,z,\zeta})^{-1} \ .$$

If $A_{g,z,\zeta}$ is differentiable in $L(H)$ then so are the right hand sides in (6.22), assuming that $A^{-1}\in L(H)$, in the second case. Q.E.D.

7. Polynomials in x and ∂_x with coefficients in ΨGX .

So far, in this chapter, we were dealing with zero order ψdo's only. This restriction now will be removed by defining the classes ψx of all polynomials in the 2n variables x, ξ with coefficients in ψx_0, $x=t,s,l$. Also, ψx_m, $m=(m_1,m_2)$, with integers $m_j\ge 0$, denotes collection of all such polynomials of degree $\le m_1$ ($\le m_2$) in ξ (in x) For general $m\in \mathbb{R}^2$ we define ψx_m , $x=l,s$, by setting

$$(7.1)\quad \psi x_m = \{a \in C^\infty(\mathbb{R}^{2n}) : a_0(x,\xi) = \langle x\rangle^{-m_2}\langle \xi\rangle^{-m_1} a(x,\xi) \in \psi x_0 \} \ .$$

Note that we have

$$(7.2)\quad \langle x\rangle = s_0(x) + \sum x_j s_j(x) \ , \ s_0(x) = \langle x\rangle^{-1} \ , \ s_j(x) = x_j/\langle x\rangle \ ,$$

and a corresponding formula for $\langle \xi\rangle$, where the $s_j(x)$ and $s_j(\xi)$ are in $\psi c_0\subset \psi x_0$, for $x=t,l,s$. Thus for integer $m_j\ge 0$ an $a\in \psi x_m$ of (7.1) may be written as $a=a_0\langle x\rangle^{m_2}\langle \xi\rangle^{m_1}$, a polynomial (of proper degrees) with coefficients in ψx_0, showing that the two definitions agree.

Note that $\psi x_m \subset \psi x$, even if m_j are not integers, since we always can write $a \in \psi x_m$ as $a = (a_0\langle \xi\rangle^{m_1-M_1}\langle x\rangle^{m_2-M_2})\langle \xi\rangle^{M_1}\langle x\rangle^{M_2}$, with $a_0\in \psi x_0$, and with integers $0<M_j\ge m_j$, so that the expression (.) is also in ψx_0. We get $\psi x=\cup\{\psi x_m : m\in \mathbb{R}^2\}$ as well. Clearly the ψx are graded algebras under pointwise multiplication; we have $\psi x_m\cdot \psi x_{m'}=\psi x_{m+m'}$. Moreover, the Frechet topology may be carried to ψx_m using (7.1). As seminorms on ψx_m use (6.16) for a_0 of (7.1).

The above definition of ψx_m uses the 'weight functions' $\langle x\rangle = (1+|\xi|^2)^{1/2}$, $\langle x\rangle=(1+|x|^2)^{1/2}$ again. The symbols of ψx_m are said to have multiplication (differentiation) (total) order m_2 (m_1) (m).

<u>Proposition 7.1</u>. The class ψs_m consist precisely of all $a\in C^\infty(\mathbb{R}^{2n})$ such that, with $\pi_m(x,\xi)$ of III,(3.1) i.e.,

$$(7.3)\qquad \pi_m(x,\xi) = \langle \xi\rangle^{m_1}\langle x\rangle^{m_2} \ , \ m = (m_1,m_2) \in \mathbb{R}^2 \ ,$$

we have the estimates (7.4) for all products $\Pi_{j=1}^N \eta_{p_j q_j}$ of η_{pq} :

$$(7.4)\qquad (\Pi_{j=1}^N \eta_{p_j q_j} a)(x,\xi) = O(\pi_m(x,\xi)) \ , \ x,\xi \in \mathbb{R}^n \ .$$

Similarly, ψl_m is described by (7.4) with $\eta_{\mu\nu}$ replaced by $\varepsilon_{\mu\nu}$.

Proof. Consider $x=s$ only. If $a\in \psi s_m$, then $a\pi_m^{-1}=b\in \psi s_0$. Hence

$$\eta_{pq}a = (\eta_{pq}b)\pi_m + (b(\eta_{pq}\pi_m)/\pi_m)\pi_m \quad . \tag{7.5}$$

Note that $\eta_{pq}\pi_m/\pi_m \in \psi c_0$. Thus $\eta_{pq}a=O(\pi_m)$. Moreover, (7.5) implies that $\eta_{pq}a\in \psi s_m$, since $\psi c_0\subset \psi s_0$, and since ψs_0 is an algebra. Hence we may iterate, for (7.4). Vice versa, let (7.4) hold for a. Then

$$\eta_{pq}b = \pi_{-m}\eta_{pq}a + b(\eta_{pq}\pi_{-m})/\pi_{-m}) = O(1) , \tag{7.6}$$

using that $\eta_{pq}\pi_{-m}/\pi_{-m} \in \psi c_0 \subset \psi s_0$. Apply η_{rs} to (7.6) for $\eta_{rs}\eta_{pq}b=O(1)$, etc. ... .It follows that $b\in \psi s_0$, so that $a\in \psi s_m$, q.e.d.

Proposition 7.2. Let $a\in \psi x_m$, $x=s,l$, $m\in \mathbb{R}^2$. Then, for all α,β,

$$a^{(\alpha)}_{(\beta)} \in \psi x_{m+r(e^1-e^2)} \text{ , as } -|\alpha| \le r \le |\beta| \text{ , } r = \text{integer.} \tag{7.7}$$

Proof. First let $|\alpha|+|\beta|=1$. Prop.7.1 shows that ψx_m is left invariant by arbitrary applications of ε_{pq} or η_{pq}, resp., hence $a^{(\alpha)}_{(\beta)}\in \psi x_m$ for $a\in \psi x_m$. Thus (7.7) holds for r=0. On the other hand, (6.8) expresses $a^{(\alpha)}_{(\beta)}$ by a_{pq} with coefficients in $\psi c_{r(e^1-e^2)}$, $r=\pm1$, correspondingly. This gives (7.7) for $r=\pm1$. Formulas (6.8) may be iterated, proving (7.7) for arbitrary α,β and r. Q.E.D.

Remark 7.3. Note prop.7.2 contains thm.6.3. Generalized classes $Z_{n,m} \supset \psi s_m$, with $Z_{n,0}=Z_n$, may be introduced, replacing O(.) in (6.12) by $O(\pi_{m+r(e^1-e^2)})$.

For a symbol $a\in \psi x$, $x=s,l$, we will define the ψdo

$$A = a(x,D) = \sum x^\alpha a_{\alpha,\beta}(x,D)D^\beta \text{ ,as } a(x,\xi) = \sum a_{\alpha,\beta}(x,\xi)x^\alpha\xi^\beta \ . \tag{7.8}$$

However, one just as well might write the same operator A as

$$A = \sum b_{\alpha,\beta}(x,D)x^\alpha D^\beta \quad , \tag{7.9}$$

with (other) coefficients $b_{\alpha,\beta}(x,\xi)\in \psi x_0$. Or also in any other order $A=\sum c_{\alpha,\beta}(x,D)D^\beta x^\alpha=\sum x^\alpha D^\beta d_{\alpha,\beta}(x,D)$, etc. This follows from (2.8):

The reason for this fact can be found in formulas (2.8) : For A = For $A=a(x,D)\in \Psi GX\subset \Psi GT$, $X=L,S,T$, we have $[x_j,A]$, $[D_j,A]\in \mathrm{Op}\psi x_0$.

In fact, for $x=s$ or l, the commutator symbols have better decay properties: Using (2.20) we get

$$\partial_{z_j}A_{0,0} = a_{|x_j}(x,D) \text{ , } \partial_{\zeta_j}A_{0,0} = a_{|\xi_j}(x,D) \text{ ,} \tag{7.10}$$

so that

(7.11) $[x_j,A]=a^j(x,D)=ia_{|\xi_j}(x,D)$, $[D_j,A]=a_j(x,D)=-ia_{|x_j}(x,\xi)$

satisfies stronger estimates of Weinstein-Zelditch type :

$$a_{j(\beta)}^{(\alpha)}(x,\xi) = O((\langle x\rangle/\langle\xi\rangle)^l) \ , \ -|\beta|-1 \le l \le |\alpha| \ , \tag{7.12}$$
$$a^{j(\alpha)}_{(\beta)}(x,\xi) = O((\langle x\rangle/\langle\xi\rangle)^l) \quad -|\beta| \le l \le |\alpha|+1 \ .$$

Similarly for higher order commutators and higher derivatives.

Hence ψx is an algebra under $a^\Delta b$ of prop.7.4, since in a product of 2 expressions (7.8) one may unite the x- (D-) powers at left (right). With calculus of ψdo's, below,one also confirms that $\psi x_m{}^\Delta \psi x_{m'} \subset \psi x_{m+m'}$, i.e., ψx is a graded algebra, under "Δ" .

Next we look at calculus of ψdo,s, for the algebra Opψs, similar as in I,6 for $\psi h_{m,\rho,\delta}$. We have $\psi x_m \subset \psi t_m$, hence may use the Leibniz-Taylor formulas I,(5.7) , I,(5.9) for product and adjoint in Opψs. The question remains whether (in what sense) the remainders decay, and about significance of the terms of the expansion.

In view of applications in ch's 9,10 we focus on

(7.13) KA , AK , [K,A] , K = k(x,D) , A = a(x,D) ,

where $a\in \psi x_m$, $k\in \psi c_{m'}$, $m,m'\in \mathbb{R}^2$, $x=s,l$. We get a Leibniz formula with controlled remainder, but decay of order not quite as fast.

<u>Proposition 7.4</u>. We have

$$KA = \sum_{j=0}^N \sum_{|\theta|=j} (-i)^j/\theta! \ (k^{(\theta)}a_{(\theta)})(x,D) + R_N^1 \ ,$$

$$AK = \sum_{j=0}^N \sum_{|\theta|=j} (-i)^j/\theta! \ (a^{(\theta)}k_{(\theta)})(x,D) + R_N^2 \ , \tag{7.14}$$

$$[K,A] = \sum_{j=1}^N (-i)^j/j! \ \langle k,a\rangle_j(x,D) + R_N^3 \quad ,$$

where $k^{(\theta)}a_{(\theta)}$,and $a^{(\theta)}k_{(\theta)}$, and the 'iterated Poisson brackets'

$$\langle k,a\rangle_j = \sum_{|\theta|=j} \binom{j}{\theta}(k^{(\theta)}a_{(\theta)}-k_{(\theta)}a^{(\theta)}) \ , \ \binom{j}{\theta} = j!/\theta! \ , \tag{7.15}$$

belong to $\psi x_{m+m'-re^1-r'e^2}$, for all $r,r'=0,1,2,\dots$, $r+r'=j$, while $R_N^j\in \psi x_{m+m'+re^1+r'e^2}$, for all $r,r'=0,1,\dots$, with $r+r'=N+1$.

<u>Proof</u>. Note that $k^{(\theta)}\in \psi c_{m'-je^1}$, $a_{(\theta)}\in \psi x_{m+r(e^1-e^2)}$,$r=0,\dots,j$, as $|\theta|=j$, by prop.7.2. Their products are $\in \psi s_{m+n'-(j-r)e^1-re^2}$, proving the first statement. In the remainder of I,(5.27), i.e.,

$$\rho_N=(N+1)\sum_{|\theta|=N+1}\frac{(-i)^{N+1}}{\theta!}\rho_{\theta,N}, \ \rho_{\theta,N}=\int_0^1(1-\tau)^N d\tau I_{\theta,N}(x,\xi,\tau), \tag{7.16}$$

with the 'finite part integrals'

$$(7.17)\qquad I=I_{\theta,N}(\tau,x,\xi) = \int d\!\!{}^-\eta d\!\!{}^-y\, e^{-i\eta y}k^{(\theta)}(x,\xi-\tau\eta)a_{(\theta)}(x-y,\xi)\ ,$$

I may be written as an improper Riemann integral, namely

$$(7.18)\ I = \int d\!\!{}^-y d\!\!{}^-\eta e^{-iy\eta}\langle y\rangle^{-2M}(1-\Delta_\eta)^M(\langle\eta\rangle^{-2M}\, k^{(\theta)}(x,\xi-\tau\eta)a_{M(\theta)}(x-y,\xi))$$

with $a_M = (1-\Delta_x)^M a$, and the usual formal partial integration.

In I,5, by a somewhat difficult interchange of limits, we proved (7.16), (7.17), (7.18). Formally this is Taylor's formula with integral remainder. Moreover, we saw that the derivatives $I^{(\alpha)}_{(\beta)}$ are obtained by differentiating (7.18) under the integral sign.

We discuss R^1_N only, noting that R^2_N, R^3_N may be treated analogously. We need an estimate for $I^{(\alpha)}_{(\beta)}$. They are in the span of integrals (7.18), with $k^{(\theta)}\cdot a_{N(\theta)}$ replaced by

$$(7.19)\qquad k^{(\alpha'+\theta)}_{(\beta')}(x,\xi-\tau\eta)\cdot a^{(\alpha''+\theta)}_{N(\beta'')}(x-y,\xi)\ ,\ \alpha'+\alpha''=\alpha\ ,\ \beta'+\beta''=\beta\ ,$$

We have $a_N\in\psi s_{m'}$ again. Hence, using estimates (6.9) for k and (7.7) for a_N , we arrive at an estimate for $I^{(\alpha)}_{(\beta)}$ by a sum of terms

$$7.20)\int dyd\eta(\langle y\rangle\langle\eta\rangle)^{-2M}\langle x\rangle^{m_2-|\beta'|}\langle\xi-\tau\eta\rangle^{m_1-N-1-|\alpha'|}\langle x-y\rangle^{m'_2-r}\langle\xi\rangle^{m1+r},$$

where $-|\alpha''|\le r\le N+1+|\beta''|$. Here we can estimate

$$(7.21)\qquad \begin{aligned}&\int dy\langle y\rangle^{-2M}\langle x-y\rangle^{m'_2-r} = O(\langle x\rangle^{m'_2-r})\ ,\\ &\int d\eta\langle\eta\rangle^{-2M}\langle\xi-\tau\eta\rangle^{m_1-N-1-|\alpha'|} = O(\langle\xi\rangle^{m_1-N-1-|\alpha'|})\ ,\end{aligned}$$

using I,(6.5). Note (7.21) holds uniformly in t, $0\le t\le 1$. The estimates remain true if $\alpha'=\alpha''=\beta'=\beta''=0$. Hence,

$$(7.22)\qquad I^{(\alpha)}_{(\beta)} = O(\pi_{m+m'-re^1-r'e^2})\ ,\ r+r' = N+1\ ,$$

for all multi-indices α,β, giving the corresponding for the symbol r^1_N of R^1_N . These estimates are weaker than the stated ones. But,

$$(7.23)\qquad r^1_N = \sum_{N+1\le|\theta|\le R}(-i)^{|\theta|}/\theta!k^{(\theta)}a_{(\theta)} + r^1_R\ ,$$

where the sum at right is $\in\psi x_{m+m'-re^1-r'e^2}$, $r+r'=N+1$, for all $R=N+1,N+2,\ldots$ For fixed r,r', $r+r'=N+1$, let $R=N+2j$, with $j>0$. Get $R+1=(r+j)+(r'+j)$, hence $r^1_R\in\psi t_{m+m'-re^1-r'e^2-je}$, by (7.22). Thus,

$$(7.24)\qquad \Pi^{N'}_{j=1}\eta_{p_jq_j}r^1_{2j}\in\psi t_{m+m'-re^1-r'e^2}\ ,\text{as } N'\le j\ .$$

Thus all products of $\le j$ factors are $O(\pi_{m+m'-re^1-r'e^2})$, for r^1_R, hence also for r^1_N. Since j is arbitrary get $r^1_N \in \psi x_{m+m'-re^1-r'e^2}$, as required. Q.E.D.

Proposition 7.5. We have

$$(7.25)\qquad Op\psi x_m = \{\langle x\rangle^{m_2} A\langle D\rangle^{m_1} : A \in \Psi GX = Op\psi x_0\} ,$$

and, more generally, for arbitrary $s \in \mathbb{R}^2$,

$$(7.26)\qquad Op\psi x_m = \{\pi_s(x,D)A\pi_{m-s}(x,D) : A \in \Psi GX\} ,$$

Theorem 7.6. The space $Op\psi x_m$ is characterized as set of all $A \in O(m)$ such that for all $T=T_{g,z,\zeta,\varphi}$ of (5.2) the operator $T^{-1}AT$ belongs to the subgroup GX of GL, $A \to T^{-1}AT$ defining a C^∞-map $gx \to O(m)$ in the Frechet topology of $O(m)$, as discussed in III,3. Moreover, a sufficient condition for $A \in Op\psi x_m$ is that (i) $A \in L(H_s, H_{s-m})$, (ii) $T^{-1}AT$ is $C^\infty(gx, L(H_s, H_{s-m})$ for just one $s \in \mathbb{R}^2$.

We leave the proofs of prop.7.5 and thm.7.6 to the reader.

8. Characterization of ΨGX by the Lie algebra.

In [B_2], [Dn_1], [C_1],IV,9-10 one introduces 'derivatives' of a linear operator $A \in L(H)$, as commutators with D_j or M_1. If all such derivatives exist (in suitable topology) then A is found to have a symbol in a class like CB^∞ or $\psi h_{0,\rho,\delta}$ - i.e., A is a ψdo.

There is a link between that approach and our present one: It points to the relation between the Lie group GH and its Lie algebra, as discussed in ch.0,sec.9. If a Lie group G is represented as group of invertible matrices on some $\mathbb{R}^N$, the 'tangent vectors' at $I \in G \subset GL(\mathbb{R}^N)$ form a linear space A of $N\times N$-matrices, containing its commutators - a representation of the Lie algebra of G. The connected component of I in G then is the set of all products of e^A, for $A \in A$, by the Campbell-Hausdorff formula.

Now focus on the 'representation' $A \to A_{z,\zeta,\varphi} = E_{z,\zeta,\varphi}{}^{-1}AE_{z,\zeta,\varphi}$ of our Lie group gh of (1.11),(1.12) on the ∞-dimensional space $L(H)$. Actually, two representations are involved: We represent gh $(=\mathbb{R}^n\times\mathbb{R}^n\times\mathbb{S}^1)$ by $(z,\zeta,\varphi) \to E_{z,\zeta,\varphi} = e^{i(\varphi+M\zeta)}T_z = e^{i(\varphi+M\zeta)}e^{izD}$ on H (as unitary operators). That representation generates the above $A \to A_{z,\zeta}$ as invertible maps $L(H) \to L(H)$. Following the lines of ch.0,9 we formally get a tangent space at $(z,\zeta,\varphi)=0$ for the first representation as the linear span (called $AT=AH$) of the folpde's

(8.1) ∂_{x_j} , i , ix_j , $j=1,\dots,n$.

The (8.1) are <u>unbounded</u> operators of H. A precise definition will specify the domain - that of the infinitesimal generator of $\{e^{iD_jt}\}$, $\{e^{it}\}$, $\{e^{ix_jt}\}$, respectively. These groups consist of unitary operators of H, thus their generators are skew-selfadjoint each having an orthogonal spectral measure.

Notice that the precise domains of the operators (8.1) are not identical. Linear combinations no longer are skew-self-adjoint. However, as easily seen, the self-adjoint relizations of 1, D_j , ix_j , and their linear combinations $zD+\zeta x+\varphi$, are unique. Declare as common domain of 1, D_j , M_j the space S. Then AH indeed is a linear space; for $D_{z,\zeta,\varphi}\in AH$ the closure (in H) is skew-self-adjoint, and $e^{iD_{z,\zeta,\varphi}^{**}}$ is unitary. Using ch.0,sec.6 we get

$$(8.2)\qquad e^{iD_{z,\zeta,\varphi}^{**}} = E_{z,\zeta,\varphi+(z.\zeta)/2} .$$

Note that (8.2) supplies an analogue to the Campbell-Hausdorf formula for our present infinite dimensional representation. GH and AH are related exactly as G and A of 0,0.

Our second representation $A\to A_{z,\zeta}$ of gh on $L(H)$ supplies another representation of the Lie algebra AH , as the (also unbounded) operators on $L(H)$, formally given as linear combinations of

$$(8.3)\qquad \mathrm{ad}\,\partial_{x_j} ,\ \mathrm{ad}\,ix_j ,\ j=1,\dots n, \text{ with } (\mathrm{ad}\,X)A = [X,A] = XA-AX .$$

Indeed, let $L=D_{z,\zeta,\varphi}^{**}$, then, formally,

$$(8.4)\qquad \frac{d}{dt}(e^{-iLt}Ae^{iLt}) = -i(\mathrm{ad}\,L)A .$$

Formula (8.4) seems to imply that $A\to A_{z,\zeta}$ is C^1 as a map $\mathbb{R}^{2n}\to L(H)$ if and only if $(\mathrm{ad}\,\partial_j)A$, $(\mathrm{ad}\,x_j)A\in L(H)$, moreover, that $A\to A_{z,\zeta}$ is C^∞ if and only if finite products of ad ∂_{x_j} , ad x_1 take A to a bounded operator. In turn $A\to A_{z,\zeta}$ is C_∞ if and only if $A\in$ Opψt$_0$, by the ψdo-theorem. Both facts suggest thm 8.1, below.

<u>Theorem 8.1</u>. An operator $A\in L(H)$ belongs to Opψt$_0$ if and only if the operators $(\mathrm{ad}\,x)^\alpha(\mathrm{ad}\,D)^\beta A$: $S\to S'$ all have their image in H and extend to bounded operators of H , where

$$(8.5)\qquad (\mathrm{ad}\,x)^\alpha=(\mathrm{ad}\,x_1)^{\alpha_1}(\mathrm{ad}\,x_2)^{\alpha_2}\dots,\ (\mathrm{ad}\,D)^\beta=(\mathrm{ad}\,D_1)^{\beta_1}(\mathrm{ad}\,D_2)^{\beta_2}\dots$$

Here $(\mathrm{ad}\,x_j)A= x_jA-Ax_j$ trivially is a map $S\to S'$. Similarly

$(\mathrm{ad}\ \partial_{x_j})A$. Higher order commutators $(\mathrm{ad}\ x)^{\alpha}A$ are sums of terms $\Pi x A \Pi x$, with only one A between products Πx of x_j. Again they are $\mathcal{S}\to\mathcal{S}'$. Thus the statement is meaningful. The proof will be a verification of above relation between differentiability and existence of commutators in $L(\mathcal{H})$. Details are left to the reader.

Clearly the map $(g,\lambda)\to T(g,\lambda)=T_{g,z,\zeta,\varphi}$ of (5.2) defines a unitary representation of the Lie group gl on $\mathcal{H}$. Note that the map $E_{-z,\zeta,\varphi}$ is just the restriction to $gt\subset gs\subset gl$ of T above.

For *GT* , *GS* , *GL* the Lie algebras *AT* , *AS* , *AL* are the (unbounded) directional derivatives of $T(g,\lambda)=T_{g,z,\zeta,\varphi}$ at $e=(1,0,0,0)$ of gl, in directions allowed by the subgroup. In this way we get *AL* and *AS* as the real linear spans of (respectively)

$$(8.6)\qquad x_l\partial_{x_j} + \delta_{jl}/2\ ,\ \partial_{x_j}\ ,\ ix_l\ ,\ i\ ,\ j,l = 1,\dots,n\ ,$$

and

$$(8.7)\qquad \textstyle\sum_{k=1}^{n} x_k\partial_{x_k} + n/2\ ,\ x_l\partial_{x_j} - x_j\partial_{z_l}\ ,\ j<l,\ \partial_{x_j},\ ix_l,\ i\ ,\ j,l=1,\dots,n\ .$$

Again declare $\mathcal{S}$ as joint domain of all operators (8.6),(8.7). The exponentiations e^{Lt}, as in 0,6, for $L\in$ *AX*, define operators of *GX*. We again get a Campbell-Hausdorff formula, of the form

$$(8.8)\qquad \begin{aligned}&\exp(iD_{\gamma,z,\zeta,\varphi})=T_{g,z^{\Delta},\zeta^{\Delta},\varphi^{\Delta}}\ ,\ \text{where } g=e^{\gamma},\ z^{\Delta}=e_1(\gamma)z,\\ &\zeta^{\Delta}=e_1(\gamma^{t})\zeta,\ \varphi^{\Delta}=\varphi+\zeta.e_2(\gamma)z\ ,\ e_j(\gamma)=\textstyle\sum_0^{\infty}\frac{\gamma}{(j+k)!}\ ,\end{aligned}$$

and where

$$(8.9)\qquad D_{\gamma,z,\zeta,\varphi} = \sum(x_q\partial_{x_p}+\tfrac{1}{2}\delta_{pq}) + \sum z_p\partial_{x_p} + i\sum\zeta_p x_p + i\varphi\ .$$

For $L\in$ *AX* we get (8.4) again. The algebras Ψ*GX* may be characterized by the Lie algebras *AX* , just as in thm.8.1 for $X=H$:

<u>Theorem 8.3</u>. For $X=T,S,L$, the class Ψ*GX* is identical with the set of $A\in L(\mathcal{H})$ allowing arbitrary finite application of ad L, $L\in$ *AX*, in the sense that $\Pi(\mathrm{ad}\ L_j)A:\mathcal{S}\to\mathcal{S}'$ maps $\mathcal{S}\to\mathcal{H}$, and extends to $L(\mathcal{H})$.

The proof is similar to that of thm.8.1, and is omitted.

<u>Problems</u>: (1) Verify formula (8.8) in details, using ch.0,sec.6. (2) Question: Can you relate the proof of the real Campbell-Hausdorff formula (0,lemma 9.5) offered in 0,9, with pbm. (1) above?

Chapter 9. PARTICLE FLOWS AND INVARIANT ALGEBRA OF A SEMI-STRICTLY HYPERBOLIC SYSTEM; COORDINATE INVARIANCE OF $Op\psi x_m$.

0. Introduction.

We use the natural symbol classes ψx, $x=l,s$, in this chapter. First we ask for invariance of $Op\psi x$ under global coordinate transforms. Each ΨGX is invariant under conjugation with $T\in GX$. For a subgroup these are linear coordinate transforms: For GT get the translations e^{zD}, for GS the 'similarities' $T_{\sigma o,z,0}$ (all distances multiplied by a constant), for GL get all linear substitutions. Another subgroup gives 'gauge transforms' (conjugation by $e^{i\zeta x}$) .

The question about more general coordinate (or gauge-) invariance of $\Psi GX=\Psi AX$ may be phrased as follows: ΨGX is the set of $A\in L(H)$ with $A_t=e^{Lt}Ae^{-Lt}\in C^\infty(\mathbb{R},L(H))$ for every $L\in AX$. In new coordinates $y=\varphi(x)$ the folpdes $L\in AX$ will transform to other folpdes $L^\sim$ forming a Lie algebra $AX^\sim$, where $Lu=f \leftrightarrow L^\sim v=g$, $v=u\circ\varphi^{-1}$, $g=f\circ\varphi^{-1}$. Clearly e^{Lt} transforms to $e^{L^\sim t}$, as solution operator of $\partial_t u=Lu \leftrightarrow \partial_t v=L^\sim v$. Thus "$A\in \Psi GX$" means $A^\sim_t=e^{tL^\sim}Ae^{-tL^\sim}\in C^\infty(\mathbb{R},L(H))$, assuming that $u(x)\to u^\sim(x)=u(\varphi(x))$ defines an isomorphism of H. Thus, in new coordinates the property "$A\in VGX$" transforms to smoothness under certain e^{Lt}, with more complicated $L\in AX^\sim$, depending on φ .

Vice versa, since a coordinate transform is invertible, "$A\in \Psi GX$" (that the transformed operator is in ΨGX) may be expressed as smoothness of $e^{Mt}Ae^{-Mt}$ for certain M - the transforms of $L\in AX$ back to the old coordinates. For coordinate invariance we must show that $e^{Mt}Ae^{-Mt}\in C^\infty(\mathbb{R},L(H))$ for the transforms M of ε_{pq} or η_{pq}.

Observe that our folpdes $L=\varepsilon_{pq}$ (or $=\eta_{pq}$) were either skew-symmetric, or, at least, have $L+L^*$ a multiplication of order 0. We shall require that condition for M, as an assumption on the coordinate transform φ. Then e^{Mt} is the solution operator of a hyperbolic equation $\partial_t u=Mu$, of the form studied in ch.VI.

We always will require $M\in Op\psi c_e$, $M+M^*\in Op\psi c_0$, so that the existence theorems of ch.VI are applicable. For a detailed discussion of coordiate invariance cf.sec.4. But observe that coordinate invariance appears as a corollary of a study of invariance of ψx_0 under the particle flows of the hyperbolic symmetric equation $\partial_t u=Mu$. For $e^{Mt}Ae^{-Mt}\in C^\infty(\mathbb{R},L(H))$ we first prove $a_t=a_\circ\nu_t\in C^\infty(\mathbb{R},\psi x_0)$, $a=symb(A)\in \psi x_0$ (sec.1 for $x=l$, sec.2 for $x=s$). The discussions are parallel to VI,6, but use a different approach: a_t solves the PDE $\partial_t a_t = \langle k,a\rangle$, $k=-i\,\mathrm{symb}(M)$. Repeated application of ε_{pq} (or η_{pq}) generates a system of PDE's. We use this for boundedness in t of $\Pi\varepsilon_{p_jq_j} a_t$ (or $\Pi\eta_{p_jq_j} a_t$), and, more generally $\Pi\varepsilon_{p_jq_j}\partial_t^l a_t$, etc. In sec.3 we apply our result to show that $A_t = e^{Mt}Ae^{-Mt} \in C^\infty(\mathbb{R},\mathcal{L}(\mathcal{H}))$.

We prove this for a rather general class of hyperbolic ψde's $\partial_t u = iKu$ with $K \in Op\psi c_e$, $K-K^* \in Op\psi c_0$. Under proper assumptions on K we get $e^{iKt}\Psi GXe^{-iKt} = \Psi GX$, as for $Op\psi c_0$ in VI,5. In fact, $Op\psi x_m = e^{iKt}Op\psi x_m e^{-iKt}$, and even $e^{iKt}Ae^{-iKt} \in C^\infty(\mathbb{R}, Op\psi x_m)$, as $A \in Op\psi x_m$.

This <u>invariance</u> <u>under</u> <u>particle</u> <u>flows</u> and <u>invariance</u> <u>of</u> $Op\psi x$ <u>under</u> <u>conjugation</u> is the second topic of the present chapter.

Thirdly, we will look at particle flows for systems - in sec's 5,6,7. focusing on the algebras $Op\psi x$. Here $K \in Op\psi c_e^\nu$ will be a $\nu\times\nu$-matrix of ψdo's. To include the Dirac equation (ch.10) we require the system $\partial_t u + iKu = 0$ to be <u>semi-strictly</u> <u>hyperbolic</u> of type e^Δ , a generalization of a concept of VI,4: The eigenvalues of (the essential part of) $k=\mathrm{symb}(K)$ must be real, of <u>constant</u> multiplicity, but need not be distinct (sec.5).

It is no longer true that ΨGX^ν (or $Op\psi x_m^\nu$) is invariant under conjugation with e^{-iKt}. Rather $e^{-iKt}Ae^{iKt}$ remains a ψdo in $Op\psi x_m^\nu$ only if A belongs to a certain subalgebra $P=P(K)$ of $Op\psi x^\nu$. The existence of this <u>invariant</u> <u>algebra</u> P gives raise to physical speculations in case of the Dirac equation, to be looked at in ch.10.

A necessary condition for a ψdo $A \in Op\psi x_m$ to belong to $P(K)$ is that its symbol matrix $a=\mathrm{symb}(A)$ commutes with $k=\mathrm{symb}(K)$, modulo terms of lower order. While investigating invariance of P we will find one particle flow for each of the distinct eigenvalues $\lambda(x,\xi)$ of $k(x,\xi)$. The flows satisfy VI,(5.19) with k replaced by λ .

For the Dirac equation we get 2 particle flows, for electrons and positrons, resp. The flows will describe the exact relativistic motion of the particle under the potentials imposed.

For 1-dimensional systems we get a theory parallel to that of VI,5,6. For $\nu>1$ the different parts of the symbol matrix $a(x,\xi)$ will propagate along different flows. Roughly, an $a(x,\xi)$ commuting with $k(x,\xi)$ will split into 'diagonal boxes' corresponding to the eigenvalues of k. Modulo lower order each box propagates along its flow, undergoing a similarity. For details cf. sec's 5,6,7.

<u>1. Flow-invariance of</u> $\psi\mathcal{I}_0$.

In this section we assume a scalar real-valued symbol $k=k_1 \in$

ψc_e. Let ψ_t be the flow of VI,(5.18), for this symbol k. In VI,6 we proved that $\nu_t:\mathbb{R}^{2n}\to\mathbb{R}^{2n}$ satisfies $a_t=a\circ\nu_t\in\psi c_m$ for $a\in\psi c_m$. Here we prove the same for ψl, under stronger assumptions on k.

From VI,(5.17) and 0,6 we know that a_t is given as solution of

$$(1.1)\qquad \partial_t a_t + \langle k,a_t\rangle = 0\ ,\ t\in\mathbb{R}\ ,\ a_0=a\ ,$$

with the Poisson bracket $\langle .,.\rangle=\langle .,.\rangle_1$ of VI,(5.16). Note that (1.1) is a Cauchy problem for a first order PDE in t,x,ξ. Our conditions on k will insure that a_t solving (1.1) gives a symbol $a_t\in\psi l$, for all $t\in\mathbb{R}$, $a\in\psi l$.

Assuming $a\in\psi l_0$ we know that $\Pi\varepsilon_{p_jq_j}a$ is bounded for finite such products. We apply ε_{pq} (or a finite product) to (1.1), for a system of equations in $a_{pqt}=\varepsilon_{pq}a_t$, $a_{pqrst}=\varepsilon_{rs}\varepsilon_{pq}a_t$, etc. Then we will derive apriori estimates showing that a_{pqt}, ... remain bounded in t,x,ξ. As a preparation we note the following.

<u>Proposition 1.1</u>. (a) The expressions ε_{pq}, of VIII,(5.17),(5.27) may be written as Poisson brackets

$$(1.2)\qquad \varepsilon_{pq}a = \{\varepsilon^{pq},a\} = \varepsilon^{pq}{}_{|\xi}\cdot a_{|x} - \varepsilon^{pq}{}_{|x}\cdot a_{|\xi}\ ,$$

with the functions

$$(1.3)\qquad \varepsilon^{pq} = -x_q\xi_p\ ,\text{as } p,q\ge 1\ ,\ \varepsilon^{p0}=\xi_p\ ,\ \varepsilon^{0q}=-x_q\ .$$

(b) For Poisson brackets we have <u>Jacobi's identity</u>

$$(1.4)\qquad \langle f,\langle g,h\rangle\rangle = \langle\langle f,g\rangle,h\rangle + \langle g,\langle f,h\rangle\rangle\ ,\ f,g,h\in C^\infty(\mathbb{R}^n)\ .$$

Using (1.1), (1.2), (1.3), and Jacobi's identity, we get

$$(1.5)\qquad \partial_t a_{pqt} + \langle k,a_{pqt}\rangle + \langle(\varepsilon_{pq}k),a_t\rangle = 0\ .$$

Suppose that the folpde $L_{pq}:a\to\langle(\varepsilon_{pq}k),a\rangle$ may be written as a linear combination of $\varepsilon_{00}=1$ and the other ε_{jl}, with coefficients in ψc_0. In other words, assume that, for $p,q=0,\dots,n$, $(p,q)\ne(0,0)$,

$$(1.6)\qquad \langle(\varepsilon_{pq}k),.\rangle = \textstyle\sum_{r,s=0}^n \lambda_{pqrs}\varepsilon_{rs}\ ,\ \text{where } \lambda_{pqrs}\in\psi c_0\ .$$

Then (1.5) assumes the form

$$(1.7)\qquad \partial_t a_{pqt} + \langle k_0,a_{pqt}\rangle + \sum_{rs}\lambda_{pqrs}a_{rst} = 0,\ p,q=0,\dots,n,\ (p,q)\ne 0.$$

Now (1.1) and (1.7) may be thought of as a linear system of $(n+1)^2$ ODE's for the unknown functions a_{pq}, along the characteristic curves of the common principal part of all equations, That is, along the particle flow ν_t, above, we get a linear homogeneous system (with $\lambda^1_{pqrs}=\lambda_{pqrs}\circ\nu_t$)

$$(1.8)\quad d/dt(a_{pqt}(x_t(x,\xi),\xi_t(x,\xi))+\sum_{rs}\lambda^1_{pqrs}a_{pqt}(x_t(x,\xi),\xi_t(x,\xi))=0.$$

Also we of course have the initial conditions

$$(1.9)\qquad a_{pqt}(x_t(x,\xi),\xi_t(x,\xi)) = a_{pq}(x,\xi)\ ,\ \text{at } t = 0\ .$$

Since the coefficient matrix $((\lambda_{pqrs}))$ is bounded in all variables $x,\xi \in \mathbb{R}^n$, $t \in \mathbb{R}$, it follows that a_{pqt} are bounded in x,ξ, over compact t-intervals (cf. the detailed discussion in VI,6). We get

$$(1.10)\quad \gamma(t) = \sum_{pp}|a_{pq}(t)|^2\ ,\ |\tfrac{d\gamma}{dt}|\le c\gamma\ ,\ \gamma(0) = \sum_{pq}|a_{pq}|^2= O(1)\ ,$$

c independent of t,x,ξ, which may be integrated for $\gamma=O(e^{ct})$. Thus

$$(1.11)\quad a_t,\ \partial_{x_p}a_t,\ \partial_{\xi_q}a_t,\ \varepsilon_{pq}a_t= \xi_p\partial_{\xi_q}a_t- x_q\partial_{x_p}a_t,\ p,q=1,\dots,n,$$

are bounded in every compact t-interval, as $x,\xi\in \mathbb{R}^n$.

Clearly the procedure may be iterated, under proper conditions on k. Once more, apply $\varepsilon_{\rho\sigma}$ to (1.7). Write $\varepsilon_{\rho\sigma}\varepsilon_{pq}a_t=a_{pq\rho\sigma t}$:

$$(1.12)\qquad \partial_t a_{pq\rho\sigma t} + \langle k,a_{pq\rho\sigma t}\rangle + \sum_{rs}\lambda_{pqrs}a_{rs\rho\sigma t}$$

$$+ \langle\varepsilon_{\rho\sigma}k,a_{pqt}\rangle + \sum_{rs}(\varepsilon_{\rho\sigma}\lambda_{pqrs})a_{rst} = 0\ .$$

In the last term we use that $\lambda_{pqrs}\in \psi c_0\subset \psi l_0$, implying $\varepsilon_{\rho\sigma}\lambda_{pqrs}\in \psi c_0$. For the fourth term we get $\sum_{rs}\lambda_{\rho\sigma rs}\varepsilon_{rs}a_{pqt}=\sum\lambda_{\rho\sigma rs}a_{pqrst}$, using (1.6) again. With this, and the relations (1.5), (1.7) get

$$(1.13)\qquad \partial_t a_{pqrst}+\langle k,a_{pqrst}\rangle+ \sum_{\pi\kappa\rho\sigma}\lambda_{pqrs\pi\kappa\rho\sigma}a_{\pi\kappa\rho\sigma t} = 0\ ,$$

$p,q,r,s=0,\dots,n$, again with $\lambda_{pqrs\pi\kappa\rho\sigma}\in \psi c_0$. Again the 'vector' $(a_{pqrst})_{p,q,r,s=0,..n}$ satisfies a system of ODE's in t along the particle flow with coefficients bounded in x,ξ,t. Again it follows that the a_{pqrst} are bounded for $x,\xi\in \mathbb{R}^n$, $t\in I\subset\subset \mathbb{R}$. We even may differentiate for t, deriving initial conditions for $(\partial_t^j a_{pqrst})_{t=0}$ from (1.13), for yet another system with bounded coefficients. Thus t-derivatives are bounded as well. We proved, for m=0 :

<u>Theorem 1.2</u>. Assume that the real-valued symbol $k\in \psi c_e$ satisfies condition (1.6). Then $a_t=a\circ\nu_t\in C^\infty(\mathbb{R},\psi l_0)$, for every $a\in \psi l_m$, $m\in \mathbb{R}^2$.

For general m we use VIII, prop.7.1: ψl_m is characterized by $\Pi\varepsilon_{p_jq_j}a=O(\pi_m)$, for all such products. For $a\in\psi l_m$, a_t as above, again get (1.1) , (1.7) , (1.13), etc. The coefficients k, λ_{pqrs} , $\lambda_{pqrs\rho\sigma}$, etc., are bounded as before. However, the initial values are $O(\pi_m(x,\xi))$. Thus, the resulting estimates now will be

$$(1.14)\qquad \partial_t^j a_{pqt}\ ,\ \partial_t^j a_{pqrst}\ ,\ \ldots\ ,\ = O(\pi_m\circ\nu_{-t}(x,\xi))\ .$$

However, in VI,6 we proved that the right hand side is $O(\pi_m(x,\xi))$. This completes the proof of thm.1.2 for general m.

<u>The class of all symbols</u> $k\in\psi c_e$ <u>satisfying (1.6) will be denoted by</u> $\psi\lambda\pi_e$. Note prop.1.3, below, describing a subset $\psi\lambda_e$ of of $\psi\lambda\pi_e$. For another such class see sec.4.

<u>Proposition 1.3</u>. Let $\psi\lambda_e$ be the class of $k\in\psi c_e$ with

$$(1.15)\qquad k(x,\xi)=\sum_{pq}k_{pq}x_p\xi_q+k_d+k_m\ ,\ k_d\in\psi c_{e^1}\ ,\ k_m\in\psi c_{e^2}\ ,\ k_{pq}\in\mathbb{C}\ .$$

Then $\psi\lambda_e\subset\psi\lambda\pi_e$ .The (real-valued) functions of $\psi\lambda_e$ form a Lie-algebra over $\mathbb{C}$ ($\mathbb{R}$) under the Poisson bracket product $\langle k_1,k_2\rangle$, containing the ε^{jl} of (1.3), and we have $\varepsilon_{pq}:\psi\lambda_e\to\psi\lambda_e$.

The proof is a calculation.

2. Invariance of ψs_m under particle flows.

In this section we will extend the results of sec.1 to ψs_m. We introduce analogues to $\psi\lambda_e$, $\psi\lambda\pi_e$: Let $\psi\sigma_e$ consist of all

$$(2.1)\qquad k(x,\xi) = \kappa_{00}x\xi + \sum\kappa_{pq}(x_p\xi_q-x_q\xi_p) + k_d + k_m\ ,$$

with $\kappa_{jl}\in\mathbb{C}$, and symbols k_d, k_m, as in (1.15). Let $\psi\sigma\pi_e$ consist of all $b\in\psi c_e$ with $b_{pq}=\eta_{pq}b$ satisfying

$$(2.2)\qquad \langle b_{pq},.\rangle = \sum_{p,q=0}^n\gamma_{pqrs}\eta_{rs}\quad ,\ \gamma_{pqrs}\in\psi c_0\ .$$

$\psi\lambda\pi_e$ is a proper subset of ψc_e: For n=1, $b(x,\xi)=\langle x\rangle^{1/2}\langle\xi\rangle\notin\psi\sigma\pi_e$ and $\notin\psi\lambda\pi_e$, as easily seen. Clearly $\psi\sigma_e\subset\psi\lambda_e$ and $\psi\sigma_e\subset\psi\sigma\pi_e$.

<u>Theorem 2.1</u>. Let the real-valued symbol $k(x,\xi)$ be in $\psi\sigma\pi_e$, and let $a\in\psi s_m$ .Then, with the characteristic flow $\nu_t=(x_t,\xi_t)$ of (0.2) and $a_t=a\circ\nu_t$, we have $a_t\in C^\infty(\mathbb{R},\psi s_m)$.

The proof of thm.2.1 is analogous to that of thm.1.2. Define

$$(2.3)\qquad \eta^{00}=-x.\xi,\ \eta^{jl}=x_j\xi_l-\xi_jx_l,\ j,l=1,\ldots,n,\ \eta^{10}=\xi_1,\ \eta^{01}=-x_1,$$

similarly than the ε^{jl} in sec.1, such that

(2.4) $\eta_{jl}a = \langle \eta^{jl}, a\rangle$, $j,l=0,\ldots,n$, $\eta_{l0} = \partial_{x_l}$, $\eta_{0l} = \partial_{\xi_l}$, $l\geq 1$ ..

Proposition 2.2. The (real) linear span of the functions η^{jl} , as well as the class $\psi\sigma_e$, are (real) Lie-algebras under $\langle .,.\rangle$.

The proof is a calculation again.

More general symbols in $\psi\sigma\pi_e$ are used in the proof of thm.4.1 For hyperbolic systems in sec.5 we discuss the following:

Proposition 2.3. Let $a\in \psi s_m$, $r=(\rho,\rho)$, $\rho\in \mathbb{R}$, $\pi_r(x,\xi)=\langle x\rangle^{\rho}\langle \xi\rangle^{\rho}$. Then

$$(2.5) \qquad \langle \pi_r, a\rangle = \rho\pi_{r-2e}\Big(\sum_{p<q}\eta^{pq}(\eta_{pq}a) + \eta^{00}(\eta_{00}a)\Big) \quad ,$$

The proof is a calculation :

$$\langle \pi_r, a\rangle = \rho\langle x\rangle^{\rho}\langle \xi\rangle^{\rho-2}\Sigma\xi_j a_{|\xi_j} - \rho\langle \xi\rangle^{\rho}\langle x\rangle^{\rho-2}\Sigma x_j a_{|\xi_j}$$

$$(2.6) \qquad = \rho\pi_{r-2e}\sum_{pq}(x_p^2 x_q a_{|x_q} - \xi_q^2 x_p a_{|\xi_p})$$

$$= \sum_{pq} x_p\xi_q(x_p a_{|x_q} - x_q a_{|x_p} + \xi_p a_{|\xi_q} - \xi_q a_{|\xi_p}) + \sum_{pq} x_q\xi_q(x_p a_{|x_p} - \xi_p a_{|\xi_p}),$$

which implies (2.5).

For a more general class of functions in $\psi\sigma\pi_e$ we introduce

$$(2.7) \qquad \zeta_{jl} = \eta^{jl}\pi_{-e} \quad , \ j,l=0,\ldots,n \ .$$

Proposition 2.4. The class $\psi\sigma\pi_0{}'$ of all functions c of the form

$$(2.8) \qquad c(x,\xi) = \Theta(\zeta_{jl}) + c_d(x,\xi) + c_m(x,\xi) \ ,$$

with a polynomial Θ in $\zeta_{00}, \zeta_{12},\ldots$, with complex coefficients, and $c_d\in \psi c_{-e^2}$, $c_m\in \psi c_{-e^1}$, is a subalgebra of ψc_0 . For $c\in \psi\sigma\pi_0{}'$ we have $b=c\pi_e\in \psi\sigma\pi_e$, and $\Pi_{j=1}^{N}\eta_{p_j q_j} b=\pi_e c^{\sim}$, $c^{\sim}\in \psi\sigma\pi_e$, for all products.

Proof: The algebra property of $\psi\sigma\pi_0{}'$ (under the pointwise product) is trivial: Clearly $\zeta_{jl}\in \psi c_0$, hence a polynomial in ζ_{jl} also belongs to c_0 (Note that $\psi c_{-e^1} + \psi c_{-e^2}$ is an ideal of ψc_0). Now let

$$(2.9) \qquad b = \pi_e\Theta(\zeta) + b_d + b_m \quad , \ b_d \in \psi c_{e^1} \ , \ b_m \in \psi c_{e^2} \ .$$

We introduce a multi-index-notation, writing

$$(2.10) \quad \Theta(\zeta) = \Sigma\gamma_\alpha\zeta^\alpha = \Sigma\gamma_\alpha\pi_{-|\alpha|e}\,\eta^\alpha \ , \ \zeta^\alpha = \Pi_{jl=0}^{n}\zeta_{jl}^{\alpha_{jl}} \ , \ \gamma_\alpha\in \mathbb{C} \ ,$$

with an $(n+1)^2$-dimensional multi-index $\alpha = (\alpha_{jl})$. The next observation is that every application $\Pi_{j=1}^N \eta_{p_j q_j} b = b^{\sim}$ again is of the form (2.9) . Indeed $\eta_{pq} b_d$ and $\eta_{pq} b_m$ have the same property than b_d and b_m, respectively. Also, calling the first term in (2.9) b_0, we get $b_0 \in \psi c_e$, hence $\eta_{p0} b_0 \in \psi c_{e^1}$, $\eta_{0q} b_0 \in \psi c_{e^2}$. One obtains

$$(2.11)\quad \eta_{00}\pi_{\mu e} = \mu(\pi_{\mu e-2e^2} - \pi_{\mu e-2e^1}) \ ,\ \eta_{jl}\pi_{\mu e} = 0 \ ,\ j,l=1,\dots,n \ ,$$

where $\mu \in \mathbb{R}$, by a calculation. Thus we get

$$(2.12)\quad \eta_{jl} b_0 = \sum_{\nu=0}^{N} \sum_{|\alpha|=\nu} \gamma_\alpha((\eta_{jl}\pi_{(1-\nu)e})\, \eta^\alpha + \pi_{(1-\nu)e}\, (\eta_{jl}\eta^\alpha)).$$

For $j,l \geq 1$ the first term vanishes ; for $j=l=0$ this term is in $\psi c_{e^1} + \psi c_{e^2}$, hence will go together with $b_d + b_m$. We must focus on

$$(2.13)\qquad \pi_{(1-\nu)e}\, \eta_{jl}\eta^\alpha = \pi_{(1-\nu)e} \sum_{pq=1}^{n} \alpha_{pq} \eta^{\alpha-\varepsilon^{pq}} \langle \eta_{jl}, \eta_{pq} \rangle \ ,$$

with ε^{pq} having (p,q)-component equal 1, all others zero. By prop. 2.2 $\langle \eta_{jl}, \eta_{pq} \rangle$ is a linear combination of the η_{rs}. Thus $\eta_{jl} b_0$ is of the same form as b_0 ,except for an additive term in $\psi c_{e^1} + \psi c_{e^2}$. Hence $\Pi\eta_{p_j q_j} b$ is of the form (2.9). We are left with showing that $\langle b,.\rangle$ is a combination of the η_{pq} with coefficients in ψc_0 .

The terms $\langle b_d,.\rangle$, $\langle b_m,.\rangle$ are handled as those in prop.1.3, using VIII,prop.6.2. For one of the terms of b_0 we get

$$(2.14)\quad \langle \pi_{(1-\nu)e}\, \eta^\alpha, a\rangle = \eta^\alpha \langle \pi_{(1-\nu)e}, a\rangle + \pi_{(1-\nu)e} \sum \alpha_{pq} \eta^{\alpha-\varepsilon^{pq}} \eta_{pq} a \ .$$

The last term indeed is a combination of the $\eta_{pq} a$, coefficients $\in \psi c_0$. For the first term apply (2.5), with $r=(1-\nu)e$. Q.E.D.

Define $\psi o\pi_m = \pi_{m-e} \psi o\pi_e$, and $\psi o\pi_m' = \pi_{m-e} \psi o\pi_e'$, for $m=\mu e$. Then $\psi o\pi_m$ consists precisely of all $c \in \psi c_m$ with $c^{\sim} = \Pi\eta_{p_j q_l} c$ satisfying

$$(2.15)\qquad \langle c^{\sim},.\rangle = \sum \gamma_{jl}^{\sim} \eta_{jl} \ ,\ \gamma_{jl}^{\sim} \in \psi c_{m-e} \ ,$$

as is easily verified, using (2.5). Moreover we note that $\psi o\pi_0$ and $\psi o\pi_0'$ are algebras, with the pointwise product. For $\psi o\pi_0'$ this was part of prop.2.4. For $c,d \in \psi o\pi_0$ one finds that

$$(2.16)\qquad \langle cd,.\rangle = c\langle d,.\rangle + d\langle c,.\rangle = \sum (c\delta^{jl} + d\gamma_{jl})\eta_{jl} \ ,$$

with evident notations, where we get $c\delta_{jl} + d\gamma_{jl} \in \psi c_{-e}$. Similarly for $(cd)^{\sim}$, which gives the algebra property.

Proposition 2.5. Every continuous root $\lambda(x,\xi)$, $|x|+|\xi|>>1$, of

$$(2.17)\qquad \Theta(\lambda) = \sum_{j=0}^{N}\theta_j\lambda^j \ , \quad \theta_N = 1 \ , \ \theta_j \in \psi\sigma\pi_0' \ , \ j=0,\dots,N-1,$$

may be extended to a symbol $\lambda \in \psi\sigma\pi_0$, with all $\Pi\eta_{p_jq_j}\lambda \in \psi\sigma\pi_0$, provided that for large $|x|+|\xi|$ all roots have constant multiplicity and their mutual distance is bounded away from zero.

Proof. For a root $\lambda(x,\xi)$ with constant multiplicity ρ we can write

$$(2.18)\qquad \lambda(x,\xi) = (2\pi i\rho)^{-1}\int_{|\tau-\lambda(x,\xi)|=\varepsilon} d\tau\ \tau\Theta'(\tau)/\Theta(\tau) \quad ,$$

where $\varepsilon>0$ is fixed -say, as half the minimum distance between two roots. But we have expansions

$$(2.19)\qquad \langle\theta_j,.\rangle = \sum\gamma_{jpq}\eta_{pq} \ , \ \gamma_{jpq} \in \psi c_{-e} \ , j=0,1,\dots,N \ .$$

Hence, with polynomials $\gamma_{pq}=\sum\gamma_{jpq}\lambda^j$,

$$(2.20)\qquad \langle\Theta(\tau),.\rangle = \sum\gamma_{pq}(\tau)\eta_{pq} \ , \ \langle\Theta'(\tau),.\rangle = \sum\gamma_{pq}'(\tau)\eta_{pq} \ .$$

Substituting into (2.18) one obtains

$$(2.21)\qquad \langle\lambda,.\rangle = \sum\lambda_{jl}\eta_{jl} \quad , \ \lambda_{jl} = (2\pi i\rho)^{-1}\int\tau d\tau(\gamma_{jl}'\Theta-\Theta'\gamma_{jl})/\Theta^2 \ ,$$

with path of (2.18). The path is kept constant during differentiation, the integral being locally independent. For t on the path,

$$(2.22)\qquad |\Theta(t)| = |\Pi(\lambda_j-t)| \ge \varepsilon^N > 0 \quad ,$$

since by assumption the path has distance $\ge\varepsilon$ from all roots λ_j . This allows to estimate λ_{jl} uniformly in x, ξ, for large $|x|+|\xi|$. Since Θ and Θ' are ψc_0 , but $\gamma_{jl} \in \psi c_{-e}$, we get $\lambda_{jl}=O(\pi_{-e})$. The formulas for λ_{jl} may be differentiated under the integral sign for $\lambda_{jl} \in \psi c_{-e}$. Also we get expansions for $\Pi\eta_{p_jq_j}\lambda$: Apply η_{pq} to (2.18), using estimates as above. Q.E.D.

Corollary 2.6. Let $k \in \psi\sigma\pi_{\mu e}$. Then we have

$$(2.23)\qquad \langle k,a\rangle \in \psi s_{m+(\mu-1)e} \ ,\text{for all } a \in \psi s_m \ , \ m \in \mathbb{R}^2 \ .$$

The proof is a consequence of (2.5) and (2.15).

3. Conjugation of Opψx with e^{iKt} , $K \in$ Opψc_e .

With invariance of the classes ψx_m $x=l,s$ under the particle flow of $k\in\psi\lambda\pi_e$ or $k\in\psi\sigma\pi_e$ established we now prove the invariance of $\mathrm{Op}\psi x_m$ under conjugation with e^{iKt}. The results will be similar in form and principle to VI,thm.5.1. Existence of the group $\{e^{iKt}\}$ of operators $S\to S$ (or $S'\to S'$, or $H_s\to H_s$, with the spaces of III,3) is insured by VI,thm.3.1. It would be easy to carry the results to the case of a solution operator $U(\tau,t)$ of $u^{\cdot}+Ku=0$, with $K=K(t)$.

In thm.3.1 we work with $k_0\in\psi\xi_e$, $\xi=\sigma,\lambda$, focusing on a simple nontrivial case. Slightly weaker conditions on k will be needed in sec.4, to prove coordinate invariance (cf. cor.3.4). A result for systems, requiring the classes $\psi\sigma\pi_e$ is discussed in sec.5.

<u>Theorem 3.1</u>. Let $K=k(x,D)$, where k is of the form $k=k_0+\kappa$, with a complex-valued $\kappa\in\psi c_0$, and a real-valued $k_0\in\psi\xi_e$, $\xi=\sigma,\lambda$. Then, for every $A=a(x,D)\in\psi l_m$, $m\in\mathbb{R}^2$, we have $A_t=e^{-iKt}Ae^{iKt}\in\mathrm{Op}\psi l_m$, $t\in\mathbb{R}$. Moreover, we get $A_t=a(t,x,D)$ with

$$(3.1)\qquad a(t,x,\xi)=(a\circ\nu_t)(x,\xi)+z(t,x,\xi)\ ,\ z\in\psi l_{-e^1}\cap\psi l_{-e^2}\ .$$

Also, $A_t\in C^\infty(\mathbb{R},\mathrm{Op}\psi x_m))$, and $\partial_t^j A_t=(-i\ \mathrm{ad}\ K)^j A$.

<u>Proof</u>. We retrace the proof of VI,thm.5.1, taking special care of the remainders, due to less perfect ψdo-calculus of VIII,7. Let

$$(3.2)\qquad P(t)=e^{iKt}a_t(x,D)e^{-iKt}\ ,$$

where $a\in\psi x_m$, $x=l,s$, $a_t=a\circ\psi_t$, $A_t=a_t(x,D)$. Here we use thm.1.2 (or thm.2.1) to assure that $a_t\in\psi x_m$, and $a_t(x,D)$ is meaningful. Note that this is the only direct reference to $x=s$ or $x=l$. We get

$$(3.3)\qquad dP(t)/dt=e^{iKt}(i[K,A_t]-\langle k,a_t\rangle(x,D))e^{-iKt}\ ,$$

where the operators are $S\to S$. By VI,thm.3.1 $\frac{d}{dt}(e^{iKt})$ exists in $L(H_s,H_{s-e})$, and by sec.1 (or sec.2) $\partial_t a_t=-\langle k_0,a_t\rangle$ exists in ψx_m. Hence dP/dt exists in $L(H_s,H_{s-m-e})$, for all s, and (3.3) is valid, using III,thm.3.1.

Now we use our calculus of ψdo's. VIII,prop.7.4 yields

$$(3.4)\quad [K,A]=c_A(x,D)\ ,\ c_A(x,\xi)=-i\langle k,a\rangle-\langle k,a\rangle_2/2+i\langle k,a\rangle_3/6+\ldots$$

where $\langle k,a\rangle\in\psi x_{m+e-re^1-r'e^2}$, $r+r'=j$, and the remainder R_N^3 has symbol in $\psi x_{m+e-re^1-r'e^2}$, $r+r'=N+1$, if the first N terms are used.

It is important that, for our present k, we even get $\langle k,a\rangle_j\in\psi x_{m-re^1-r'e^2}$, $r+r'=j-1$, hence $R_N^3\in\mathrm{Op}\psi x_{m-re^1-r'e^2}$, $r+r'=N$. For $x=l$, $j=1$ get $\langle x_p\xi_q,a\rangle_1=-\langle\varepsilon_{qp},a\rangle=-\varepsilon_{qp}a\in\psi x_m$, while $\langle k_d,a\rangle$, $\langle k_m,a\rangle$ belong to $\psi x_{m+e^1-re^1-r'e^2}$, and $\psi x_{m+e^2-re^1-r'e^2}$, resp. With r=1 (r=0) get all terms $\in\psi s_m$, as stated. Similarly for j=1, $x=s$. For j>1 the first sum (1.15) gives 0, since $\varepsilon^{pq}{}_{|xx}=\varepsilon^{pq}{}_{|\xi\xi}=\eta^{pq}{}_{|xx}=\eta^{pq}{}_{|\xi\xi}=0$. The other terms in (1.15) go as for j=1. To summarize:

Proposition 3.2. Under the assumptions of thm.3.1 we have

$$c_A=\sum_{j=1}^{N}(-i)^j/j!\ \langle k,a\rangle_j+r_N^3,\ r_N^3\in \psi x_{m-re^1-r'e^2},\ r+r'=N, \tag{3.5}$$

where the j-th term belongs to $\psi x_{m-re^1-r'e^2}$, r+r'=j-1.

Applying prop.3.2 for N=1 we get

$$\gamma(t,x,\xi) = ic_{A_t}(x,\xi) - \langle k,a_t\rangle \in \psi x_{m-e^1} \cap \psi x_{m-e^2}\ . \tag{3.6}$$

Note that (3.6)implies $\gamma(t,x,\xi)=O((\pi_m(x,\xi)(\langle\xi\rangle\langle x\rangle)^{-1/2})$.In fact, we get $\gamma\in\psi t_{m-e/2}$. Moreover, we know from sec.1 or 2, and the calculus of ψdo's between ψc_e and ψx that γ is continuous as a map $\mathbb{R}\to\psi x_{m-e^1}$ (or ψx_{m-e^2}). Actually, we even get it C^∞: One knows from sec.1 or 2 that a_t is C^∞ , and the asymptotic expansion (3.4) (or Leibniz-Taylor-expansion (3.5)) may be differentiated arbitrarily for t. Then prop.3.2 may be applied to the differentiated expansion. Hence we even get $\gamma\in C^\infty(\mathbb{R},\psi t_{m-e/2})$. Again this gives

$$P(t) = A + \int_0^t d\tau e^{iK\tau}\gamma(\tau,x,D)e^{-iK\tau}\ . \tag{3.7}$$

Or,

$$e^{-iKt}Ae^{iKt} = a_t(x,D) - \int_0^t d\tau e^{iK(\tau-t)}\gamma(\tau,x,D)e^{iK(t-\tau)}\ . \tag{3.8}$$

This is a decomposition like (3.1), but the reminder (second term) is not in ψdo-form. It is norm continuous in $L(H_s,H_{s-m-f})$, f $=e^1$, e^2, e/2. Also, $\frac{d}{dt}(e^{-iKt}Ae^{iKt})$ exists in $L(H_s,H_{s-m})$, by (3.8).

(3.8) may be iterated: Noting γ satisfies (3.6), and even $\gamma\in C^\infty(\mathbb{R},\psi x_{m-f})$, $f=e^1$, e^2 , we get, setting $\gamma_\kappa(t,x,\xi)=\gamma\circ\nu_\kappa(t,x,\xi)$,

$$e^{-iK\kappa}\gamma(\tau,x,D)e^{iK\kappa}=\gamma_\kappa(\tau,x,D)-\int_0^\kappa d\kappa' e^{iK(\kappa'-\kappa)}W(\kappa',\tau)e^{iK(\kappa-\kappa')}, \tag{3.9}$$

with a ψdo $W = w(\kappa',\tau,x,D)$ obtained by substituting γ for a into (3.7) . Clearly it follows that $w\in C^\infty(\mathbb{R}^2,\psi x_{m-re^1-r'e^2})$, r+r'=2 , using (3.6) . Also, $\gamma_\kappa\in C^\infty(\mathbb{R}^2,\psi x_{m-f})$, $f=e^1$, e^2 .

With (3.9), for $\kappa=t-\tau$, in the integrand of (3.8) we get

$$e^{-iKt}Ae^{iKt} = a_t(x,D) + z_{1t}(x,D) + W_{1t}\ , \tag{3.10}$$

where $W_{1t}\in L(H_s,H_{s-m+re^1+r'e^2})$, while the symbol ζ_{1t} is in $C^\infty(\mathbb{R},\psi x_{m-f})$, $f=e^1$,e^2 . An N-fold iteration then yields

$$e^{-iKt}Ae^{iKt} = a_t(x,D) + \sum_{j=1}^{N}z_{jt}(x,D) + W_N\ , \tag{3.11}$$

with symbols $z_j \in C^\infty(\mathbb{R}, \psi x_{m-re^1-r'e^2})$, $r+r'=j$, and with a remainder in $O(m-re^1-r'e^2)$, $r+r'=N+1$. Moreover, the remainder will have all derivatives for t of order up to N+1 existing in norm convergence of every $L(H_s, H_{s-re^1-r'e^2})$, since it is an N+1-fold iterated integral over a bounded, strongly continuous integrand.

Finally we take the asymptotic sum

$$(3.12) \qquad z_t = \sum_{j=1}^{\infty} z_{jt} \pmod{\psi t_{-\infty}} .$$

It exists, because $\zeta_{jt} \in \psi t_{-je/2}$ follows as above. Then $e^{-iKt} A e^{iKt} - a_t(x,D) - z_t(x,D) = V_t$ clearly is of order $-\infty$. Its t-derivatives of all orders exist and are of order $-\infty$ again. Thus $V_t = v_t(x,D)$, $v_t \in C^\infty(\mathbb{R}, S(\mathbb{R}^{2n}))$. Also all t-derivatives of the asymptotic sum z_t exist in ψx_{m-f}, $f=e^1, e^2$. Writing $z_t + v_t = z(t,x,\xi)$ we get $e^{-iKt} A e^{iKt} = a_t(x,D) + z(t,x,D)$, which amounts to (3.1). In particular, we have verified all the properties stated for z in thm.3.1.

With differentiability established we clearly get $\partial_t A_t = -i[K, A_t] = -i(\mathrm{ad}\, K) A_t$, where $[K, A_t] \in Op\psi x_m$ again. This may be iterated, to complete the proof of thm.4.1.

It will be useful for sec.4 to reexamine the proof of thm.4.1 for a possible generalization to the case where the first sum of (1.15) (or (2.1)) is replaced by a linear combination of terms $\langle x\rangle \sigma(s(x)) \xi^q$, with $s(x) = \frac{x}{\langle x\rangle}$, and a function $\sigma \in C^\infty(B_1^c)$, with the closed unit ball $B_1^c \subset \mathbb{R}^n$, as in IV. In other words, let

$$(3.13) \quad k = \langle x\rangle \sum_q \sigma_q(s(x)) \xi_q + k_m + k_d,\ k_m \in \psi c_{e^2},\ k_d \in \psi c_{e^1},\ \sigma_q \in C^\infty(B_1^c),$$

Let $\psi\lambda\kappa_e$ be the class of such functions. Clearly $\psi\sigma_e \subset \psi\lambda_e \subset \psi\lambda\kappa_e$.

<u>Proposition 3.3</u>. We have $\psi\lambda\kappa_e \subset \psi\lambda\pi_e$. In fact, not only $\langle \varepsilon^{pq} k, .\rangle$, but also $\langle k, .\rangle$ is a combination of the ε_{jl}, with ψc_0-coefficients.

<u>Proof</u>. It suffices to focus on the case $k(x,\xi) = \langle x\rangle \sigma(s(x)) \xi_q = \alpha \xi_q$. A calculation shows that $\varepsilon^{pq} : \psi\lambda\kappa_e \to \psi\lambda\kappa_e$. Thus we only must show that $\langle \alpha\xi_q, .\rangle = \sum \gamma_{pq} \varepsilon^{pq}$. But $\langle \alpha\xi_q, .\rangle = \alpha\partial_{x_q} - \sum \alpha_{|x_l} \xi_q \partial_{\xi_l} = \alpha\partial_{x_q} - \sum \alpha_{|x_l} \varepsilon_{ql}$ $+ \sum \alpha_{|x_l} x_l \partial_{x_q}$. Here $\alpha_{|x_l} \in \psi c_0$, so look at $\alpha\partial_{x_q} - \sum \alpha_{|x_l} x_l \partial_{x_q} = \langle x\rangle^{-1} \sigma \partial_{x_q}$ $+ \sum x_l (s_l \sigma(s) - (\langle x\rangle \sigma(s))_{|x_l}) \partial_{x_q}$. Finally, $(\langle x\rangle \sigma(s(x)))_{|x_l} = (s_l\sigma)(s(x))$ $+ \sum \sigma_{|x_m}(s(x))(\delta_{lm} - s_l s_m(x))$, so that $\sum_l x_l (s_l \sigma(s) - (\langle x\rangle \sigma(s))_{|x_l}) =$ $- \sum_{m,l} x_l \sigma_{|x_m}(s(x))(\delta_{lm} - \sigma_l s_m) = \langle x\rangle^{-1} \sum s_l \sigma_{|l}(s(x))$. Q.E.D.

Looking at prop.3.3 and the proof of thm.3.1: In the case $x=1$ we clearly get a_t well defined again, and $a_t \in C^\infty(\mathbb{R},\psi l_m)$. Also, $\langle k,a_t\rangle=\sum\gamma_{pq}\varepsilon_{pq}a_t\in\psi l_m$ again, if we weaken the assumptions, requiring $k_0\in\psi\lambda\kappa_e$ only. Formula (3.6) remains true as well: We need $\langle k,a\rangle_2=-\sum_{jl}k_{|x_jx_l}a_{|\xi_j\xi_l}\sum_{jl}\alpha_{|x_jx_l}\xi_q a_{|\xi_j\xi_l}$, for $k=\alpha\xi_q$ again. First observe that $\alpha_{|x_j}=s_j\sigma(s)+s_j\sigma_{|j}(s)-s_j\sum\sigma_{|m}(s)s_m$ is a sum of function $\sigma(s(x))$. Thus we may look at $\sum\sigma(s(x))_{|x_l}\xi_q\partial_{x_l}\partial_{\xi_j}=\sum\sigma(s(x))_{|x_l}\varepsilon_{ql}\partial_{\xi_j}$ $+\sum\sigma(s(x))_{|x_l}x_l\partial_{x_q}\partial_{\xi_j}$. We get $\sigma(s(x))_{|x_l}=\sum\sigma_{|m}(s)(\delta_{lm}-s_ls_m)/\langle x\rangle\in$ ψc_{-e^2} , hence the first term is in $\psi l_{m-e_1}\cap\psi l_{m-e_2}$. For the second term we get $\sum s_l\sigma_{|m}(s)(\delta_{lm}-s_ls_m)=\sum\sigma_m\sigma_{|m}(s)/\langle x\rangle^2$, Using VIII,(6.12) we get the proper estimate for (3.6) for the second term as well.

We have proven:

Corollary 3.4. In the case $x=1$ the statement of thm.3.1 is still valid, if "$k_0\in\psi\lambda_0$" is replaced by "$k_0\in\psi\lambda\kappa_0$" .

Indeed, the iteration and asymptotic sum leading to (3.1) no longer depend on "$k_0\in\psi\lambda_0$".

4. Coordinate and gauge invariance; extension to *S*-manifolds.

We now will look into coordinate invariance of the two algebras $\Psi\mathcal{AL}$ and $\Psi\mathcal{AS}$,and ,more generally, $Op\psi l$, $Op\psi s$. Given a global transform $x=\varphi(y)$,with a diffeomorphism $\varphi:\mathbb{R}^n\to\mathbb{R}^n$, inverted by ψ: $\mathbb{R}^n\to\mathbb{R}^n$:We have $\varphi(\chi(x))=\chi(\varphi(x))=x$, $x\in\mathbb{R}^n$. Assume (the components of) $\varphi(x)$, $\chi(x)$ are $C^\infty(\mathbb{R}^n)$. The Jacobian matrix $((\partial\varphi_j/\partial x_l(x)))$ = $J(x)$ is nonsingular, together with $((\partial\chi_j/\partial x_l(x)))=J^{-1}(\varphi(x))$, $x\in\mathbb{R}^n$

Consider the substitution operators T_φ, T_χ defined by

$$(4.1)\qquad (T_\varphi u)(x)=u(\varphi(x))=(u\circ\varphi)(x)\ ,\ (T_\chi u)(x)=u(\chi(x))\ ,$$

for functions u over $\mathbb{R}^n$. Assume $0<c_1\le|\det J(x)|\le c_2$, with constants, to insure that $T_\varphi\ T_\chi : H\to H$.

If a folpde

$$(4.2)\qquad L=\sum_j a_j(x)\partial_{x_j}+a_0(x)\quad ,\ a_j\in C^\infty(\mathbb{R}^n)\ ,j=0,1,\dots,n\ ,$$

is given, then we get a transformed folpde $L^{\sim} = T_\varphi L T_\chi$. We get

$$(4.3)\qquad L^{\sim} = \sum_j a^{\sim}_j(x)\partial_{x_j} + a^{\sim}_0(x) \ , \ a^{\sim}_0 = T_\varphi a_0 \ ,$$

$$a^{\sim}_j(x) = \sum_l (\partial\chi_j/\partial\xi_l)(\varphi(x))a_l(\varphi(x)) \ .$$

The exponentiations e^{Lt} and $e^{L^{\sim}t}$ are related by

$$(4.4)\qquad e^{L^{\sim}t} = T_\varphi e^{Lt} T_\chi \ ,$$

since the PDE $\partial_t u = Lu$ transforms into $\partial_t v = L^{\sim} v$, with $v = T_\varphi u$.

To prove coordinate invariance of ΨGL or ΨGS we will use thm.3.1 or cor.3.4, with the ψdo $K=k(x,D)=-iM$, M as $L^{\sim}$ in (4.3), φ replaced by $\chi=\varphi^{-1}$, with $L=ix_q$, ∂_{x_p} , $x_q\partial_{x_p}+\frac{1}{2}\delta_{pq}$, in case of ΨGL.

For ΨGS we use $L=ix_q$, ∂_{x_p} , $x_p\partial_{x_q}-x_q\partial_{x_p}$, and $\sum x_j\partial_{x_j}+\frac{n}{2}$. We must verify that the transformed expressions satisfy the assumptions. as outlined in sec.0. We consider S-admissible transforms, as in IV,3, only: With $s(x)=\frac{x}{\langle x\rangle}$, $s:\mathbb{R}_n\to B_1$, of IV,(1.5), assume that

$$(4.5)\qquad \varphi = s^{-1}\circ\psi\circ s \ , \ \chi=\varphi^{-1} = s^{-1}\circ\theta\circ s \ ,$$

with diffeomorphisms $\psi: B_1^c \Leftrightarrow B_1^c$, $\theta: B_1^c \Leftrightarrow B_1^c$, inverting each other. Under this general assumption, if it can be shown that $T_\varphi \Psi GX T_\chi \subset \Psi GX$, for $X=L$ or $X=S$, it follows at once that also $T_\varphi Op\psi x_m T_\chi \subset Op\psi x_m$, because we already know from IV,thm.3.3 that $T_\varphi Op\psi c_m T_\chi \subset Op\psi c_m$. For $A\in Op\psi x_m$ we have $B=A\Pi_{-m}\in Op\psi x_0 = \Psi GX$. Then $T_\varphi B T_\chi \in \Psi GX$ follows. Also, $T_\varphi B T_\chi = (T_\varphi A T_\chi)(T_\varphi P_{-\mu} T_\chi)$, where the second factor belongs to $Op\psi c_{-m} \subset Op\psi x_{-m}$. Also $T_\varphi P_{-m} T_\chi$ is inverted by $T_\varphi P^*_m T_\chi \in Op\psi x_m$. Hence $T_\varphi A T_\chi = (T_\varphi B T_\chi)(T_\varphi P^*_m T_\chi) \in Op\psi x_m$.

For thm.4.1 below we must focus on the expressions $M=M_{pq}$,

$$(4.6)\qquad M_{0q} = i\chi_q(x) \ , \ M_{p0} = \sum_j \varphi_{j|x_p}(\chi(x))\partial_{x_j} \ ,$$

$$M_{pq} = \sum_j \chi_q(x)\varphi_{j|x_p}(\chi(x))\partial_{x_j} + \tfrac{1}{2}\delta_{pq} \ , \ p,q=1,\dots n \ ,$$

in case of $X=L$, and M_{0q} , M_{p0} , $\sum_j M_{jj}$, $M_{pq}-M_{qp}$, $1\le p=q\le n$, for $X=S$.

Here we first apply IV,prop.2.1: With our assumption (4.5) on φ, we have $\varphi_{j|x_p}\circ s^{-1}\in C^\infty(B_1^c)$, and $\chi=s^{-1}\circ\theta\circ s$, hence $\varphi_{j|x_p}(\chi(x)) = =\varphi_{j|x_p}\circ(s^{-1}\circ\theta\circ s)(x) = ((\varphi_{j|x_p}\circ s^{-1})\circ\theta)\circ s(x) = \sigma_{jp}(s(x))$, with

$\sigma_{jp}=(\varphi_{j|x_p}\circ s^{-1})\circ\theta\in C^\infty(Bc_1)$.

On the other hand, $\langle x\rangle^{-1}\varphi(x)=\psi(s(x))\{\langle x\rangle\sqrt{1-|\psi(s(x))|^2}\}^{-1}$, where the second factor may be written as $\{\frac{1-s^2}{1-\psi(s)^2}\}^{1/2}\in C^\infty(B_1^c)$, just as in IV,(2.8),(2.9). This shows that $M_{pq}=ik_{pq}(x,D)+\kappa_0$, where

(4.7) $k_{0q}=\langle x\rangle\sigma_{0q}\circ s(x)$, $k_{p0}=\sum\sigma_{jp}\circ s(x)\xi_j$, $k_{pq}=\langle x\rangle\sum\sigma_{pqj}\circ s(x)\xi_j$,

are symbols of $\psi\lambda\kappa_e$ of sec.3, so that cor.3.4 applies. We proved

Theorem 4.1. Let the diffeomorphism $\varphi:\mathbb{R}^n\leftrightarrow\mathbb{R}^n$ and its inverse function $\chi:\mathbb{R}^n\leftrightarrow\mathbb{R}^n$ be of the form (4.5) with $s(x)=\frac{x}{\langle x\rangle}$, and diffeomorphisms $\psi:B_1^c\to B_1^c$, $\theta:B_1^c\to B_1^c$, $\psi\circ\theta=id$. Then the transforms T_φ and T_χ (4.1) leave ΨGL and every $Op\psi l_m$, $m\in\mathbb{R}^2$, invariant. That is,

(4.8) $T_\varphi\Psi GLT_\chi=T_\chi\Psi GLT_\varphi=\Psi GL$, $T_\varphi Op\psi l_mT_\chi=T_\chi Op\psi l_mT_\varphi=Op\psi l_m$, $m\in\mathbb{R}^2$.

Theorem 4.2. Assume that the diffeomorphisms φ,χ satisfy

(4.9) $\varphi(x)=gx+\omega(x)$, $\chi(x)=g^{-1}(x)+\upsilon(x)$, $g\in GL(\mathbb{R}^n)$, $\omega,\upsilon\in\psi c_0$.

Then (4.8) is valid - i.e., T_φ and T_χ leave $Op\psi l_m$ invariant. Moreover, if in addition we have $g=\sigma o$, $0<\sigma\in\mathbb{R}$, o orthogonal, then also ΨGS and $Op\psi s_m$ are invariant under T_φ and T_χ .

Next let us turn to gauge transforms. A calculation gives

(4.10) $$L^{\sim}=e^{-i\mu(x)}Le^{i\mu(x)} = L+i\sum_p a_p(x)\mu_{|x_p}(x) .$$

Applying our arguments to the generators ix_q, ∂_{x_p}, $x_q\partial_{x_p}+\frac{1}{2}\delta_{pq}=L_{pq}$ again we now must prove $e^{Mt}A\varepsilon^{-Mt}\in C^\infty(\mathbb{R},L(H))$, for $A\in\Psi GX$, where $M=M_{pq}=L_{pq}+\mu_{pq}$, $\mu_{0q}=0$, $\mu_{p0}=i\mu_{|x_p}$, $\mu_{pq}=ix_q\mu_{|x_p}$.

Theorem 4.3. Let the 'gauge function' $\mu(x)$ be real-valued and let $\mu(x)\in\psi c_{e^2}$. Then $Op\psi x_m$, $m\in\mathbb{R}^2$, $x=s,l$, are invariant under

(4.11) $$A\to e^{-i\mu(x)}Ae^{i\mu(x)} .$$

The proof follows from thm.3.1, for m=0. For general m we still must show that $e^{-i\mu}\langle D\rangle^s e^{i\mu}\in\psi c_s$, $s\in\mathbb{R}$. We leave this to the reader; the proof is not difficult.

Finally, after discussing coordinate invariance, it is clear that the class $Op\psi l_m$ of natural ψdo's may be considered on a manifold with conical ends, just as earlier $Op\psi h_{m,\rho,\delta}$ (or $Op\psi c_m$).

Repeating the discussions of IV,4 with ψl_m instead of $\psi h_{m,\rho,\delta}$ generates classes called LL_m of 'natural ψdo's' on an S-manifold Ω .

On the other hand, we do not expect invariance of $Op\psi s_m$ under general S-admissible transforms (4.5), although, evidently, thm.4.2 implies invariance under local transforms extended as a multiple of the identity outside a compact set.

Our use of $Op\psi s$ in Dirac theory might profit from introduction of $Op\psi s(\Omega)$ for an Ω coinciding with $\mathbb{R}^n$ outside $K\subset\subset\Omega$.

Problem: Give a definition of $Op\psi s(\Omega)$, for Ω as above, using a partition $1=\Sigma\chi_j$ with $\chi_1=1$ near ∞, and also, by smoothness properties.

5. Conjugation with e^{iKt}, for a matrix-valued $K=k(x,D)$.

In this section we take up the question of extending thm.3.1 to the case of a $\nu\times\nu$-matrix $K = k(x,D)=((k_{jl}(x,D)))_{j,l=1,\dots,\nu}$ of of ψdo's $K_{jl} = k_{jl}(x,D)$. We shall give preference to the class ψs , from now on, but note that most arguments will also work for ψl. Generally we assume $k_{jl}\in\psi\sigma_e$ (cf. sec.2). Instead of requiring

$$(5.1)\qquad \partial_t u - ik(x,D)u = 0 \ , \ u=\varphi \text{ at } t=0 \ ,$$

to be strictly hyperbolic of type e^Δ, as in VI,4, we will accomodate the Dirac equation by allowing multiple eigenvalues, but of constant multiplicity for $k(x,\xi)$, similarly as in [CD] .

With comparison operators $\Lambda=\pi_{e^\Delta}(x,D)$, $e^\Delta=e,e^1,e^2$, as in VI,4, we call equation (5.1) semi-strictly hyperbolic (of type e^Δ) if

(i) We have $k(x,\xi)=k_{e^\Delta}(x,\xi)+\kappa(x,\xi)$, $k_{e^\Delta}=\pi_{e^\Delta}k_0$, $\kappa\in\psi c_{e^\Delta-e}$, with entries of the $\nu\times\nu$-matrix k_{e^Δ} of the form (2.1):

$$(5.2)\quad k_{e^\Delta jl}=\sum_{p,q=0}^n\kappa_{pq}^{jl}\eta^{pq}+k_{djl}+k_{mjl},\ \kappa_{pq}^{jl}\in\mathbb{C},\ k_{djl}\in\psi c_{e^1},\ k_{mjl}\in\psi c_{e^2},$$

where all but k_{djl} or k_{mjl} vanishes in case of $e^\Delta=e^1$, e^2, resp.

(ii) The $\nu\times\nu$-matrix $k_{e^\Delta}(x,\xi)$ is diagonalizable for all $f_2|x|+f_1|\xi|\ge\eta_0$, $e^\Delta=(f_1,f_2)$, sufficiently large η_0. All eigenvalues λ_j of k_{e^Δ} are real, and of constant multiplicity ν_j (independent of x,ξ), as $f_2|x|+f_1|\xi|\ge\eta_0$.

Well known perturbation arguments imply that the λ_j are C^∞-functions of x and ξ . The matrices k_{e^Δ} and $k_0=k_{e^\Delta}\pi_{-e^\Delta}$ have the same spectral projections, and the eigenvalues of k_0 are given by $\mu_j = \lambda_j\pi_{-e^\Delta}$. We may arrange for

$$(5.3)\qquad \mu_1(x,\xi)<\mu_2(x,\xi)<\dots<\mu_\rho(x,\xi),\ \nu_1+\dots\nu_\rho=\nu,\ f_2|x|+f_1|\xi|\ge\eta_0.$$

As our third condition, we assume that

(iii) $|\mu_j(x,\xi)-\mu_l(x,\xi)| \ge \delta_0$, $f_2|x|+f_1|\xi| \ge \eta_0$, $j=1,\dots,\rho-1$, with some positive constant δ_0 independent of x,ξ .

We only will admit the weight functions π_{e^Δ} mentioned, although others, such as $\pi_{e^1}+\pi_{e^2}$ might be useful as well. Mainly $e^\Delta=e^1$ will be of interest, for the Dirac equation.

<u>Theorem 5.1</u>. If equation (5.1) is semi-strictly hyperbolic of type e^Δ then the solution operator e^{iKt} of (5.1) exists as an operator in $O(0)$. Moreover, the partial derivatives (5.4) exist in norm topology of $L(H_s,H_{s-je^\Delta})$. For every $s\in\mathbb{R}^2$, and $j=0,1,2,\dots$, we get

$$(5.4) \qquad \partial_t^j(e^{iKt}) = K^j e^{iKt} \in O(jf) .$$

<u>Proof</u>. We use VI,thm.4.2 with a symmetrizer of the form

$$(5.5) \qquad r(x,\xi) = \sum_{j=1}^{\rho} \bar{p}_j^{\,T}(x,\xi)p_j(x,\xi) + r_{-e}(x,\xi) \ , \ r_{-e}\in\psi c_{-e} \ ,$$

using the spectral projections $p_j(x,\xi)$ to $\mu_j(x,\xi)$ of $k_0(x,\xi)$, $f_2|x|+f_1|\xi| \ge 2\eta_0$, extended to symbols (with coeffivients) in ψc_0 . The discussion follows the proof of VI,thm.4.4, with small amendments, left to the reader. The point is formula VI,(4.19), i.e.,

$$(5.6) \quad p_j(x,\xi)=i/2\pi\int d\lambda_{|\lambda-\mu_j(x,\xi)|=\rho_0/2}\,(k_0(x,\xi)-\lambda)^{-1}, \ |x|+|\xi|>\eta_0,$$

describing the spectral projections in $f_2|x|+f_1|\xi|>\eta_0$. Then the p_j must be properly extended to $\mathbb{R}^{2n}$, and r_{-e} must be chosen to satisfy the cdn's of VI,4.

<u>Proposition 5.2</u>. The eigenvalues λ_{e^Δ} , $f_2|x|+f_1|\xi|>2\eta_0$, of $k_{e^\Delta}(x,\xi)$ extend ty symbols $\lambda_j\in\psi\sigma\pi_e$ (cf. sec.2).

<u>Proof</u>. The eigenvalues μ_j solve the algebraic equation

$$(5.7) \qquad \Theta(\mu,x,\xi) = \det(k_0(x,\xi)-\mu) = \sum_{j=0}^{\nu} \theta_j(x,\xi)\mu^j = 0 \ ,$$

where the coefficients θ_j are polynomials in the coefficients of k_0 . If $e^\Delta\ne e$ then we just get $\mu_j\in\psi c_0$, $\lambda_j=\mu_j\pi_{e^\Delta}\in\psi c_{e^\Delta}$, by an argument as for VI,prop.4.3. For $e^\Delta=e$ apply prop.2.4 for $\theta_j\in\psi\sigma\pi_0'$, hence also $\mu_j\in\psi\sigma\pi_0'$, by prop.2.5. Note that the assumptions hold, by (i)-(iii) above. Thus we get $\lambda_j=\pi_e\mu_j\in\psi\sigma\pi_e$, as stated. Q.E.D.

The following example shows that the class $Op\psi s_m^\nu$ of all $\nu\times\nu$-matrices $A=((A_{jl}))$ of ψdo's $A_{jl}\in Op\psi s_m$ is not invariant under the conjugation $A\to A_t=e^{-iKt}Ae^{iKt}$. Let

$$(5.8) \qquad k(x,\xi) = \begin{pmatrix}\langle\xi\rangle & 0\\ 0 & -\langle\xi\rangle\end{pmatrix} \ , \quad a(x,\xi) = \begin{pmatrix}0 & 1\\ 0 & 0\end{pmatrix} \ .$$

Then clearly $k \in \psi\sigma_e^2$, and $a \in \psi s_0^2$. But a calculation shows that

(5.9) $$e^{-iK}Ae^{iK} = \begin{pmatrix} 0 & P \\ 0 & 0 \end{pmatrix} \text{ , with } P = e^{-2i\langle D\rangle} .$$

Clearly $P = p(D)$, $p(\xi) = e^{-2i\langle\xi\rangle} \in \psi t_0$,but $p \notin \psi s_0$.

On the other hand, as a simple consequence of thm.3.1 we get $e^{-iKt}Ae^{iKt} \in Op\psi s_m$ if $a(x,\xi)$ and $k_0(x,\xi)$ are diagonal matrices for large $|x|+|\xi|$. If only a and k commute for large $|x|+|\xi|$ we can expect (3.8) with an error term Z_t (the integral at right) of lower order, not necessarily a ψdo (cf.also [Tl_3]).

Thm.5.4, below, shows that even the error Z_t becomes a ψdo $\in Op\psi s_{m-e^\Delta}$, $e^\Delta = e^1$, e^2 , if only a suitable 'correction' of lower order is added to the symbol $a(x,\xi)$ commuting with $k_0(x,\xi)$ (that is, if we let $a=q+z$, with $[k_0,q]=0$, and suitable lower order z).

From now on we always assume that (5.1) is semi-strictly hyperbolic of type e^Δ. We shall see that a subalgebra $P=P^K \subset Op\psi s^\nu$ can be defined, essentially by the property that its ψdo's remain ψdo's, of the same symbol type, when conjugated with e^{iKt} .

For $m \in \mathbb{R}^2$ we define $P_m = P_m^K$ as the class of all $A=a(x,D) \in Op\psi s_m^\nu$ such that A_t of (5.10) belongs to $Op\psi s_m^\nu$, and, moreover that

$(5.10)_e$ $$A_t = e^{-iKt}Ae^{iKt} \in C^\infty(\mathbb{R}, Op\psi s_m^\nu) \text{ , in case of } e^\Delta = e.$$

For $e^\Delta = e^1$, or $e^\Delta = e^2$ we replace $(5.10)_e$ by

$(5.10)_{e^\Delta}$ $$\partial_t^j A_t \in C^\infty(\mathbb{R}, Op\psi s^\nu_{m-j(e-e^\Delta)}) \text{ , } j = 0,1,\ldots .$$

Then we define, for $e^\Delta = e$, e^1 , or e^2 ,

(5.11) $$P = P^K = \bigcup_{m \in \mathbb{R}^2} P_m^K .$$

<u>Proposition 5.3</u>. The class P is a graded algebra invariant under ad iK (as in VIII,(8.4)), i.e., $(\text{ad } iK)A = \partial_t(e^{iKt}Ae^{-iKt})|_{t=0}$). Especially, if $A \in P_m$, $B \in P_{m'}$, then $AB \in P_{m+m'}$. Moreover, if $A \in P_m$, then $(\text{ad } K)A \in P_{m-(e-\varepsilon^\Delta)}$. Moreover P_0, with the Frechet topology of VIII,6, is an (adjoint invariant) ψ^*-subalgebra of $L(H)$.

Proof evident. Clearly $B=(\text{ad } iK)A$ exists for $A \in P_m$, and $B_t = e^{-iKt}Be^{iKt} = -\dot{A}_t \in C^\infty(\mathbb{R}, Op\psi s^\nu_{m-(j+1)(e-e^\Delta)})$.

<u>Theorem 5.4</u>. Let (5.1) be semi-strictly hyperbolic of type e^Δ.

(1) For each ν×ν-matrix-valued symbol q with

(5.12) $$q \in \psi s_m^\nu \text{ , } [k_0(x,\xi), q(x,\xi)] = 0 \text{ , } |x|+|\xi| \geq \eta_1 \text{ ,}$$

there exists a symbol $z \in \psi s^\nu_{m-e}$ such that $A = a(x,D)$, with $a=q+z$, is an operator in P_m . That is, in particular,

$$(5.13)\quad \begin{array}{l} A_t = e^{-itK} Ae^{itK} \in Op\psi s^{\nu}_{m} , t \in \mathbb{R} , A_t = a_t(x,D) , \\ \partial^j_t a_t \in C^{\infty}(\mathbb{R},\psi s^{\nu}_{m-j(e-f)}) , j=0,1,2,\ldots . \end{array}$$

(2) Vice versa, if $A = a(x,D) \in P^K_m$ is given , then there exists a decomposition (valid for all $t \in \mathbb{R}$)

$$(5.14)\quad a_t=q_t+z_t , z_t\in \psi s^{\nu}_{m-e} , [k_0(x,\xi),q_t(x,\xi)]=0 , |x|+|\xi| \geq\eta_0 .$$

Moreover, the decomposition (5.14) may be differentiated for t , for the corresponding decomposition $\partial^j_t a_t=\partial^t_t q_t+\partial^j_t z_t$ of the symbol $\partial^j_t a_t$ of $((ad_{iK})^j A)_t$. In particular, $\partial^j_t z_t \in \psi s^{\nu}_{m-j(e-f)-e}$.

(3) If, for any q with (5.12), the symbols z_1, z_2 both satisfy (1) then z_1-z_2 satisfies (5.13) for m-e , instead of m .

We shall lay out the proof of thm.5.4 in the remainder of this section, and will finish it in sec.7, after discussing some auxiliary results on a commutator-differential equation in sec.6.

The particle flows of the eigenvalues $\lambda_j(x,\xi)$ will be important. Even in the strictly hyperbolic case, when all eigenvalues are distinct, we will have to solve a succesive sequence of matrix commutator equations, for careful alignment of the correction z. Moreover, for multiple eigenvalues, not only will the symbol flow along these flows, but, in addition, there will be a 'similarity action' within the eigenspaces of $k(x,\xi)$. For details see sec.6. For a different interpretation for the Dirac equation see ch.10.

<u>Discussion of (3)</u>. Prop.5.3 implies that $z_3(x,D)=(z_1-z_2)(x,D)\in$ $\in P_{m-e}$. Thus, indeed, z_3 allows a decomposition (5.13) for m-e.

<u>Discussion of (2)</u>. First let $e^{\Delta}=e$. For an $A\in P_m$ we get $A_t=a_t(x,D)$, $a_t\in C^{\infty}(\mathbb{R},\psi s^{\nu}_m)$. Then again (5.1) yields

$$(5.15)\qquad A^{\cdot}_t +i[K,A_t] = 0 , t \in \mathbb{R} , A_0 = A .$$

(We first get (5.15), applied to $u\in S$, due to differentiability of $e^{\pm iKt}$ in $L(H_s,H_{s-e})$ (thm.5.1). Then, since all operators involved are ψdo's, we get (5.15) as an equation for ψdo's.) Using calculus of ψdo's we may translate (5.15) into symbols :

$$(5.16)\qquad 0 = a^{\cdot}_t + i[k,a_t] + \langle k,a_t\rangle -i/2\langle k,a_t\rangle_2 + \ldots(\mathrm{mod}\ S) .$$

Here the Poisson brackets are formed with matrix multiplication :

$$(5.17)\ \langle a,b\rangle =a_{|\xi}\cdot b_{|x}-b_{|\xi}\cdot a_{|x} , \langle a,b\rangle_j= \sum_{|\theta|=j} \theta!(a^{(\theta)}b_{(\theta)}-b^{(\theta)}a_{(\theta)}).$$

The commutator $[k,a_t]$ normally does not vanish. However, the discussion of VIII,prop.7.4, carries over with the same proofs, except there will be an extra term $[k,a]$ in the third formula VIII,(7.14). Moreover, as in sec.3, we have $\langle k,a_t\rangle_j \in \psi s_{m-re^1-r'e^2}$, $r+r'=j-1$, under our assumptions on k and a_t. Thus (5.16) implies

$$(5.18)\qquad [k_f,a_t] = r_t \in \psi s^{\vee}_m .$$

We regard (5.18) as a commutator equation for the unknown matrix function a_t. Here we know that (5.18) admits a solution. With λ_j and p_j as constructed, we get

$$(5.19)\qquad a_t = \sum_{j=1}^{\rho} p_j a_t p_j + \sum_{j\neq l} p_j r_t p_l/(\lambda_j-\lambda_l) = q_t + z_t .$$

(Just note that $a_t=\sum p_j a_t p_l$, and that $p_j[k_0,a_t]p_l=(\lambda_j-\lambda_l)p_j a_t p_l$, leading to $p_j a_t p_l = p_j r_t p_l/(\lambda_j-\lambda_l)$, as $j\neq l$, and the solvability condition (5.20) (which here holds automatically).)

$$(5.20)\qquad p_j r_t p_j = 0 ,\ j=1,\dots,\rho .$$

Notice that (5.19), with q_t and z_t equal to the first and second sum, respectively, gives a decomposition (5.14), as desired, using that $p_j\in \psi c_0$, $(\lambda_j-\lambda_l)^{-1}\in \psi c_{-e}$, (possibly after a correction for small $f_2|x|+f_1|\xi|$), using (iii) above. Moreover, since p_j, λ_j are independent of t, we may differentiate (5.19) for a decomposition of $\partial_j^t a_t$, the symbol of B^j_t, $B_j=(\mathrm{ad}\ iK)^j A\in \mathrm{Op}\psi s^{\vee}_{m-j(e-e^{\Delta})}$, using (5.10). This proves (2) for $e^{\Delta}=e$.

Next consider $e^{\Delta}=e^1$, noting that the case $e^{\Delta}=e^2$ handles similarly, with x and ξ reversed. Again get (5.16), where now we have $k_f=k_d\in \psi c^{\vee}_{e^1}$, however. Also, $a_t^{\cdot}\in \psi s^{\vee}_{m-e^2}$, by $(5.10)_{e^1}$. Hence,

$$(5.21)\qquad \langle k,a_t\rangle_j \in \psi s^{\vee}_{m+e^1-re^1-r'e^2} ,\ r,r'\geq 0 ,\ r+r'=j ,\ j=0,1,\dots .$$

The proof runs exactly parallel to the proof of prop.3.2. (One gains the slight advantage of an improved multiplication order because the term k_m is missing.)

Now, assuming that $k=k_{e^{\Delta}}+\kappa$, $\kappa\in \psi c^{\vee}_{-e^2}$, as required by (i) of our e^1-semi-strict hyperbolicity, we get (5.18), with $r_t\in \psi s^{\vee}_{m-e^2}$. But we now also have $\lambda_j \in \psi c_{e^1}$, instead of $\lambda_j \in \psi c_e$. Therefore the second term at right in (5.9) still is a symbol in $\psi s^{\vee}_{m-e}$ again, and (2) follows for $f=e^j$ as well.

<u>Discussion of (1)</u>. First assume $e^{\Delta}=e$. For a q satisfying (5.12),

if $z \in \psi s^{\nu}_{m-e}$ exists with $A=q(x,D)+z(x,D) \in P_m$, then get (5.16) for for $a_t=q_t+z_t$. Separating symbols in (5.16) of order m from those of lower order (with (5.12), and cor.10.5), get

$$(5.22)\quad i[k_f,z_t] + q_t^{\cdot} + \langle k_f,q_t\rangle + i[\kappa,q_t] = 0 \ (\mathrm{mod}\ \psi s^{\nu}_{m-e^1} \cap \psi s^{\nu}_{m-e^2}).$$

(In particular we know from (2), already proven, that $z_t^{\cdot} \in \psi s^{\nu}_{m-e}$. Also, the only additional term of (5.16) to be taken into (5.22) for a relation mod ψs^{ν}_{m-e} will be $-i/2\langle k,q_t\rangle_2$. We will not use this term, at the expense of a weaker (5.21), resulting in slower asymptotic convergence of the series for a_t, similar as in thm.3.1.)

Vice versa, we start the construction of a symbol z with the attempt to solve (5.22) exactly, not mod lower order, assuming $q_t \in \psi s^{\nu}_m$ given. Rewrite (5.22) as commutator equation for z_t,

$$(5.23)\qquad [k_f,z_t] = -[\kappa,q_t] + i(q_t^{\cdot} + \langle k_f,q_t\rangle) = \varphi_t .$$

Assuming that φ_t is known, for a moment, (5.23) implies

$$(5.24)\quad z_t = \Sigma_1^{\rho} p_l z_t p_l + i \sum_{j\neq l} p_j(q_t^{\cdot}+\langle k_f,q_t\rangle+i[\kappa,q_t])p_l/(\lambda_j-\lambda_l) .$$

This solution is valid only under the condition (5.20), i.e.,

$$(5.25)\qquad p_l(q_t^{\cdot} + \langle k_f,q_t\rangle + i[\kappa,q_t])p_l = 0 \ ,\ l = 1,\dots,\nu' .$$

In sec.6 we will investigate commutator equations of the form (5.18), with solvability condition (5.20) as in (5.25). In sec.7 construction of z_t and the proof of thm.5.4 will be completed.

6. A technical discussion of commutator equations.

In this section we consider an equation of the form

$$(6.1)\qquad [b,w] = d \qquad ,$$

for general $\nu\times\nu$-square matrix valued symbols $b(x,\xi)$, $d(t,x,\xi)$, $w(t,x,\xi)$, where b,d are given and w is to be found. Assume that b, d are C^{∞} for $(x,\xi) \in \Omega \subset \mathbb{R}^{2n}$, and $(t,x,\xi) \in \mathbb{R}\times\Omega$. Let the matrix b be diagonalizable, have real eigenvalues, and eigenspaces of dimension independent of x,ξ. As used before (sec.5) the eigenvalues $\lambda_j(x,\xi)$ may be arranged as $C^{\infty}(\Omega)$-functions: λ_j is a simple root of $q^{(\rho-1)}(\lambda)=0$, with $q(\lambda)=\det(b(x,\xi)-\lambda)$, and the multiplicity ρ of λ. Hence $\partial_{\lambda} q^{(\rho-1)}(\lambda_j)\neq 0$. The implicit function theorem yields $\lambda_j \in C^{\infty}$, locally. The real distinct eigenvalues may be ordered by size, giving globally defined $C^{\infty}(\Omega)$-functions. For $\lambda=\lambda_j$,

$$(6.2)\qquad p = i/2\pi\int_{|\mu-\lambda(x,\xi)|=\varepsilon} (b(x,\xi) - \mu)^{-1}\, d\mu \in C^{\infty}(\Omega)$$

with $\varepsilon=\varepsilon(x,\xi) > 0$ sufficiently small, is the eigenprojection.

For solvability of (6.1) it is necessary and sufficient that $p(x,\xi)d(t,x,\xi)p(x,\xi)=0$, $x,\xi \in \Omega$, for each eigenprojection p of b. (Indeed, if (6.1) is solvable, then $pdp=[pbp,pwp]=[\lambda p,pwp]=0$. Vice versa, if the latter holds for all $p_j=p$, then (6.1) is solved by (5.19) with λ_j, p_j, $j=1,\dots,\rho$, the distinct eigenvalues an projections of w (and a_t replaced by w). The first sum in (5.19) is an arbitrary matrix commuting with b. All solutions of (6.1) are given by (5.8).)

We plan to solve (6.1) for d of the form (6.3), with given (t-depedent) symbols g, $d^{\wedge}$, and $d^{\sim}$. Assume $g=g(t,x,\xi)$ commutes with $b(x,\xi)$, for all $(t,x,\xi) \in \mathbb{R}\times\Omega$, while $d^{\wedge}(t,x,\xi)$, $d^{\sim}(t,x,\xi)$ are C^{∞}-matrix functions, and $\langle .,.\rangle=\langle .,.\rangle_1$ is the Poisson bracket.

$$(6.3)\qquad d = g^{\cdot} + \langle b,g\rangle + [d^{\wedge},g] + d^{\sim} \quad , \quad g^{\cdot} = \partial_t g .$$

For given λ, p let $\varphi_1(x,\xi),\dots,\varphi_\rho(x,\xi)$, $\psi_1(x,\xi),\dots,\psi_\rho(x,\xi)$ be a bi-orthogonal pair of bases of the eigenspace $S=S(x,\xi)=\mathrm{im}\, p(x,\xi)$ (such that $\psi_j^*\varphi_l=\delta_{jl}$, for all $(x,\xi)\in\Omega$, and that also $\varphi_j,\psi_l\in C^{\infty}(\Omega)$) Consider the local matrices $\gamma=((\gamma_{jl}))$, $\delta^{\wedge}=((\delta^{\wedge}_{jl}))$, $\delta^{\sim}=((\delta^{\sim}_{jl}))$ of g, $pd^{\wedge}p$, $pd^{\sim}p$, (all leaving S invariant): In detail, $\gamma=\gamma(t,x,\xi)$, $\delta^{\wedge}(t,x,\xi)$, $\delta^{\sim}(t,x,\xi)$ are defined by

$$(6.4)\qquad pgp = \sum_{j,l}\gamma_{jl}p_{jl} \;,\; pd^{\wedge}p = \sum_{j,l}\delta^{\wedge}_{jl}p_{jl} \;,\; pd^{\sim}p = \sum_{j,l}\delta^{\sim}_{jl} \;,$$

with $p_{jk}u = \varphi_j(\psi_k^* u)$, $u \in \mathbb{C}^{\nu}$, $\psi_k^* = \overline{\psi}_k^T$, for all $(x,\xi) \in \Omega$.

<u>Proposition 6.1</u>. The condition $pdp=0$ with d of (6.3), and $[b,g]=0$ for all $(t,x,\xi) \in \mathbb{R}\times\Omega$, translates into a differential equation for the restriction $g^0(t,x,\xi) = g(t,x,\xi)|S(x,\xi)$ of g onto the eigenspace S of λ, of the form

$$(6.5)\qquad p(g^0p)'p|S + [v,g^0] + d^{\sim 0} = 0 \quad , \quad d^{\sim 0} = pd^{\sim}p|S \; ,$$

where

$$(6.6)\qquad " \, ' \, " = \partial_t + \lambda_{|\xi}\partial_x - \lambda_{|x}\partial_\xi \;,\; v = (pd^{\wedge}p + p(p_{|\xi}b_{|x})p)|S \; .$$

Moreover, with the matrices γ, $\delta^{\wedge}$, $\delta^{\sim}$ of γ^0, $d^{\wedge 0}=pd^{\wedge}p|S$, $\delta^{\sim 0}$,

$$(6.7)\qquad d\gamma/ds + [\theta,\gamma] + \delta^{\sim}=0 \;,\; d/ds = \partial_t + \lambda_{|\xi}\partial_x - \lambda_{|x}\partial_\xi \;,$$

with

$$(6.8)\qquad \theta=((\theta_{jl}))\ ,\ \theta_{jl}=-\psi_j^{*\prime}\varphi_l + \psi_j^*(pd^{\wedge}p+p_{|\xi}b_{|x})\varphi_l\ ,\ \text{"}'\text{"}=d/ds.$$

The restrictions g^0 , $d^{\wedge 0}$, $d^{\vee 0}$ clearly are linear operators of S . Prop.6.1 implies that, under above assumptions, cdn. pdp=0 involves only g^0, not the restrictions of g to the other eigenspaces, hence a 'decoupled' system of ρ^2 DE,s results for each eigenspace of b, where $\rho=\dim S(x,\xi)$. This is a system of PDE with same principal part, translating into a first order system of ODE along the flow of the common principal part, in the sense of sec.1.

Let us first focus on (6.5). In this discussion we again consider the full set of eigenvalues $\lambda_1,\dots,\lambda_\rho$ of multiplicities ρ_j, with projections $p=p_j$, and $g_j=gp_j=p_jg$. From $p_kg_r=0$, $k\neq r$, we get

$$(6.9)\qquad p_k\dot{g}_rp_k = -\dot{p}_kg_rp_k = 0\ ,\ p_kg_{r|x}p_k = p_kg_{r|\xi}p_k = 0\ ,\ k\neq r\ .$$

We have $g=\sum g_k$, $b=\sum\lambda_rp_r$, hence

$$(6.10)\qquad \begin{aligned} p_k\langle b,g\rangle p_k &= p_k\left\{\sum\lambda_{r|\xi}p_rg_{|x} - g_{|\xi}\lambda_{r|x}p_r\right\}p_k \\ &\quad + p_k\sum\lambda_r\langle p_r,g\rangle p_k \ = T_1 + T_2\ , \end{aligned}$$

where

$$(6.11)\qquad T_1 = p_k\langle\lambda_k,g\rangle p_k = p_k\langle\lambda_k,g_k\rangle p_k\quad ,$$

using (6.9). Also, as $j\neq k\neq r\neq j$, using $g_{j|x}p_k = -g_jp_{k|x},\dots,$

$$(6.12)\qquad p_k\langle p_r,g_j\rangle p_k = p_kp_rg_{j|\xi}p_{k|x} - p_{k|\xi}g_{j|x}p_rp_k = 0\ .$$

Similarly, as $j\neq k=r$, using $p_kp_{k|\xi} = p_{k|\xi}(1-p_k)$,..... ,

$$(6.13)\qquad p_k\langle p_k,g_j\rangle p_k = -p_{k|\xi}(1-p_k)g_jp_{k|x} + p_{k|\xi}g_j(1-p_k)p_{k|x} = 0.$$

On the other hand, for $j=k\neq r$, and $j=k=r$,

$$p_k\langle p_r,g_k\rangle p_k = p_{k|\xi}p_{k|x}g_k - g_kp_{r|\xi}p_{k|x} = [p_kp_{k|\xi}p_{r|x}p_k,g_k]\ ,$$

$$(6.14)\qquad \begin{aligned} p_k\langle p_k,g_k\rangle p_k &= p_{k|\xi}(1-p_k)g_{k|x}p_k - p_kg_{k|\xi}(1-p_k)p_{k|x} \\ &= p_{k|\xi}p_{k|x}g_k - g_kp_{k|\xi}p_{k|x} = [p_kp_{k|\xi}p_{k|x}p_k,g_k]\ . \end{aligned}$$

Accordingly, using (6.9) again,

$$(6.15)\qquad T_2 = \sum\lambda_r\,[p_kp_{k|\xi}p_{r|x}p_k,g_k] = [p_kp_{k|\xi}b_{|x}p_k,g_k].$$

Then (6.10), (6.11), (6.15), and (6.9) imply that indeed (6.5) is equivalent to pdp=0 with d of (6.3), and [b,g] = 0.

Next a calculation, not given in details, will show that the "covariant derivative" $p_k g_k' p_k$ of g_k in the space S_k has the matrix

$$(6.16) \qquad \gamma^{k'} + [\theta_1, \gamma^k] \ , \ \theta_1 = ((\psi_j^* \varphi_r')) = -((\psi_j'^* \varphi_r)) \quad ,$$

with respect to any biorthogonal base $\{\varphi_j\}$, $\{\psi_j\}$ of S_k, where γ^k is the matrix of g_k . Accordingly, (6.5) is equivalent to (6.7), with θ defined by (6.8) , q.e.d.

Clearly the system (6.7) of ρ_k ODE's has a unique local solution satisfying $\gamma=\gamma^0$, at t=0, defined near each (x,ξ), for small t. Let it be assumed that $\Omega = \mathbb{R}^{2n}$,but that only a local base pair φ_j , ψ_l exists, in subsets $\Omega_l \subset \mathbb{R}^{2n}$, covering $\mathbb{R}^{2n}$. Then solutions in overlapping sets Ω_j, for different bases, remain compatible as long as they are jointly defined. From the well known properties of ODE's we conclude existence of a unique solution $g = \sum g_j$ of our matrix commutator problem, assuming a given value $g^0(x,\xi)$, at t=0. Note that d/ds just is the derivative along the Hamiltonian flow (1.3) of the function $\lambda(x,\xi)$ (instead of k_0).This flow is well defined for all t ,if we assume that $\lambda \in \psi\sigma_{e^\Delta}$, for example (cf. prop.2.5). We then get existence and uniqueness of $g=g(t,x,\xi)$ for all $(t,x,\xi) \in \mathbb{R}^{2n+1}$.

<u>Theorem 6.2</u>. Let the assumptions of prop.6.1 hold for $f_2|x|+f_1|\xi| > \eta_0$, and let $b \in \psi\sigma^{\vee}_{e^\Delta}$, $d^\wedge \in C^\infty(\mathbb{R}, \psi c^{\vee}_{e^\Delta - e})$, $d^- \in C^\infty(\mathbb{R}, \psi s^{\vee}_{m+e^\Delta - e})$. For $e^\Delta = e$ assume in addition that the coefficients of b are $\in \psi\sigma\pi_e$. For $e^\Delta \neq e$ even $\partial_t^l d^\wedge \in C^\infty(\mathbb{R}, \psi c^{\vee}_{(l+1)(e^\Delta - e)})$, $\partial_t^l d^- \in C^\infty(\mathbb{R}, \psi s^{\vee}_{m+(l+1)(e^\Delta - e)})$. Then, for every $g^0 \in \psi s^{\vee}_m$ a unique $g_t \in C^\infty(\mathbb{R}, \psi s^{\vee}_m)$ exists such that for all $p=p_j$ of b we have (for large $f^2|x|+f_1|\xi|$)

$$(6.17) \quad [b,g]=0 \ , \ p(g^{\cdot} + \langle b,g\rangle + [d^\wedge, g] + d^-)p = 0 \ , \ t \in \mathbb{R} \ , \ g=g_0 \text{ at } t=0.$$

We also have $\partial_t^l g_t \in C^\infty(\mathbb{R}, \psi s^{\vee}_{m+l(e^\Delta - e)})$, $l = 0,1,2,\ldots$.

<u>Proof</u>.(Assume $f_2|x|+f_1|\xi|$ large,where needed.) We know existence of a unique solution g of (6.17).In order to show that g has the stated symbol properties we note that (6.5) and $g_j = p_j g_j p_j$ imply

$$(6.18) \quad g_j' + [v_j, g_j] = p_j' g_j + g_j p_j' - p_j d^- p_j \ , \ v_j = p_j(d^\wedge + p_{j|\xi} b_{|x}) p_j \ .$$

This may be interpreted as another set of ODE's for the ν^2 coefficients of g_j. There is a unique solution of (6.18) under the initial conditions of (6.17), defined in $\mathbb{R}^{2n+1}$. This must be the same

g_j , also solving (6.5).

We have $p_j^{'}=\langle\lambda_j,p_j\rangle\in\psi c^{\nu}_{f-e}$, $v_j=p_j(d^{\wedge}+p_{j|\xi}b_{|x})p_j\in C^{\infty}(\mathbb{R},\psi c^{\nu}_{f-e})$. Therefore (6.18) may be interpreted as a $\nu^2\times\nu^2$-system of first order ODE's for $g_j^{\sim}=g_\circ v_t^j$ of the form

$$(6.19)\qquad dg_j^{\sim}/ds+c_j^1(s)g_j^{\sim}=c_j^2(s)\ ,\ s\in\mathbb{R}\ ,\ g_j^{\sim}=g^0p_j\ ,\ \text{at } s=0,$$

$g_j^{\sim}$ being interpreted as a ν^2-component vector with entries $\in\psi s^{\nu}_m$, while $c_j^1\in C^{\infty}(\mathbb{R},\psi c^{\nu^2}{}_{e^{\wedge}-e})$,and the ν^2-vector c_j^2 has entries in $C^{\infty}(\mathbb{R},\psi s_{m+e^{\wedge}-e})$. By our remarks a global solution of (6.19) exists for all t. Also, prop.2.5 was used, for the above. It is a matter of deriving suitable apriori estimates to show that $g_j^{\sim}\in C^{\infty}(\mathbb{R},\psi s_m)$ hence also $g_j\in C^{\infty}(\mathbb{R},\psi s^{\nu}_m)$, etc., proving the theorem.

For example, let us introduce the norm

$$(6.20)\qquad |a|_m^2=\sum_{j,r}|a_{jr}(x,\xi)|^2(1+|x|^2)^{-m_2}(1+|\xi|^2)^{-m_1}\ .$$

From (6.19) one obtains the differential inequality

$$(6.21)\qquad |d|g_j^{\sim}|_m^2/ds|\le\eta_T|g_j^{\sim}|_m^2+2|c_j^2|_m^2\ ,\ |s|\le T\ ,$$

with a constant $\eta_T=\sup\{2\|c_j^1\|+1:|s|\le T\}$, with a matrix norm $\|.\|$. Then (6.21) may be integrated, using the initial conditions, for

$$(6.22)\qquad |g^{j-}|_m\le\eta_1(s)\ ,\ s\in\mathbb{R}\ ,$$

with locally bounded $\eta_1(s)$. Then (6.19) implies

$$(6.23)\qquad |\partial_t g^k|_{m-e+e^{\wedge}}\le\eta_2(s)\quad .$$

One may differentiate (6.19) for t to derive estimates for $dg_j^{\sim}/dt$ recursively. Also apply (a finite number of) η_{jl} of (2.3) to (6.18) to obtain a similar system of ODE's for $\Pi_{p=1}^{Ne}\eta_{j_pl_p}$ with coefficcients in ψs_0 , following the proof of thm.1.2. This will give estimates for all expressions $|d/ds\Pi_{p=1}^{N}\eta_{j_pl_p}g_j^{\sim}|_m^2$, completing the proof of thm.6.2. Details are left to the reader.

7. Completion of the proof of theorem 5.4.

In all discussions of sec.7 a restriction $f_2|x|+f_1|\xi|\ge\eta_0$, with sufficiently large η_0 , is assumed wherever needed.

Returning to the proof of thm.5.4,(1) , we observe that (5.24), together with $[k_f,q_t]=0$, i.e., (5.14) are of the form of

(6.17), so that thm.6.2 applies (presently with $e^{\Delta}=e$). Thus q_t is determined by its initial value q_0, by property $[q,k_e]=0$ and cdn. (5.20) (i.e., (5.24) for (5.22)). By thm.6.2 such symbol exists, belongs to $C^\infty(\mathbb{R},\psi s^\nu_m)$. We also get $z_t \in \psi s^\nu_{m-e}$ as solution of (5.22), i.e., in the form (5.23). First set $p_1 z_t p_1=0$. (5.23) implies $z_t \in C^\infty(\mathbb{R},\psi s^\nu_{m-e})$, similarly as in the proof of thm.5.4 (2).

Note however, that $a_t=q_t+z_t$ does not satisfy (5.16) precisely, but only mod $\psi s^\nu_{m-e}j$, j=1,2. To improve our choice of z_t we will set up a recursion, for a sequence of improvement symbols. An asymptotic sum will give the total correction symbol. First let

$$(7.1) \qquad W_t = -e^{-iKt}q(x,D)e^{iKt} + q_t(x,D) + z_t(x,D) ,$$

and note that

$$(7.2) \qquad \begin{gathered} W_t^{\cdot} + i[K,W_t] = R_t = r_t(x,D) \ , \ t\in\mathbb{R} \ , \ W_0 = Z_0 = z_0(x,D) \ , \\ r_t=z_t^{\cdot}+i[\kappa,z_t]+\langle k,z_t\rangle -i/2\langle k,z_t\rangle_2+\ldots+\langle\kappa,q_t\rangle -i/2\langle k,q_t\rangle_2+\ldots \ . \end{gathered}$$

In particular, we have used (6.22) and $[k_f,q_t] = 0$. Note that $r_t \in C^\infty(\mathbb{R},\psi s_{m-e}j)$, j=1,2, and $z_0 \in \psi s_{m-e}$.

We now may solve the linear inhomogeneous differential equation (7.2): It follows that $d/dt(e^{iKt}W_t e^{-iKt})=e^{iKt}R_t e^{-iKt}$, hence,

$$e^{iKt}W_t e^{-iKt} = -Q + e^{iKt}(Q_t + Z_t)e^{-iKt} = Z_0 + \int_0^t e^{iK\tau}R_t e^{-iK\tau}\,d\tau ,$$

which implies

$$(7.3) \qquad e^{-iKt}(Q+Z_0)e^{iKt} = Q_t+Z_t - \int_0^t e^{-iK\tau}R_{t-\tau}e^{iK\tau}\,d\tau' .$$

Start the recursion by attempting to write $[k_f,z_t^2]=ir_t$, with z_t^2 to be found. This is a commutator relation as studied. It can be solved only if r_t satisfies a cdn. (5.20), generally not to be expected. But recall that the first term of (5.23) was set 0 so far. We may replace z_t by $z_t^1=z_t^0+z_t$, where z_t^0 commutes with k_f but otherwise is arbitrary. Substituting this into r_t of (7.2) we get $r_t=\bar r_t+z_t^{0\cdot}+i[\kappa,z^0]+\langle k_f,z_t^0\rangle+\langle\kappa,z_t^0\rangle-i/2\langle k,z_t^0\rangle_2+\ldots$, with $\bar r_t$ denoting our old r_t formed with z_t. Here we write $r_t = r_t^1 + r_t^2$, with $r_t^2 = \langle\kappa,z_t^0\rangle -i/2\langle k,z_t^0\rangle_2 + \ldots$, and improve the above 'recursion ansatz' by attempting to find z_t^2 solving the commutator equation

$$(7.4) \qquad [k_f,z_t^2] = ir_t^1 \ , \ r_t^1=\bar r_t+z_t^{0\cdot}+\langle k_f,z_t^0\rangle+i[\kappa,z^0] \ , \ [k_f,z_t^0]=0 \ ,$$

exactly of the form (6.1), (6.3) again. For j=1,2 we get $\bar r_t \in \psi s^\nu_{m-e}$. With initial value $z_0^0=0$ use thm.6.2 to get $z_t^0 \in C^\infty(\mathbb{R},\psi s^\nu_{m-e})$ to solve the compatibility conditions $p_j r_t^1 p_j=0$. Then (5.19), with the first sum set 0, gives a solution of (7.4) of the form

$$(7.5)\quad z_t^2 = \sum_{j\neq 1} p_j i r_t^1 p_1/(\lambda_j - \lambda_1) \in C^\infty(\mathbb{R}, \psi s^\nu_{m-e^1} - e) \cap C^\infty(\mathbb{R}, \psi s^\nu_{m-e^2} - e) \ .$$

After fixing the symbol $z_t^0 \in C^\infty(\mathbb{R}, \psi s^\nu_{m-e} j)$ we also have a well defined $r_t^2 \in C^\infty(\mathbb{R}, \psi s^\nu_{m-je^1-j'e^2})$, $j+j'=2$, using prop.5.3.

Set $r_t = [k_f, z_t^2] + r_t^2$ in (7.3), and 'integrate by parts': Write

$$(7.6)\qquad r_t = r_t^3 - i([k, z_t^2] - i\langle k, z_t^2\rangle + (-i)^2/2\langle k, z_t^2\rangle + \ldots),$$

with $r_t^3 = r_t^2 + i([\kappa, z_t^2] + i\langle k, z_t^2\rangle + - \ldots) \in C^\infty(\mathbb{R}, \psi s^\nu_{m-je^1-j'e^2})$, $j+j'=2$. Get

$$\int_0^t e^{-iK\tau} R_{t-\tau} e^{-K\tau} d\tau = \int_0^t e^{-iKt} R^3_{t-\tau} e^{iK\tau} d\tau + i\int_0^t e^{-iK\tau} [K, Z^2_{t-\tau}] e^{iK\tau} d\tau$$

$$(7.7)\qquad = \int_0^t e^{-iK\tau} (R^3_{t-\tau} - Z^{2\cdot}_{t-\tau}) e^{iK\tau} d\tau + \int_0^t d/dt(e^{-iK\tau} Z^2_{t-\tau} e^{iK\tau}) d\tau$$

$$= e^{-iKt} Z_0^2 e^{iKt} - Z_t^2 + \int_0^t e^{-iK\tau} R^2_{t-\tau} e^{iK\tau} d\tau \ ,$$

where we introduced $r_t^2 = r_t^3 - z_t^{2\cdot}$ instead of the old r_t^2. Substitute in (7.3) (we now have $Z_t^0 + Z_t^1$ instead Z_t but $Z_0^1 = 0$ by construction). Get

$$(7.8)\qquad e^{iKt}(Q + Z_0^0 + Z_0^2) e^{iKt} = Q_t + Z_t^0 + Z_t^1 + Z_t^2 - \int_0^t e^{-iK\tau} R^2_{t-\tau} e^{iK\tau} d\tau \ .$$

In particular, we have $r_t^2 \in C^\infty(\mathbb{R}, \psi s^\nu_{m-je^1-j'e^2})$, $j+j'=2$, and $z_t^j \in C^\infty(\mathbb{R}, \psi s^\nu_{m-je^1-j'e^2})$, $j=0,1,2$.

It is clear that this proceedure may be iterated. In the next step we seek for a representation of r_t^2 as a commutator. Again this requires (a) redefining the symbol z_t^2 of (7.5), adding a suitable $C^\infty(\mathbb{R}, \psi s^\nu_{m-je^1-j'e^2})$-function z_t^{2-}, commuting with k_{e^Δ}, and (b) in the redefined (7.8), using $z_t^2 + z_t^{2-}$, splitting r_t^2 into a lower order term r_t^{2-}, and another term $r_t^{2\wedge}$ fitting into thm.6.2.

The point is this: The z_t^{2-} again will have to be of order $m-je^1-j'e^2$, $j+j'=2$, since it must solve (6.5). The r_t^{2-} will be of this order, with $j+j'=3$, and the z_t^3, solving $[k_{e^\Delta}, z_t^3] = r_t^{2\wedge}$, will be of order $m-je^1-j'e^2-e$, $j+j'=2$, which is even lower.

After solving that commutator equation the integral in (7.8) is treated as that in (7.3) earlier. The splitting corresponding to (7.6) gives symbols of the proper order, and $z_t^{3\cdot}$ (going to the next remainder), is of proper order $m-je^1-j'e^2$, $j+j'=3$. Thus get

$$(7.9)\quad e^{-iKt}(Q + Z_0 + \ldots + Z_0^N) e^{iKt} = (Q_t + Z_t^0 + \ldots + Z_t^N) - \int_0^t d\tau\, e^{-iK\tau} R^N_{t-\tau} e^{iK\tau} \ ,$$

where generally

$$(7.10)\qquad z_t^l \ ,\ r_t^l \in C^\infty(\mathbb{R}, \psi s^\nu_{m-je^1-j'e^2}) \ , j+j'=l \ , l=0,1,2,\ldots \ .$$

Clearly the integral in (7.9) is $\in C^\infty(\mathbb{R},\psi s^\vee_{m-je^1-j'e^2})$, $j+j'=N$.

Next, the asymptotic sum $z_t=\sum z^j_t$ is well defined, and we get

$$(7.11) \qquad e^{-iKt}(Q+Z_0)e^{iKt} = Q_t + Z_t + W_t^{\sim} ,$$

with $W_t^{\sim}\in O(-\alpha)$, $W_0^{\sim}=0$, and $W_t^{\sim}\in C^\infty(\mathbb{R},Op\psi s^\vee_m)$, using (7.9). We get $A=Q+Z_0\in P_m$, $A_t=Q_t+Z_t+W_t^{\sim}$. We have proven (1) of thm.5.4, for $e^\Delta=e$.

Notice that we were selecting all the initial conditions for the successive 'commuting parts' of the z^j_t as 0 . This means that the decomposition $a = q+z_0$, for t=0, coincides with the initial decomposition of thm.5.4,(2). In particular, we get q_0 of (5.14) equal to our initially given symbol q of (5.12) .

On the other hand the lower order commuting parts of the z^j are not all zero,in general. Hence the decomposition $q_t+z_t+w_t^{\sim}=a_t$ suggested by (7.11) in general is not of type (5.14). In fact, for general $q \in \psi s^\vee_m$, the symbol $z_t + w_t^{\sim}$ may not be of order m-e but only will be of order $m-e^j$, j=1,2. The reason is, that the 'commuting part' $\sum p_l z_t p_l$ of z_t will have to go together with q_t , to make a decomposition (5.14) .

This suggests that the commuting part of a_t is no 'clean propagation' of q , according to the differential equations (6.5).

The above discussion of thm.5.4 (1), $e^\Delta=e$, works again (with the following amendments) for $e^\Delta=e^1$ (and hence also for $e^\Delta=e^2$). First, z_t of (5.23) again is in $C^\infty(\mathbb{R},\psi s^\vee_{m-e})$, but for a different reason: The right hand side of (5.22) now is in ψs_{m-e_2} , but division by $\lambda_j-\lambda_l$ gets this into ψs_{m-e} again. q_t of (5.24) has $q_t^{\cdot}\in \psi s^\vee_{-e^2}$. Moreover, $\partial^1_t z_t\in C^\infty(\mathbb{R},\psi s^\vee_{m-1e^2-e})$, and, $\partial^1_t q_t\in C^\infty(\mathbb{R},\psi s^\vee_{m-1e^2})$, by a calculation similar to that in [CE].

The corresponding effect is observed on higher iterations. For example, the symbol r_t of (7.2) now is of one order $-e^2$ better since $z_t^{\cdot}$, and $[\kappa,z_t]$, and $\langle k_f,z_t\rangle$ all are better. Therefore the result follows just as well. Details left to the reader. Q.E.D.

For a possible later application we summarize the special observations in the proof of thm.5.4 (1) as follows.

<u>Corollary 7.1</u>. For any $A \in P^K_m$ a decomposition (5.14), valid for $t\in \mathbb{R}$, may be constructed as follows: Start from any such decomposition $a = a_0 = q_0 + z_0$ at t=0 ,and then set

$$(7.12) \qquad q_t = \sum_{j=1}^{\rho} q_{jt}p_j , \quad q_{jt}:S_j \to S_j , \ S_j = \mathrm{im}\, p_j .$$

Here the linear operators q_{jt} of the j-th eigenspace S_j of k_{e^Δ} are given as unique solutions of the first order system of PDE (7.13), with covariant derivatives in $S(x,\xi)$, changing with x and ξ ,

$$(7.13)\quad p_j((q_{jt}p_j)_{|t}-\lambda_{j|\xi}\cdot(q_{jt}p_j)_{|x}+\lambda_{j|x}\cdot(q_{jt}p_j)_{|\xi})|S_j+[v_j,q_{jt}]=0,$$

satisfying the initial condition

$$(7.14)\qquad q_{jt} = q_t|S_j \ , \text{ at } t = 0 \ ,$$

and $v_j : S_j \to S_j$ are defined by

$$(7.15)\qquad v_j = (ip_j\kappa p_j - p_j p_{|\xi}k_{f|x}p_j)|S_j \ .$$

In "local coordinates"-that is if $\varphi_1,\ldots,\varphi_{\rho_j}$, and $\psi_1,\ldots,\psi_{\rho_j}$ is a local bi-orthogonal basis of the eigenspace $S_j(x,\xi)$, defined and C^∞ in some open set Ω of (x,ξ)-space - if $\kappa j_t = ((\kappa^{jt}_{rs}(x,\xi)))$ is the matrix of q_{jt}, with respect to that basis, then, for $(x,\xi)\in\Omega$, (7.13) and (7.14) are equivalent to

$$(7.16)\qquad \kappa^j_t|t + \lambda_{j|\xi}\kappa^j_{t|x} - \lambda_{j|x}\kappa^j_{t|\xi} + [\theta_j,\kappa^j_t] = 0 \ , \quad \kappa^1_0 = \kappa^1 \ ,$$

where θ_j is the sum of the matrix of v_j and $((-\langle\lambda_j,\psi^*_r\rangle\varphi_s))$.

For the q_t thus obtained one next writes

$$(7.17)\quad z_t=z^0_t+i\sum_{j\neq 1}p_j(q^{\cdot}_t+\langle k_f,q_t\rangle+i[\kappa,q_t])p_1/(\lambda_j-\lambda_1) \ , \ [k_f,z^0_t]=0.$$

Here $z^0_0=0$, $z^0_t=\sum z^0_{jt}p_j$, $z^0_{jt}:S_j\to S_j$ given as solution of

$$(7.18)\quad \begin{aligned}&(p_j(C_{\lambda_j}(z^0_{jt}p_j)))|S_j + [v_j,z^0_{jt}] +\\ &+ (p_j(z^{\cdot}_t+i[\kappa,z_t]+\langle k_f,z_t\rangle+\langle\kappa,q_t\rangle-i/2\langle k_f,q_t\rangle_2))|S_j = 0 \ ,\end{aligned}$$

with the differential expression of the "j-th particle flow"

$$(7.19)\qquad C_{\lambda_j} = \partial_t + \lambda_{j|\xi}\partial_x - \lambda_{j|x}\partial_\xi \ .$$

The above holds only for large $f_2|x|+f_1|\xi|$, but the symbols obtained, defined for large $f_2|x|+f_1|\xi|$, may be extended to $\mathbb{R}^{2n}$, to get symbols in $\psi s^{\vee}_m$, $\psi s^{\vee}_{m-e}$, $\psi s^{\vee}_{m-e^1}\cap\psi s^{\vee}_{m-e^2}$, respectively, and,

$$(7.20)\qquad a_t = (q_t+z^0_t) + (z_t+w_t) \quad ,$$

a decomposition (5.14), where $w_t\in\psi s^{\vee}_{m-je^1-j'e^2}$, $j+j'=3$ may be determined up to order $-\infty$ by a recursion, similar as above.

Chapter 10. THE INVARIANT ALGEBRA OF THE DIRAC EQUATION

0. Introduction. Consider the motion of a charged particle in an electromagnetic field with potentials $V(x)$, $A(x)=(A_1, A_2, A_3)(x)$ The particle is thought as a small charged sphere of mass m and total charge e spinning about an axis through its center of gravity, and moving along some orbit $x(t)=(x_1, x_2, x_3)(t)$.

In the sense of classical Physics the particle experiences forces from the fields E and B determining its motion, once an initial location x^0 and velocity v^0 are given.

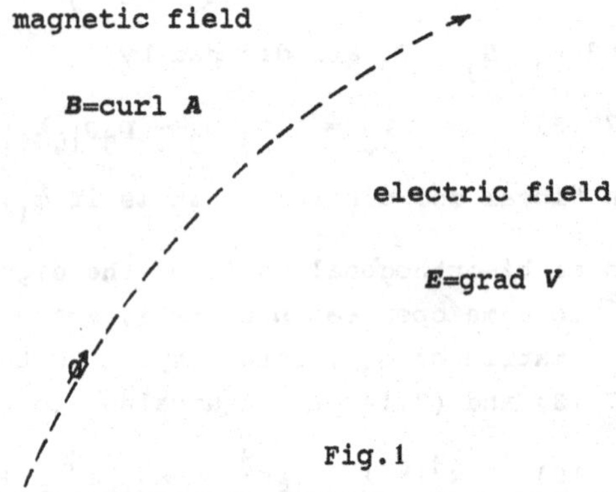

Fig.1

The spinning charge makes the particle a magnet (generates a magnetic moment) which experiences a twist from the fields B (and E) Thus the particle is acting like a spinning top with an angular momentum trying to turn it.

Assume the velocities large to implicate special relativity. Let c denote the speed of light.

The motion is described by a pair of systems of ODE. With location x(t) and magnetic moment $j(t)$ at time t we get (with "x" denoting the cross product of 3-vectors, also in the following)

(0.1) $$\pm\left(\frac{mx^{\cdot}}{\sqrt{1-x^2/c^2}}\right)^{\cdot} = -E + \frac{e}{c}(x^{\cdot}\times B)$$

as equation of motion for the particle, while $j(t)$ satisfies

(0.2) $$j^{\cdot} \pm M\sqrt{1-x^{\cdot 2}/c^2}\; B^{\sim}\times j = 0 ,$$

where

(0.3) $$B^{\sim} = B + (1/ec)(1+\sqrt{1-x^{\cdot 2}/c^2})^{-1}x^{\cdot}\times E .$$

denotes the magnetic field 'seen by the particle' (even if $B=0$, the relatively moving electrostatic field E is experienced as a magnetic field). Clearly $E = E(x(t))$, $B=B(x(t))$. We work in terms of special relativity. x' in (0.2) and (0.3) may be substituted with $x'(t)$ from (0.1), so that the combination (0.1),(0.2) is a nonlinear system in the variables $(x_j(t)\ ,\ \jmath_j(t))_{j=1,\dots,3}$.

Also M is a coupling constant, involving the quotient of the magnetic moment and the mechanical moment of the rotation. It is given only after making some assumption about the geometric distribution of charges and masses of the little sphere.

In somewhat greater precision one will have to also expect an effect of the magnetic moment on the motion of the particle: If B changes rapidly there will be different forces acting on north and south pole of the spinning magnet, resulting in a non-zero combined force on the particle. We have reason to first ignore this "Stern-Gerlach-effect", as a lower order quantity.

In terms of quantum mechanics, assuming that the particle has 'spin 1/2' - such as an electron or a positron (or certain mesons) , the physical description is very different, given by the Dirac equation, a system of first order PDE .

The physical state of the system is no longer described by initial position and velocity coordinates and initial spin, but rather by a unit vector

$$\psi(x) = (\psi_1(x),\psi_2(x),\psi_3(x),\psi_4(x))^T \ ,\ x=(x_1,x_2,x_3)$$

in the Hilbert space $H = L^2(\mathbb{R}^3,\mathbb{C}^4)$ of squared integrable 4-vector functions with complex coefficients. (The inner product is written as $(\psi,\omega) = \int_{\mathbb{R}^3} \sum_1^4 \overline{\psi_j}\omega_j dx$, and the norm as $\|\psi\|=(\psi,\psi)^{1/2}$.)

The (bounded or unbounded) self-adjoint operators acting on a dense subdomain of H are called <u>observables</u>. For example the x_1-coordinate of the particle is an observable, given by the unbounded self-adjoint operator $\psi(x)\to x_1\psi(x)$ with domain $\{\psi\in H : x_1\psi\in H\}$.

'Measuring' an observable A for a physical state ψ will in general not produce a precisely predictable result. Rather one will measure one of the eigenvalues of A with a certain probability (assuming that A has discrete spectrum). If $\{\varphi_j : j=1,2,\dots\}$ denotes an orthonormal base of eigenvectors with corresponding eigenvalues $\lambda_1,\lambda_2,\dots$, then we may look at the expansion

$$(0.4)\qquad \psi = \sum_{j=1}^{\infty} a_j\varphi_j \ ,\ a_j = (\varphi_j,\psi) \ ,\ \sum_{j=1}^{\infty} |a_j|^2 = \|\psi\|^2 = 1 \ .$$

The measurement of A in the state ψ will produce the result

λ_j with probability $|a_j|^2$. Observe that the sum of all probabilities is 1, as expected, and that the statistical expectation value of the measurement is given by

$$(0.5) \qquad \bar{A} = \sum_{j=1}^{\infty}\lambda_j|a_j|^2 = (\psi,A\psi) \text{ , for } \psi \in \text{dom } A \text{ .}$$

If A has continuous spectrum one can talk about the probability of measuring a value in a given interval Δ=[a,b]. This probability is given by $\|E_\Delta\psi\|^2$ with E_Δ the value of the spectral measure on the interval Δ . The expectation value still is given by (0.5).

After measuring the observable A , the physical state ψ in general has changed: If the eigenvalue λ_j has been measured, then the new physical state will be the corresponding eigenvector φ_j . A successive measurement of the same observable A will produce the same result λ_j with certainty (probability 1), as the expansion (0.4) for $\psi=\varphi_j$ shows. However, measuring another observable B will transfer the state into an eigenvector of B, hence a following measurement of A no longer will produce λ_j with certainty, except if the operators A and B commute. Then the expansion of φ_j into eigenvectors of B will lead to values of B corresponding to simultaneous eigenvectors of A and B for which both may be measured in any order, always producing the same result.

The first momentum coordinate p_1 is given as (the unique self-adjoint realization of) the differential operator $-i\hbar\partial_{x_1}$. Clearly thus the observables x_1 and p_1 do not commute, we have

$$(0.6) \qquad x_1p_1-p_1x_1 = i\hbar \quad ,$$

the Heisenberg uncertainty relation: Successive measurement of the observables cannot eliminate a certain minimal uncertainty.

The total energy of the system is given as such an observable, denoted by H. This observable has a special significance. It determines the 'time propagation' -substitute for (0.1) and (0.2).

The law of conservation of energy requires that a measurement of H must produce the same result at any time. The representation of states and observables by vectors and operator of $\mathcal{H}$ is not unique, of course. The choice has been a matter of convenience as well as of individual preference.

Time propagation may be described either by the <u>Schroedinger representation</u>: Observables are constant in time, but physical states change. Then the change of the state ψ_0 in time will be determined by the differential-equation-initial-value-problem

$$(0.7) \qquad \hbar d\psi/dt + i\, H\psi = 0 \text{ , } 0\le t<\infty \text{ , } \psi(0) = \psi_0 \text{ .}$$

Or, equivalently, by the <u>Heisenberg</u> <u>representation</u>: Physical states are constant in t, but observables change, according to

$$A \to A_t = e^{itH}Ae^{-itH} \quad . \tag{0.8}$$

Observe that $U(t)=e^{itH}$ is a unitary operator, defined by the spectral resolution, and is the solution operator of (0.7).

The quantum mechanical problem corresponding to our above classical problem is described by the following energy observable:

$$H = V(x) + mc^2\beta + c\sum_{j=1}^{3}\alpha_j(-i\hbar\partial/\partial x_j - \tfrac{e}{c}A_j(x)) \quad . \tag{0.9}$$

Here $V(x)$ and $A_j(x)$ are the potentials as above, m,e,c,ħ are mass, charge, speed of light and Planck constant. Also β , α_1, α_2, α_3 are certain constant 4×4-matrices. We are used to work with

$$\alpha_j = i\begin{pmatrix} 0 & \sigma_j \\ -\sigma_j & 0 \end{pmatrix} , \; j=1,2,3 \; , \; \beta = \begin{pmatrix} I & 0 \\ 0 & -I \end{pmatrix} , \tag{0.10}$$

with the 'Pauli matrices' σ_j and I defined by

$$\sigma_1 = \begin{pmatrix} 0 & i \\ -i & 0 \end{pmatrix} , \; \sigma_2 = \begin{pmatrix} 0 & 1 \\ 1 & 0 \end{pmatrix} , \; \sigma_3 = \begin{pmatrix} 1 & 0 \\ 0 & -1 \end{pmatrix} , \; I = \begin{pmatrix} 1 & 0 \\ 0 & 1 \end{pmatrix} . \tag{0.11}$$

For convenience we also define the 4×4-matrices

$$\mu_j = \begin{pmatrix} 0 & \sigma_j \\ \sigma_j & 0 \end{pmatrix} , \; \rho_j = \begin{pmatrix} \sigma_j & 0 \\ 0 & \sigma_j \end{pmatrix} , \; j=1,2,3 \; , \tag{0.12}$$

and observe that the following (anti)-commutator relations hold:

$$\sigma_j^2=1, \; \sigma_1\sigma_2=-\sigma_2\sigma_1=i\sigma_3, \; \sigma_2\sigma_3=-\sigma_3\sigma_2=i\sigma_1, \; \sigma_3\sigma_1=-\sigma_1\sigma_3=i\sigma_2 \; , \tag{0.13}$$

hence

$$\begin{gathered} \sigma_j\sigma_k + \sigma_k\sigma_j = 2\delta_{jk}, \; \rho_j\rho_k+\rho_k\rho_j = 2\delta_{jk}, \; \beta\alpha_j+\alpha_j\beta = 0 \; , \\ \alpha_j\alpha_k + \alpha_k\alpha_j = 2\delta_{jk}, \; \mu_j\mu_k+\mu_k\mu_j = 2\delta_{jk}, \; \beta^2 = 1 \; , \\ \beta\alpha_j= i\mu_j \; , \; \alpha_1\alpha_2= i\rho_3 \; , \; \alpha_2\alpha_3= i\rho_1 \; , \; \alpha_3\alpha_1= i\rho_2 \; . \end{gathered} \tag{0.14}$$

Note (0.15) below, for the formal 3-vector $\sigma = (\sigma_1,\sigma_2,\sigma_3)$, and arbitrary formal vectors $\boldsymbol{a},\boldsymbol{b}$, components commuting with σ :

(0.15)

(a) $\sigma(\sigma\boldsymbol{a}) = \boldsymbol{a} + i\boldsymbol{a}\times\sigma \; , \; (\sigma\boldsymbol{a})\sigma = \boldsymbol{a} - i\boldsymbol{a}\times\sigma \; ,$

(b) $(\sigma\boldsymbol{a})(\sigma\boldsymbol{b}) = \boldsymbol{a}\cdot\boldsymbol{b} + i\sigma\cdot(\boldsymbol{a}\times\boldsymbol{b}) \; ,$

(c) $[(\boldsymbol{a}\times\sigma),(\boldsymbol{b}\cdot\sigma)] =(\boldsymbol{a}\times\sigma)(\boldsymbol{b}\cdot\sigma)-(\boldsymbol{b}\cdot\sigma)(\boldsymbol{a}\times\sigma) =2i\{(\boldsymbol{a}\sigma)\boldsymbol{b}-(\boldsymbol{ab})\sigma\}.$

The (system of) differential equation(s) (0.7), with H of (0.9), is called the Dirac equation. Clearly the Dirac equation is a symmetric hyperbolic 4×4-system, in the sense of VI,2. Moreover, we shall see that it is semi-strictly hyperbolic of type e' , under proper assumptions on V and A_j . Accordingly our results of IX,5 apply. We are getting an invariant algebra of ψdo-s.

From now on let us assume the potentials V, A_j to be bounded $C^\infty(\mathbb{R}^3)$-functions. Note, we exclude singularities like that of a Coulomb potential. This appears as a serious handicap of the theory, but may, at the contrary, prove to be one of its major points.

It then follows easily that the differential expression H is formally self-adjoint, and, moreover, that the minimal operator (with domain $C_0^\infty(\mathbb{R}^3)$) admits precisely one self-adjoint realization In other words, there is a unique self-adjoint energy observable (the Hamiltonian) induced, and our theory is well set.

The above scheme indeed is capable of explaining physical observations around our model of fig.1 with such amazing accuracy and detail, that it survived, in spite of the fact that some strange contradictions or paradoxa also were noted. In earlier work ([CD], [CF]) we attempted to show that our invariant algebra of IX,5 may explain some of the paradoxa. One might even go beyond this and ask the question whether the 'perturbation symbols' we discuss in sec.5 could explain other physical phenomena so far only derived from quantum field theory or gauge theory.

In sec. 1 we propose a modified observable concept. In sec.2 we discuss a link between invariant algebra and Foldy-Wouthuysen transform. Sec's 3,4,5 give details of the invariant algebra in case of the Dirac system (0.7),(0.9). In sec's 5-9 we focus on the 'correction symbol' first for 'standard observables' then in general, discussing some finer details.

1. A refinement of the concept of observable.

To motivate an application of the theory of ψdo's, let us focus on the concept of observable introduced in sec.0 . So far, any bounded or unbounded self-adjoint operator of $H = L^2(\mathbb{R}^3, \mathbb{C}^4)$ (having a spectral resolution) qualifies as observable. On the other hand, the observables of real interest normally turn out to be either multiplications (like the x_1-coordinate, we mentionned) or differential operators - such as H above, or the momentum observable, given as $p_1 = -i\hbar\partial/\partial_{x_1}$, etc. . We will meet other "standard observables" later on, and mention that multiplication by a (proper) matrix may occur as well, such as the spin observable, commonly defined as multiplication by ρ_j , j=1,2,3, with ρ_j of (0.12) .

On the other hand, in many respects the admission of all unbounded self-adjoint operators as observables appears inconveniently large and the question can be asked whether it is useful to get restricted to a special class of observables.

We choose to suggest a certain condition of continuity or smoothness on observables, to be discussed now:

Note that for the remainder of this section it is not significant that physical states are 4-vectors. We might just as well assume that $\mathcal{H} = L^2(\mathbb{R}^3)$ (with complex-valued functions), or, more generally, that $\mathcal{H} = L^2(\mathbb{R}^n, \mathbb{C}^\nu)$, for arbitrary $n, \nu = 1,2,\ldots$.

First consider a bounded operator $A \in \mathcal{L}(\mathcal{H})$. We propose to admit A as observable only if it is norm continuous under translations. In more detail, it is well known that the operator family $A_h = T_{-h}AT_h$, $h \in \mathbb{R}^3$, with the translation operator $T_h = e^{ihD}$, $(T_h u)(x) = u(x+h)$, $u \in \mathcal{H}$, in general is not norm continuous in h in the Banach algebra $\mathcal{L}(\mathcal{H})$, but only 'strongly continuous' : For each fixed $u \in \mathcal{H}$ the family $u_h = A_h u$ is continuous as a map from $\mathbb{R}^3$ to $\mathcal{H}$. However, this continuity may degenerate if u varies over a bounded set of $\mathcal{H}$, it may not be uniform. Example: let A = multiplication by a discontinuous function a(x) ,like $a(x) = x_1/|x_1| = \mathrm{sgn}(x_1)$.

If we ask for differentiability in the parameter h the situation gets worse: A derivative $\partial_{h_j} A_h u$ as limit of a difference quotient exists only under special assumptions on A and u .

We impose as condition (τ) for a bounded observable A that its family A_h is not only norm continuous, but even norm-C^∞ : All partial derivatives $\partial_h^\alpha A_h$ exist in norm convergence.

Observe that this condition may be very natural: Given the inherent inaccuracy of space measurements an observable can have little significance, if it changes strongly under very small translations. (Although of course this may be an idealization just as the concept of derivative - interpreting velocity as time derivative amounts to a strong simplification, but the built-in limit in praxis never can be carried out, but will have to be replaced by a difference quotient).

Next, we impose the same condition also in phase space: The Fourier transform

$$(1.1) \qquad u^\wedge(\xi) = Fu(\xi) = \int u(x)e^{-ix\xi}đx \; , \; đx = (2\pi)^{-3/2}dx \; .$$

defines a unitary operator of H 'diagonalizing' the momentum obser vable (i.e., $p_j = -i\partial_{x_j}$ is transformed into the multiplication operator $u^\wedge(x) \to x_j u^\wedge(x)$).

Our quantum mechanical measuring process is invariant under unitary transforms: If ψ is a state and A is an observable, as above, and if $U:H\to H$ is a unitary operator then we get exactly the same probabilities and expectation values if we replace ψ and A by $\psi^\sim = U\psi$ and $A^\sim = UAU^*$, respectively. In sec.0 the position observable was diagonal (was a multiplication operator). If we apply the Fourier transform then the momentum operator gets diagonal. Position and momentum play a dual role in much of the theory. Hence it appears natural to impose the same "translation smoothness condition" (i.e.,cdn.(τ)) also in the momentum-diagonal form. A simple calculation shows that this amounts to the following gauge-smoothness condition (γ): The operator family $A^h = e^{ihx}Ae^{-ihx}$ is norm-C^∞ in the parameter h (i.e., $A^h \in C^\infty(\mathbb{R}^3, L(H))$).

[Indeed, $(FAF^*)_h = T_{-h}FAF^*T_h = FA^hF^*$, as easily checked] .

Remark 1.1. Both conditions (τ) and (γ) (i.e. translation and gauge smoothness together) amount to the condition of norm-smoothness of conjugation within the Heisenberg group HG : The map $g \to A(g) = g^*Ag$ is $C^\infty(HG, L(H))$. Recall that the Heisenberg group (or its standard representation within the unitary group $U(H)$ of H) consists of all unitary operators of the form $g = e^{i\varphi}e^{i\zeta x}e^{izD} = e^{i(\varphi+\zeta x)}T_z$, $\varphi \in \mathbb{R}$, $z,\zeta \in \mathbb{R}^3$ with operator multiplication as group operation.

At this time we recall the concept of pseudo-differential operator (ψdo) with symbol in $CB^\infty(\mathbb{R}^3)$:

The space of symbols $\psi t_0 = CB^\infty(\mathbb{R}^3)$ is defined as

$$(1.2) \qquad \psi t_0 = \{a(x,\xi) \in C^\infty(\mathbb{R}^6): \text{all derivatives are bounded}\}$$

(we mean all partial derivatives in the 6 variables x and ξ and of all orders). For a "symbol" $a(x,\xi) \in \psi t_0$ define the operator $A = op(a) = a(x,D)$ by setting

$$(1.3) \qquad (Au)(x) = (2\pi)^{-3}\int d\xi \int dy e^{i\xi(x-y)}a(x,\xi)u(y) .$$

The integrals at right exist in the order stated whenever $u \in S = S(\mathbb{R}^3)$, the space of rapidly decreasing $C^\infty(\mathbb{R}^3)$-functions, and the function $v = Au$ then is in S again, as seen by a technical but straight-forward calculation (cf. I,1, VIII,1). Thus A of (1.3) is defined as a (continuous) map $S \to S$, whenever $a \in CB^\infty = \psi t_0$.

The significance of the last remarks become clear if we

look at the following result:

Theorem 1.2. Every pseudodifferential operator A = a(x,D) with symbol $a \in \psi t_0$ extends to a continuous operator of $L(H)$ (from the dense subspace S of H), also denoted by A . Moreover the class of all such L^2-bounded pseudo-differential operators precisely coincides with the class of all bounded observables satisfying cdn's (τ) and (γ) above (i.e., the class of A with $A(g) \in C^\infty(HG, L(H))$) . Furthermore, for the latter type of A the corresponding symbol is uniquely defined by the formula

$$(1.4) \qquad a(z,\zeta) = (2\pi)^n \mathrm{trace}\{Q_-^* P(\partial_z, \partial_\zeta) A_{z,\zeta}\} \quad ,$$

where $A_{z,\zeta} = e^{-i\zeta x} e^{izD} A e^{-izD} e^{i\zeta x}$, and where $P(\partial_z, \partial_\zeta)$ is any differential polynomial (constant coefficient differential expression) in $\partial_{z_j} = \partial/\partial_{z_j}$ and ∂_{ζ_j} with the property that it admits a fundamental solution $q(z,\zeta)$ of polynomial growth, and such that the pseudodifferential operator $Q_- = q(-x,-D)$ is of trace class in the Hilbert space H .(Such a polynomial is given by $p(z,\zeta) = (1+z^2)^n(1+\zeta^2)^n$, for example, as may be seen. The corresponding fundamental solution q then is given as the product of Bessel potentials inverting $(1-\Delta_z)^n(1-\Delta_\zeta)^n$).

This result was discussed in [CL]. Actually, it coincides with VIII, thm.2.1, of course, except for some variations in the symbol formula.

The conclusion we derive from thm.1.2 is that, with our above restriction of translation and gauge smoothness imposed on bounded observables, we automatically arrive at the consequence that a bounded observable must be a ψdo , and that a self-adjoint ψdo $A \in \mathrm{Op}\psi t_0$ automatically is a bounded observable satisfying the translation smoothness and gauge-smoothness condition.

Notice that we are lead into a set of axioms selecting better observables which seems elegant, but has several defects. First, most standard observables are not bounded (such as energy, position and momentum, for example) . Second the class $\mathrm{Op}\psi t_0$ of ψdo's with symbol in ψt_0 has severe shortcomings, as far as standard theory of ψdo's is concerned: These ψdo's need not to obey the asymptotic calculus of ψdo's valid for classes of symbols with derivatives decaying at ∞ .

We address the second objection first: If in addition we also impose a "rotation smoothness condition" (cdn.(ρ), below) , and also a "dilation smoothness condition" (cf. cdn.(δ), below).

then we get a smaller class of bounded observables, and a class of ψdo's allowing a global calculus of ψdo's, in a form to be specified.

<u>Condition</u> (ρ) (<u>rotation smoothness</u>): For an orthogonal 3×3-matrix o define the unitary operator $S_o \in L(H)$ by $u(x) \to u(ox)=S_o u(x)$. Then the family $A(o) = S_o^* A S_o$ is norm-C^∞ over the orthogonal group O^3 of $\mathbb{R}^3$.

<u>Condition</u> (δ) (<u>dilation smoothness</u>): For a positive number δ>0 define $R_\delta \in L(H)$ by setting $R_\delta u(x) = \delta^{1/2} u(\delta\xi)$. Then the family $R_\delta^* A R_\delta$ is norm-C^∞ over $\mathbb{R}^+$.

Note that cdn's (ρ) and (δ) are Fourier invariant: They mean the same, if imposed on the position-diagonalized or the momentum diagonalized representation, because the Fourier transform leaves the corresponding subgroups of $U(H)$ invariant. All above cdn's (τ),(γ),(ρ),(δ) together may be expressed by using the Lie-subgroup GS of $U(H)$ of all operators of the form

$$g = e^{i\varphi} e^{izD} e^{i\zeta x} R_\delta S_o \ , \ \varphi \in \mathbb{R} \ , \ z,\zeta \in \mathbb{R}^3 \ , \ o \in O^3 \ . \tag{1.5}$$

<u>Proposition 1.3</u>. For a bounded observable cdn's (τ),(γ),(ρ),(δ) together are equivalent to the property that the function $A(g) = gAg^*$, defined over GS, is $C^\infty(GS, L(H))$, with norm topology used in $L(H)$.

We denote the class of operators used in prop.1.3 by ΨGS, and then have the result ,below (cf. VIII, thm.5.4).

<u>Theorem 1.4</u>. The class ΨGS precisely consists of all ψdo's A = a(x,D) with symbol in the class ψs_0, where ψs_0 denotes the class of symbols in ψt_0 such that application of a finite number of differential expressions η_{jl} of (1.6), below, repeated in arbitrary orders always gives a function of ψt_0.

$$\begin{aligned} \eta_{00} &= \textstyle\sum_{j=1}^n (\xi_j \partial_{\xi_j} - x_j \partial_{x_j}) \ , \\ \eta_{jl} &= (\xi_j \partial_{\xi_l} - \xi_l \partial_{\xi_j}) + (x_j \partial_{x_l} - x_l \partial_{x_j}) \ , \ j,l=1,\dots,n \ . \end{aligned} \tag{1.6}$$

One confirms easily that the functions in ψs_0 have the "classical symbol property" on compact sets: For any compact set $K \subset \mathbb{R}^n$ and all multi-indices α , β we have

$$\partial_x^\alpha \partial_\xi^\beta a(x,\xi) = O(\langle\xi\rangle^{-|\beta|}) \ , \ x,\xi \in \mathbb{R}^n \ , \ x \in K \ . \tag{1.7}$$

There is an analogous condition with x and ξ reversed. In fact, global such conditions (degenerating at infinity) hold cf. VIII,6.

Next we introduce the class ψs of all polynomials

$$(1.8) \qquad a(x,\xi) = \sum_{\alpha,\beta} a_{\alpha,\beta}(x,\xi) x^{\alpha}\xi^{\beta} \ , \ a_{\alpha,\beta} \in \psi s_0 \ .$$

in the 6 variables x,ξ and observe that ψs as well as the class $\Psi S = \mathrm{Op}\psi s = \{a(x,D) : a \in \psi s\}$ are algebras (cf. VIII,7). For a of the form (1.8) the operator $A=a(x,D)$ is of the form

$$(1.9) \qquad A = \sum_{\alpha,\beta} x^{\alpha} A_{\alpha,\beta} D^{\beta} \ , \text{ with } A_{\alpha,\beta} = a_{\alpha,\beta}(x,D) \ .$$

However, the commutators $[D_j,A]$ and $[x_j,A]$ for $A \in \mathrm{Op}\psi s$, handle rather well (cf. VIII,7), and we also may write

$$(1.10) \qquad A = \sum a^1_{\alpha,\beta}(x,D) x^{\alpha} D^{\beta}_x = \sum x^{\alpha} D^{\beta}_x a^2_{\alpha,\beta}(x,D) \ ,$$

with different coefficient symbols a^j_{α}. A Leibniz-Taylor formula as in I,5 holds in $\mathrm{Op}\psi t$, with the class ψt of polynomials with coefficients in ψt_0, but the remainder may not be well behaved. On the other hand, for symbols in ψs, estimates for the remainder can be derived, resulting in asymptotic expansions, and a calculus of ψdo's, (although degenerating for $|x|=|\xi|=\infty$, cf. VIII,7).

<u>In the following we shall require all observables considered to be members of</u> $\Psi S = \mathrm{Op}\psi s$. (Note, however, that further restrictions will be introduced in sec.2, below.

Introducing the classes ψs_m (for $m = (m_1,m_2)$, $m_j=0,1,2,..,$) as the polynomials (1.8) of order m_1 in ξ and order m_2 in x, and the operator classes $\Psi S_m = \mathrm{Op}\psi s_m$, we observe that $\Psi S = \cup_m \Psi S_m$, and that a characterization of ΨS_m by conditions like our above 'smoothnesses' $(\tau),(\gamma),(\rho),(\delta)$ is possible with the (polynomially weighted) L^2-Sobolev spaces of III,3, i.e.,

$$(1.11) \qquad H_s = H_{s_1,s_2} = \{u \in S' : \langle x\rangle^{s_2} \langle D\rangle^{s_1} u \in H\} \ .$$

<u>Theorem 1.5</u>. Let $s=(s_1,s_2)$ be given fixed. The class ΨS_m (for any $m=(m_1,m_2)$) consists precisely of all operators in $L(H_s,H_{s-m})$ with the norm-C^{∞} properties (τ), (γ), (ρ), (δ), all with respect to the operator norm of $L(H_s,H_{s-m})$ instead of $L(H)$.

The proof will be omitted.

<u>2. The invariant algebra and the Foldy-Wouthuysen transform.</u>

As stated above, an <u>observable</u> will be a (self-adjoint) operator of ΨS, from now on. But it will be seen later on that the algebra ΨS is <u>not</u> in general invariant under time translation.

Assuming states constant and observables to change by conjugation with e^{-iHt}, we need $A_t=e^{iHt}Ae^{-iHt}$ to remain in ΨS for all t before we can admit $A \in \Psi S$ as an observable. In fact, we tend to require a norm smoothness of the time translation of an observable similar as posted for space and momentum translation.

To be precise: For an L^2-bounded A we require that

$$(2.1) \qquad \langle x\rangle^j \partial_t^j A_t \in C^\infty(\mathbb{R},L(H)) \quad , \quad j=1,2,\ldots,$$

and, moreover, (2.1) is not only required for A_t , but also for every g-derivative of $A_t(g)$, as in sec.1. (over the group GS) . For general $A \in \Psi S_m$ we require (2.1) (for all the derivatives mentioned) with $L(H)$ replaced by $L(H_s,H_{s-m})$, as in thm.1.5.

The above condition will be referred to as 'cdn.(t)'. Clearly cdn.(t) implies that $A_t \in \Psi S$ for all t. Thus A_t has a symbol $a_t(x,\xi)$. Equivalently we also may express the same cdn.(t) in terms of the symbol a_t . For $a\in \psi s_m$ this takes the following form. <u>Condition</u> (t'): We have $A_t = e^{iHt}Ae^{-iHt} = a_t(x,D)$, with $a_t \in \psi s_m$ and (with $e^2 = (0,1)$) we have

$$(2.2) \qquad \partial_t^j a_t \in \psi s_{m-je^2} \text{ for all } t \in \mathbb{R} \text{ , } j=0,1,2,\ldots .$$

Moreover, if we apply any combination (of arbitrary order) of the differential expressions η_{jl} of (1.6) to a_t we obtain a symbol $b_t \in \psi s_m$ still satisfying (2.2) .

The class of all $A \in \Psi S_m$ (for a given m) satisfying cdn (t) (or equivalently (t')) will be called $P_m = P_m^H$ (since it depends on the 'Hamiltonian' H). One trivially notes that (for fixed H) $P_m \subset P_{m'}$ as $m_j \le m_j'$, j=1,2 . The union $P = \cup_m P_m$ is a graded algebra, left invariant under conjugation with e^{-iHt} . It will be called <u>the invariant algebra of the Dirac equation</u> (or, more generally of the Hamiltonian H).

Let us now analyze the significance of cdn (t). The special form (0.9) of the Dirac operator is of crucial importance, in this respect. We know that $\psi(t) = e^{-itH}\psi^0$ is given as the solution of the Dirac equation (0.7) with $\psi(0)=\psi^0$. Explicitly we are given an initial-value problem for the partial differential equation

$$(2.3) \qquad \partial\psi/\partial t + i\{V(x)+mc^2\beta+ c\textstyle\sum_{j=1}^3\alpha_j(-i\partial/\partial x_j-(e/c)A_j(x))\}\psi(x)=0 .$$

Note that (2.3) is a <u>symmetric hyperbolic first order system</u> of 4 partial differential equations in 4 unknown functions. Its principal part has constant coefficients. The existence and uniqueness of the solution of such a system is a well known fact (discussed in detail in VI,2). We will require as additional condition that (for j=1,2,3, and all multi-indices α) we have

$$(2.4) \qquad V^{(\alpha)}(x) = O(\langle x\rangle^{-|\alpha|}) \ , \ A_j^{(\alpha)}(x) = O(\langle x\rangle^{-|\alpha|}) \ .$$

Then $H = h(x,D)$ is a pseudodifferential operator in $Op\psi c_{e^1}$ ($e^1=(1,0)$) with symbol h and pricipal symbol $h^{\sim}$ given by

$$(2.5) \quad h(x,\xi)=V(x)+mc^2\beta+c\alpha\cdot\pi(x,\xi),\ h^{\sim}(\xi)=c\alpha\cdot\xi,\ \pi(x,\xi)=\xi-(e/c)A(x).$$

Both the symbol h and principal symbol $h^{\sim}$ are hermitian symmetric 4×4-matrix-valued functions, confirming the symmetric hyperbolic property of (2.3). Under (2.4) we have $h\in\psi c_{e^1}$ so that $H\in Op\psi c_{e^1}$.

On the other hand, the hyperbolic system (2.3) is <u>not</u> strictly hyperbolic: The matrix $h^{\sim}(\xi)$ of course has real eigenvalues, but they are not distinct. A quick check shows that the eigenvalues of $h^{\sim}(\xi)$ are $\lambda^{\sim}(\xi)=\pm|\xi|$. For $\xi\neq 0$ they both are double, hence have constant multiplicity.

The hermitian matrix-function $h(x,\xi)$ has the diagonalization

$$(2.6) \qquad \varphi^{+}(x,\xi)h(x,\xi)\varphi(x,\xi)=l(x,\xi) = ((l_j(x,\xi)\delta_{jl}))\ ,$$

with a unitary 4×4-matrix $\varphi(x,\xi)$, and eigenvalues

$$(2.7) \quad l_1=l_2=\lambda_+\ ,\ l_3=l_4=\lambda_-\ ,\ \lambda_{\pm}=V(x)\pm cf_r\ ,\ \text{where } f_r=\sqrt{m^2c^2+\pi^2}\ ,$$

for all $x,\xi\in\mathbb{R}^n$. ('+' denotes the adjoint 4×4-matrix, 'a^*' the Hilbert space adjoint's symbol - $a(x,D)^*=a^*(x,D)$, in all of ch.X.)

In I,7 we defined the class of symbols ψc_m, $m=(m_1,m_2)\in\mathbb{R}^2$, by

$$(2.8) \quad \psi c_m = \{a\in C_0^{\infty}:\ a_{(\tau)}^{(\theta)} = O(\langle\xi\rangle^{m_1-|\theta|}\langle x\rangle^{m_2-|\tau|}),\theta,\tau\in\mathbb{Z}_+^n\}\ ,$$

with $\langle\xi\rangle = \sqrt{1+|\xi|^2}$, $a_{(\tau)}^{(\theta)} = \partial_\xi^\theta\partial_x^\tau a$. Now we claim that,under the conditions (2.4), we have $\lambda_{\pm}\in\psi c_{e^1}$, $e^1 = (1,0)$. Moreover, the entries of φ are global symbols in ψc_0. In fact the 2-fold eigenvalues $\lambda_{\pm}$ have the two-dimensional orthogonal spectral projections

$$(2.9) \quad p_{\pm}=1/2(1\pm a_0\beta\pm\alpha\cdot\zeta),\ a_0=mc/(m^2c^2+\pi^2)^{1/2},\ \zeta=\xi/(m^2c^2+\pi^2)^{1/2},$$

A calculation confirms that $\lambda_{\pm}$,as well as the entries of $p_{\pm}$, are in ψc_0 . Let ψ_1, ψ_2 be the first two columns of p_+ , let ψ_3, ψ_4 be the last two columns of p_- , and let $\varphi_j = \sqrt{2}\,\psi_j/(1+a_0)^{1/2}$. Then

(2.10) $\varphi(x,\xi) = (\varphi_1(x,\xi),\varphi_2(x,\xi),\varphi_3(x,\xi),\varphi_4(x,\xi))$

is the desired unitary matrix satisfying (2.5) and again has coefficients in ψc_0, by a calculation. We write ψc^r_m for the class of all r×r-matrices with entries in ψc_m, so that p±, $\varphi \in \psi c^4_0$.

In matters of the invariant algebra *P* it should first be noted that for a single first order hyperbolic equation we get an Egorov-type result, saying that cdn (t) (or (t')) holds for <u>every</u> A ∈ ΨS (cf. VI,6). For a semi-strictly hyperbolic system of the type discussed in IX,5 we found that the invariant algebra *P* is properly contained in ΨS , however. In essence, membership in *P* depends on the symbol of A to commute with the symbol of H , at least modulo lower order terms. Such a result was proven in IX,5.

In order to apply these results we just have to confirm that the Dirac equation (2.3) is semi-strictly hyperbolic of type e' . Indeed, we have the explicit eigenvalues (2.7) of the (complete) symbol of H , and it is readily checked that IX,5,(i),(ii),(iii) are satisfied, if only the potentials *V* and A_j satisfy (2.4). (Note that the classes $\psi\sigma\kappa_e$,... simplify to just $\psi c_{e'}$, as e^{Δ} =e' .)

One finds that most standard dynamical observables (such as position coordinates x_j, momentum coordinates $p_j=-i\partial_{x_j}$, angular momentum coordinates $(x\times p)_j$, etc. all belong to ΨS .

Surprisingly none of these operators belong to *P*. They all need (normally small) additive corrections to become members of *P*. To illuminate this fact we proceed stating a pair of theorems.

From now on we will distinguish in notation between complex-valued and 4×4-matrix-valued symbols: Let ψt^4_m , ψs^4_m , etc. be the 4×4-matrices of symbols in ψt_m, ψs_m , etc. Note that (2.4) implies that $\in \psi c^4_{e'}$, hence H ∈ Op$\psi c^4_{e'}$, e' =(1,0), for h and H of (2.5).

Evidently the algebra *P* contains the identity, and the Hamiltonian H as well as all polynomials in H. In fact, it is clear that every A∈ ΨS commuting with H - in the sense that $e^{iHt}A-Ae^{iHt}$ =0 , using the operator product of *O*(∞) , satisfies A_t=A=const. hence will belong to *P* . For example, the resolvent $R(\mu)=(H-\mu)^{-1}$ can be shown to belong to $\Psi S_{-e'}$, as μ is non-real: Clearly the symbol $h(x,\xi)-\mu = \varphi(\lambda-\mu)\varphi^+$ is md-elliptic of order e' - one easily finds that $(h(x,\xi)-\mu)^{-1}=O(\langle\xi\rangle^{-1})$. It has a special Green inverse, in the sense of III,thm.4.2 (or V,thm.1.3). Since it is invertible in *H* - due to self-adjointness of H - it follows that R(μ) must be that special Green inverse. Hence R(μ)∈ $\Psi S_{-e'}$, implying that

$R(\mu)\in P_{-e^1}$. On the other hand, the operator e^{iHt} does not belong to ΨS , hence not to P either . The operator $e^{iHt}\chi$, for $0\neq\chi\in C_0^\infty$, shifts singularities (cf.VI,7) while $A\chi\in Op\psi c$, for $A\in\Psi S$, leaves wave front sets invariant, by II, thm.5.4.

The result, below, in essence links the algebra P to the ψdo's with symbol matrix commuting with $h(x,\xi)$.

<u>Theorem 2.1.</u> (1) For every $A\in P_m$, of the form $A=a(x,D)$, the symbol a allows a decomposition

$$(2.11)\qquad a = q + z\ ,\ z\in\psi s^4_{m-e},\ q\in\psi s^4_m\ ,\ [h(x,\xi),q(x,\xi)] = 0\ ,$$

the latter for all $x,\xi\in\mathbb{R}^3$.

(2) Vice versa, if a matrix $q\in\psi s^4_m$ of symbols commutes with the symbol $h(x,\xi)$, at least for all $x,\xi\in\mathbb{R}^3$ satisfying $|x|+|\xi|\geq\gamma>0$, then there exists a symbol $z\in\psi s^4_{m-e}$ such that $A = a(x,D) = q(x,D) + z(x,D)\in P_m$.

(3) Suppose A_1 , $A_2\in P_m$ (which must have a decomposition (2.11) both) have the same q, but different z ($=z_1,z_2$), resp. Then $b=z_1-z_2$ allows a decomposition (2.11) with m replaced by m-e.

Thm.2.1 is just IX,thm.5.4, in case of the Dirac equation.

In the next result we are going to construct a unitary operator of H ,and a pseudo-differential operator which decouples the Dirac equations modulo a term of order $-\infty$. The fact that this can be done for finite orders - not order$-\infty$ - is well known. The corresponding unitary operator is called <u>Foldy-Wouthuysen</u> <u>transform</u>.

<u>Theorem 2.2</u>. There exists a unitary operator $X:H\to H$ of the Hilbert space $H = \mathbb{C}^4\times L^2(\mathbb{R}^3)$, which also is a ψdo with symbol in ψc^4_0 , such that

$$(2.12)\qquad X=\chi(x,D)\ ,\ \chi=\varphi+\omega\ ,\ \omega\in\psi c^4_{-e}\ ,\ \varphi \text{ as in } (2.10)\ ,\ e=(1,1),$$

and that the substitution $u = Xv$ (and multiplication from left by X^{-1}) brings the Dirac equation to the form

$$(2.13)\qquad \partial u/\partial t+ i(\Lambda + \Gamma)v = 0\ .$$

Here Γ is a (4×4-matrix of) ψdo(-s) in $O(-\infty)$, while Λ vanishes in its upper right and lower left 2×2-corners. Moreover, we have

$$(2.14)\qquad \mathrm{symb}(\Lambda) = \mathrm{diag}\ (\lambda_+,\lambda_+,\lambda_-,\lambda_-)\quad (\mathrm{mod}\ \psi c_{-e^2}\ )\ ,\ e^2 = (0,1)\ .$$

with diag( ) denoting the diagonal matrix with entries listed.

A proof of thm.2.2 will be discussed in sec.6, below.

Thm.2.1 and thm.2.2 are related by the following argument: First, for a hyperbolic system of the form (2.13) consider the operator $Z_t=e^{-i(\Lambda+\Gamma)t}e^{i\Lambda t}$. Confirm that $Z_t{}^\cdot=e^{-i(\Lambda+\Gamma)t}\Gamma e^{i\Lambda t}\in O(-\infty)$, and, in fact, $Z_t{}^\cdot\in C^\infty(\mathbb{R},O(-\infty))$. Accordingly $Z_t-1=W_t\in C^\infty(\mathbb{R},O(-\infty))$, by integration, and we get $e^{-i(\Lambda+\Gamma)t}=e^{-i\Lambda t}(1+W_t)=e^{-i\Lambda t}+V_t$, where $V_t\in C^\infty(\mathbb{R},\psi s_{-\infty})$. It follows that $A_t=e^{i(\Lambda+\Gamma)t}Ae^{-i(\Lambda+\Gamma)t}=e^{i\Lambda t}Ae^{-i\Lambda t}+$ $+Q_t = B_t + Q_t$, where $Q_t\in C^\infty(\mathbb{R},\psi s_{-\infty})$.

This shows that $P_\Lambda=P_{\Lambda+\Gamma}$, while thm.2.2 and a simple argument imply that $A\in P_H \Leftrightarrow X^*AX\in P_{\Lambda+\Gamma}$, i.e., $P_H=XP_{\Lambda+\Gamma}X^*$. It follows that

$$P_H = XP_\Lambda X^* . \tag{2.15}$$

On the other hand, it is clear that P_Λ contains <u>all</u> matrices

$$A = \begin{pmatrix} B & 0 \\ 0 & C\end{pmatrix} \text{ with } B,C\in \psi s^2 = \mathrm{Op}\,\psi s^2 , \tag{2.16}$$

while $A = \begin{pmatrix} 0 & P \\ Q & 0\end{pmatrix}\in P_\Lambda$ requires that

$$e^{i\lambda_+(x,D)t}Pe^{-i\lambda_-(x,D)t} , e^{i\lambda_-(x,D)t}Qe^{-i\lambda_+(x,D)t}\in \psi s , \tag{2.17}$$

to satisfy differentiability conditions of type (2.1). Using (2.1) for $A\in O(m)$ and $j=1$ yields $\langle x\rangle(\lambda_+-\lambda_-)(x,D)P-\langle x\rangle[P,\lambda_-(x,D)]\in O(m)$, with md-elliptic $\langle x\rangle(\lambda_+-\lambda_-)(x,D)$ having a K-parametrix Ξ, while $\langle x\rangle[P,\lambda_-(x,D)]\in O(m)$. Thus $P\in \Xi O(m)\subset O(m-e)$. Continuing with (2.1), for $j>1$, get $P\in O(-\infty)$. A similar argument shows that also $Q\in O(-\infty)$.

In other words, we have

$$P_H = \{XAX^*+F: A \text{ of the form (2.16), } F\in \mathrm{Op}\psi s^4_{-\infty}\} . \tag{2.18}$$

<u>3. The geometrical optics approach for the Dirac algebra</u> P.

We mention again that thm.2.1 is a special case of IX, thm. 5.4. the proof of which is discussed in IX,5,6,7. In the present section we want to reexamine the formal geometrical optics of this proof in the special case of the Dirac equation. The point will be to derive explicit formulas for the successive perturbation terms, to be evaluated for their physical significance. We will settle for explicit formulas of the first and second step, since calculations for higher orders transcend into the unmanageable.

Changing notation, we write $A=e^{itH}A_0e^{-itH}$, for some $A_0\in P_m$, (with some $m\in \mathbb{R}^2$). The function A=A(t) with values in $O(m)$ assumes values in $Op\psi s_m = \Psi S_m$, by assumption. Differentiating $u(t)=A(t)u$,

for $u \in S \subset \cap\{\mathrm{dom}\ H^k : k=1,2,\ldots\}$ yields $\partial_t u(t) = i[H,A]u$. Accordingly the ψdo-family $A(t)=a_t(x,D)$ satisfies

$$(3.1)\qquad A^{\boldsymbol{\cdot}} = dA/dt = \partial_t a(t,x,D) \quad = i[H,A] \ ,$$

with the time derivative existing in $Op\psi s^4_{m-e^2}$, and in norm convergence of $L(H_s, H_{s-m+e^2})$, H_s of III,3, using III, thm.3.1. Similarly it is seen that time derivatives of all orders of A(t) are in P. In fact, we get $\partial_t^k A(t) \in P_{m-ke^2}$, as $A \in P_m$, for all k.

In terms of symbols, equation (3.1) assumes the form

$$(3.2)\qquad a^{\boldsymbol{\cdot}} = i[h,a] + \langle h,a\rangle - i/2\langle h,a\rangle_2 + \ldots \ (\mathrm{mod}\ \psi c_{-\infty}) \ .$$

Actually, for all common observables we examined, the symbol a belongs to $\psi c \subset \psi s$. A subalgebra $PC \subset P$ of $Op\psi c^4$, and spaces $PC_m \subset P_m$ of $Op\psi c^4_m$ may be defined by imposing a condition (t_c), like cdn.(t) above, but with ψs_m replaced by ψc_m. Then a result like thm.2.1 holds for $A\in PC_m$, with all perturbation symbols in $\psi c_\mu \subset \psi s_\mu$, $\mu=m-e$, etc. Actually, all proofs are the same, but simplified since ψdo-calculus in $Op\psi c$ is somewhat simpler. The corresponding result is discussed in detail in [CD], cf. thm.2.1 there.

Since we have $\psi c_m \subset \psi s_m$ (and the topology of ψc_m is stronger than that of ψs_m) it is clear that $PC_m \subset P_m$, for all m. For a $q \in \psi c_m$ commuting with h any sequence of correction symbols to obtain an $A=Q+Z\in PC_m$ also is a valid sequence of correction symbols in the sense of thm.2.1, above, -i.e., for P_m. This explains why we choose here to assume that all correction symbols are in ψc_μ, instead of ψs_μ, for the corresponding μ. Of course, a more general choice of $z\in \psi s_{m-e} \backslash \psi c_{m-e}$ is possible, we note that in (3.10), below, the choice of $c_\pm$, at t=0, is arbitrary.

This explains why we will work in the algebra PC, for our symbol calculations. In particular, instead of (2.2) we assume

$$(3.3)\qquad \partial_t^j a_t \in \psi c_{m-je^2} \ , \ t \in \mathbb{R} \ , \ j = 0,1,2,\ldots \quad ,$$

restricting the choice of z, but not the possibility of choice.

The first term at right of (3.2) has order $m+e^1$. all others have order $m-e^2$ or less (assuming $A\in PC_m$). It follows that $p=[h,a] \in \psi c^4_{m-e^2}$, for all t. This gives (2.11) with

(3.4) $q = p_+ap_+ + p_-ap_-$, $z = p_+ap_- + p_-ap_+ = z_+ + z_-$,

where clearly $[h,q]=0$, $[h,z_\pm] = (\lambda_\pm - \lambda_\mp)z_\pm = 2cf_r z_\pm = p_\pm p p_\mp \in \psi s^4_{m-e^2}$,

hence $z = (p_+pp_- - p_-pp_+)/2cf_r \in \psi c^4_{m-e}$, as stated in (1).

Note that we have achieved a decomposition (2.11) for all $t \geq 0$, not only for t=0. Moreover, since $p_\pm$ are time independent it follows that $a^\cdot = q^\cdot + z^\cdot$ is the decomposition (2.11) for the operator $A^\cdot \in P_{m-e^2}$. Therefore we get $z^\cdot \in \psi c^4_{m-e-e^2}$. Using this and (2.11) in (3.2) already proves (1) : we get

(3.5) $[h,z] = -i(q^\cdot - \langle h,q\rangle)$ (mod $\psi s^4_{m-e-e^2}$) .

Vice versa, in order to discuss the iteration leading to a proof of thm.2.1 (2), let $q \in \psi s_m$ be a given symbol commuting with $h(x,\xi)$ in the sense stated. If an A (of thm.2.1(2)) exists, then A_t is well defined for all t, and we must have the decomposition (2.11) in the form (3.4), for all t as well. Moreover, also the commutator relation (3.5) will follow for all t, $z=z_t$, $q=q_t$.

To obtain a first approximation for q_t and z_t let us regard (3.5) as a sharp equation, not an equation modulo ψc_{m-e-e^2} . Such an equation is solvable if and only if its right hand side $w=w(q) = i(\langle h,q\rangle - q^\cdot)$ satisfies the two relations

(3.6) $p_\pm w(q_t)p_\pm = 0$, for "+,+" and "-,-" , and all t .

If (3.6) holds then we get

(3.7) $z = h(w)/2cf_r + c_+ + c_-$, $h(w) = p_+wp_- - p_-wp_+$,

where $c_\pm$ are arbitrary symbols leaving $S_\pm = \operatorname{im} p_\pm$ invariant and equal to 0 on $S_\mp$ (i.e., $c_+ = p_+c_+p_+$, $c_- = p_-c_-p_-$) .

Let us first assume that $q(x,\xi)$ is a multiple of the identity in each of the two spaces $S_\pm$. This means that

(3.8) $q(x,\xi) = \kappa_+(x,\xi)p_+(x,\xi) + \kappa_-(x,\xi)p_-(x,\xi)$

with complex-valued symbols $\kappa_\pm \in \psi c_m$. Apparently this is the case if we seek for a z fitting the position or momentum observable. We then may assume that (3.8) holds for all t because that 'Ansatz' will supply a valid symbol $z(t,x,\xi)$. From (3.8) we get $q^\cdot = \kappa_+^\cdot p_+ + \kappa_-^\cdot p_-$,since $p_\pm$ are time independent. Accordingly,

(3.9) $p_+q^\cdot p_- = p_-q^\cdot p_+ = 0$,

and we may replace $w(q)$ in (3.7) by $w_0 = i\langle h,q\rangle$, for

$$(3.10)\qquad z = ih(\langle h,q\rangle)/2cf_r + c_+ + c_- = z_0 + c_+ + c_- \ .$$

Note that (3.10) no longer contains time derivatives.

Anticipating a later discussion, note that the symbols $c_\pm$ at t=0 remain arbitrary, subject to restrictions stated. They must satisfy certain differential equations, but initial values are free.

Note that (3.6) (for all t) leads into a pair of Cauchy problems of first order PDE, where $\kappa_\pm$ are given for t=0:

$$(3.11)\qquad \kappa_\pm^{\cdot} - \langle\lambda_\pm,\kappa_\pm\rangle = 0 \ , \ \kappa_\pm(0,x,\xi) = \kappa_\pm(x,\xi) \ .$$

We have discussed this in IX, prop.6.1, in more general setting. For our present special case it may be derived from (4.1) and prop.4.1, below - we have $q=\kappa_+p_++\kappa_-p_-=g_++g_-$; for "+" in (3.6) we get $0=g_+^{\cdot}-p_+\langle h,g_+\rangle p_+-p_+\langle h,g_-\rangle p_+$, where the last term vanishes, by $(4.2)_2$, while the second term equals $p_+\langle\lambda_+,g_+\rangle p_+$, by $(4.2)_1$, using that the commutator in $(4.2)_2$ vanishes. Finally (4.1) may be used to convert into $(3.11)_1$. Similarly one obtains $(3.11)_2$.

Solving (3.11) means solving the Cauchy problems for the characteristic equations (cf. ch.0, sec.6)

$$(3.12)\qquad x^{\cdot} = -\lambda_{\pm|\xi}(x,\xi) \quad , \ \xi^{\cdot} = \lambda_{\pm|x}(x,\xi) \quad .$$

Clearly (3.12) is a Hamiltonian system of 6 first order ODE in 6 unknown functions $x(t)$, $\xi(t)$ with Hamilton funtion $-\lambda_\pm$. The field of solutions of (3.12) defines a Hamiltonian flow

$$(3.13)\qquad \nu^\pm_{-t}:\mathbb{R}^{2n} \to \mathbb{R}^{2n}, \ \nu^\pm_t(x,\xi) = (x^\pm_t(x,\xi),\xi^\pm_t(x,\xi))$$

of diffeomorphisms $\mathbb{R}^{2n}\Leftrightarrow\mathbb{R}^{2n}$ where $f(t)=x^\pm_t(x^0,\xi^0)$, $\varphi(t)=\xi^\pm_t(x^0,\xi^0)$ is defined as the unique solution of the Cauchy problem

$$(3.14)\qquad f^{\cdot} = \lambda_{\pm|\xi}(f,\varphi) \ , \ \varphi^{\cdot}=-\lambda_{\pm|x}(f,\varphi) \ , \ f(0)=x^0 \ , \ \varphi(0)=\xi^0 \ .$$

The solutions of (3.11) simply are the functions $\kappa_\pm(t,x,\xi)$ constant along the solution curves of (3.12). Hence, for a given $\kappa_\pm$ define $\kappa_{\pm;t} = \kappa_\pm\circ\nu^\pm_t$ (that is, $\kappa_{\pm;t}(x,\xi)=\kappa_\pm(x^\pm_t(x,\xi),\xi^\pm_t(x,\xi))$). Then $\kappa_\pm(t,x,\xi)=\kappa_{\pm;t}(x,\xi)$ solves (3.11), since $\kappa_{\pm;t}\circ\nu^\pm_{-t}=\kappa_\pm\circ\nu^\pm_{-t}\circ\nu^\pm_t$ $=\kappa_\pm$ is indeed constant (independent of t). Also $\kappa_{\pm;0}=\kappa_\pm$, so that $\kappa_{\pm;t}$ indeed solves (3.11).

In other words, the initially (for t=0) defined symbols $\kappa_\pm(x,\xi)$ "flow" along their Hamilton flow (with Hamilton function $+\lambda^\pm(x,\xi)$) defining the symbols $\kappa_{\pm;t}=\kappa_\pm(t,x,\xi)$: We have

(3.15) $\kappa_{\pm,t}=\kappa\circ v^{\pm}_t$, $\kappa_{\pm}(t,x,\xi)=\kappa(x^{\pm}_t(x,\xi),\xi^{\pm}_t(x,\xi))$.

Clearly this determines a function $q(t,x,\xi)$ for all times t. In VI,6 we have shown that not only the flow ψ_t as a family of homeomorphisms $\mathbb{R}^6\to\mathbb{R}^6$ is well defined, but also that (3.8) and (3.15) indeed define a family of symbols in ψc^4_m with the proper dependence on t (i.e. $C^\infty(\mathbb{R},\psi c^4_m)$) .

After determining q for all t we obtain a first correction symbol z , for all t , given in the form (3.10). Since we were neglecting lower order terms the symbol q+z will not be a precise solution of (3.2), but we now may go into (3.2) with

(3.16) $a=q+z+s$, $a^{\cdot}=q^{\cdot}+z^{\cdot}+s^{\cdot}$,

with q , z of (3.8), (3.10), and (3.15), and with $s\in\psi c_{m-2e}$ to be determined. Using $[h,q]=0$ and (the sharp) (3.5), we derive that

(3.17) $[h,s]=-i(z^{\cdot}-\langle h,z\rangle)+(1/2)\langle h,q\rangle_2 \pmod{\psi c_{m-2e-e^2}}$,

as another commutator equation of the general form (3.5), again to be considered as a sharp equation, not only mod ψc_{m-2e-e^2} .

As a condition for solvability we now get (3.6) with

(3.18) $w=w(q,z)=-i(z^{\cdot}-\langle h,z\rangle)+\frac{1}{2}\langle h,q\rangle_2$.

In (3.18) we have $z=z_0+c_++c_-$, with z_0 of (3.10), where $p_\pm z_0 p_\pm=0$, hence $p_\pm z^{\cdot}_0 p_\pm=0$. Thus the solvability condition assumes the form

(3.19) $p_\pm w(c_+)p_\pm+p_\pm w(c_-)p_\pm+p_\pm(i\langle h,z_0\rangle+\frac{1}{2}\langle h,q\rangle_2)p_\pm=0$

with $w(c)=-i(c^{\cdot}-\langle h,c\rangle)$.

As discussed in IX,thm.5.4, under more general assumptions, (3.19) amounts to a pair of first order PDE - equivalently, ODE's along the flow (3.15) - for the symbols $c_\pm$, so far undetermined. After setting the initial values of $c_\pm$, at t=0, the first correction z of (3.10) is fully determined. From (3.17) we then get

(3.20) $s=s_0+d_++d_-$, $s_0=h(\frac{1}{2}\langle h,q\rangle_2-i(z^{\cdot}-\langle h,z\rangle))/(2cf_r)$,

with $d^\pm$ so far free. For the next iteration carry a=q+z+s+r into (3.2) with above q,z,s and a new correction $r\in\psi c_{m-3e}$. The construction of a sequence of corrections and asymptotic sum a=q+z+r+s+..., leading into a valid $A=a(x,D)\in PC_m\subset P_m$ was discussed in IX,5,6,7.

<u>Presently we want to look into explicit calculations of only the full first correction symbol</u> $z=z(t,x,\xi)$ <u>for the special case</u>

of the Dirac equation (2.3), assuming that the potentials V, A_j are symbols in ψc_0, first focusing on $q(x,\xi)= x_k$, ξ_k, (of position and momentum), and other (scalar) standard observables.

Note that our above iteration ignores self-adjointness. Even if we start with a self-adjoint $q(x,D)$, the above construction will not necessarily give a self-adjoint $a(x,D)$. Basing our calculations on the Weyl-representation of ψdo's (cf. I,3), rather than than the left-multiplying would be a cure. However, the algebra PC and its spaces PC_m are adjoint invariant. After obtaining an $A\in PC_m$ by our iteration we just use the selfadjoint $\frac{A+A^*}{2}\in PC_m$.

We close here with the observation that the Hamiltonian flows $v_{\pm t}$ of (3.13) indeed are the particle flows of electron and positron in the following sense: If we eliminate ξ from the 6 equations (3.12) we get the system of 3 second order ODE's

$$(3.21)\qquad \pm\left(\frac{mx^{\cdot}}{\sqrt{1-\dot{x}^2/c^2}}\right)^{\cdot} = -\text{grad } V + \frac{e}{c}(x^{\cdot}\times\text{curl } A) = -E + \frac{e}{c}(x^{\cdot}\times B),$$

in 3 unknown functions. These are the relativistic equations of motion (0.1) we departed from, for a particle of negative and positive charge, respectively.

Indeed, with $\lambda_\pm=V(x)\pm cf_r$ equations (3.12) assume the form

$$(3.22)\qquad x^{\cdot} = \pm c\pi/\sqrt{m^2c^2+\pi^2}\quad,\ \xi^{\cdot} = -(E(x) \pm f_{r|x})$$

Solving the first equation for π we get

$$(3.23)\qquad \pi = \xi-\frac{e}{c}A = \pm mx^{\cdot}/\sqrt{1-x^{\cdot 2}/c^2}\quad.$$

For $A=0$ we get $f_{r|x}=0$ and (3.21) arises by differentiating (3.23) and equating the result with the second relation (3.22). For general A an additional calculation is required which we skip.

Note that the momentum variable for the case "-" is $-\xi$, not ξ. In fact, one of the classical difficulties of Dirac theory arises if one tries to link (3.11), (3.12), for "-", to a variational principle in the sense of classical Hamilton Jacobi theory.

We shall see later that the equations of motion (0.2) for the spin also can be "derived" from (2.3), if a general matrix q not reducing to 1 in $S_\pm$ is allowed.

4. Some identities for Dirac matrices.

First we provide a list of useful identities around the Dirac equation. To simplify notation we write $p_+= p$ $p_-=q$ (while the primarily given symbol q of thm.2.1.(2) will be written as κ). Also $\lambda_{\mp}^{=}$ λ (but we may fall back to the old notations if they are clearer).

For a summary of conventions used cf. (4.3), (4.4), (4.5), below. First we discuss some matrix results of IX,6, in the present special case. Formulas have analogues with p and q interchanged.

For "'"= "$_{|x}$" or "$_{|\xi}$" we have $pp'p=-pq'p=p'qp=0$, etc., hence

$$(4.1) \qquad pp'p = qq'q = pq'p = qp'q = 0 \ .$$

<u>Proposition 4.1</u>. For any symbol g with $pg = gp = pgp = g$ we have

$$p\langle h,g\rangle p = p\langle \lambda,g\rangle p - (\lambda-\lambda_-)[g,(pp_{|\xi}p_{|x}p)]$$

$$(4.2) \qquad = \langle \lambda,g\rangle + gp\langle \lambda,q\rangle q + q\langle \lambda,q\rangle pg - (\lambda-\lambda_-)[g,(pp_{|\xi}p_{|x}p)] ,$$

$$q\langle h,g\rangle q = 0 \ .$$

Note that (4.2) holds the derivation of (3.11) from (3.6): For (q of (3.6) now called κ) $\kappa = \kappa_+p + \kappa_-q$ with scalars $\kappa_\pm$ we get

$$0 = p(\kappa^{\cdot}-\langle h,\kappa\rangle)p = \kappa_+{}^{\cdot}p - p\langle h,\kappa_+p\rangle p - p\langle h,\kappa_-q\rangle p \ .$$

The last term at right vanishes, using (4.2), with p and q interchanged, since $g=\kappa^-q$ satisfies the (reversed) assumptions. For the second term we use the first formula (4.2) with $g=\kappa_+p$, noting that the commutator vanishes, while $pg'p=\kappa_+{}'p$. The second form of the first (4.2) follows since (3.19) amounts to an ODE along the classical orbits: We get $pw(c_-)p = 0$ by the (reversed) second relation while $pw(c_+)p = (\partial_t + \lambda_{|\xi}\partial_x - \lambda_{|x}\partial_\xi)c_+$ + terms not involving derivatives of c_+. The above term is the directional derivative along the classical orbits. Thus (3.19) indeed translates into an inhomogeneous linear first order ODE along those orbits.

<u>Proof of prop.4.1</u>. For the second relation (4.2) we have

$$q\langle h,g\rangle q = q(\lambda_{|\xi}p+\lambda_{-|\xi}q+\lambda p_{|\xi}+\lambda_-q_{|\xi})g_{|x}q$$
$$-qg_{|\xi}(\lambda_{|x}p+\lambda_{-|x}q+\lambda p_{|x}+\lambda_-q_{|x})q \ .$$

Here the first term produced by each of the parantheses vanishes since it contains the factor $pq=qp=0$. The second terms also vanish since they contain the factors $qg_{|x}q=-qgq_{|x}$ and $qg_{|\xi}q=-qgq_{|\xi}$ (note that $qg=gq=0$, hence $q'g=-qg'$, $qg'q=-qgq'=0$). We organize the remaining terms at right in the form $\lambda T + \lambda_-T_-$, where

$$T = qp_{|\xi}g_{|x}q - qg_{|\xi}p_{|x}q = qp_{|\xi}pg_{|x}q - qg_{|\xi}pp_{|x}q \text{ (using (4.1))}$$
$$=-qp_{|\xi}pgpq_{|x}q + qq_{|\xi}pgpp_{|x}q \quad \text{(using } qg'=-q'g)$$
$$= 0 \ , \text{ using } p'+q'=0, \text{ due to } p+q=1 \ .$$ The term T_- vanishes

for analogous reason, proving the second relation (4.2).

For the first relation we have

$$p\langle h,g\rangle p = p(\lambda_{|\xi}p+\lambda_{-|\xi}q+\lambda p_{|\xi}+\lambda_{-}q_{|\xi})g_{|x}p$$
$$- pg_{|\xi}(\lambda_{|x}p+\lambda_{-|x}q+\lambda p_{|x}+\lambda_{-}q_{|x})p ,$$

where the second term in each paranthesis produces 0 , due to the factor $pq=qp=0$. For the third and fourth term we use $p'=-q'$. These terms of both parantheses give

$$(\lambda-\lambda_{-})(pp_{|\xi}qg_{|x}p - pg_{|\xi}qp_{|x}p) \qquad \text{(using (4.1))}$$
$$=(\lambda-\lambda_{-})(-pp_{|\xi}q_{|x}pg + gpq_{|\xi}p_{|x}p) \qquad \text{(using } qg'=-q'g)$$
$$=(\lambda-\lambda_{-})[pp_{|\xi}p_{|x}p,g] \qquad \text{(using } q'=-p') .$$

Therefore the first form of (4.2) follows. For the second form we write $pg'p = g' - pg'q - qg'p$ (noting that $qg'q = 0$)
$= g' + g(pq'q) + (qq'p)g$ (using $g'q=-gq'$). This achieves the desired form, proving prop.4.1.

Now we discuss a list of identities for 'Dirac matrices'. For α_j , β , μ_j , ρ_j cf. (0.10) f. We use the abbreviations

$$\pi = \xi - \tfrac{e}{c}A \ , \ f_r = \sqrt{m^2c^2+\pi^2} \ , \ a_0=mc/f_r \ , \ \zeta = \pi/f_r \ ,$$

(4.3) $$h_0=h/c-V = mc\beta + \alpha\cdot\pi \ , \ \eta_0 = h_0/f_r = a_0\beta + \alpha\cdot\zeta = p - q \ ,$$

$$p = \tfrac{1}{2}(1+\eta_0) = \tfrac{1}{2}(1+a_0\beta+\alpha\cdot\zeta) \ , \ q = \tfrac{1}{2}(1-\eta_0) = \tfrac{1}{2}(1-a_0\beta - \alpha\cdot\zeta).$$

where p , q are orthogonal projections, and form a decomposition of 1 , the spectral resolution of h and h_0:

(4.4) $$p+q=1 \ , \ p^+=p \ , \ q^+=q \ , \ p^2=p \ , \ q^2=q \ , \ pq = qp = 0 \ ,$$

$$h_0 = f_rp - f_rq \ , \ h = \lambda p + \lambda_{-}q \ , \ \lambda_{\pm} = V \pm cf_r \ .$$

For a general 4×4-matrix κ we write $\boldsymbol{f}=\boldsymbol{f}_+$, $\boldsymbol{h}=\boldsymbol{h}_-$, where

(4.5) $$\boldsymbol{f}_{\pm}(\kappa) = p\kappa p \pm q\kappa q \ , \ \boldsymbol{h}_{\pm}(\kappa) = p\kappa q \pm q\kappa p \ .$$

The following formal 3-vectors of 4×4-matrices are useful:

(4.6) $$\omega_c = \alpha - \eta_0\zeta = \alpha - (a_0\beta+\alpha\cdot\zeta)\zeta \ , \ \theta_c = a_0\mu + \rho\times\zeta \ ,$$

$$\omega = f_r^2\omega_c = (m^2c^2+\pi^2)\alpha + h_0\pi \ , \ \theta = mc\mu + \rho\times\pi \ .$$

<u>We discuss the following list of identities</u> (4.7)-(4.19):

(4.7) $$p\alpha q = \tfrac{1}{2}(\omega_c + i\theta_c) \ , \ q\alpha p = \tfrac{1}{2}(\omega_c - i\theta_c) \ ,$$

(4.8) $$p\alpha p = p\zeta \ , \ q\alpha q = - q\zeta \ ,$$

(4.9) $$p\beta q = \tfrac{1}{2}(\beta - a_0\eta_0) - \tfrac{i}{2}\mu\zeta = (q\beta p)^+ \ , \ p\beta p = a_0 p \ ;$$

(4.10) $$p\mu q = \tfrac{1}{2}(a_0\theta_c - (\zeta\cdot\mu)\zeta) + \tfrac{i}{2}(\beta\zeta - a_0\alpha) \ ,$$

(4.11) $$p\rho q = \tfrac{1}{2}(\zeta^2\rho - \zeta(\rho\cdot\zeta) - a_0\mu\times\zeta + i\alpha\times\zeta) = (q\rho p)^+ \ ,$$

(4.12) $$(\alpha\cdot\zeta)\alpha(\alpha\cdot\zeta) = -\alpha\zeta^2 + 2(\alpha\cdot\zeta)\zeta \ .$$

The 'operators' $\boldsymbol{f}_+$ and $\boldsymbol{h}_+$ clearly act on the 4×4-matrices as projections (onto the spaces of matrices commuting and anti-commuting with h_0, respectively), and we get

(4.13) $$\boldsymbol{f}_+ + \boldsymbol{h}_+ = 1 \ , \text{ i.e., } \boldsymbol{f}_+(\kappa) + \boldsymbol{h}_+(\kappa) = \kappa \ , \text{ for all } \kappa \ .$$

In the same sense (regarding $\boldsymbol{f}, \boldsymbol{h}$, etc. as operators on matrices),

(4.14) $$\boldsymbol{f}_+^2 = \boldsymbol{f}_2^2 = \boldsymbol{f}_+ \ , \ \boldsymbol{h}_+^2 = \boldsymbol{h}_-^2 = \boldsymbol{h}_+ \ .$$

It illuminates the role of ω_c and θ_c above to note that

(4.15) $$\boldsymbol{h}_+(\theta_c) = \theta_c \ , \ \boldsymbol{h}_+(\omega_c) = \omega_c \ ,$$

while

(4.16) $$\boldsymbol{h}(\alpha) = i\theta_c \ , \ \boldsymbol{h}(\theta_c) = -i\omega_c \ , \ \boldsymbol{h}(\omega_c) = \boldsymbol{h}(\alpha) = i\theta_c \ .$$

Also note

(4.17) $$\boldsymbol{h}(\mu) = i(\beta\zeta - a_0\alpha) \ , \ \boldsymbol{h}(\rho) = i\alpha\times\zeta \ , \ \boldsymbol{h}(\beta) = -i\mu\cdot\zeta \ .$$

Furthermore:

(4.18) $$p\omega_c = p\omega_c q = p\alpha q = \tfrac{1}{2}(\omega_c + i\theta_c),\ q\omega_c = q\omega_c p = q\alpha p = \tfrac{1}{2}(\omega_c - \theta_c),$$
$$p\theta_c = p\theta_c q = \tfrac{1}{2}(\theta_c - i\omega_c) \ , \ q\theta_c = q\theta_c p = \tfrac{1}{2}(\theta_c + i\omega_c) \ .$$

Or, in terms of ω and θ :

(4.19) $$p\omega = \tfrac{1}{2}(\omega + if_r\theta) \ , \ q\omega = \tfrac{1}{2}(\omega - if_r\theta) \ ,$$
$$p\theta = \tfrac{1}{2}(\theta - i\omega/f_r) \ , \ q\theta = \tfrac{1}{2}(\tau + i\omega/f_r) \ .$$

First we verify (4.7)-(4.8), using (0.15):

$$2p\alpha_1 = \alpha_1 + a_0\beta\alpha_1 + \zeta_1\alpha_1{}^2 + \zeta_2\alpha_2\alpha_1 + \zeta_3\alpha_3\alpha_1$$

$$= \alpha_1 + \zeta_1 + a_0\mu_1 + i(\rho_2\zeta_3 - \rho_3\zeta_2) ,$$

confirming the first component of the vector equation

$$(4.20) \qquad 2p\alpha = \alpha + \zeta + i(a_0\mu + \rho\times\zeta) = \alpha + \zeta + i\theta_c$$

(the other components follow analogously).

In (4.20) we may take the self-adjoint part:

$$(4.21) \qquad p\alpha + \alpha p = \alpha + \zeta .$$

Applying p from left and right, using (4.4), confirms the first relation (4.8) (the second follows similarly). Then (4.20) yields

$$2p\alpha q = 2p\alpha - 2\pi\alpha p = \alpha + \zeta + i\theta_c - \zeta(1+\eta_0) = \alpha - \eta_0\zeta + i\theta_c = \omega_c + i\theta_c,$$

confirming (4.7) (the second relation is the adjoint of the first)

From (4.7) we conclude

$$(4.22) \qquad \boldsymbol{h}_+(\alpha) = \omega_c \quad , \ \boldsymbol{h}(\alpha) = i\theta_c .$$

Then the evident identity

$$(4.23) \qquad \boldsymbol{h}(\boldsymbol{h}(\kappa)) = \boldsymbol{h}_+(\kappa)$$

implies $\boldsymbol{h}(i\theta_c) = \boldsymbol{h}(\boldsymbol{h}(\alpha)) = \boldsymbol{h}_+(\alpha) = \omega_c$, or, (4.16). Also, (4.15) is a consequence of the projection properties of $\boldsymbol{h}$, i.e., $\boldsymbol{h}^3 = \boldsymbol{h}$, eic.

For a confirmation of (4.12) focus on its first component:

$$(\alpha\cdot\zeta)\alpha_1(\alpha\cdot\zeta) = (\alpha_1\zeta_1 + (\alpha\zeta)^{\check{}})\alpha_1(\alpha_1\zeta_1 + (\alpha\zeta)^{\check{}}) \quad \text{(with } (\alpha\zeta)^{\check{}} = \alpha_2\zeta_2 + \alpha_3\zeta_3)$$

$$= \zeta_1^2\alpha_1 - \alpha_1((\alpha\zeta)^{\check{}})^2 + 2\zeta_1(\alpha\zeta)^{\check{}} \qquad \text{(where } ((\alpha\zeta)^{\check{}})^2 = \zeta_2^2 + \zeta_3^2)$$

$$= 2\zeta_1(\alpha\cdot\zeta) - \alpha_1\zeta^2 , \qquad \text{(using (0.15) again)} .$$

Again the other components follow similarly.

Next for (4.9) proceed as for (4.7)-(4.8) :

$$2p\beta = (\beta + a_0) - i\zeta\cdot\mu , \ p\beta + \beta p = \beta + a_0 ,$$

onto which p may be applied from left and right for the second relation (4.9) . Then also the first follows.

For (4.10): Multiply (4.20) from right by β for

$$(4.24) \qquad 2p\mu = (\mu - (\rho\beta)\times\zeta) + i(\beta\zeta - a_0\alpha) .$$

This implies

$$(4.25) \qquad 2\boldsymbol{h}(\mu) = 2p\mu - 2\mu p = 2i(\beta\zeta - a_0\alpha) .$$

Applying "$\boldsymbol{h}$" to this:

$$(4.26) \qquad \boldsymbol{h}_+(\mu) = \boldsymbol{h}(\boldsymbol{h}(\mu)) = i\zeta\boldsymbol{h}(\beta) - ia_0\boldsymbol{h}(\alpha) = \zeta(\mu\cdot\zeta) + a_0\theta_c .$$

Using that

$$(4.27)\qquad p\kappa q = \tfrac{1}{2}(\boldsymbol{h}_+(\kappa) + \boldsymbol{h}(\kappa)) \ , \text{ for any } \kappa \ ,$$

we get the required formula (4.9).

Next, for (4.11): $2p\rho = s + i\alpha\times\zeta$, with a self-adjoint matrix s. Thus the second formula (4.17) follows. As above we then get

$\boldsymbol{h}_+(\rho) = i\boldsymbol{h}(\alpha)\times\zeta = -\theta_c\times\zeta$, and (4.11) follows.

Finally, (4.18),(4.19) follow from (4.7),(4.8), and (4.27).

5. The first correction z_0 for standard observables.

Using the identities of sec.4 we now are equipped for an explicit discussion of the 'anti-commuting part' $z_0(t,x,\xi)$ of the first perturbation symbol, as defined by (3.10). It will be more difficult to also comment on the propagation of $c_\pm$ of (3.10). These symbols may be set zero for t=0, but the might propagate into nonvanishing quantities.

We get restricted to the case of a symbol $\kappa(x,\xi)\in\psi c_m^4$ commuting with $h(x,\xi)$ which equals multiples of the identity $\kappa_\pm$ in the two eigenspaces $\boldsymbol{S}_\pm$ of h . Such symbol may be written as $\kappa=\kappa_+p+\kappa_-q$, with complex-valued $\kappa_\pm$. Introduce $\sigma_\kappa=\frac{1}{2}(\kappa_++\kappa_-)$, $\delta_\kappa=\frac{1}{2cf_r}(\kappa_+-\kappa_-)$.

<u>Proposition 5.1</u>. The symbol $z_0=\frac{i}{2cf_r}\boldsymbol{h}(\langle h,\kappa\rangle)$ of (3.10) is given by

$$(5.1)\qquad z_0 = i\boldsymbol{h}(\langle p,\sigma_\kappa\rangle) - icf_r\boldsymbol{h}_+(\langle p,\delta_\kappa\rangle_+) - i\delta_\kappa \boldsymbol{E}.\boldsymbol{h}(p_{|\xi}) \ ,$$

where we have written $\langle p,\delta\rangle_+ = p_{|\xi}.\delta_{|x} + p_{|x}.\delta_{|\xi}$, $\boldsymbol{E} = \boldsymbol{V}_{|x}$.

<u>Proof</u>. First assume that $\kappa_-=0$. Write κ as κp , assume $V=0$. Then $h=ch_0=cf_r(p-q)$, by (4.3). We must compute $z_0=\frac{i}{2f_r}\boldsymbol{h}(\langle f_r(p-q),\kappa p\rangle)$. Calculating $\langle f_r(p-q),\kappa p\rangle$ we get $(f_r(p-q))'=f_r'(p-q)+f_r(p-q)'$, and $(\kappa p)'=\kappa'p+\kappa p'$. Substituting this we need only consider terms with one derivative on p or p-q and the other on f_r or κ , due to $\langle p-q,p\rangle=0$ and $\boldsymbol{h}((p-q)p)=0$. Thus we get

$$(5.2)\qquad \langle f_r(p-q),\kappa p\rangle = T_1 + T_2 =$$

$$=\{f_{r|\xi}(p-q)p_{|x}\kappa-\kappa p_{|\xi}(p-q)f_{r|x}\} + \{f_r(p-q)_{|\xi}p\kappa_{|x}-\kappa_{|\xi}p(p-q)_{|x}f_r\},$$

where

$h((p-q)p_{|x})=p(p-q)p_{|x}q-q(p-q)p_{|x}p=h_+(p_{|x})$, $h(p_{|\xi}(p-q))=-h_+(p_{|\xi})$, so, $h(T_1) = f_{r|\xi}\kappa h_+(p_{|x})+f_{r|x}\kappa h(p_{|\xi})$. Also, $h((p-q)_{|\xi}p)=-2qp_{|\xi}p=(h-h_+)(p_{|\xi})$, $h(p(p-q)_{|x})=2pp_{|x}q=(h+h_+)(p_{|x})$, so, $h(T_2) =-f_r\kappa_{|x}h_+(p_{|\xi}) -f_r\kappa_{|\xi}h_+(p_{|x}) + f_r h(\langle p,\kappa\rangle)$. Summing we get $h(T_1+T_2) =-f_r^2(\frac{\kappa}{f_r})_{|\xi}h_+(p_{|x})-f_r^2(\frac{\kappa}{f_r})_{|x}h(p_{|\xi})+f_r h(\langle p,\kappa\rangle)$. In our present case we get $2\sigma_\kappa=\kappa=cf_r\delta_\kappa$, hence

$$z_0=\frac{i}{2f_r}h(T_1+T_2) = -icf_r h_+(\langle p,\delta_\kappa\rangle_+) +ih(\langle p,\sigma_\kappa\rangle) , \tag{5.3}$$

in agreement with formula (5.1). Now we may do the same calculation with κq instead of κp . It is left to the reader to confirm that the same formula (5.3) follows, now for $2\sigma_\kappa=\kappa=-cf_r\delta_\kappa$. Summing over the two formulas we get (5.3) for $\kappa=\kappa_+p+\kappa_-q$, still with $V=0$.

Finally, to account for a nontrivial $V(x)$ we evaluate $\langle V,\kappa p\rangle=-E(\kappa_{|\xi}p+\kappa p_{|\xi})$, implying $h(\langle V,\kappa p\rangle) = -E\kappa h(p_{|\xi})$. Similarly, $h(\langle V,\kappa q\rangle) =-E\kappa h(q_{|\xi})=+E\kappa h(p_{|x})$. For $k=k_+p+\kappa_-q$ we thus get

$$\frac{i}{2cf_r}h(\langle V,\kappa\rangle) =-\frac{iE}{cf_r}(\kappa_+-\kappa_-)h(p_{|\xi}) =-iE\delta_\kappa h(p_{|x}) . \tag{5.4}$$

Adding (5.3) and (5.4) we get (5.1) in the general case. Q.E.D.

It is useful to offer (5.1) in a different form. Recalling (4.3) we get $2\langle p,\sigma\rangle = \langle \eta_0,\sigma\rangle = \langle a_0,\sigma\rangle\beta + \langle\zeta,\sigma\rangle.\alpha$, where $\langle\zeta,\sigma\rangle = \frac{1}{f_r}\langle\pi,\sigma\rangle + \frac{\zeta}{a_0}\langle a_0,\sigma\rangle$. Thus $2\langle p,\sigma\rangle=\langle a_0,\sigma\rangle\{\beta+\alpha\frac{\zeta}{a_0}\} + \frac{1}{f_r}\langle p,\sigma\rangle$. Or,

$$\langle p,\sigma\rangle =\frac{1}{2a_0}\langle a_0,\sigma\rangle(p-q) +\frac{1}{2f_r}\langle\pi,\sigma\rangle.\alpha , \quad h(\langle p,\sigma\rangle) =\frac{1}{2f_r}\langle\pi,\sigma\rangle\theta_c .$$

Conclusion:

$$ih(\langle p,\sigma_\kappa\rangle) =\frac{1}{2f_r}\langle\sigma_\kappa,\pi\rangle.\theta_c , \tag{5.5}$$

With exactly the same argument we get

$$-i\delta_\kappa Eh(p_{|\xi}) = -\frac{i\delta_\kappa}{2f_r}E.\alpha , \quad -icf_r h_+(\langle p,\delta_\kappa\rangle_+)=-\frac{ic}{2}\langle\pi,\delta_\kappa\rangle_+.\omega_c . \tag{5.6}$$

For the last formula we used

$$h_+(\alpha)=ih(\theta_c)=\omega_c=\alpha-(a_0\beta+\alpha\zeta)\zeta . \tag{5.7}$$

Also, we get $\langle p,\delta_\kappa\rangle_+=\frac{1}{a_0}\langle a_0,\delta_\kappa\rangle_+(p-q)+\frac{1}{f_r}\langle\pi,\delta_\kappa\rangle_+$ and $h_+(p-q)=0$, etc.

Summarizing:

Corollary 5.1. Formula (5.1) may also be expressed in the form

(5.8) $$z_0 = \frac{1}{2f_r}\langle\sigma_\kappa,\pi\rangle.\theta_c + \frac{\delta}{2f_r}E.\theta_c - \frac{ic}{2}\langle\delta_\kappa,\pi\rangle_+.\omega_c\ ,\quad E = V_{|x}\ ,$$

θ_c and ω_c of (4.6). The hermitian symmetric part of this symbol is

(5.9) $$z_{0,s} = \frac{1}{2f_r}\{\langle\sigma_\kappa,\pi\rangle + \delta_\kappa E\}.\theta_c\ .$$

Let us now look at the explicit form of this perturbation for some common dynamical observables. To get an idea of orders of magnitude we reintroduce $\hbar\neq 1$: The Dirac equation (0.7) with H of (0.9) coincides with that for $\hbar=1$ if we replace m, V, A by $m/\hbar, V/\hbar, A/\hbar$, respectively. Accordingly the corresponding substitution in (5.8) and (5.9) will lead to the corresponding terms for general $\hbar$. One finds that the very same formulas will remain intact, except that now ξ must be replaced by $\hbar\xi$. In all details, we get

(5.10) $$z_{0,s} = \frac{1}{2f_r}\{\langle\sigma_\kappa,\pi\rangle + \hbar\delta_\kappa E\}.\theta_c\ ,\quad \theta_c = a_0\mu + \rho\times\zeta\ ,\ \text{where now}$$

$$\pi = \hbar\xi - \frac{e}{c}A\ ,\ f_r = \sqrt{m^2c^2+\pi^2}\ ,\ a_0 = \frac{mc}{f_r}\ ,\ \zeta = \frac{\pi}{f_r},\ \sigma_\kappa = \frac{1}{2}(\kappa_+ + \kappa_-),\ \delta_\kappa = \frac{1}{2cf_r}(\kappa_+ - \kappa_-)$$

with $\kappa_\pm$ denoting the scalar symbols in the two eigenspaces of h . For space, momentum, and angular momentum, electrostatic potential mechanical momentum, relativistic mass, energy, particle density we all have initially $\kappa_+ = \kappa_- = \sigma_\kappa$, $\delta=0$ - the initial observable is a multiple of the 4×4-identity. However, as time progresses, this will not remain true: We know that $\kappa_\pm$ will propagate along different particle flows. Still it is interesting to note that an initial correction is needed. (We decide for the moment to set $c_\pm=0$, at t=0). We get the following list of corrected observables. All of them are stated for t=0 only. In all of them we use the term

$$\lambda_c = \frac{\hbar}{2f_r}\,\theta_c = \frac{1}{2f_r}(a_0\mu + \rho\times\zeta)\ ,\ f_r\ ,\ a_0\ ,\ \zeta \text{ of (5.10)}.$$

(5.11) Position $x = (x_1, x_2, x_3)$: $x_{corr} = x - \lambda_c$.

(5.12) Momentum $p = -i\hbar\partial_x$: $p_{corr} = \hbar\xi - \frac{e}{c}\sum_j \lambda_{cj} A_{j|x}$.

(5.13) Angular momentum $k = x\times p$: $k_{corr} = \hbar x\times\xi - \hbar\lambda_c\times\xi$

$$-\frac{e}{c}\sum_j \lambda_{cj}(x\times A_{j|x}).$$

(5.14) <u>Electrostatic</u> <u>potential</u> $V(x)$: $V_{corr}=V-\lambda_c\cdot E$.

(5.15) <u>Mechanical</u> <u>momentum</u> $p-\frac{e}{c}A$: $\pi_{corr}=\pi-\frac{e}{c}B\times\lambda_c$, $\pi=\xi-\frac{e}{c}A$.

(5.16) <u>Relativistic</u> <u>mass</u> $M=H-V$: $M_{corr}=h(x,\xi)-V(x)+\lambda_c\cdot E$.

(5.17) <u>Energy</u> H : $H_{corr} = h(x,\xi)$.

(5.18) <u>Particle</u> <u>density</u> $\delta=\delta(x-x^0)$: $\delta_{corr}=\delta(x-x^0)-\lambda_c\cdot\delta_{|x}(x-x^0)$.

The last observable (5.18) only has a formal meaning since the Dirac function δ is not a symbol in ψc . In (5.11) -(5.18) x_{corr} , ..., δ_{corr} each denote the corrected symbol, not the ψdo. We have added only the first correction, while an infinite number of lower and lower order corrections remain to be worked in.

Perhaps it might be time well spent if we now look at a possible physical interpretation of our mathematical results. From the view point outlined in sec.0 this will amount to a study of the spectrum - or the spectral theoretical properties - of the self-adjoint operator $A=a(x,D) = \kappa(x,D)+z_s(x,D)+\ldots$, where $\kappa(x,D)$ is one of the classical observables of (5.11)-(5.18). These $\kappa(x,D)$ are self-adjoint operators of H , with their domain properly defined. The newly proposed observables consist of $\kappa(x,D)$ with an additive perturbation $(z_s(x,D)+..)$. We thus should study the influence of the perturbation on the spectrum, the (generalized) eigenfunctions, even a properly introduced spectral density function, etc.

The mathematical difficulties liable to occur in such investigations are well known. Moreover, it is clear that our mathematical approach does not define a unique corrected observable - an additive self-adjoint operator of order $-\infty$ remains completely arbitrary. Such additional perturbation $P\in O(-\infty)$ clearly is a compact operator of H . Any $P=\sum\gamma_j^1\rangle\langle\gamma_j^2$, with a finite sum and $\gamma_j^1\in S(\mathbb{R}^n)$, will qualify. An addition of such perturbation could produce any finite number of arbitrarily prescribed eigenvalues and eigenvectors. There are examples of compact perturbations with very singular effects on the spectral resolution of a self-adjoint operator. Such effects are physically undesirable or impossible. In fact, the cut-off procedure used in constructing the asymptotic sums of I,lemma 6.4 appears physically undesirable.

Our hope to derive significant physical information perhaps should depend on isolating spectral properties invariant under perturbation by $P\in \mathcal{O}(-\infty)$ - or even by $P\in \mathcal{K}(\mathcal{H})$, since we will have to leave higher order perturbations (beyond $z_S(x,D)$) unattended anyhow. As a first such concept the <u>essential spectrum</u> comes to mind - the set $\sigma_{ess}(A)$ of ∞-dimensional eigenvalues and cluster points of the spectrum. It is known that $\sigma_{ess}(A+C)=\sigma_{ess}(A)$, for all compact C. (cf.[C_1], p.279, thm.6.2., for example). More refined discussions might be interesting but are beyond the scope of this book.

In this section we confine ourselves to more general observations about the operator norm of the first perturbations listed. Note that in all cases (5.11)-(5.18) the correction to the symbol carries the factor λ_c which has the dimension of a length. From the differential equations (3.11) of the particle flow we get

$$(5.19) \qquad \zeta = \frac{\pi}{f_r} = \frac{x^{\cdot}}{c} \ , \quad a_0 = \sqrt{1-x^{\cdot 2}/c^2} \ .$$

Using this we get

$$\lambda_c = \frac{\hbar}{2f_r}\{\rho\times\frac{x^{\cdot}}{c} + \mu\sqrt{1-x^{\cdot 2}/c^2}\} \approx \frac{\hbar}{2mc}\mu = \iota_c\frac{\mu}{2} \ , \text{ as } x^{\cdot}<< c \ .$$

The number $\iota_c=\frac{\hbar}{mc}\approx 4\times10^{-11}$ cm is known as the Compton wave length It often is regarded as a lower limit of measurable length. Our correction term $-\lambda_c$ for the observable x of (5.11) - the position observable - agrees well with this empirical fact: when measuring position, there will be a term $\approx\lambda_c(x,D)$ not in diagonal form, with expectation value depending on the physical state, but of the order of magniture of ι_c .

More precisely, ρ_j, μ_j have matrix norm 1. With ι_c as above, and the '<u>classical electron radius</u>' $\rho_c=\frac{e^2}{mc^2}\approx 3\times10^{-13}$ cm write

$$(5.20) \quad \lambda_c=\iota_c\mu_c,\ \mu_c=\tfrac{1}{2}\langle\iota_c\xi-\rho_c A/e\rangle^{-2}\{\mu+\rho\times(\iota_c\xi-\rho_c A/e)\},\ \langle z\rangle=\sqrt{1+|z|^2}\ .$$

It is not hard to estimate the L^2-operator norm of $\mu_c(x,D)$, using III, thm.1.1. There we were not interested in a numerically useful bound. Estimates given in [CC] on the basis of our proof in VIII,2 might be handier, for present needs. There we prove that

$$(5.21) \quad \|\mu_c(x,D)\|_{L^2} \le c_0\|(1-\Delta_x)(1-\Delta_\xi)\mu_c(x,\xi)\|_{L^\infty} \ , \quad c_0=\tfrac{1}{4}\kappa_1[[L]] \ ,$$

with the trace norm [[.]] of the integral operator L on $\mathbb{R}^3$ with kernel $L(x,y)=e^{-ixy}\tau(x)\tau(y)$, $\tau(x)=e^{-|x|}/|x|$ (cf. [CC],p.124) . The trace norm may be easily estimated. For example, we obtain L=

$\frac{1}{4\pi}MN$, with Schmidt class integral operators M,N , having kernel $\frac{1}{|x|}e^{\langle x\rangle/2-|x|}\ \tau(x-y)$, and $\tau(y)(1-\Delta_x)e^{-ixy-\langle x\rangle/2}$, respectively. Thus $[[L]]\le\frac{1}{2\pi}[[[M]]][[[N]]]$ with Schmidt norms at right. By a crude estimate, $[[[M]]]\le4\pi\sqrt{\frac{e}{2}}$, $[[[N]]]\le10\pi$, so $c_0\le\frac{5}{4}\sqrt{\pi e}<4$. This and formula (5.20) (which allows easy calculation and estimates of x and ξ-derivatives of μ_c) shows that, indeed, the norm $\|\lambda_c\|$ is of the order of ι_c. In fact, for $A=0$ we get the simpler estimate $\|\lambda_c(x,D)\|_{L^2}=\|\lambda_c(\xi)\|_{L^\infty}$; for general A we still note invariance under change of unit of lenth: In (5.21) we may replace $\|(1-\Delta_x)(1-\Delta_\xi)\mu_c(x,\xi)\|_{L^\infty}$ by $\inf\{\|(1-\rho\Delta_x)(1-\frac{1}{\rho}\Delta_\xi)\mu_c(x,\xi)\|_{L^\infty}:0<\rho<\infty\}$, invariant under change of unit of length, as a calculation shows.

The above assumes that $\|\rho_c(A_{|x})/e\|_{L^\infty}$, $\|(\rho_c A_{|xx})/e\|_{L^\infty}$ have order of magnitude ≤1 . Then indeed the correction ζ_0 of the position observable should be of the order of ι_c.

Looking at the correction of momentum in (5.12): For vanishing potentials A the first order correction z_0 vanishes. Generally we will get a first perturbation $-\hbar\rho_c\sum\mu_{cj}(A_{j|x}/e)$ with norm of order ι_c, assuming the above condition, regarding $\|A_{j|x}/e\|_{L^\infty}$. Note, $p=\frac{\hbar}{i}\partial_x$ is the <u>canonical</u> <u>momentum</u>, defined as variable canonical to x , in the sense of Legendre theory. As $A\neq0$ the <u>mechanical</u> <u>momentum</u> of (5.15) is the standard physical observable. It also has the more meaningful correction term $\frac{e}{c}B\times\lambda_c$ of order ι_c whenever $\frac{e}{c}\|B\|_{L^\infty}$ $\frac{e}{c}\|B_{|x}\|_{L^\infty}$, $\frac{e}{c}\|B_{|xx}\|_{L^\infty}$ are of order $\frac{1}{c_0}$. Similarly for (5.14), (5.16), with the magnetic field strength B replaced by the electric E.

In other words, the norms of all above correction terms are small or negligible as long as the potentials $V(x)$, $A(x)$ <u>and their gradients</u> do not assume excessive values. On the other hand, a potential wall, such as $V(x)=V_0\operatorname{sgn}(x_1)$, or suitable approximation, having slope near $x_1=0$ comparable to $1/\iota_c$ produces a perturbation $E.\lambda_c$ of the observable $V(x)$ which seems no longer negligible. This fact might serve as an explanation for the so-called Klein paradox of Dirac theory (cf. [Sm]) .

Similarly, a Coulomb potential $V(x)=-e^2/|x|$ is inadmissible of course, due to its singularity, but will become admissible, after "capping" the singularity, in an ever so small neighbourhood of zero. Then, however, one will have to account for the perturation term $\mathbf{E}.\lambda_c$ of the observable $V(x)$. To make it negligible, one would have to arrange for a "capping" - replacing $V(x)$ by a smooth function in $|x|\le\varepsilon$, ε small, in such a way that $V_{|x}=\mathbf{E} \ll \lambda_c$. Physically one might argue that the potential V of a hydrogen atom might not well be measurable very close to the nucleus - the first electron orbit is about 100 times larger than ι_c . Actually since the mathematical requirement for the observable $V(x)$ seems to require the presence of the term $\mathbf{E}.\lambda_c$ near 0 , one even might ask about the forces such an additional potential might produce. - Note that forces other than that produced by $E = \text{grad } V$ are known to act on a particle getting very close to the nucleus. It might be interesting to investigate whether such forces can be explained from our above correction of the potential V.

So far we have not looked at the angular momentum observable k_{corr} of (5.13). Again, this observable is of particular interest in case of $A=0$. In particular it leads to the common definition of the <u>spin</u> <u>observable</u>. It is known that the operator $J=k+\frac{\hbar}{2}\rho$ commutes with the Hamiltonian H whenever $A=0$, and V is rotationally symmetric. Indeed, look at $J_1=k_1+\frac{\hbar}{2}\rho_1=\frac{\hbar}{i}(x_2\partial_{x_3}-x_3\partial_{x_2})+\frac{\hbar}{2}\rho_1$. If V is a function of $|x|$ only, then we get $[k_1,V]=0$, since k_1 contains only angular derivatives. Thus $[J_1,H]=c[J_1,H_0]$, $H_0=h_0(x,D)$.

Clearly $[J_1,\beta]=0$, hence $[J_1,H]=\sum_{k=1}^3\{\frac{\hbar}{i}c\alpha_k[k_1,\partial_{x_k}]+\frac{\hbar^2}{2i}c[\rho_1,\alpha_k]\partial_{x_k}\}$.

Here $[k_1,\partial_{x_k}]$ and $[\rho_1,\alpha_k]$ may be calculated, using (0.10), (0.12) and (0.13), and $A=0$. One finds that $[J_1,H]=0$. Similarly for J_2 , J_3 . This formal commutativity indeed implies a spectral theoretical one (i.e., the resolvents and spectral families all commute), as we will not discuss in detail.

Commonly this is used as a motivation to interpret the (constant-matrix-valued) vector $\frac{\hbar}{2}\rho$ as the <u>spin</u> <u>observable</u>, so that J becomes the total angular momentum, satisfying a conservation law: Since J and H commute, $J_t=e^{itH}Je^{-itH}=J$ is independent of t.

Note, we have $J\in \mathrm{Op}\psi c_e$, and $J\in PC_e$ follows trivially, since $J_t=$ const. With k_{corr} of (5.13), $j=\mathrm{symb}(J)$ we get

$$(5.22)\qquad j=k_{corr}+\frac{\hbar}{2}\tau \ , \ \tau = f(\rho)=f_r^{-2}(m^2c^2\rho + (\pi\cdot\rho)\pi + mc\mu\times\pi),$$

assuming $A=0$, but not necessarily rotational symmetry of V. Indeed

$j(x,\xi)=\frac{\hbar}{i}x\times\xi+\frac{\hbar}{2}\rho=k_{corr}+(\frac{\hbar}{2}\rho+\hbar\lambda_c\times\xi)=k_{corr}+\frac{\hbar}{2}\tau$, where $\tau=\rho+2\lambda_c\times\xi$. Here $2\lambda_c\times\xi=\frac{1}{f_r^2}(mc\mu\times\xi+(\rho\times\xi)\times\xi)=\frac{1}{f_r^2}(mc\mu\times\xi+\xi(\rho x)-\rho|\xi|^2)=a_0\mu\times\zeta+(\rho\zeta)z-\rho|\zeta|^2$. Using $1=|z|^2+a_0^2$ we get $\tau=\rho+2\lambda_c\times\xi=a_0^2\rho+(r\zeta)\zeta+\alpha_0\mu\times\zeta$, in agreement with (5.22). Finally, to show that $\tau=f(\rho)$, note that $f(\rho)=\rho-h_+(\rho)$. From (4.11) get $h_+(\rho)=p\rho q+(p\rho q)^+=|\zeta|^2\rho-(\rho\zeta)\zeta-a_0\mu\times\zeta$. Hence $\tau=f(\rho)$.

In the frame of our present discussion it thus is natural to interpret the matrix-valued symbol $s=\frac{\hbar}{2}\tau\in\psi c_0$ as the (symbol of the) <u>spin</u> <u>obervable</u>. We then get the angular momentum conservation law for $J=k_{corr}(x,D)+\frac{\hbar}{2}\tau(x,D)$, as above.

Note that $\tau=f(\rho)=p_+\rho p_++p_-\rho p_-$ commutes with h. Thus the new spin $\frac{\hbar}{2}\tau$ will be a <u>basic</u> <u>observable</u> (def.7.3), qualifying for the correction procedure of thm.2.1: A correction $z\in\psi c_{-e}$ will give a symbol of a $z(x,D)\in PC_0$. On the other hand the classical spin symbol $\frac{\hbar}{2}\rho$ does not commute with $h(x,\xi)$, hence cannot be corrected.

There might be a physical objection: The spin is supposed to measure always $\pm\frac{\hbar}{2}$, while the operator $\tau(x,D)$ -or $\frac{1}{2}(\tau(x,D)+\tau(x,D)^*)$ has continuous spectrum, different from ± 1. To counter this argument with physical reasoning: For $\zeta=\frac{x^\bullet}{c}\ll 1$ we will have $\tau\approx\rho$, so that one still might expect to measure $\pm\frac{\hbar}{2}$, except for relativistically large velocities. (There is a similar contradiction with the Stark effect's Schroedinger operator. It has purely continuous spectrum, while, of course, spectral lines are observed.)

The above first perturbations are only those encountered for t=0. As time progresses, the symbol $\kappa_\pm$ will each flow along their corresponding particle flows. Clearly they no longer will assume the same value at each x,ξ , even though they did initially. Thus we now must focus on the second term of (5.10). Write it as

(5.23) $T_2=\frac{1}{2}\Delta\kappa.V_{pt}/E_t$, $\Delta\kappa=\kappa_+-\kappa_-$, $V_{pt}=E.\lambda_c$, $E_t=cf_r=\sqrt{m^2c^4+c^2\pi^2}$.

Clearly V_{pt} is the correction of the observable V of (5.14), containing the 'noncommuting Compton wave length' λ_c. Again this may be estimated in norm and found negligibly small, assuming $E\ll 1/\iota_c$.

As a common experience, many negligibly small quantities have non-negligible time derivatives. This prompts us to consider time-derivatives of our new observables. (3.6) and (3.9) implies

(5.24) $\kappa^\bullet=p\langle h,\kappa\rangle p+q\langle h,\kappa\rangle q=f(\langle h,\kappa\rangle)$.

Here $\kappa=\kappa_+p+\kappa\_q$. By $(4.2)_3$ we get $p\langle h,\kappa\_q\rangle p=q\langle h,\kappa_+q\rangle q=0$, hence

(5.25) $\kappa^\bullet=p\langle h,\kappa_+p\rangle p+q\langle h,\kappa\_q\rangle q=p\langle\lambda,\kappa_+p\rangle\pi+q\langle\lambda\_,\kappa\_q\rangle q$, where $(4.2)_2$ was used, noting that $[\kappa_+ p,(pp_{|\xi}p_{|x}p)]=0$, and similar for q, since $\kappa_\pm$ are scalars. Finally, using (4.1), (4.3), we get

(5.26) $$\kappa^{\cdot}= \langle\lambda,\kappa_+\rangle p+\langle\lambda_-,\kappa_-\rangle q=\tfrac{1}{2}(\langle\lambda,\kappa_+\rangle+\langle\lambda_-,\kappa_-\rangle)+\tfrac{1}{2}(\langle\lambda,\kappa_+\rangle-\langle\lambda_-,\kappa_-\rangle)\eta_0 .$$

We reintroduce σ_κ, δ_κ with $\kappa_\pm=\sigma_\kappa\pm c\delta_\kappa f_r$, $\lambda_\pm=V\pm cf_r$, and get

(5.27) $$\begin{aligned}\hbar\kappa^{\cdot} &= -\tfrac{1}{2}E.\sigma_{\kappa|\xi} +\frac{c}{2f_r^2}\pi.\langle\pi,\sigma_\kappa\rangle(mc\beta+\alpha.\pi)\\ &+ \frac{c^2}{2}\pi.\langle\pi,\delta_\kappa\rangle -\frac{c}{2f_r}E.(f_r\delta_\kappa)_{|\xi}(mc\beta+\alpha.\zeta) .\end{aligned}$$

In (5.26) we had $\hbar=1$, but (5.27) checks for general $\hbar$, if the quantities of (5.10) are used.

Let us use (5.27) to calculate $A^{\cdot}$ of (3.1) for $\kappa=x$ - clearly this should be the velocity observable $v = A^{\cdot}$ We get (at t=0)

(5.28) $$v = \frac{\pi}{f_r^2}(mc^2\beta+c\alpha\cdot\pi) = \frac{\pi c}{f_r}(p-q) .$$

Clearly the components of the vector v commute with each other and with $h(x,\xi)$ as well. As eigenvalues one obtains

(5.29) $$\pm\frac{\pi c}{f_r} \approx \pm\frac{\pi}{m} , \text{ as } x^{\cdot} << c .$$

In turn, for the velocity (5.28) we have $\kappa_\pm=\pm\frac{\pi c}{f_r}$, $\sigma=0$, $\delta= 2\pi\frac{1}{f_r}$. Substituting this into (5.27) one finds (at t=0)

(5.30) $$\begin{aligned} a = v^{\cdot} &=-\frac{c}{f_r}(eB\times\zeta + (E-(E\zeta)\zeta)(p-q)) = \kappa_+p + \kappa_-q ,\\ &\text{with } \kappa_\pm = -\frac{c}{f_r}(eB\times\zeta \pm (E-(E\zeta)\zeta)) ,\end{aligned}$$

with the electric and magnetic field vectors E and B, as a calculation shows. In particular, using (5.19), we get

(5.31) $$\kappa_\pm =-\frac{1}{m}\sqrt{1-x^{\cdot 2}/c^2}\{\tfrac{e}{c}B\times x^{\cdot}\pm(E- \tfrac{1}{c^2}(Ex^{\cdot})x^{\cdot})\} ,$$

corresponding to the classical law of motion of a particle of (relativistic) mass $m/\sqrt{1-x^{\cdot 2}/c^2}$ under the electromagnetic force $\pm(E-\frac{1}{c^2}(Ex^{\cdot})x^{\cdot})-\frac{e}{c}B\times x^{\cdot}$.

Actually, the above could have been derived more directly: We know that $\kappa=\kappa_+p+\kappa_-q$, where $\kappa=\kappa_0\circ v^\pm_t$ with $v^\pm$ of (3.12). We get (3.11) as propagation law. For $\kappa_\pm=x$ this gives (5.26). In fact, (5.31) follows for all t, not only t=0. We offer this derivation to show that there is no "Zitterbewegung", and that the paradox of velocity components, in classical Dirac theory, does not exist.

A similar argument shows that there no longer is a difficulty with the current observable.

It was argued before that the difficulties are resolvable by means of the Foldy-Wouthuysen transform of thm.2.2. Possible our axiomatic necessity for corrections could be a more natural way.

6. Proof of the Foldy-Wouthuysen theorem.

Let us next discuss a proof of thm.2.2. One might expect that the desired unitary operator $\mathcal{X}$ will be constructed as asymptotic sum $\sum \mathcal{X}_j$, $\mathcal{X}_j=\chi_j(x,D)\in \mathrm{Op}\psi c^4_{-je}$, by an iteration similar to that of the proof of thm.2.1. Assume that $\mathcal{X}_j= \chi_j(x,D)\in \mathrm{Op}\psi c^4_{-je}$, $j=0,\dots,N$, have been constructed, such that $\sum_{j=0}^N \mathcal{X}_j = \mathcal{X}$ satisfies

(6.1) $\mathcal{X}^*\mathcal{X}=1-\Gamma$, $\mathcal{X}^*H\mathcal{X}=\Lambda-\Delta$,$\Gamma,\Delta,\Lambda\in \mathrm{Op}\psi c^4$,order $-Ne-e,-Ne-e^2$,e^1 ,resp.,

with a pseudo-diagonal matrix Λ. Here and in the following 'pseudo-diagonal' (or, ψ-diagonal) means that the upper right and lower left 2×2-corners - called Λ_{UR} and Λ_{LL} - of the matrix Λ vanish. We shall construct $\Omega=\mathcal{X}_{N+1}\in \mathrm{Op}\psi c^4_{-Ne-e}$ such that $Y=\mathcal{X}+\Omega$ satisfies (6.1) for N+1. By induction we then get an infinite sequence $\{\mathcal{X}_j\}$ with $\mathcal{X}_0,\dots,\mathcal{X}_N$ satisfying (6.1) for every $N=0,1,\dots$. The asymptotic sum $\sum\mathcal{X}_j$ then satisfies (6.1) for $N=\infty$. It will provide the desired unitary operator after a final correction in $\mathrm{Op}\psi c^4_{-\infty}$.

For N=0 let $\chi_0=\varphi$, $\mathcal{X}_0= \Phi=\varphi(x,D)$, with φ of (2.10). Then $\mathcal{X}_0^*H\mathcal{X}=\mathrm{diag}(\lambda_+,\lambda_+,\lambda_-,\lambda_-)-\Delta_0$, $\Delta_0\in \mathrm{Op}\psi c^4_{-e^2}$, by calculus of ψdo's, confirming (6.1) for $\{\mathcal{X}_0\}$. For the above Ω we get $\Omega^*\Omega\in \mathrm{Op}\psi c_{-2Ne-2e}$ $\subset \mathrm{Op}\psi c^4_{-Ne-e}$, and $\mathcal{X}^*\Omega=\Phi^*\Omega$, mod $O(-Ne-2e)$. Hence (6.1) for X at N and Y at N+1 implies

$$(6.2) \qquad \mathcal{X}^*\Omega + \Omega^*\mathcal{X} = \Gamma \quad (\mathrm{mod}\ \mathrm{Op}\psi c^4_{-(N+2)e}) ,$$

and, in terms of symbols, using $(\mathcal{X}-\Phi)^*\Omega\in O(-Ne-2e)$, this yields

$$(6.2') \qquad \varphi^+\omega+(\varphi^+\omega)^+ = \varphi^+\omega+\omega^+\varphi = \gamma \quad (\mathrm{mod}\ O(-(N+2)e)) .$$

It is possible to satisfy (6.2') sharply - not only modulo $O(-(N+2)e)$: Just choose $\varphi^+\omega = \frac{\gamma}{2}+i\theta$, $\tau=\frac{1}{2i}(\varphi^+\omega-\omega^+\varphi)=\tau^+$. That is,

$$(6.3) \qquad \omega = \varphi(\gamma/2+i\theta) , \ \theta^+=\theta, \ \theta\in \psi c^4_{-Ne-e} .$$

Having thus $Y^*Y-1\in Op\psi c^4_{-Ne-2e}$ satisfied, we seek to determine θ to also get the second relation (6.1) for Y and N+1 - i.e., $Y^*HY= X^*HX+\Omega^*HX+X^*H\Omega+\Omega^*H\Omega = \Lambda_{N+1}-\Delta_{N+1}$ with ψ-diagonal $\Lambda_{N+1}\in O(e^1)$, and general $\Delta_{N+1}\in O(-(N+1)e-e^2)$. Here we have $\Omega^*H\Omega\in O(-2Ne-e-e^2)$, and $X^*HX=\Lambda-\Delta=\Lambda_N-\Delta_N$, $\Delta_N\in O(-Ne-e^2)$. One may assume that the operators Λ_N, Δ_N both are hermitian, implying that the symbol δ_N of Δ_N satisfies $\delta_N^+-\delta\in O(-(N+1)e-e^2)$. Also, again $\Omega^*H(X-\Phi)\in O(-(N+1)e-e^2)$, hence the symbol of Ω^*HX equals $\omega^+h\varphi$ (modulo $\psi c^4_{-(N+1)-e^2}$). All together, we obtain the relation

$$\Omega^*H\Phi + \Phi^*H\Omega = (\Lambda_{N+1}-\Lambda_N)+\Delta_N \pmod{O(-(N+1)e-e^2)},$$

and, for symbols, we get - with $\lambda_j = \mathrm{symb}(\Lambda_j)$,

$$(6.4)\qquad \varphi^+h\omega + \omega^+h\varphi = (\lambda_{N+1}-\lambda_N) + \delta_N \pmod{\psi c^4_{-(N+1)e-e^2}}.$$

Here we substitute from (6.3), noting that $\gamma=\gamma_N$, δ_N are given, but the hermitian symbol $\theta\in\psi c^4_{-(N+1)e}$ still is arbitrary. We get

$$(6.5)\quad \varphi^+h\omega+\omega^+h\varphi = \mathit{1}(\tfrac{\gamma}{2}+i\theta)+(\tfrac{\gamma}{2}-i\theta)\mathit{1} = \tfrac{1}{2}(\mathit{1}\gamma+\gamma\mathit{1})+i[\mathit{1},\theta]=(\lambda_{N+1}-\lambda_N)+\delta_N,$$

mod $\psi c^4_{-(N+1)e-e^2}$, using (2.6), where λ_{N+1} and θ are to be found. Again we regard (6.5) as a sharp equation, after still replacing δ_N by $\frac{1}{2}(\delta_N+\delta_N^+)$. That is we seek to solve

$$(6.6)\qquad [\mathit{1},\theta] = \tfrac{i}{2}\{(\mathit{1}\gamma_N-\delta_N)+(\mathit{1}\gamma_N-\delta_N)^+\} - i(\lambda_{N+1}-\lambda_N).$$

A hermitian symbol θ solving (6.6) will exists if and only if λ_{N+1} is chosen according to $\lambda_{N+1}-\lambda_N=\begin{pmatrix}\upsilon^1 & 0\\ 0 & \upsilon^4\end{pmatrix}$, where υ^1, υ^4 are the ψ-diagonal corners of $Z=\frac{1}{2}(W+W^+)$, $W=\mathit{1}\gamma_N-\delta_N$, i.e., $Z=\begin{pmatrix}\upsilon^1 & \upsilon^2\\ \upsilon^3 & \upsilon^4\end{pmatrix}$.

Then we will have $\theta=\begin{pmatrix}\theta^1 & \theta^2\\ \theta^3 & \theta^4\end{pmatrix}$, with arbitrary hermitian θ^1, θ^4, and $\theta^2=\theta^{3+}$ satisfying $(\lambda_+-\lambda_-)\theta^2 = \upsilon^2$. In other words, we get

$$(6.7)\qquad \lambda_{N+1}=\lambda_N+\begin{pmatrix}\upsilon^1 & 0\\ 0 & \upsilon^2\end{pmatrix},\quad \chi_{N+1}=\varphi(\tfrac{1}{2}\gamma_N+\tfrac{i}{2cf_r}\begin{pmatrix}0 & \upsilon^2\\ \upsilon^3 & 0\end{pmatrix}+\begin{pmatrix}\theta^1 & 0\\ 0 & \theta^4\end{pmatrix}),$$

with arbitrary skew-hermitian θ^1, $\theta^4\in\psi c^2_{-(N+1)e}$, and υ^j from the matrix Z above. In particular, θ obtained this way is hermitian since the matrix iZ clearly is skew-hermitian, giving $\upsilon^3=\upsilon^{2+}$. Evidently we have $W=\mathit{1}\gamma_N-\delta_N\in O(-Ne-e^2)$, giving $\upsilon^j\in O(-Ne-e^2)$, hence $\upsilon^2=\upsilon^{2+}\in O(-(N+1)e)$, exactly as required.

It is now clear that the construction of X_j is possible for all j=0,1,2,..., and we may form the asymptotic sum $X=\sum_{j=0}^{\infty}X_j$.

Then it follows at once that (6.1) is valid for all N, for the redefined $\mathcal{X}$. In other words, we now have Δ,Γ of order $-\infty$.

We plan to make another correction of $\mathcal{X}$ such that it actually becomes unitary. First we conclude that $\Phi \in \mathrm{Op}\psi c_0^4$ is a (bounded) Fredholm operator of $\mathcal{H}$, by II, thm.6.2 and III, thm.4.2, for any choice of potentials A_j and V, since the symbol φ of Φ is uniformly non-singular at $|x| + |\xi| = \infty$. Moreover, if the potentials are zero, then Φ is unitarily equivalent to a multiplication operator by the Fourier transform F. It then follows that Φ is invertible (even unitary). In particular, Φ has its Fredholm index equal to zero. For general potentials we introduce a family $\varphi(s)$ of symbols corresponding to the potentials sA_j , sV , $0\le s\le 1$. A calcultion shows that $\varphi(s)$ is continuous in $0\le s\le 1$, as a map into the Frechet space ψc_0^4 , with the semi-norms induced by (6.6). Hence the corresponding operator family $\Phi(s)$ is norm continuous, by III, thm.1.1, and index(Φ)=0 for potentials satisfying our general assumptions.

Observe that $\mathcal{X}=\Phi+C$, with $C=\sum_{j=1}^{\infty} \mathcal{X}_j \in \mathrm{Op}\psi c_{-e}$ compact in $L(\mathcal{H})$, by III, thm.5.1. Thus $\mathcal{X}$ is Fredholm as well, and index($\mathcal{X}$)=0. Hence dim ker $\mathcal{X}^*\mathcal{X}$=dim ker $\mathcal{X}\mathcal{X}^*$. We get $1-\mathcal{X}\mathcal{X}^*=\Gamma\in O(-\infty)$, $1-\mathcal{X}\mathcal{X}^*=\Gamma^{\sim} \in O(-e)$.

<u>Proposition 6.1</u>. The null spaces of $\mathcal{X}$ and $\mathcal{X}^*$ are contained in S .
<u>Proof</u>: For $u \in S' = \cap\{\mathcal{H}_s : s\in \mathbb{R}^2\}$ with $\mathcal{X}u=0$ we get $u =-\Gamma u \in S$, since Γ is of order $-\infty$, hence takes S' to S (cf. III, sec.4). For $u \in S'$ with $\mathcal{X}^*u = 0$ we get $u = -\Gamma u = (-\Gamma)^r u$, $r = 1,2,\dots$, hence $u\in \mathcal{H}_{s-Ne}$ for some s and all N=1,2,..., which implies $u \in S$, q.e.d.

Let us choose a basis $\omega_1,\dots,\omega_r$ of ker $\mathcal{X}^*\mathcal{X}$, and a basis $\psi_1,\dots,\psi_r$ of ker $\mathcal{X}\mathcal{X}^*$ (both orthonormal), and let

$$(6.8)\qquad \mathcal{X}^{\sim} = \mathcal{X}+\Omega\ ,\ \Omega=\sum \omega_j\rangle\langle\psi_j\ ,\ \text{with } \{\omega\rangle\langle\psi\}u=(\psi,u)\,\omega\ ,\ u\in \mathcal{H}.$$

Clearly we have $\Omega \in \mathrm{Op}\psi c_{-\infty}^4$. In fact,

$$(6.9)\qquad \omega\rangle\langle\psi u(x)=\int(\omega(x)\overline{\psi}(\xi)e^{-ix\xi})e^{ix\xi}u^{\wedge}(\xi)d\xi\quad ,$$

where indeed the symbol is in $\psi c_{-\infty}^4$.

Note that $\mathcal{X}^{\sim}$ of (6.8) is an invertible operator,since

$$(6.10)\qquad \mathcal{X}^{\sim *}\mathcal{X}^{\sim} = \mathcal{X}^*\mathcal{X} + \Omega^*\Omega + \Omega^*\mathcal{X} + \mathcal{X}^*\Omega = \mathcal{X}^*\mathcal{X} + \sum\psi_j\rangle\langle\psi_j\ ,$$

where the two operators at right are self-adjoint and positive definite in $(\ker \mathcal{X}^*\mathcal{X})^{\perp}$ and ker $\mathcal{X}^*\mathcal{X}$, and zero in ker $\mathcal{X}^*\mathcal{X}$ and $(\ker \mathcal{X}^*\mathcal{X})^{\perp}$, respectively. It follows that $\mathcal{X}^{\sim *}\mathcal{X}^{\sim}$ is self-adjoint and positive definite in $\mathcal{H}$, hence it is invertible. Similarly one concludes that $\mathcal{X}\mathcal{X}^*$ is invertible, hence $\mathcal{X}$ also is invertible. Since $\mathcal{X}$ is determined only up to operators of order $-\infty$ we may regard Ω as another correction of $\mathcal{X}$, and hence we may assume without loss of generality that $\mathcal{X}$, as constructed in this section, is invertible in $\mathcal{H}$. We then may introduce

$$(6.11) \qquad \Gamma^{\wedge} = (\mathcal{X}^*\mathcal{X})^{-1/2}-1 = (1-\Gamma)^{-1/2} -1 = -2\Gamma/\pi \int_0^\infty (1-\Gamma+\lambda^2)d\lambda$$

<u>Proposition 6.2</u>. $\Gamma^{\wedge}$ is an operator of $\mathrm{Op}\psi c^4_{-\infty}$.

<u>Proof.</u> With another change μ =arctan λ of integration variable get

$$(6.12) \qquad \Gamma^{\wedge} = -2\Gamma/\pi \int_0^{\pi/2} (1-\Gamma\cos^2\mu)^{-1} \cos^2\mu \, d\mu \, .$$

This is now an integral over a finite interval with norm continuous integrand in $\mathcal{H}$. In fact, we observe that $1-\Gamma\cos^2\mu$ is continuous in norm convergence in every $\mathcal{H}_s$, for $0\le\mu\le\pi/2$. Since it is invertible in $\mathcal{H}$ it also is invertible in every $\mathcal{H}_s$ - actually, that inverse is a Green inverse in the sense of III,4. Hence the integrand is norm continuous in $\mathcal{H}_s$,and the integral exists in every $\mathcal{H}_s$,in norm convergence. Thus the integral is in $O(0)$. Due to the factor Γ in (6.11) we then get $\Gamma^{\wedge}\in O(-\infty)$, q.e.d.

We have $(\mathcal{X}+\mathcal{X}\Gamma^{\wedge})^*(\mathcal{X}+\mathcal{X}\Gamma^{\wedge})=(1+\Gamma^{\wedge})\mathcal{X}^*\mathcal{X}(1+\Gamma^{\wedge})=(\mathcal{X}^*\mathcal{X})^{1/2}\mathcal{X}^*\mathcal{X}(\mathcal{X}^*\mathcal{X})^{1/2}=1$. Hence $\mathcal{X}\Gamma^{\wedge}$ is another correction $\mathcal{X}\Gamma^{\wedge}\in\mathrm{Op}\psi c^4_{-\infty}$ of $\mathcal{X}$, which will effect that $\mathcal{X}$ is invertible and $\mathcal{X}^*\mathcal{X}=1$. In other words, $\mathcal{X}$ now is unitary. The proof is complete.

Let us ask for an explicit form of the first terms of the resulting asymptotic expansions

$$(6.13) \qquad \chi = \mathrm{symb}(\mathcal{X}) = \textstyle\sum_{j=0}^\infty \chi_j \, , \; \mathrm{symb}(\Lambda) = \lambda_0 + \sum_{j=0}^\infty(\lambda_{j+1}-\lambda_j) \, .$$

We started setting $\chi_0=\varphi$, making, as a possible choice,

$$(6.14) \qquad \chi_0 = \varphi = \frac{1}{\sqrt{2+2a_0}}(1+a_0-i\mu\zeta) \, , \; \lambda_0 = \mathcal{l} = V+cf_r\beta \, .$$

Looking for λ_1 , χ_1 , we calculate symbols γ , δ of (6.1) for N=0. By I,(5.7), (5.9) we get (modulo $O(-2e)$ or $O(-e-e^2)$, resp.)

$$\gamma=1-\mathrm{symb}(\varphi(x,D)^*\varphi(x,D)) = i(\varphi^+_{|x\xi}\varphi+\varphi^+_{|\xi}\varphi_{|x}) \, ,$$

(6.15)

$$\delta=\mathcal{l}-\mathrm{symb}(\varphi(x,D)^*H\varphi(x,D)) = i(\varphi^+_{|x\xi}h\varphi+\varphi^+_{|\xi}(h\varphi)_{|x}+\varphi^+h_{|\xi}\varphi_{|x}) \, .$$

Recall, that we agreed to choose δ hermitian. A calculation shows that δ of (6.15) satisfies $\delta-\delta^+=i(\varphi^+h\varphi)_{|x\xi}=i\mathcal{l}_{|x\xi}$, using $h_{|x\xi}=0$. This has the effect that we still have (6.6) and (6.7) true, but must replace the second formula (6.14) by

$$(6.16)\qquad \lambda_0 = \mathit{l} + \tfrac{i}{2}\mathit{l}_{|x\xi} = V + cf_r\beta - \tfrac{ie}{2f_r}\Sigma A_{k|x_j}\zeta_j\zeta_k\,\beta\ .$$

Confirm that γ of (6.15) is hermitian - we have $\gamma-\gamma^+=i(\varphi^+\varphi)_{|x\xi}=0$.

Calculating γ and $W=\gamma\mathit{l}-\delta$, introduce $X=\varphi^+\varphi_{|x}$, $\Xi=\varphi^+\varphi_{|\xi}$, and get

$$(6.17)\qquad \gamma=i\Xi_{|x}\ ,\ W=\gamma\mathit{l}-\delta = -i\{\mathit{l}_{|\xi}X-\Xi\mathit{l}_{|x}+[\Xi,\mathit{l}]X\}\ .$$

Indeed, the expression for γ follows from (6.15). With $\mathit{l}=\varphi^+h\varphi$ get $W=i\{\varphi^+_{|\xi}\varphi_{|x}\mathit{l}-\varphi^+_{|\xi}(h\varphi)_{|x}-\varphi^+h_{|\xi}\varphi_{|x}\}=i\{\varphi^+_{|\xi}\varphi_{|x}\mathit{l}-\varphi^+_{|\xi}(\varphi\mathit{l})_{|x}-\varphi^+h_{|\xi}\varphi_{|x}\}=$ $=-i\{\varphi^+_{|\xi}\varphi\mathit{l}_{|x}+\varphi^+h_{|\xi}\varphi_{|x}\}$. Notice that X and Ξ are skew-hermitian, due to $X+X^+=(\varphi^+\varphi)_{|x}=1_{|x}=0$, $\Xi+\Xi^+=1_{|\xi}=0$. We get $h_{|\xi}=(\varphi\mathit{l}\varphi^+)_{|\xi}=\varphi_{|\xi}\mathit{l}\varphi^+ +\varphi\mathit{l}_{|\xi}\varphi^+ +\varphi\mathit{l}\varphi^+_{|\xi}$. Substituting this we get (6.17) for W .

For the further evaluation it now will be necessary to calculate the two matrix symbols $X=\varphi^+\varphi_{|x}$ and $\Xi=\varphi^+\varphi_{|\xi}$. The proof of prop.6.3, below, is postponed to the end of this section.

Proposition 6.3. We have

$$(6.18)\qquad X=-\tfrac{e}{c}\Sigma A_{k|x}\Xi_k\ ,\ \Xi=\tfrac{-i}{2f_r}\begin{pmatrix}\varepsilon & \tau\\ \tau & \varepsilon\end{pmatrix}\ ,$$

$$\text{where}\qquad \varepsilon=\frac{1}{1+a_0}\zeta\times\sigma\ ,\ \tau=\sigma-\frac{\zeta\sigma}{1+a_0}\zeta\ .$$

First calculate $(\tfrac{\gamma}{2})_{UR}=\tfrac{i}{2}(\Xi_{|x})_{UR}=\tfrac{1}{4}(\tau/f_r)_{|x}=\tfrac{-e}{4c}\Sigma(\tau_k/f_r)_{|\pi_j}A_{j|x_k}$.

Here $(\tau_k/f_r)_{|\pi_j}=\{\tfrac{1}{f_r}(\sigma_k-\tfrac{\sigma\zeta}{1+a_0}\zeta_k)\}_{|\pi_j}=-(\zeta_j/f_r^2)(\sigma_k-\tfrac{\sigma\zeta}{1+a_0}\zeta_k)$

$$\tfrac{1}{f_r}\{\tfrac{\sigma\zeta}{f_r(1+a_0)^2}a_0\zeta_j\zeta_k-\tfrac{\sigma\zeta}{f_r(1+a_0)}(\delta_{kj}-\zeta_k\zeta_j)-\tfrac{1}{f_r(1+a_0)}\Sigma_l\sigma_l(\delta_{jl}-\zeta_j\zeta_l)\zeta_k\}$$
$$=\tfrac{1}{f_r^2}\{\zeta_j\sigma_k+\tfrac{\sigma\zeta}{1+a_0}\zeta_j\zeta_k-\tfrac{a_0}{f_r(1+a_0)^2}(\sigma\zeta)\zeta_j\zeta_k+\tfrac{\sigma\zeta}{1+a_0}\delta_{jk}-2\tfrac{\sigma\zeta}{1+a_0}\zeta_j\zeta_k+\zeta_k\sigma_j\}$$
$$=\tfrac{1}{f_r^2}\{(\zeta_k\sigma_j-\zeta_j\sigma_k)+\tfrac{\sigma\zeta}{1+a_0}\delta_{jk}-\tfrac{1+2a_0}{(1+a_0)^2}(\sigma\zeta)\zeta_j\zeta_k\}$$

. It follows that

$$(6.19)\qquad (\tfrac{\gamma}{2})_{UR}=(\tfrac{\gamma}{2})_{LL}=\tfrac{e}{4cf_r^2}\{\sigma.(B\times\zeta)+\tfrac{1+2a_0}{1+a_0}\tfrac{\sigma\zeta}{1+a_0}\Sigma A_{j|x_k}\zeta_j\zeta_k\}$$

We also need $(\tfrac{\gamma}{2})_{UL}=(\tfrac{\gamma}{2})_{LR}=\tfrac{1}{4}(\varepsilon/f_r)_{|x}=\tfrac{-e}{4c}\Sigma(\varepsilon_k/f_r)_{|\pi_j}A_{j|x_k}$, for the hermitian ψ-diagonal part of $\varphi^+\chi_1$. Here we get $(\varepsilon_k/f_r)_{|\pi_j}=$ $=\{(\zeta\times\sigma)_k/(mc+\sqrt{m^2c^2+\pi^2})\}_{|\pi_j}=\tfrac{1}{f_r^2(1+a_0)}((e^j-\zeta_j\zeta)\times\sigma)_k-\tfrac{\zeta_j(\zeta\times\sigma)_k}{f_r^2(1+a_0)^2}=$ $=\tfrac{1}{f_r^2(1+a_0)}(e^j\times\sigma)_k-\tfrac{2+a_0}{f_r^2(1+a_0)^2}\zeta_j(\zeta\times\sigma)_k$. Accordingly,

(6.20) $$(\tfrac{\gamma}{2})_{UL}=(\tfrac{\gamma}{2})_{LR}=\frac{-e}{4cf_r^2}\frac{1}{(1+a_0)}(\sigma B)+\frac{e}{4cf_r^2}\frac{2+a_0}{(1+a_0)^2}\sum\zeta_j A_{j|x}(\zeta\times\sigma)\ .$$

Next we calculate $W=\gamma\mathit{1}-\delta=-i\mathit{1}_{|\xi}X+i\Xi\mathit{1}_{|x}-i[\Xi,\mathit{1}]X=W_1+W_2+W_3$.

Get $W_1=\frac{e}{2f_r}\sum A_{k|x_j}\zeta_j\begin{pmatrix}\varepsilon & \tau\\ -\tau & -\varepsilon\end{pmatrix}_k$, hence $Z_1=\frac{1}{2}(W_1+W_1^*)=\frac{e}{2f_r}\sum A_{k|x_j}\zeta_j\begin{pmatrix}\varepsilon & 0\\ 0 & -\varepsilon\end{pmatrix}_k$

$=\frac{e}{2f_r}\sum A_{k|x_j}\zeta_j\varepsilon_k\beta$. Similarly, $W_2=\frac{1}{2f_r}E.\begin{pmatrix}\varepsilon\tau\\ \tau\varepsilon\end{pmatrix}-\frac{e}{2f_r}\sum A_{j|x_k}\zeta_j\begin{pmatrix}\varepsilon & -\tau\\ \tau & -\varepsilon\end{pmatrix}_k$, hence

$Z_2=\frac{1}{2f_r}E.\varepsilon-\frac{e}{2f_r}\sum A_{j|x_k}\zeta_j\varepsilon_k\beta+\frac{1}{2f_r}(E\tau)\begin{pmatrix}0 1\\ 1 0\end{pmatrix}$. Thus $Z_1+Z_2=\begin{pmatrix}\kappa^1 & \kappa^2\\ \kappa^3 & \kappa^4\end{pmatrix}$, where

$$\kappa^2=\kappa^3=\frac{1}{2f_r}(E\tau)=\frac{1}{2f_r}(E\sigma)-\frac{1}{2f_r(1+a_0)}(E\zeta)(\zeta\sigma)\ ,\ \text{while}$$

$$\begin{pmatrix}\kappa^1 & 0\\ 0 & \kappa^4\end{pmatrix}=\frac{1}{2f_r(1+a_0)}E.(\zeta\times\sigma)+\frac{e}{2f_r(1+a_0)}\sum(A_{k|x_j}-A_{j|x_k})\zeta_j(\zeta\times\sigma)_k\beta$$

(6.21) $$=\frac{1}{2f_r(1+a_0)}\{E.(\zeta\times\sigma)+e\zeta.((\zeta\times\sigma)\times B)\beta\}$$

$$=\frac{1}{2f_r(1+a_0)}\{E.(\zeta\times\sigma)+e(\zeta\sigma)(B\zeta)\beta\}-\frac{e}{2f_r}(1-a_0)(B\sigma)\beta\ .$$

Finally, $[\Xi,\mathit{1}]=ic\begin{pmatrix}0 & \tau\\ -\tau & 0\end{pmatrix}$, hence $W_3=\frac{ie}{2f_r}\sum A_{k|x_j}\ \tau_j\begin{pmatrix}\tau & \varepsilon\\ -\varepsilon & -\tau\end{pmatrix}_k$. Therefore,

$(Z_3)_{UL}=-(Z_3)_{LR}=\frac{ie}{4f_r}\sum(A_{k|x_j}-A_{j|x_k})\tau_j\tau_k=\frac{ie}{4f_r}B.(\tau\times\tau)=\frac{ie}{4f_r}\{B_1[\tau_2,\tau_3]+$ $+B_2[\tau_3,\tau_1]+B_3[\tau_1,\tau_2]\}$, where $[\tau_2,\tau_3]=2i\{\sigma_1-\frac{1}{1+a_0}\sigma.(\zeta\times(e^2\zeta_3-e^3\zeta_2))\}$, etc., so that

(6.22) $$(Z_3)_{UL}=-(Z_3)_{LR}=\frac{-e}{2f_r}\{a_0(B\sigma)+\frac{1}{1+a_0}(\sigma\zeta)(B\zeta)\}\ .$$

Also, $(Z_3)_{UR}=-(Z_3)_{LL}=\frac{ie}{4f_r}\sum A_{k|x_j}[\tau_j,\varepsilon_k]_+$. A calculation yields

(6.23) $$[\tau_j,\tau_k]_+=\tau_j\tau_k+\tau_k\tau_j=2(\delta_{jk}-\zeta_j\zeta_k)\ ,\ [\tau_j,\varepsilon_k]_+=\frac{1}{1+a_0}\zeta.(e^j\times e^k)\ .$$

(We have used (0.15), in this connection.) Therefore,

(6.24) $$(Z_3)_{UR}=-(Z_3)_{LL}=\frac{ie}{2f_r}z.\sum A_{k|x_j}e_j\times e_k=\frac{ie}{2f_r}\zeta.\mathrm{curl}\,A=\frac{ie}{2f_r}(B\zeta)\ .$$

From (6.21) an (6.22) we first get an expression for the matrix $\begin{pmatrix}\upsilon^1 & 0\\ 0 & \upsilon^4\end{pmatrix}=U$ of (6.7) :

(6.25) $$U=\frac{1}{2f_r}\{\frac{1}{1+a_0}E.(\zeta\times\rho)-e(B\rho)\beta\}\ .$$

Accordingly, we get formula (6.27), below, for λ_1, using (6.16). On the other hand, for χ_1 we get the (more complicated) expression (6.26). Indeed, looking at (6.7), with $\theta^1=\theta^4=0$, we first assemble the hermitian part $\frac{\gamma}{2}$ of $\varphi^+\chi_1$: (6.19) and (6.20) yield

$$\frac{\gamma}{2}=\frac{e}{4cf_r^2}\{(B\times\zeta).\mu-\frac{1}{1+a_0}(B\rho)+\frac{2+a_0}{(1+a_0)^2}D(\zeta)\rho+\frac{1+2a_0}{(1+a_0)^2}C(\zeta)(\mu\zeta)\} .$$

Next, for the skew-hermitian part, using (6.7), (6.21), (6.23),

$\frac{i}{2cf_r}\begin{pmatrix}0 & \upsilon^2\\ \upsilon^3 & 0\end{pmatrix}=\frac{i}{4cf_r^2}(E\mu-\frac{E\zeta}{1+a_0}(\zeta\mu))-\frac{1}{4cf_r^2}\begin{pmatrix}0 & e\\ -e & 0\end{pmatrix}(B\zeta)$. Taking the sum we get (6.26). Let us summarize:

<u>Theorem.6.4</u>. We have - setting $C(\zeta)=\sum\zeta_j A_{j|x}\zeta$, $D(\zeta)=\sum\zeta_j A_{j|x}\times\zeta$ -

$$\varphi^+\chi_1=\frac{1}{4cf_r^2}\{i(E\mu)+e\mu(B\times\zeta)-\frac{1}{1+a_0}(i(E\zeta)(\zeta\mu)+e(B\rho))\}$$

(6.26)

$$+\frac{e}{4cf_r^2}\{C(\zeta)\frac{1+2a_0}{(1+a_0)^2}(\mu\zeta)-(B\zeta)\begin{pmatrix}0 & 1\\ -1 & 0\end{pmatrix}+\frac{2+a_0}{(1+a_0)^2}(\rho D(\zeta))\}+\begin{pmatrix}\theta^1 & 0\\ 0 & \theta^4\end{pmatrix},$$

with arbitrary skew-hermitian $\theta^j\in\psi c^2_{-e-e^2}$, $j=1,4$, and

(6.27) $$\lambda_1=V+cf_r\beta+\frac{1}{2f_r}\{\frac{1}{1+a_0}\rho.(E\times\zeta)-e\beta(B\rho)\}-\frac{ie}{2f_r}C(\zeta) .$$

The last term in (6.27) is skew-hermitian, hence will not appear in the corrected observable. Up to terms in $O(e^1-2e)$ we get

(6.28) $$\Lambda=V+2cf_r(M_W,D)\beta+(\frac{1}{2f_r}\{\frac{1}{1+a_0}\rho.(E\times\zeta)-e\beta(B\rho)\})(M_W,D) .$$

Physically, one should regard Λ as a corrected energy observable, cf. [IZ], (2-82), for example, where such expression is obtained with the third kind Foldy-Wouthuysen transform, evaluating terms. Note, the symbol $\varphi+\chi_1$ belongs to (the expansion of) a unitary but nonselfadjoint operator, not an observable. Its physical interpretation might be more meaningful in Weyl representation.

Some ingredients of the symbol χ_1 of (6.25) will show up in the coefficients of the differential equation governing time propagation of "corrections" - as seen in sec.7, below.

Finally we discuss a proof of prop.6.3. First we neglect real multiples of the 4×4-identity matrix I_4, and write $b\equiv c \Leftrightarrow b-c=\upsilon I_4$, $\upsilon\in\mathbb{R}$. From (6.14) get $\varphi^+\varphi_{\pi_\ell}\equiv\frac{1}{2+2a_0}\chi^+\chi_{|\pi_\ell}$, $\chi=1+a_0-i\mu\zeta$, where

$\chi^+\chi_{|\pi_1}=\frac{1}{f_r}(1+a_0+i\mu\zeta)(-a_0\zeta_1-i\sum\mu_j(\delta_{j1}-\zeta_j\zeta_1))\equiv\frac{1}{f_r}\{-a_0\zeta_1 i\mu\zeta-i(1+a_0)\mu_1$ $+i(1+a_0)(\mu\zeta)\zeta_1+(m\zeta)(\mu_1-(\mu\zeta)\zeta_1)\}$. Using (0.12) and (0.15)(b) conclude that $(\mu\zeta)^2=|\zeta|^2$ is a multiple of I_4. By the same formulas, $(\mu\zeta)\mu_1=i(\rho\times\zeta)_1$, so that $\chi^+\chi_{|\pi_1}\equiv\frac{i}{f_r}\{-(1+a_0)(\mu_1-\frac{\mu\zeta}{1+a_0}\zeta_1)+(\rho\times\zeta)_1\}$, confirming (6.18) modulo a real multiple of 1. Both, $\varphi^{+}\varphi_{|\pi_1}$, and $i\begin{pmatrix}\varepsilon & \tau\\ \tau & \varepsilon\end{pmatrix}_1$ are skew-hermitian. Hence we have equality, not only "≡", since no real multiple of I_4 except 0 is skewhermitian. Q.E.D.

7. Non-scalar symbols in diagonal coordinates of h(x,ξ).

We return to consideration of observables in our algebra *PC*= *PC*$_H$, for the operator H=h(x,D). Looking at the first perturbation symbol z(x,ξ) of (3.7) we have calculated z_0 in sec.5, for a variety of well known observables, for t=0, as well as for general t.

We even developed a general formula (5.10), valid for $\kappa=\kappa_+p_+ +\kappa_-p_-$ with scalars $\kappa_\pm$. The calculation of z_0 for the general case where $\kappa_\pm$ are operators on $S_\pm$ remains to be done. Also, nothing was worked out for the two symbols $c_\pm$ of (3.7). We know of course that one may set $\kappa_\pm$=0 at t=0 - as perhaps physically indicated. On the other hand, for t≥0, $c_\pm$ will have to satisfy cdn.'s (3.19), i.e., a pair of differential equations. They will be determined as solutions assuming the given initial conditions at t=0. We will now look explicitly at these equations.

The formulas will look more transparent if we work out the form of the quantities of sec.4 in a coordinate system of $\mathbb{C}^4$ having the symbol matrix h(x,ξ) diagonal at all x,ξ. A unitary matrix symbol $\varphi\in\psi c_0^4$ diagonalizing h(x,ξ) was defined in (2.10). We get

(7.1) $$\varphi(x,\xi)=\frac{1}{\sqrt{2+2a_0}}(1+a_0-i\mu\cdot\zeta) .$$

In effect we are performing a "zero-order-Foldy-Wouthuysen transform at the symbol level".

According to (2.6) we get

(7.2) $$h^{\sim}=\varphi^{+}h\varphi=\begin{pmatrix}\lambda & 0\\ 0 & \lambda_-\end{pmatrix},\quad p^{\sim}=\varphi^{+}p\varphi=\begin{pmatrix}1 & 0\\ 0 & 0\end{pmatrix},\quad q^{\sim}=\varphi^{+}q\varphi=\begin{pmatrix}0 & 0\\ 0 & 1\end{pmatrix},$$

and for matrices g,f satisfying the assumptions of prop.4.1 in the form pg = gp = p , qf = fq = f generally $g^{\sim}$, $f^{\sim}$ are of the form

(7.3) $$g^{\sim}=\begin{pmatrix}u & 0\\ 0 & 0\end{pmatrix},\quad f^{\sim}=\begin{pmatrix}0 & 0\\ 0 & v\end{pmatrix} .$$

All matrices in (7.2), (7.3) are 4×4 , with 0 , 1 , u , v , all denoting 2×2-blocks (0-matrix, identity matrix).

For a 4×4-matrix κ let generally $\kappa^{\sim}=\varphi^{+}\kappa\varphi$. We ask for the explicit form of the transformed matrices of the quantities of (4.5)-(4.19). To simplify the calculation note that $\chi = 1+a_0-i\mu\cdot\zeta$ and $\chi^{-1} =(2+2a_0)^{-1}\chi^{+}$, differing from φ by a scalar factor, may be used instead of φ : We have

(7.4) $$\kappa^{\check{}} = \chi^{-1}\kappa\chi = \frac{1}{2+2a_0}\chi^{+}\kappa\chi \quad .$$

Thus, using (0.15) again, we get

$$(2+2a_0)\alpha^{\check{}} = (1+a_0)^2\alpha + (1+a_0)\beta((\alpha\cdot\zeta)\alpha + \alpha(\alpha\cdot\zeta)) + \beta(\alpha\cdot\zeta)\alpha(\alpha\cdot\zeta)\beta \ .$$

Here we employ (4.12) for

$$\begin{aligned}(2+2a_0)\alpha^{\check{}} &= ((1+a_0)^2 + \zeta^2)\alpha + 2\beta(1+a_0))\zeta - 2(\alpha\cdot\zeta)\zeta \\ &= (2+2a_0)\alpha + 2\beta(1+a_0)\zeta - 2(\alpha\cdot\zeta)\zeta \ .\end{aligned}$$

Dividing by $(2+2a_0)$ we get

(7.5) $$\alpha^{\check{}} = \alpha - \frac{\alpha\zeta}{1+a_0}\zeta + \beta\zeta \ .$$

Next, $(2+2a_0)\beta^{\check{}} = (1+a_0)^2\beta - 2(1+a_0)(\alpha\cdot\zeta) - \zeta^2\beta$. Or,

(7.6) $$\beta^{\check{}} = a_0\beta - \alpha\cdot\zeta \ .$$

With control of $\alpha^{\check{}}$ and $\beta^{\check{}}$ we first get

(7.7) $$\eta_0^{\check{}} = p^{\check{}} - q^{\check{}} = \beta \ .$$

Then,

(7.8) $$\omega_c^{\check{}} = \alpha^{\check{}} - \eta_0^{\check{}}\zeta = \alpha - \frac{\alpha\zeta}{1+a_0}\zeta \quad .$$

Reintroducing the matrices

(7.9) $$\tau = (\tau_1,\tau_2,\tau_3) \ , \ \tau = \sigma - \frac{\sigma\zeta}{1+a_0}\zeta$$

of (6.18) we may write

(7.10) $$\omega_c^{\check{}} = \begin{pmatrix} 0 & i\tau \\ -i\tau & 0 \end{pmatrix} .$$

Then we get

(7.11) $$\theta_c^{\check{}} = -i(h(\omega_c))^{\check{}} = \begin{pmatrix} 0 & \tau \\ \tau & 0 \end{pmatrix} \ ,$$

and, using (4.7),

(7.12) $$(p\alpha q)^{\check{}} = \tfrac{1}{2}(\omega_c + i\theta_c) = \begin{pmatrix} 0 & i\tau \\ 0 & 0 \end{pmatrix} \ , \ (q\alpha p)^{\check{}} = \tfrac{1}{2}(\omega_c - i\theta_c) = \begin{pmatrix} 0 & 0 \\ -i\tau & 0 \end{pmatrix} .$$

Next we translate (4.2) of prop.4.1 to the new coordinates. Using the identities of sec.4, we get (using pp'p=0)

(7.13) $$\lambda - \lambda_- = 2cf_r \ , \ pp_{|\xi}p_{|x}p = (pp_{|\xi}q)(qp_{|x}p) \ .$$

Recalling that $\eta_0 = h_0/f_r = p-q$, hence $ph_0q=0$, using (4.7), we get

$$p\eta_0'q = (\tfrac{1}{f_r})'ph_0q + \tfrac{1}{f_r}ph_0'q = \tfrac{\pi'}{f_r}p\alpha q = \tfrac{\pi'}{2f_r}\cdot(\omega_c + i\theta_c) \ .$$

Using (4.3), we get

(7.14) $pp_{|\xi}q = \frac{1}{4f_r}\Sigma_j\pi_{j|\xi}(\omega_c+i\theta_c)_j$, $pp_{|x}q = \frac{1}{4f_r}\Sigma_j\pi_{j|x}(\omega_c+i\theta_c)_j$.

Looking at the transformed quantities

(7.15) $(pp_{|\xi}q)^{\check{}} = \frac{i}{2f_r}\Sigma_j\pi_{j|\xi}\begin{pmatrix}0 & \tau_j\\ 0 & 0\end{pmatrix}$, $(pp_{|x}q)^{\check{}} = \frac{i}{2f_r}\Sigma_j\pi_{j|x}\begin{pmatrix}0 & \tau_j\\ 0 & 0\end{pmatrix}$,

we get, setting $T_{jl}=\begin{pmatrix}\tau_j\tau_l & 0\\ 0 & 0\end{pmatrix}$,

(7.16) $r_1 = (\lambda-\lambda_-)(pp_{|\xi}p_{|x}p)^{\check{}} = \frac{c}{2f_r}\Sigma(\pi_{j|\xi}\cdot\pi_{l|x})T_{jl} = \frac{-e}{2f_r}\Sigma A_{l|x_j}T_{jl}$.

Calculating as for (6.17), for any symbol $g^{\check{}} = \begin{pmatrix}u & 0\\ 0 & 0\end{pmatrix}$, we get

(7.17) $\langle\lambda,g\rangle^{\check{}} = \langle\lambda,g^{\check{}}\rangle + [r_3,g^{\check{}}]$, $r_3 = X\lambda_{|\xi} - \Xi\lambda_{|x}$,

with the matrices $X=\varphi^+\varphi_{|x}$, $\Xi=\varphi^+\varphi_{|\xi}$ of (6.18). With this information we change coordinates in (4.2). Using (7.2) and (7.3) we get

(7.18) $(p\langle h,g\rangle p)^{\check{}} = \begin{pmatrix}1 & 0\\ 0 & 0\end{pmatrix}\langle h,g\rangle^{\check{}}\begin{pmatrix}1 & 0\\ 0 & 0\end{pmatrix}$,

a matrix having only the upper left (2×2-) corner different from zero. Accordingly, in transforming (4.2), we are only interested in the upper left corner b_{UL} of every matrix b occurring. We get

(7.19) $(p\langle h,g\rangle p)^{\check{}}_{UL} = (\langle\lambda,g\rangle^{\check{}})_{UL} + ([r_1,g^{\check{}}])_{UL}$

Cor.7.1, below, is the result of a transformation of (4.2) - as well as its analogue for q and $f^{\check{}} = \begin{pmatrix}0 & 0\\ 0 & v\end{pmatrix}$ - i.e.,

(7.20) $q\langle h,f\rangle q = q\langle\lambda_-,f\rangle q + (\lambda-\lambda_-)[f,(qq_{|\xi}q_{|x}q)]$,

to diagonal coordinates, using (7.16), (7.17), and corresponding formulas for q . We also have linked the "coefficients" of (7.19) to the matrix Z of (6.7) , as occurring in thm.6.3.

<u>Corollary 7.1</u>. For matrices g, f with $g^{\check{}} = \begin{pmatrix}u & 0\\ 0 & 0\end{pmatrix}$, $f^{\check{}} = \begin{pmatrix}0 & 0\\ 0 & v\end{pmatrix}$ we have

(7.21) $\langle h,g\rangle^{\check{}}_{UL} = \langle\lambda,u\rangle + i[Z_{UL},u]$, $\langle h,f\rangle^{\check{}}_{LR} = \langle\lambda_-,v\rangle + i[Z_{LR},v]$,

where $Z=\frac{1}{2}(W+W^+)$ denotes the hermitian part of the matrix W of (6.17). That is, using formula (6.24), we have

(7.22) $Z_{UR} = \frac{1}{2f_r}\{\frac{1}{1+a_0}E.(\zeta\times\sigma)-e(B\sigma)\}$, $Z_{LR} = \frac{1}{2f_r}\{\frac{1}{1+a_0}E.(\zeta\times\sigma)+e(B\sigma)\}$.

Only the second formula (7.22), and the occurrence of W of (6.17) with its hermitian part, still need to be discussed. We get (4.2) as $\langle h,g\rangle^{\sim}{}_{UL}=(p\langle h,g\rangle p)^{\sim}{}_{UL}=\langle\lambda,u\rangle + [r_{UL},u]$, with $r=r_1+r_3$, r_j as defined above, and notice that $r_{3UL}=i(W_1+W_2)_{UL}$, with W_j as defined after (6.19). Similarly, comparing (7.16) with W_3, below (6.20), we get $r_{1UL} = iW_{3UL}$. Accordingly $r_{UL}=iW_{UL}$, with W of (6.17). This confirms $(7.21)_1$, but with Z replaced by W. Observe also that $(W_1+W_2)_{UL}$ is hermitian, while the skew-hermitian part of W_{3UL} is a multiple of the identity, hence may be omitted in the commutator of (7.21).- Confirm that $[\tau_j,\tau_l]_+=2(\delta_{jl}-\zeta_j\zeta_l)$, whence

$\sum A_{j|x_l}[\tau_j,\tau_l]_+$ is a multiple of 1. This confirms $(7.21)_1$. For

$(7.21)_2$ we first look at (7.20), where $r_1^{\sim}=-(\lambda-\lambda_-)(qq_{|\xi}q_{|x}q)^{\sim} =$

$=-2cf_r(qp_{|\xi}p_{|x}q)^{\sim} = -2cf_r(qp_{|\xi}p)^{\sim}(qp_{|x}q)^{\sim}=-2cf_r(pp_{|\xi}q)^{\sim+}(pp_{|x}q)^{\sim}$. Application of (7.15) will give a formula for $r_{1LR}^{\sim}$, confirming $r_{1LR}^{\sim}=iW_{3LR}$. Similary, a commutator with $r_{3LR}^{\sim}=i(W_1+W_2)_{LR}$ occurs in the formula for $\langle\lambda_-,f\rangle^{\sim}$ corresponding to (7.17). We get $(7.21)_2$.

Now we return to our solvability condition (3.19). We noted that $pw(c_-)p = qw(c_+)q=0$. With (7.21), recalling that $c_+^{\sim}=\begin{pmatrix}\gamma & 0\\ 0 & 0\end{pmatrix}$,

$c_-^{\sim}=\begin{pmatrix}0 & 0\\ 0 & \gamma_-\end{pmatrix}$, and that after "$^{\sim}$" all information is in "UR" or "LL",

(7.23)
$$\gamma^{\bullet}-\langle\lambda,\gamma\rangle-i[Z_{UL},\gamma] = \langle h,z_0\rangle^{\sim}{}_{UL}-\tfrac{i}{2}\langle h,\kappa\rangle_2^{\sim}{}_{UL} ,$$
$$\gamma_-^{\bullet}-\langle\lambda_-,\gamma_-\rangle-i[Z_{LR},\gamma_-] = \langle h,\zeta_0\rangle^{\sim}{}_{LR}-\tfrac{i}{2}\langle h,\kappa\rangle_2^{\sim}{}_{LR} .$$

We still may substitute from (7.22). Also, recall that

(7.24)
$$\gamma^{\grave{}} = \gamma^{\bullet}- \langle\lambda,\gamma\rangle = \partial_t(\gamma\circ\nu_{-t})$$

is the derivative of γ along the classical orbits, in negative time direction. Similarly for "$\partial_t-\langle\lambda_-,.\rangle$"="$^{\grave{}}$" in the second equation, with the positron orbits. Hence we get (7.23) in the form

(7.25)
$$\gamma^{\grave{}}+i[(G_+.\sigma),\gamma] = \gamma^{\bullet} - \langle\lambda,\gamma\rangle + i[(G_+\cdot\sigma),\gamma] = U_{UL} ,$$
$$\gamma_-^{\grave{}}+i[(G_-.\sigma),\gamma_-]= \gamma_-^{\bullet}-\langle\lambda_-,\gamma_-\rangle + i[(G_-\cdot\sigma),\gamma_-] = U_{LR} ,$$

with

(7.26)
$$G_{\pm} = \frac{1}{2f_r}\{e\mathcal{B} \pm\frac{1}{1+a_0}(\zeta\times\mathcal{E})\} ,$$

and

(7.27) $$U=\langle h,z_0\rangle^- - \tfrac{i}{2}\langle h,\kappa\rangle_2^-$$

It will require a lengthy calculation, to obtain U (cf. sec.8).

Here we focus on a case, where U=0 : The initial symbol $\kappa=0$ (for all x,ξ) satisfies all assumptions. It may be regarded of order m. We get (3.8) with $\kappa_\pm=0$, and $\kappa_t=0$, all t. Also $z_\circ=z_{0,t}=0$, hence the right hand side of (7.25) vanishes identically. Prescribing $\gamma^0_\pm\in\psi c^2_{m-e}$ at t=0 starts an iteration, for a perturbed symbol z+s+..., where z and h commute, since $z_\circ=0$.

In effect we obtain a symbol a=0+z+s+... =z+s+... of $A\in P_{m-e}$ with highest order symbol z satisfying all assumptions of our earlier $q=\kappa$, for m-e instead of m. However, we no longer impose (3.8). In other words, if we replace m-e by m then we have a construction of z according to thm.2.1 (2) as before, but without the condition (3.8) imposed on the starting symbol κ. The time propagation $\kappa_t=g_t+f_t$ of κ satisfies $\kappa_t^-=\begin{pmatrix}\gamma_t & 0\\ 0 & \gamma_{-t}\end{pmatrix}$, γ,γ_- satisfying (7.25) with U=0.

It is interesting to invoke one further transformation: Every 2×2-matrix $b=((b_{jk}))$ has a unique representation in the form

(7.28) $$b = b_0 + \sum b_j\sigma_j \ ,\ b_0=\tfrac{1}{2}\text{trace } b \ ,\ b_3=\tfrac{1}{2}(b_{11}+b_{22})$$
$$b_2=\tfrac{1}{2}(b_{12}+b_{21}),\ b_1=\tfrac{1}{2i}(b_{12}-b_{21}) \ .$$

In fact, b is hermitian if and only if b_j are real. It is unitary if and only if $b_0\in\mathbb{C}$ and the 3-vector $b^\wedge=(b_1,b_2,b_3)\in\mathbb{C}^3$ satisfy

(7.29) $$|b_0|^2+|b^\wedge|^2=1 \ ,\ \mathrm{Re}(\bar{b}_0b^\wedge)+\tfrac{i}{2}\,\bar{b}^\wedge\times b^\wedge=0 \ .$$

Actually, we are dealing with a well known representation of the group $U^2/U^1=U^2/\{e^{i\theta}\}$ in $\mathbb{R}\times\mathbb{R}^3$:

<u>Lemma 7.2</u>. Every unitary 2×2-matrix is of the form $e^{i\theta}(ib_0+b^\wedge\cdot\sigma)$, where $\theta\in\mathbb{R}$, $b_0\in\mathbb{R}$, $b^\wedge\in\mathbb{R}^3$, $b_0^2+|b^\wedge|^2=1$.

Let us write our symbols γ,γ_- in the form

(7.30) $$\gamma=\gamma_0+\gamma^\wedge.\sigma \ ,\ \gamma_-=g_0^-+\gamma_-^\wedge.\sigma \ ,\ \gamma_0,\gamma_0^-\in\mathbb{C} \ ,\ \gamma^\wedge \ ,\ \gamma_-^\wedge\in\mathbb{C}^3 \ .$$

Introducing this into the differential equations (7.25) yields
$\gamma_0^` + \gamma^{\wedge`}\cdot\sigma - i[(G\cdot\sigma),(\gamma^\wedge\cdot\sigma)] = 0$, where (0.15)(b) implies
$i[G\cdot\sigma,\gamma^\wedge\cdot\sigma] = 2\sigma\cdot(G\times\gamma^\wedge)$, i.e., (assuming U=0),

(7.31) $$\gamma_0^` = 0 \ ,\ \gamma^{\wedge`} - 2G\times\gamma^\wedge = 0 \ ,$$
$$\gamma_0^{-`} = 0 \ ,\ \gamma_-^{\wedge`} - 2G_-\times\gamma_-^\wedge = 0 \ .$$

This yields the time propagation of a general symbol, not necessarily a multiple of 1 on $S_\pm$, and also the full connection to the classical theory, as described initially in sec.0, as follows:

<u>Definition 7.3</u>. A <u>basic observable</u> is defined by assigning a point wise self-adjoint symbol $\kappa(x,\xi) \in \psi s_m^4$ (usually in ψc_m) which commutes with the symbol $h(x,\xi)$, for all x,ξ .

For a basic observable we have $\kappa^\sim = \begin{pmatrix} \gamma & 0 \\ 0 & \gamma_- \end{pmatrix}$. Also, by thm.2.1 there exists a perturbation symbol z such that $a=\kappa+z$ is a symbol of an operator in P_m resulting in a 'time propagated' symbol $a_t = \kappa_t+z_t$, where κ_t may be defined by setting $\kappa_t^\sim = \begin{pmatrix} \gamma_t & 0 \\ 0 & \gamma_{-,t} \end{pmatrix}$, with the solutions γ_t , $\gamma_{-,t}$ of the differential equations (7.25) with U=0, and initial values γ , γ_- at t=0.

Then we have proven the following result:

<u>Theorem 7.4</u>. Let $\kappa \in \psi c_m^4$ be a basic observable, not necessarily scalar in the spaces $S_\pm$. Then, to obtain the time propagation κ_t of κ without perturbations write $\kappa^\sim = \begin{pmatrix} \gamma & 0 \\ 0 & \gamma_- \end{pmatrix}$ with γ , $\gamma_- \in \psi c_m^2$. Expand γ , γ_- in the form (7.30). Then set $\gamma_{0t}=\gamma_0$, $\gamma_t^-=g_0^-$ constant along the particle orbits (of electron and positron, respectively). Also, let the 3-vectors $\gamma^\wedge$ and $\gamma_-^\wedge$ propagate along these particle orbits as a magnetic moment does - i.e. let $\gamma^\wedge{}_t$, $\gamma_t^{-\wedge}$ satisfy

$$(7.32) \qquad \gamma^{\wedge\cdot} + \frac{e}{mc}\sqrt{1-x^{\cdot 2}/c^2}\,\{B+\frac{1}{ec}(1+\sqrt{1+x^{\cdot 2}/c^2}\;)^{-1}x^\cdot \times E\}\gamma^\wedge = 0 ,$$

and a corresponding equation for $\gamma_-^\wedge$, namely ,

$$(7.33) \qquad \gamma_-^{\wedge\cdot} + \frac{e}{mc}\sqrt{1-x^{\cdot 2}/c^2}\,\{B-\frac{1}{ec}(1+\sqrt{1+x^{\cdot 2}/c^2}\;)^{-1}x^\cdot \times E\}\gamma_-^\wedge = 0 .$$

Then κ_t is given by $\kappa_t^\sim$ in the same form as $\kappa^\sim$, with γ_t , $\gamma_{-,t}$.

Note the coupling constant $M = \frac{e}{mc}$. It involves the so-called Bohr Magneton $\frac{\hbar}{mc}$ in view of the fact that we are setting $\hbar=1$ here.

Note that, in sec.5 we even established the electron spin as a basic observable. - In classical theory, it was thought that the spin is given by the matrix-valued vector ρ , while in (5.21) we got the commuting symbol $\frac{\hbar}{2}\tau$ to represent the spin, in view of the correction needed for the angular momentum. Note, the components of $\tau = p\rho p+q\rho q$ are not scalar in the spaces $S_\pm$, hence form nontrivial examples for thm.7.4.

One might thus ask whether <u>all</u> physically reasonable observables must be (related to) basic observables. Looking at the need of classical Dirac theory to "fill" all states belonging to λ_- and to regard positrons as "holes" there might be an advantage to this interpretation: Simple electron-positron transitions, with noncommuting part of lower order, are OK. Others are 'unobservable'.

8. The full symmetrized first correction symbol $z_S(x.\xi)$.

We return to our task of obtaining the full first correction symbol $z(x,\xi)$ of (3.7). Rather, as noted in sec.5, <u>only</u> its symmetric part $z_S(x,\xi)$ is of interest, so we shall focus on z_S .

In sec.7 we have reduced the conditions for $c_\pm$ of (3.7) to the form (7.25), in our diagonal coordinates. Then we investigated only the case of a vanishing right hand side, to obtain thm.7.4. From the time-propagated operator $A=\kappa(x,D)+z(x,D)+s(x,D)+\ldots$ of thm.2.1(2) we only need $A_S = \frac{1}{2}(A+A^*)$, where

$$A_S= \tfrac{1}{2}(\kappa(x,D) + \kappa(x,D)^*) + z_S(x,D) + s_S(x,D) + \ldots .$$

Thus we are only interested in $z_S(x,\xi) = \frac{1}{2}\{z(x,\xi)+z^+(x,\xi)\}$, called the <u>symmetrized</u> first correction symbol. By (3.10) we have

$$(8.1) \qquad z_S = z_{0,S} + c_{+,S} + c_{-,S} \quad ,$$

where $c_{+,S}^{\,\check{}} = \begin{pmatrix}\gamma_S & 0\\ 0 & 0\end{pmatrix}$, $c_{-,S}^{\,\check{}} = \begin{pmatrix}0 & 0\\ 0 & \gamma_{-,S}\end{pmatrix}$, with $\gamma_{\pm,S}=\frac{1}{2}(\gamma_{\pm,S}+\gamma_{\pm+S})$, $\gamma_\pm$ solving (7.25). Taking adjoints in (7.25) get

$$(8.2) \qquad \gamma^{+\backprime} -i[(G_+\cdot\sigma),\gamma^+] = U^+_{UL} \quad , \quad \gamma_-^{+\backprime} -i[(G_-\cdot\sigma),\gamma_-^+] = U^+_{LR} \, .$$

Taking the average of (7.25) and (8.2),

$$(8.3) \qquad \gamma_S^{\backprime} -i[(G_+\cdot\sigma),\gamma_S] = (U_S)_{UL} \quad , \quad \gamma_S^{-\backprime} -i[(G_-\cdot\sigma),\gamma_S^-] = (U_S)_{LR} \, .$$

<u>Proposition 8.1</u>. For z_S of (8.1) we choose $z_{0,S}$ according to (5.9). Then $c_{\pm,S}$ are described by conditions on the 2×2-matrix functions $\gamma_S=\gamma_S^+=c_{+,S}^{\,\check{}}{}_{UL}$, and $\gamma_S^-=c_{-,S}^{\,\check{}}{}_{LR}$, in diagonal coordinates of $h(x,\xi)$, as in sec.7: $\gamma^\pm_S$ are solutions of the differential equations (8.3) with arbitrarily given self-adjoint initial data $\gamma_0^\pm$, and

$$(8.4) \qquad U_S= \langle h,z_0\rangle_S^{\check{}} \; -(\tfrac{i}{2}\langle h,\kappa\rangle^2)_S^{\check{}} \quad .$$

Next we will calculate U_S of (8.4). As a preparation we state:

<u>Proposition 8.2</u>. For any (4×4-matrix) function $b(x,\xi)$ we have

$$(8.5)\qquad (\partial_\xi b)^{\check{}} = \partial_\xi b^{\check{}} + [\Xi,b^{\check{}}]\ ,\ (\partial_x b)^{\check{}} = \partial_x b^{\check{}} + [X,b^{\check{}}]\ ,$$

with $X=\varphi^+\varphi_{|x}$, $\Xi=\varphi^+\varphi_{|\xi}$ of (6.18).

The proof is a calculation, similar as for (7.17).

We shall use (8.5) to calculate $\langle h,z_0\rangle^{\check{}}{}_S$, for $h^{\check{}}=\begin{pmatrix}\lambda & 0\\ 0 & \lambda_-\end{pmatrix}$:

$$(8.6)\qquad \langle h,\kappa\rangle^{\check{}} = (h_{|\xi})^{\check{}}(z_{0|x})^{\check{}} - (z_{0|\xi})^{\check{}}(h_{|x})^{\check{}}\ .$$

For $z_0^{\check{}}$ we use (5.8) and (7.10), (7.11):

$$(8.7)\qquad z_0^{\check{}} = \tfrac{1}{2f_r}(\delta_\kappa E + \langle\sigma_\kappa,\pi\rangle)\begin{pmatrix}0 & \tau\\ \tau & 0\end{pmatrix} + \tfrac{c}{2}\langle\delta_\kappa,\pi\rangle_+\begin{pmatrix}0 & \tau\\ -\tau & 0\end{pmatrix} = \begin{pmatrix}0 & \iota\\ \upsilon & 0\end{pmatrix}\ ,$$

where we introduce $\iota = \tau\cdot P = z_0^{\check{}}{}_{UR}$, $\upsilon = \tau\cdot Q = z_0^{\check{}}{}_{LL}$. Note that

$$(8.8)\qquad \begin{gathered}\iota = \tau\cdot P=\tau\cdot(R+S)\ ,\ \upsilon = \tau\cdot Q=\tau\cdot(R-S)\ ,\ \text{where}\\ R=\tfrac{1}{2f_r}(\delta_\kappa E+\langle\sigma_\kappa,\pi\rangle)\ ,\ S = \tfrac{c}{2}\langle\delta_\kappa,\pi\rangle_+\ .\end{gathered}$$

Using (8.8) on the expression $\langle h,z_0\rangle^{\check{}}$, we get , setting

$$(8.9)\qquad \delta=-\tfrac{e}{c}\textstyle\sum A_{k|x}\varepsilon_k\ ,\ \nu=-\tfrac{e}{c}\sum A_{k|x}\tau_k\ ,\ \Sigma=-2if_r\ ,$$

$$(8.10)\qquad \begin{gathered}\partial_\xi z_0^{\check{}} = \tfrac{1}{\Sigma}\begin{pmatrix}\tau\upsilon-\iota\tau & \Sigma\,\iota_{|\xi}+[\varepsilon,\iota]\\ \Sigma\,\upsilon_{|\xi}+[\varepsilon,\upsilon] & \tau\iota-\upsilon\tau\end{pmatrix}\ ,\\ \partial_x z_0^{\check{}} = \tfrac{1}{\Sigma}\begin{pmatrix}\nu\upsilon-\iota\nu & \Sigma\,\iota_{|x}+[\delta,\iota]\\ \Sigma\,\upsilon_{|x}+[\delta,\upsilon] & \nu\iota-\upsilon\nu\end{pmatrix}\ .\end{gathered}$$

On the other hand, by a calculation,

$$(8.11)\qquad \begin{gathered}(\partial_\xi h)^{\check{}} = \begin{pmatrix}\lambda_{|\xi} & ic\tau\\ -ic\tau & \lambda_{-|\xi}\end{pmatrix} = c\begin{pmatrix}\zeta & i\tau\\ -i\tau & -\zeta\end{pmatrix}\ ,\\ (\partial_x h)^{\check{}} = \begin{pmatrix}\lambda_{|x} & ic\nu\\ -ic\psi & \lambda_{-|\xi}\end{pmatrix} = E - e\textstyle\sum A_{k|x}\begin{pmatrix}\zeta & i\tau\\ -i\tau & -\zeta\end{pmatrix}_k\ .\end{gathered}$$

Hence,

$$(8.12)\qquad \begin{gathered}\langle h,z_0\rangle^{\check{}}{}_{UL}=\tfrac{1}{\Sigma}\lambda_{|\xi}(\nu\upsilon-\iota\nu)+ic\tau(\upsilon_{|x}+\tfrac{1}{\Sigma}[\delta,\upsilon])-\tfrac{1}{\Sigma}\lambda_{|x}(\tau\upsilon-\iota\tau)-ic(\iota_{|\xi}+\tfrac{1}{\Sigma}[\varepsilon,\iota])\nu,\\ \langle h,z_0\rangle^{\check{}}{}_{LR}=-ic\tau(\iota_{|x}+\tfrac{1}{\Sigma}[\delta,\iota])+\lambda_{-|\xi}(\tfrac{\nu}{\Sigma}\iota-\upsilon\tfrac{\nu}{\Sigma})+ic(\upsilon_{|\xi}+\tfrac{1}{\Sigma}[\varepsilon,\upsilon])\nu+\lambda_{-|x}(\tfrac{\tau}{\Sigma}\iota-\upsilon\tfrac{\tau}{\Sigma})\end{gathered}$$

Focus on " UL " , and write

$$(8.13)\qquad \begin{gathered}\langle h,z_0\rangle^{\check{}}{}_{UL} = T_1+T_2+T_3\ ,\ T_1=\tfrac{1}{\Sigma}\{\lambda_{|\xi}(\nu\upsilon-\iota\nu)-\lambda_{|x}(\tau\upsilon-\iota\tau)\},\\ T_2=i\tfrac{c}{\Sigma}\{\tau[\delta,\upsilon]-[\varepsilon,\iota]\nu\}\ ,\ T_3=ic(\tau\upsilon_{|x}-\iota_{|\xi}\nu)\ .\end{gathered}$$

First we look at

$$(8.14)\qquad T_1 = \frac{-i}{2f_r}(Z\upsilon - \iota Z)\ ,\ Z = \lambda_{|\xi}\nu - \lambda_{|x}\tau = -\tau\cdot(E + eB\times\zeta) = -\tau\cdot Y\ ,$$

by a calculation. For $V = (V_1, V_2, V_3)$, possibly operator-valued, let

$$(8.15)\qquad V^{\Delta} = V - \frac{V\zeta}{1+a_0}\zeta\ .$$

In particular $\tau = \sigma^{\Delta}$, and $V\cdot W^{\Delta} = V^{\Delta}\cdot W$, for arbitrary V,W. We get $2if_r T_1 = -(Y\tau)(Q\tau) + (P\tau)(Y\tau) = -(Y^{\Delta}\sigma)(Q^{\Delta}\sigma) + (P^{\Delta}\sigma)(Y^{\Delta}\sigma) = 2Y^{\Delta}.S^{\Delta} - 2i\sigma.Y^{\Delta}\times R^{\Delta}$, using (0.15)(b). Only the symmetric part is of use. Get

$$(8.16)\qquad T_{1,S} = -\frac{1}{f_r}\sigma.(E + eB\times\zeta)^{\Delta}\times R^{\Delta}\ .$$

Next look at $T_2 = \frac{c}{2f_r}\{\tau[\delta,\upsilon] - [\varepsilon,\iota]\nu\}$. From (0.15), $(1+a_0)\varepsilon = \zeta\times\sigma$ get

$T_2^1 = (1+a_0)\tau[\delta,\upsilon] = -\frac{e}{c}\Sigma A_{k|x_j}\tau_j[(\zeta\times\sigma)_k, (Q^{\Delta}\sigma)] = -2i\frac{e}{c}\Sigma A_{k|x_j}\tau_j\{(\zeta\sigma)(Q^{\Delta})_k - (\zeta Q^{\Delta})\sigma_k\}$. Here we use the formulas

$$(8.17)\qquad (b^{\Delta}\zeta) = a_0(b\zeta)\ ,\ \text{for all } b\ ,$$

and

$$(8.18)\qquad A_{k|x_j}\sigma_j\sigma_k = i\sigma\cdot \boldsymbol{B} + \operatorname{div}\boldsymbol{A}\ ,$$

easily derived. Again we neglect skew-hermitian summands, and write $a \doteq b$ if $a-b$ is skew-hermitian. We get $i\tau_j(\zeta\sigma) = i(e^{j\Delta}\sigma)(\zeta\sigma) \doteq -\sigma.(e^{j\Delta}\times\zeta) = -(\zeta\times\sigma)_j$, and $i\tau_j\sigma_k - i\sigma_j\sigma_k = \frac{i}{1+a_0}\zeta_j(\zeta\sigma)\sigma_k \doteq \frac{-1}{1+a_0}\zeta_j\sigma.(\zeta\times e^k) = \frac{1}{1+a_0}\zeta_j(\zeta\times\sigma)_k$. Conclusion:

$$(8.19)\qquad T_2^1 \doteq 2\frac{e}{c}\Sigma A_{k|x_j}(Q^{\Delta})_k(\zeta\times\sigma)_j + 2a_0\frac{e}{c}(\zeta Q)\Sigma A_{k|x_j}\{i\sigma_j\sigma_k + \frac{1}{1+a_0}\zeta_j(\zeta\times\sigma)_k\}.$$

With (8.18) we get

$$(8.20)\qquad \frac{c}{2e}T_2^1 \doteq (1+a_0)\Sigma A_{k|x_j}Q_k\varepsilon_j - (Q\zeta)\Sigma A_{k|x_j}\{\zeta_k\varepsilon_j - a_0\zeta_j\varepsilon_k\} - a_0(Q\zeta)(B\sigma).$$

Similarly,

$$\frac{c}{2e}T_2^2 = \frac{c}{2e}(1+a_0)[\varepsilon,\iota]\nu = -i\Sigma A_{k|x_j}P_j^{\Delta}(\zeta\sigma)(e^{k\Delta}\sigma) + ia_0(P\zeta)\Sigma A_{k|x_j}\sigma_j\tau_k$$

$$(8.21)\qquad \doteq -(1+a_0)\Sigma A_{k|x_j}P_j^{\Delta}\varepsilon_k + ia_0(P\zeta)\Sigma A_{k|x_j}\sigma_k\sigma_k - a_0(P\zeta)\Sigma A_{k|x_j}\zeta_k\varepsilon_j\ ,$$

implying

$$(8.22)\qquad \frac{c}{2e}T_2^2 \doteq -(1+a_0)\Sigma A_{k|x_j}P_j\varepsilon_k - a_0(P\zeta)(B\sigma) + P\zeta\Sigma A_{k|x_j}\{\zeta_j\varepsilon_k - a_0\zeta_k\varepsilon_j\}\ .$$

Subtracting (8.22) from (8.20) we get

(8.23) $$(1+a_0)\frac{f_r}{e}T_2=\frac{c}{2e}(T_2^1-T_2^2)=2a_0(S\zeta)(B\zeta)+(1+a_0)\sum A_{k|x_j}\{P_j\varepsilon_k+Q_k\varepsilon_j\}$$

$$-\sum A_{k|x_j}\{(P\zeta)\zeta_j\varepsilon_k+(Q\zeta)\zeta_k\varepsilon_j\}+a_0\sum A_{k|x_j}\{(Pz)\zeta_k\varepsilon_j+(Q\zeta)\zeta_j\varepsilon_k\} \quad .$$

To write this more conveniently, interchange indices and use that

(8.24) $$\sum(A_{k|x_j}-A_{j|x_k})M_jN_k = M.(N\times B) \quad ,$$

for arbitrary formal 3-vectors M,N, with commuting components. Get

(8.25) $$T_2=\frac{e}{f_r}\{2a_0\frac{S\zeta}{1+a_0}(B\sigma)+\sum A_{k|x_j}(P_{j}\varepsilon_k+Q_k\varepsilon_j)$$

$$+\frac{1}{1+a_0}\sum\{((a_0Q-P)\zeta)A_{k|x_j}+((a_0P-Q)\zeta)A_{j|x_k}\}\zeta_j\varepsilon_k\} \quad .$$

One may entirely avoid explicit magnetic potentials A, and only use $B=\mathrm{curl}A$ and $\delta=-\frac{e}{c}\sum A_{k|x_j}\varepsilon_k$. By a calculation not offered in detail (8.25) then assumes the form

(8.26) $$T_2= -\frac{c}{f_r}\{2(R\delta)-2(R\zeta)(\zeta\delta)\}$$
$$-\frac{1}{1+a_0}\frac{e}{f_r}\{(B\zeta)(Q\sigma)-a_0^2(P\zeta)(B\sigma)-\frac{1}{2}(B\zeta)(\zeta\sigma)\{2(R\zeta)-2\frac{1-a_0}{1+a_0}(S\zeta)\}\}.$$

Let us observe that $T_2=0$ if the magnetic potential A is constant.

We next focus on

(8.27) $$T_3 = ic\tau\upsilon_{|x} - ic\iota_{|\xi}\nu = T_4 + T_5$$
$$=i\{c((Q^{\Delta}{}_j)_{|x_k}{}^{\Delta} + e((A_{j|x_l}(P_k{}^{\Delta})_{|\xi_l})^{\Delta}\}\sigma_k\sigma_j = iZ_{jk}\sigma_k\sigma_j \quad .$$

Therefore,

(8.28) $$(T_3)_S = \sigma_1(Z_{23}-Z_{32}) + \sigma_2(Z_{31}-Z_{13}) + \sigma_3(Z_{12}-Z_{21}) = \sigma\cdot T(Z) \quad .$$

First examine the (symbolic) 3-vector $T(X)$ defined by (8.28) for a general (symbolic) 3×3-matrix $X=((X_{jk}))$. We will abbreviate $T(((X_{jk})))=T(X_{jk})$. Note that

(8.29) $$T(b_jc_k) = b\times c \quad , \quad T(b_{j|x_k}) = -\mathrm{curl}_x b \quad ,$$

for 3-vectors b , c, and that $T(X)=0$ for any symmetric matrix X.

Let us first focus on $X = W = ((W_{jk}))$,

(8.30) $$W_{jk}=(((Q^{\Delta}{}_j)_{|x})^{\Delta})_k=(Q^{\Delta}{}_j)_{|x_k}-\frac{1}{1+a_0}(Q_j-\frac{Q\zeta}{1+a_0}\zeta_j)_{|x_l}\zeta_l\zeta_k \quad .$$

Observe that

(8.31) $$T(W) = -\mathrm{curl}_x Q^\Delta - \frac{1}{1+a_0} T(Q_{j|x_1}\zeta_1\zeta_k) + \frac{Q\zeta}{(1+a_0)^2} T(\zeta_{j|x_1}\zeta_1\zeta_k) ,$$

where we used that $T((\frac{Q\zeta}{1+a_0})_{|x_1}\zeta_j\zeta_1\zeta_k) = 0$, since the argument matrix is symmetric. We introduce

(8.32) $$\zeta_{j|x_k} = -\frac{e}{cf_r}\{A_{j|x_k} - \zeta_j\zeta_1 A_{1|x_k}\}$$

into the last term of (8.31), noting that the second term of (8.32) again yields a symmetric matrix, thus $T=0$:

$$T(W) = -\mathrm{curl}_x Q^\Delta - \frac{1}{1+a_0} T(Q_{j|x_1}\zeta_1\zeta_k) - \frac{e}{cf_r}\frac{Q\zeta}{(1+a_0)^2} T(A_{j|x_1}\zeta_1\zeta_k) .$$

Or,

(8.33) $$T(W) = \underbrace{-\mathrm{curl}_x Q^\Delta}_{S_1} + \underbrace{\frac{1}{1+a_0}\zeta_1(\zeta\times Q_{|x_1})}_{S_2} + \underbrace{\frac{e}{cf_r}\frac{Q\zeta}{(1+a_0)^2}\zeta_1(\zeta\times A_{|x_1})}_{S_3} .$$

We will seek to simplify this formula, using the terms intro duced earlier. Remember, we will need $\sigma\cdot T(W) = S_1+S_2+S_3$, with S_j generated by the j-the term at right of (8.33). Here we get

(8.34) $$\frac{e}{cf_r}\zeta_1\sigma\cdot(\zeta\times A_{|x_1}) = -\frac{e}{cf_r}(1+a_0)\zeta_1 A_{k|x_1}\varepsilon_k = +\frac{1}{f_r}(1+a_0)(\zeta\delta) ,$$

hence

(8.35) $$S_3 = \frac{1}{f_r}\frac{Q\zeta}{1+a_0}(\zeta\delta) .$$

Next focus on S_1 :

(8.36) $$\mathrm{curl}_x Q^\Delta = \mathrm{curl}_x Q - \frac{Q\zeta}{1+a_0}\mathrm{curl}_x\zeta - (\frac{Q\zeta}{1+a_0})_{|x}\times\zeta = R_1+R_2+R_3 .$$

From (8.29) and (8.32) we get

(8.37) $$\mathrm{curl}_x\zeta = -\frac{e}{cf_r}B + \frac{e}{cf_r}\zeta_1(A_{1|x}\times\zeta) = -\frac{e}{cf_r}B - \frac{e}{cf_r}\zeta_1(\zeta\times A_{|x_1}) - \frac{e}{cf_r}\zeta_1(\zeta\times(A_{1|x} - A_{|x_1})) .$$

Application of (8.24) and (8.34) thus yields

$$\sigma\cdot(\mathrm{curl}_x\zeta) = -\frac{e}{cf_r}(B\sigma) - \frac{1}{f_r}(1+a_0)(\zeta\delta) - \frac{e}{cf_r}\{(\zeta B)(\zeta\sigma) - (1-a_0^2)(\sigma B)\} .$$

Or,

(8.38) $$\sigma\cdot(\mathrm{curl}_x\zeta) = -\frac{e}{cf_r}\{a_0^2(B\sigma) + (B\zeta)(\zeta\sigma)\} - \frac{1}{f_r}(1+a_0)(\zeta\delta) ,$$

and,

(8.39) $$\sigma\cdot R_2=\frac{e}{cf_r}\frac{Q\zeta}{1+a_0}\{a_0^2(B\sigma)+(B\zeta)(\zeta\sigma)\}+\frac{1}{f_r}(Q\zeta)(\zeta\delta)\ .$$

For R_3 we have

(8.40) $$R_3=\frac{Q\zeta}{(1+a_0)^2}(a_{0|x}\times\zeta)-\frac{1}{1+a_0}\zeta_j(Q_{j|x}\times\zeta)-\frac{1}{1+a_0}Q_j(\zeta_{j|x}\times\zeta)\ .$$

We have

(8.41) $$a_{0|x}=\frac{e}{cf_r}a_0\zeta_1 A_{1|x}\ ,\quad \zeta_{j|x}=\frac{e}{cf_r}(\zeta_j\zeta_1 A_{1|x}-A_{j|x})\ ,$$

hence,

$$R_3=\frac{Q\zeta}{(1+a_0)^2}\frac{e}{cf_r}a_0(\zeta_1 A_{1|x}\times\zeta)-\frac{1}{1+a_0}\zeta_j(Q_{j|x}\times\zeta)-\frac{e}{cf_r}\frac{Q\zeta}{1+a_0}(\zeta_1 A_{1|x}\times\zeta)$$
$$+\frac{1}{1+a_0}\frac{e}{cf_r}(Q_j A_{j|x}\times\zeta)\ .$$

The sum of the first and third term equals

(8.42) $$R_4=\frac{Q\zeta}{(1+a_0)^2}\frac{e}{cf_r}(\zeta\times\zeta_1 A_{1|x})\ .$$

Recall that

(8.43) $$\frac{e}{cf_r}(\zeta\times(\zeta_1 A_{1|x}))\cdot\sigma=(1+a_0)\frac{1}{f_r}(\zeta\delta)+\frac{e}{cf_r}\{(B\zeta)(\zeta\sigma)-(1-a_0^2)(B\sigma)\}.$$

It follows that

(8.44) $$\sigma\cdot R_4=\frac{1}{f_r}\frac{Q\zeta}{1+a_0}(\zeta\delta)-\frac{e}{cf_r}\frac{1-a_0}{1+a_0}(Q\zeta)(B\sigma)+\frac{e}{cf_r}\frac{Q\zeta}{(1+a_0)^2}(B\zeta)(\zeta\sigma)\ ,$$

and,

(8.45) $$\sigma\cdot R_3=-\frac{1}{1+a_0}\sigma\cdot(\zeta_j Q_{j|x}\times\zeta)+\frac{1}{1+a_0}\frac{e}{cf_r}(\sigma\cdot Q_j A_{j|x}\times\zeta)+\sigma\cdot R_4\ .$$

For the second term at right of (8.45) (called $\sigma\cdot R_5$) we get

$$\sigma\cdot R_5=\frac{1}{1+a_0}\frac{e}{cf_r}(\sigma\cdot Q_j A_{|x_j}\times\zeta)-\frac{1}{1+a_0}\frac{e}{cf_r}Q_j\sigma\cdot(\zeta\times(A_{j|x}-A_{|x_j}))$$
$$=-\frac{1}{f_r}(Q\delta)+\frac{e}{cf_r}(A_{j|x_k}-A_{k|x_j})Q_j\varepsilon_k$$
$$=-\frac{1}{f_r}(Q\delta)+\frac{1}{1+a_0}\frac{e}{cf_r}\{(B\sigma)(Q\zeta)-(Q\sigma)(B\zeta)\}\quad\text{(using (8.24))}$$

Summarizing we get

$$\sigma\cdot R_3=-\frac{1}{1+a_0}\sigma\cdot(\zeta_j Q_{j|x}\times\zeta)-\frac{1}{f_r}(Q\delta)+\frac{1}{1+a_0}\frac{e}{cf_r}\{(B\sigma)(Q\zeta)-(Q\sigma)(B\zeta)\}$$

(8.46) $$+\frac{1}{f_r}\frac{Q\zeta}{1+a_0}(\zeta\delta)-\frac{e}{cf_r}\frac{1-a_0}{1+a_0}(Q\zeta)(B\sigma)+\frac{e}{cf_r}\frac{Q\zeta}{(1+a_0)^2}(B\zeta)(\zeta\sigma)\ ,$$

where the third and fifth term still allow a simplification: They give the sum $+\frac{1}{1+a_0}\frac{e}{cf_r}\{a_0(B\sigma)(Q\zeta)-(B\zeta)(Q\sigma)\}$. All together we get

$$(T_4)_S = c(S_3+S_2+S_1) = \frac{c}{f_r}\frac{Q\zeta}{1+a_0}(\zeta\delta) + \frac{c}{1+a_0}\sigma\cdot(\zeta_1(\zeta\times Q_{|x_1})) - c\sigma\cdot(\mathrm{curl}_x Q)$$

$$(8.47)\quad + \frac{c}{1+a_0}\sigma\cdot(\zeta_j Q_{j|x}\times\zeta) + \frac{c}{f_r}(Q\delta) - \frac{1}{1+a_0}\frac{e}{f_r}\{a_0(B\sigma)(Q\zeta)-(Q\sigma)(B\zeta)\}$$
$$-\frac{c}{f_r}\frac{Q\zeta}{1+a_0}(\zeta\delta) - \frac{e}{f_r}\frac{Q\zeta}{(1+a_0)^2}(B\zeta)(\zeta\sigma) - \frac{c}{f_r}(Q\zeta)(\zeta\delta)$$
$$-\frac{e}{f_r}\frac{Q\zeta}{1+a_0}\{a_0^2(B\sigma) - (B\zeta)(\zeta\sigma)\} .$$

In this sum term 1 and 7 cancel, and terms 8 and 10 simplify. Also denoting $U = \mathrm{curl}_x Q$, for a moment, the sum of terms 2 and 4 give
$$-\frac{c}{1+a_0}\sigma\cdot\zeta_1(\zeta\times(Q_{|x_1}-Q_{1|x}) = -\frac{c}{1+a_0}((U\zeta)(\zeta\sigma)-(1-a_0^2)(U\sigma)) .$$

It follows that
$$(T_4)_S = -ca_0\,\sigma\cdot(\mathrm{curl}_x Q) - \frac{c}{1+a_0}(\zeta\sigma)\;\zeta\cdot(\mathrm{curl}_x Q)$$

$$(8.48)\quad + \frac{c}{f_r}\{(Q\delta)-(Q\zeta)(\zeta\delta)\} - \frac{1}{1+a_0}\frac{e}{f_r}(Q\zeta)(B\sigma)\{a_0+a_0^2\}$$
$$+\frac{1}{1+a_0}\frac{e}{f_r}(Q\sigma)(B\zeta) + \frac{e}{f_r}(Q\zeta)(B\zeta)(\zeta\sigma)\{\frac{1}{1+a_0}-\frac{1}{(1+a_0)^2}\} .$$

Or,

$$(8.49)\quad (T_4)_S = -ca_0\sigma\cdot(\mathrm{curl}_x Q) - \frac{c}{1+a_0}(\zeta\sigma)\;\zeta\cdot(\mathrm{curl}_x Q)$$
$$+\frac{c}{f_r}\{(Q\delta)-(Q\zeta)(\zeta\delta)\}-\frac{e}{f_r}\{a_0(Q\zeta)(B\sigma)-\frac{Q\sigma}{1+a_0}(B\zeta)-a_0\frac{Q\zeta}{(1+a_0)^2}(B\zeta)(\zeta\sigma)\}$$

Next we turn to the expression
$$(T_5)_S = -e\sigma\cdot T((A_{j|x_1}(P_k^{\Delta})_{|\xi_1})^{\Delta}) = -e\sigma\cdot(M_1+M_2)$$

$$(8.50)\quad = -e\sigma\cdot T(A_{j|x_1}(P_k^{\Delta})_{|\xi_1}) + \frac{e}{1+a_0}\sigma\cdot T(\zeta_m A_{m|x_1}(P_k^{\Delta})_{|\xi_1}\zeta_j).$$

Here (8.29) implies that

$$(8.51)\quad M_1 = A_{|x_1}\times P_{|\xi_1} - A_{|x_1}\times(\frac{P\zeta}{1+a_0}\zeta)_{|\xi_1} = M_3 + M_4 ,$$

while

$$(8.52)\quad M_2 = \frac{1}{1+a_0}A_{m|x_1}\zeta_m P_{|\xi_1}\times\zeta - \frac{1}{1+a_0}A_{m|x_1}\zeta_m(\frac{P\zeta}{1+a_0}\zeta)_{|\xi_1}\times\zeta = M_5+M_6 .$$

Here,
$$\sigma\cdot M_4 = -(\frac{P\zeta}{1+a_0})_{|\xi_1} A_{|x_1}\cdot(\zeta\times\sigma) - \frac{P\zeta}{1+a_0}\sigma\cdot A_{|x_1}\times\zeta_{|\xi_1} ,$$
$$= \frac{c}{e}(1+a_0)(\frac{P\zeta}{1+a_0})_{|\xi_1}\delta_1 - \frac{1}{f_r}\frac{P\zeta}{1+a_0}\sigma\cdot A_{|x_1}\times\{e^1-\zeta\zeta_1\} .$$

One finds that
$$\sigma\cdot A_{|x_1}\times e^1 = -(B\sigma) ,\quad \sigma\cdot A_{|x_1}\times\zeta\zeta_1 = -\frac{c}{e}(1+a_0)(\zeta\delta) .$$

Also,

$$\frac{c}{e}\left(\frac{P\zeta}{1+a_0}\right)_{|\xi_1}\delta_1=\frac{c}{ef_r}\frac{P\zeta}{(1+a_0)^2}(\zeta\delta)+\frac{c}{e}\frac{1}{1+a_0}P_{k|\xi_1}\zeta_k\delta_1+\frac{c}{ef_r}\frac{1}{1+a_0}P_k(e^k-\zeta\zeta_k)\cdot\delta$$
$$=\frac{c}{ef_r}\frac{P\zeta}{(1+a_0)^2}(\zeta\delta)+\frac{c}{ef_r}\frac{1}{1+a_0}\{(P\delta)-(P\zeta)(\zeta\delta)\}+\frac{c}{e}\frac{1}{1+a_0}P_{k|\xi_1}\zeta_k\delta_1 .$$

Together we get

$$\sigma\cdot M_1=\sigma\cdot A_{|x_1}\times P_{|\xi_1}-\frac{1}{f_r}\frac{P\zeta}{1+a_0}\{(-B\sigma)+\frac{c}{e}(1+a_0)(\zeta\delta)\}$$
$$+\frac{c}{ef_r}\frac{P\zeta}{1+a_0}(\zeta\delta)+\frac{c}{ef_r}\{(P\delta)-(P\zeta)(\zeta\delta)\}+\frac{c}{e}P_{k|\xi_1}\zeta_k\delta_1 . \tag{8.53}$$

Using that $\zeta\times\zeta=0$ we get

$$M_6=-\frac{P\zeta}{(1+a_0)^2}A_{k|x_1}\zeta_k(\zeta_{|\xi_1}\times\zeta)=-\frac{1}{f_r}\frac{P\zeta}{(1+a_0)^2}A_{k|x_1}\zeta_k(e^1-\zeta\zeta_1)\times\zeta$$
$$=-\frac{1}{f_r}\frac{P\zeta}{(1+a_0)^2}A_{k|x_1}\zeta_k(e^1\times\zeta) .$$

So,

$$\sigma\cdot M_6=-\frac{1}{f_r}\frac{P\zeta}{(1+a_0)^2}A_{k|x_1}\zeta_k e^1\cdot(\zeta\times\sigma)=-\frac{1}{f_r}\frac{P\zeta}{1+a_0}A_{1|x_k}\zeta_k\varepsilon_1$$
$$-\frac{1}{f_r}\frac{P\zeta}{1+a_0}(A_{k|x_1}-A_{1|x_k})\zeta_k\varepsilon_1=\frac{c}{ef_r}\frac{P\zeta}{1+a_0}(\zeta\delta)-\frac{1}{f_r}\frac{P\zeta}{(1+a_0)^2}\{(1-a_0^2)(B\sigma)-(B\zeta)(\zeta\sigma)\}.$$

Summarizing we get

$$(T_5)_S=-e\sigma\cdot M_1-e\sigma\cdot M_2=-e\sigma\cdot M_1-e\sigma\cdot M_6-\frac{e}{1+a_0}A_{k|x_1}\zeta_k\sigma\cdot(P_{|\xi_1}\times\zeta)$$

$$=-e\sigma\cdot(A_{|x_1}\times P_{|\xi_1})-cP_{k|\xi_1}\zeta_k\delta_1-\frac{e}{1+a_0}A_{k|x_1}\zeta_k\sigma\cdot(P_{|\xi_1}\times\zeta)$$
$$+\frac{e}{f_r}\frac{P\zeta}{1+a_0}\{-(B\sigma)+\frac{c}{e}(1+a_0)(\zeta\delta)\}-\frac{c}{f_r}\{(P\delta)-(P\zeta)(\zeta\delta)\}$$
$$-\frac{c}{f_r}\frac{P\zeta}{1+a_0}(\zeta\delta)-\frac{c}{f_r}\frac{P\zeta}{1+a_0}(\zeta\delta)+\frac{e}{f_r}\frac{P\zeta}{1+a_0}\{(1-a_0)(B\sigma)-\frac{1}{1+a_0}(B\zeta)(\zeta\sigma)\} . \tag{8.54}$$

Or, simplifying some terms (there are 4 terms with $(P\zeta)(\zeta\delta)$),

$$(T_8)_S=K_1-e\sigma\cdot(A_{|x_1}\times P_{|\xi_1})$$
$$-\frac{e}{f_r}\frac{P\zeta}{1+a_0}\{a_0(B\sigma)+(B\zeta)\frac{\zeta\sigma}{1+a_0}\}-\frac{c}{f_r}\{(P\delta)-2a_0\frac{\zeta\delta}{1+a_0}(P\zeta)\} , \tag{8.55}$$

with K_1 = sum of the second and third term at right of (8.54).
We simplify K_1:

$$K_1=\frac{e}{1+a_0}\{P_{k|\xi_1}\zeta_k A_{m|x_1}(\zeta\times\sigma)_m-A_{k|x_1}\zeta_k P_{m|\xi_1}(\zeta\times\sigma)_m\}$$
$$=\frac{e}{1+a_0}P_{k|\xi_1}A_{m|x_1}\{\zeta_k(\zeta\times\sigma)_m-\zeta_m(\zeta\times\sigma)_k\}$$

Setting $Y_{jk}=\frac{e}{1+a_0}P_{j|\xi_1}A_{k|x_1}$ and $\eta=\zeta\times\sigma$ for a moment we get

$$K_1=Y_{jk}\{\zeta_j\eta_k-\zeta_k\eta_j\}=(Y_{23}-Y_{32})(\zeta_2\eta_3-\zeta_3\eta_2)$$
$$+(Y_{31}-Y_{13})(\zeta_3\eta_1-\zeta_1\eta_3)+(Y_{12}-Y_{21})(\zeta_1\eta_2-\zeta_2\eta_1) .$$

This shows that

$$K_1=\frac{e}{1+a_0}(P_{|\xi_1}\times A_{|x_1}).(\zeta\times(\zeta\times\sigma))$$
$$=\frac{e}{1+a_0}(\zeta\sigma)\zeta\cdot(P_{|\xi_1}\times A_{|x_1})-e(1-a_0)\sigma\cdot(P_{|\xi_1}\times A_{|x_1}) \ .$$

Summarizing we get

$$(8.56)\qquad \begin{aligned}(T_5)_S=&-ea_0\sigma\cdot(A_{|x_1}\times P_{|\xi_1})-\frac{e}{1+a_0}(\zeta\sigma)\zeta\cdot(A_{|x_1}\times P_{|\xi_1})\\&-\frac{e}{f_r}\frac{P\zeta}{1+a_0}\{a_0(B\sigma)+\frac{\zeta\sigma}{1+a_0}(B\zeta)\}-\frac{c}{f_r}\{(P\delta)-2a_0\frac{P\zeta}{1+a_0}(\zeta\delta)\} \ .\end{aligned}$$

Finally then, combining (8.49) and (8.56):

$$(8.57)\qquad \begin{aligned}(T_3)_S=(T_4)_S+(T_5)_S\ =&-\frac{e}{1+a_0}(\zeta\sigma)\zeta\cdot(A_{|x_1}\times P_{|\xi_1})\\&-ca_0\sigma\cdot(\mathrm{curl}_x Q)-\frac{c}{1+a_0}(\zeta\sigma)(\zeta\cdot\mathrm{curl}_x Q)-ea_0\sigma\cdot(A_{|x_1}\times P_{|\xi_1})\\&+\frac{c}{f_r}\{((Q-P)\delta)-(Q\zeta)(\zeta\delta)+2a_0\frac{P\zeta}{1+a_0}(\zeta\delta)\}\\&+\frac{e}{f_r}\{a_0((P+Q)\zeta)(B\sigma)-\frac{P\zeta}{(1+a_0)^2}(B\zeta)(\zeta\sigma)+a_0\frac{Q\zeta}{(1+a_0)^2}(B\zeta)(\zeta\sigma)+\frac{1}{1+a_0}(Q\zeta)(B\sigma)\}\end{aligned}$$

We then finally can combine formulas (8.16), (8.26), and (8.57) :

<u>Proposition 8.4</u>. We have

$$(8.58)\qquad \begin{aligned}(\langle h,z_0\rangle^{\sim}{}_{UL})_S=&\ T_{1,S}+T_{2,S}+T_{3,S}\\=&-\frac{1}{f_r}\sigma.(E+eB\times\zeta)^{\wedge}\times R^{\wedge}\ -2\frac{c}{f_r}\{(R\delta)-(R\zeta)(\zeta\delta)\}\\&-\frac{1}{1+a_0}\frac{e}{f_r}\{(B\zeta)(Q\sigma)-a_0^2(P\zeta)(B\sigma)-(B\zeta)(\zeta\sigma)\{(R\zeta)-\frac{1-a_0}{1+a_0}(S\zeta)\}\}\\&-ca_0\sigma\cdot(\mathrm{curl}_x Q)-\frac{c}{1+a_0}(\zeta\sigma)(\zeta\cdot\mathrm{curl}_x Q)-ea_0\sigma\cdot(A_{|x_1}\times P_{|\xi_1})\\&-\frac{e}{1+a_0}(\zeta\sigma)(\zeta\cdot(A_{|x_1}\times P_{|\xi_1})-\frac{c}{f_r}\{(2S\delta)-((Q\zeta)-2a_0\frac{P\zeta}{1+a_0})(\zeta\delta)\}\\&+\frac{e}{f_r}\{2a_0(R\zeta)(B\sigma)+\frac{\zeta\sigma}{(1+a_0)^2}(B\zeta)\{a_0(Q\zeta)-(P\zeta)\}+\frac{1}{1+a_0}(Q\zeta)(B\sigma)\} \ .\end{aligned}$$

Finally we are left with the second term in (8.4). Looking at the second Poisson bracket $\langle h,\kappa\rangle_2$, note that $h_{|xx}=0$, hence

$$(8.59)\qquad \langle h,\kappa\rangle_2=-\kappa_{|\xi_j\xi_k}h_{|x_jx_k} .$$

In diagonal coordinates we get (assuming u,v multiples of 1, abbreviating $u;j=u_{|\xi_j}$, $u;jk=u_{|\xi_j\xi_k}$, etc., inside the matrices):

$$\kappa_{|\xi_j\xi_k}^{\sim}=(\partial_{\xi_j}+\mathrm{ad}\ \Xi_j)(\partial_{\xi_k}+\mathrm{ad}\ \Xi_k)\begin{pmatrix}u&0\\0&v\end{pmatrix}$$
$$=(\partial_{\xi_j}+\mathrm{ad}\ \Xi_j)\begin{pmatrix}u;k & ic\delta_\kappa\tau_k\\ -ic\delta_\kappa\tau_k & v;k\kappa\tau_k\end{pmatrix}$$

$$= \begin{pmatrix} u;jk & ic(\delta_\kappa\tau_k);j \\ -ic(\delta_\kappa\tau_k);j & v;jk \end{pmatrix} + \frac{1}{2if_r}[\begin{pmatrix} \varepsilon & \tau \\ \tau & \varepsilon \end{pmatrix}_j, \begin{pmatrix} u;k & ic\delta_\kappa\tau_k \\ -ic\delta_\kappa\tau_k & v;k \end{pmatrix}]$$

$$= \begin{pmatrix} \upsilon\lambda & \upsilon\rho \\ \lambda\lambda & \lambda\rho \end{pmatrix} ,$$

where (using anti-commutator brackets $[b,c]_+=bc+cb$)

$$\upsilon\lambda = u;jk-\frac{c}{2f_r}\delta_\kappa[\tau_j,\tau_k]_+, \quad \upsilon\rho = ic(\delta_\kappa\tau_k);j+i\frac{c}{f_r}\tau_j(f_r\delta_\kappa);k+\frac{c}{2f_r}\delta_\kappa[\varepsilon_j,\tau_k]$$

$$\lambda\lambda = -ic(\delta_k\tau_\kappa);j-i\frac{c}{f_r}\tau_j(f_r\delta_\kappa);k-\frac{c}{2f_r}\delta_\kappa[\varepsilon_j,\tau_k], \quad \lambda\rho = v;jk+\frac{c}{2f_r}\delta_\kappa[\tau_j,\tau_k]_+.$$

To evaluate this we use (6.23). Also,

$$\tau_{k|\xi_j} = \frac{1}{f_r}\frac{1}{1+a_0}\{(\frac{2+a_0}{1+a_0}\zeta_j\zeta_k - \delta_{jk})(\sigma\zeta) - \sigma_j\zeta_k\} , \tag{8.60}$$

and, using (0.15)(c),

$$(1+a_0)[\varepsilon_j,\tau_k] = 2i\{(\zeta\sigma)(\delta_{jk}-\frac{1}{1+a_0}\zeta_j\zeta_k) - a_0\zeta_k\sigma_j\} . \tag{8.61}$$

It follows that

$$\upsilon\lambda = u_{|\xi_j\xi_k} - \frac{c}{f_r}\delta_\kappa(\delta_{jk}-\zeta_j\zeta_k) , \quad \lambda\rho = v_{|\xi_j\xi_k} + \frac{c}{f_r}\delta_\kappa(\delta_{jk}-\zeta_j\zeta_k) ,$$

$$\upsilon\rho = - \lambda\lambda = ic\{\tau_k\delta_{\kappa|\xi_j}+\tau_j\delta_{\kappa|\xi_k}\} . \tag{8.62}$$

In particular we notice that

$$ic\delta_\kappa\tau_{k|\xi_j} +\frac{c}{2f_r}\delta_\kappa[\varepsilon_j,\tau_k] + i\frac{c}{f_r}\tau_j\delta_\kappa f_{r|\xi_k} = 0 ,$$

by a calculation.

Next we need

$$h_{|x_jx_k}^{\sim} = V_{|x_jx_k} + ch_{0|x_jx_k}^{\sim} \tag{8.63}$$

where h_0 involves the variables x and ξ only in form of $\pi=\xi-\frac{e}{c}A(x)$, In fact, $h_0=mc\beta+\alpha\pi$ is a linear function of π. Accordingly,

$$h_{0|x_j} = -\frac{e}{c}A_{1|x_j}h_{0|\pi_1} = -\frac{e}{c}A_{1|x_j}\alpha_1 , \tag{8.64}$$

and

$$h_{0|x_jx_k} = -\frac{e}{c}\alpha\cdot A_{|x_jx_k} . \tag{8.65}$$

Using (7.5) we get

$$h_{|x_jx_1}^{\sim} = V_{|x_jx_1} - eA_{|x_jx_1}\cdot\begin{pmatrix} \zeta & i\tau \\ -i\tau & -\zeta \end{pmatrix} . \tag{8.66}$$

Notice that both matrices $F = \kappa_{\xi_j\xi_k}^{\sim}$ and $J = h_{|x_jx_k}^{\sim}$ are of the general form

$$(8.67) \qquad F = \begin{pmatrix} e & if\tau \\ -if\tau & g \end{pmatrix}, \quad J = \begin{pmatrix} p & iq\tau \\ -iq\tau & r \end{pmatrix}$$

with real scalars e,g,p,r and $f,q \in \mathbb{R}^3$. We need the expressions

$$(8.68) \qquad \begin{aligned} (-\tfrac{i}{2}\langle h,\kappa\rangle_2{}^{\sim}{}_{UL})_S &= \Sigma_{jl}(+\tfrac{i}{2}FJ_{UL})_S = \Sigma_{jl}(\tfrac{i}{2}(ep+(f\tau)(q\tau))_S = N_{UL}, \\ (-\tfrac{i}{2}\langle h,\kappa\rangle_2{}^{\sim}{}_{LR})_S &= \Sigma_{jl}(\tfrac{i}{2}(gr+(f\tau)(q\tau))_S = N_{LR}. \end{aligned}$$

Using (0.15)(b) we get

$$(8.69) \qquad N_{UL} = \Sigma\tfrac{1}{2}\sigma\cdot q^{\Delta}\times f^{\Delta}, \quad N_{LR} = \Sigma\tfrac{1}{2}\sigma\cdot q^{\Delta}\times f^{\Delta}.$$

Checking with (8.62) and (8.66) we find that

$$(8.70) \qquad f^{\Delta} = ce^{k^{\Delta}}\delta_{\kappa|\xi_j} + ce^{j^{\Delta}}\delta_{\kappa|\xi_k}, \quad q^{\Delta} = -e\, A_{|x_j x_l}{}^{\Delta},$$

hence

$$(8.71) \quad N_{UL} = N_{LR} = -2ec\delta_{\kappa|\xi_j}\sigma\cdot\{e^k\times A_{|jk} - \tfrac{1}{1+a_0}(\zeta_k\zeta\times A_{|jk} + (\zeta\cdot A_{|jk})e^k\times\zeta)\},$$

where we write $A_{|x_j} = A_{|j}$, etc., for a moment.

Here the first term at right, equals

$$T_1 = -2ec\delta_{\kappa|\xi_j}\sigma\cdot B_{|x_j},$$

while the sum of the remaining terms is evaluated as

$$T_2 = 2ec\delta_{\kappa|\xi_j}\sigma\cdot\{(1-a_0)B_{|x_j} - \tfrac{\zeta|}{1+a_0}(\zeta B_{|x_j}\}.$$

It follows that

$$(8.72) \qquad N_{UL} = N_{LR} = -2ec\delta_{\kappa|\xi_j}\{a_0(\sigma B)_{|x_j} + \tfrac{\zeta\sigma}{1+a_0}(\zeta B_{|x_j})\}.$$

Combining (8.58) and (8.72) we finally get

<u>Theorem 8.4</u>. For a basic observable $\kappa = \kappa_+ p + \kappa_- q$, scalar in $S_\pm$, the first symmetric correction symbol z_S is given in the form $z_S = z_{0,S} + c_{+,S} + c_{-,S}$, where $c_{+,S}{}^{\sim} = \begin{pmatrix} \gamma_S & 0 \\ 0 & 0 \end{pmatrix}$, with γ_S solving the differential equation

$$(8.73) \quad \begin{aligned} \gamma_S{}' - i[(G_+\cdot\sigma),\gamma_S] &= -2ec\Sigma\,\delta_{\kappa|\xi j}\{a_0(B\sigma)_{|x_j} + \tfrac{\zeta\sigma}{1+a_0}\zeta B_{|x_j})\} \\ &-\tfrac{1}{f_r}\sigma.(E+eB\times\zeta)^{\Delta}\times R^{\Delta} \; -2\tfrac{c}{f_r}\{(R\delta)-(R\zeta)(\zeta\delta)\} \\ &-\tfrac{1}{1+a_0}\tfrac{e}{f_r}\{(B\zeta)(Q\sigma) - a_0^2(P\zeta)(B\sigma) - (B\zeta)(\zeta\sigma)\{(R\zeta) - \tfrac{1-a_0}{1+a_0}(S\zeta)\}\} \end{aligned}$$

$$-ca_0\sigma\cdot(\mathrm{curl}_x Q) - \frac{c}{1+a_0}(\zeta\sigma)(\zeta\cdot \mathrm{curl}_x Q) - ea_0\sigma\cdot(A|_{x_1}\times P|_{\xi_1})$$
$$-\frac{e}{1+a_0}(\zeta\sigma)(\zeta\cdot(A|_{x_1}\times P|_{\xi_1}) - \frac{c}{f_r}\{(2S\delta)-((Q\zeta)-2a_0\frac{P\zeta}{1+a_0})(\zeta\delta)\}$$
$$+\frac{e}{f_r}\{2a_0(R\zeta)(B\sigma)+\frac{\zeta\sigma}{(1+a_0)^2}(B\zeta)\{a_0(Q\zeta)-(P\zeta)\} +\frac{1}{1+a_0}(Q\zeta)(B\sigma)\} .$$

The matrix $c_{-,S}$ satisfies a similar relation, along the line of $(8.3)_2$, where we will not work out $(U_S)_{LR}$ in detail.

9. Some final remarks.

We return to the differential equation (7.25) (or rather (8.3)(or (8.73))) where now also the inhomogeneous term (the right hand side) is under control. In particular, $G_\pm = \frac{1}{2f_r}(eB \pm \frac{1}{1+a_0}(\zeta\times E))$ is given in terms of E and B , while $(U_S)_{UL}$ is given as in (8.73), with P,Q,R,S of (8.8), δ,ν of (8.9).

As a first remarkable fact we record that

$$(9.1) \qquad u = (U_S)_{UL} = \sigma\cdot u^\wedge \quad .$$

In other words in the expansion (7.28) of $(U_S)_{UL}$ in terms of Pauli matrices and 1 the term b_0 is absent: it vanishes identically.

Accordingly, writing $(8.3)_1$ in the form (7.31), we get

$$(9.2) \qquad \gamma_0{}' = 0 \quad , \gamma^{\wedge}{}' - 2G\times\gamma^\wedge = u^\wedge \quad .$$

The first equation still is homogeneous, hence we may set $\gamma_0=0$, i.e., we do not need another correction which is a multiple of 1 in $S_\pm$. The initial γ_0 will stay constant along the particle orbits

On the other hand, the equation for the classical motion of the $\mathbb{R}^3$-vector $\gamma^\wedge$ satisfies the <u>inhomogeneous</u> differential equation $(9.2)_2$. Note that - under our present assumptions - the basic observable $\kappa=\sigma_\kappa \pm icf_r\delta_\kappa$ is a multiple of 1 in $S_\pm$, for all t, hence its vectors $\kappa_\pm{}^\wedge$, in the expansion (7.31), vanish for all t. However, as $(9.2)_2$ indicates, the corrresponding perturbations $\gamma_\pm{}^\wedge$ in general will not remain zero, even though the theory allows us to prescribe them arbitrarily at t=0 - or at any other t.

A question of stability arises: If we prescribe $\gamma^\wedge=0$ at t=0, can it happen that $\gamma^\wedge$ grows large as t becomes large? This may be looked at under two aspects. First, we already noted in sec.7, in the case of $u^\wedge=0$, the law of propagation of $\gamma^\wedge$ coincides with the classical motion of a magnetic moment under the given fields, along the particle orbits. We now should have such a motion under some "magnetic space charge", described by the vector $u^\wedge$. Second, we might return to the 2×2-matrix-equation (8.73). Write it as

$$(9.3)\qquad \gamma' - i[g,\gamma] = u \ , \ g=\sigma\cdot G_+ \ , \ u=\sigma\cdot u^\wedge \ .$$

Introduce the evolution matrix $v(t,\tau)$ of the 2×2-matrix-system

$$(9.4)\qquad v' = igv \ , \ t,\tau\in\mathbb{R} \ , \ v(\tau,\tau)=1 \ .$$

It then is trivial to confirm that solutions of (9.3) satisfy

$$(9.5)\qquad \gamma(t) = v(t,\tau)\gamma(\tau)v(\tau,t) + \int_\tau^t d\theta\, v(t,\theta)u(\theta)v(\theta,t) \ .$$

Note that $v(t,\tau)$ is a unitary 2×2-matrix, for all t,τ, since g is hermitian: Just confirm that $v^{-1+}(t,\tau)=w(t,\tau)$ satisfies (9.4) as well, and apply uniqueness of the solution of (9.4) . Thus the matrix norm of the first term in (9.5) stays constant, while the matrix norm of the second term is bounded by $\int_\tau^t |u(\theta)|d\theta$, an average of the inhomogeneity along the particle orbit, from τ to t. Generally one will expect the unitary matrix $v(\tau,t)$ to oscillate along these orbits, hence the second term should remain smaller.

It would be a matter of investigating special well studied observables now - like those of sec.5 - to check about physical consequences of our present approach. Of course we must remember that we only look at the first correction, with an infinity of other corrections to follow.

The expression at the right hand side of (8.73) appears to be quite complicated, and might be hard to interpret. Thus, perhaps it is useful to look at it in the special case where the magnetic potentials $A(x)$ vanish identically, just to have a less complicated formula. We then get $B=\delta=0$, hence

$$(9.6)\qquad u^\wedge = -\frac{1}{f_r}E^\Delta\times R^\Delta - ca_0\mathrm{curl}_x Q - \frac{c}{1+a_0}(\zeta\cdot\mathrm{curl}_x Q)\zeta \ ,$$

with R,P,Q of (8.8). Specifically,

$$(9.7)\qquad 2R = -\frac{1}{f_r}(\sigma_{\kappa|x} - \delta_\kappa E) \ ,$$

and, using that f_r is independent of x, hence $\lambda_{|x}=E$, and that the curl of a gradient vanishes,

$$(9.8)\qquad \mathrm{curl}_x Q = -\frac{1}{2f_r}\mathrm{curl}_x(u_{|x} - \delta_\kappa\lambda_{|x}) = \frac{1}{2f_r}\delta_{\kappa|x}\times E \ .$$

Therefore,

$$(9.9)\qquad \begin{aligned} u^{\wedge} &= \frac{1}{2f_r^2} E^{\Delta} \times (\sigma_{\kappa}|x - \delta_{\kappa} E)^{\Delta} - \frac{c}{2f_r} a_0\, \delta_{\kappa}|x \times E - \frac{c}{2f_r} \frac{\zeta}{1+a_0} \zeta \cdot (\delta_{\kappa}|x \times E) \\ &= \frac{1}{2f_r^2} E^{\Delta} \times \sigma_{\kappa}|x^{\Delta} - \frac{c}{2f_r} a_0 \delta_{\kappa}|x \times E - \frac{c}{2f_r} \frac{\zeta}{1+a_0} \zeta \cdot (\delta_{\kappa}|x \times E) \ . \end{aligned}$$

Here we use another general formula valid for arbitrary $a, b \in \mathbb{C}^3$ (or also symbolic 3-vectors with commuting components):

$$(9.10)\qquad a^{\Delta} \times b^{\Delta} = a_0 a \times b + \frac{\zeta}{1+a_0} \zeta \cdot (a \times b) \ .$$

Indeed, we have

$$a^{\Delta} \times b^{\Delta} = a \times b - \frac{1}{1+a_0} \{ (a\zeta)(\zeta \times b) - (b\zeta)(\zeta \times a) \} = P_1 - P_2 \ ,$$

where

$$\begin{aligned} (1+a_0) P_2 &= \zeta \times \{ (a\zeta) b - (b\zeta) a \} = \zeta \times (\zeta \times (b \times a)) \\ &= \zeta \times (\zeta \times c) = (c\zeta)\zeta - (1 - a_0^2) c \ , \quad \text{setting } c = b \times a \ . \end{aligned}$$

Thus we get

$$P_1 - P_2 = a \times b - (1 - a_0) a \times b + \frac{\zeta}{1+a_0} \zeta \cdot (a \times b) \ , \text{ q.e.d.}$$

Combining (9.9) and (9.10) we get (with $\kappa_+ = \sigma_{\kappa} + c f_r \delta_{\kappa}$, as in sec.5)

$$(9.11)\qquad u^{\wedge} = \frac{1}{2f_r^2} E^{\Delta} \times \sigma_{\kappa}|x^{\Delta} + \frac{c}{2f_r} E^{\Delta} \times \delta_{\kappa}|x^{\Delta} = \frac{1}{2f_r^2} E^{\Delta} \times \kappa_+|x^{\Delta} \ .$$

Here we might be tempted again to look at the observables (5.11)-(5.18), to obtain results on their corresponding first order correction symbol c_+ . Note that (9.2) with $u^{\wedge}$ of (9.11) is fairly simple now. We will not pursue this further, since it might deviate to far from the subject of this book.

REFERENCES

[Ar$_1$] N.Aronszajn, A unique continuation theorem for solutions of elliptic partial differential equations or inequalities of second order; J.Math.Pures Appl.36(1957)235-249.

[B$_1$] R.Beals, A general calculus of pseudo-differential operators; Duke Math.J.42 (1975) 1-42.

[B$_2$]________, Characterization of pseudodifferential operators and applications; Duke Math.J. 44 (1977) ,45-57 ; correction, Duke Math.J.46 (1979),215 .

[B$_3$]________, On the boundedness of pseudodifferential operators; Comm.Partial Diff.Equ. 2(10) (1977) 1063-1070.

[B$_4$]________, L^p and Hoelder estimates for Pseudodifferential operators; Ann.Inst.Fourier Grenoble 29,3(1979),239-260

[BA$_1$] M.Ben-Artzi and A.Devinatz, the limiting absorption principle for partial differential operators; Memoirs AMS vol. 66 (1987) Providence.

[Bz$_1$] I.M.Berezanski, Expansions in eigenfunctions of self-adjoint operators; AMS transl. of math. monographs, vol.17, Providence 1968.

[BS] A.Boettcher and B. Silbermann, Analysis of Toeplitz operators Akademie-Verlag Berlin and Springer-Verlag Heidelberg 1989.

[BdM$_1$] L.Boutet de Monvel, Boundary problems for pseudo-differential operators; Acta Math. 126 (1971) 11-51.

[BS$_1$] J.Bruening and R.Seeley, Regular singular asymptotics; Adv. Math. 58 (1985) 133-148.

[BC$_1$] M.Breuer and H.O.Cordes,On Banach algebras with σ-symbol; J.Math.Mech. 13 (1964) 313-324 .

[BC$_2$]__________, part 2 ; J.Math.Mech.14 (1965) 299-314 .

[Ca] A.Calderon, Intermediate spaces and interpolation,the complex method; Studia Math.24 (1964) 113-190.

[Ca$_1$] T.Carleman, Sur la theorie mathematique de l'equation de Schroedinger; Arkiv f. Mat., Astr.,og Fys. 24B N11 (1934).

[CV$_1$] A.Calderon and R.Vaillancourt,On the boundedness of pseudo-differential operators; J.Math.Soc.Japan 23 (1971) 374-378.

[CV$_2$] __________, A class of bounded pseudodifferential operators;

Proc.Nat.Acad.Sci.USA 69 (1972) 1185-1187.

[Chu_1] J.Chaillou, Sur les ensembles bornés A de distributions polynomes inversibles dans $\mathcal{D}'(\Gamma)$ at d'inverse A^{-1} borné et sur les hyperfaces Γ-hyperboliques. Thesis, Univ. of Paris 1969.

[Che_1] J.Cheeger, Spectral geometry of spaces with cone-like singularities;Proc.Nat.Acad.Sci. USA (1979) 2103.

[Ch] P.Chernoff, Essential selfadjointness of powers of generators of hyperbolic equations ; J.Functional Analysis 12 (1973) 402-414 .

[CM] R.Coifman and Y.Meyer, Au dela des operateurs pseudodifferentielles; Asterisque 57 (1979) 1-184.

[CC_1] P.Colella and H.O.Cordes,The C^*-algebra of the elliptic boundary problem;Rocky Mtn.J.Math. 10 (1980) 217-238 .

[Cd_1] E.A.Coddington,The spectral representation of ordinary selfadjoint differential operators;Ann.of Math.60 (1954) 192-211.

[CdLv] _____ and N.Levinson, Theory of ordinary differential equations,McGraw Hill,New York 1955.

[C_1] H.O.Cordes,Elliptic pseudo-differential operators,an abstract theory.Springer Lecture Notes in Math.Vol.756, Berlin,Heidelberg,New York 1979.

[C_2] __________, Spectral theory of linear differential operators; London Math. Soc. Lecture Notes vol.76, Cambridge .Univ. Press Cambridge 1987.

[C_3] __________, On the technique of comparison algebra for elliptic boundary problems on noncompact manifolds. Proceedings Symposia Pure Math. 51(1990),part 1, 113-130.

[C_4] __________, On some C^*-algebras and Frechet-*-algebras of pseudodifferential operators; Proc.Sympos.P.Appl. Math.43 (1985) 79-104.

[CC] __________, On compactness of commutators of multiplications and convolutions,and boundedness of pseudodifferential operators; J.Functional Analysis,18 (1975) 115-131.

[CD] __________, A pseudo-algebra of observables for the Dirac equation; Manuscripta Math.45 (1983) 77-105.

[CE] __________, A version of Egorov's theorem for systems of hyperbolic pseudo-differential equations;J.of Functional Analysis 48 (1982) 285-300.

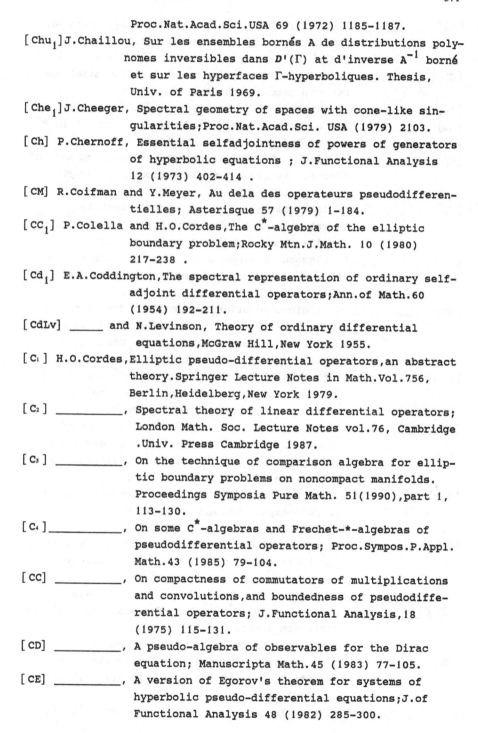

[CF] ________, A pseudodifferential Foldy-Wouthuysen transform; Comm.in Partial Differential Equations,8(13) (1983) 1475-1485.

[CFo] ________, On maximal first order partial differential operators; Amer.J.Math.82 (1960) 63-91.

[CG] ________, On geometrical optics;lecture notes,Berkeley,1982

[CI] ________, The algebra of singular integral operators in $\mathbb{R}^n$; J.Math.Mech.14 (1965) 1007-1032.

[CL] ________, On pseudodifferential operators and smoothness of special Lie-group representations;Manuscripta math. 28 (1979) 51-69 .

[CP] ________, A global parametrix for pseudodifferential operators over $\mathbb{R}^n$;Preprint SFB 72 (Bonn) (1976) (available as preprint from the author).

[CT] ________, An algebra of singular integral operators with two symbol homomorphisms;Bulletin AMS ,75 (1969) 37-42 .

[CU] ________, Ueber die eindeutige Bestimmtheit der Loesungen elliptischer Differentialgleichungen durch Anfangsvorgaben. Nachr. Akad. Wiss. Goettingen Math.-Phys. Kl.IIa, No.11, 239-258 (1956).

[CPo] ________, On the two-fold symbol chain of a C^*-algebra of singular integral operators on a polycylinder; Revista Mat. Iberoamericana 2 (1986) 215-234.

[CDg_1] ____, and S.H.Doong, The Laplace comparison algebra of a space with conical and cylindrical ends; Proceedings, Topics in pseudodifferential operators; Oberwolfach 1986; Springer LN, vol. 1256.

[CEr] ____,and A.Erkip, The N-th order elliptic boundary problem for non-compact boundaries;Rocky Mtn.J.of Math. 10 (1980) 7-24 .

[CHe_1] ____,and E.Herman , Gelfand theory of pseudo-differential operators; American J.of Math. 90 (1968)681-717.

[CHe_2] ____________,Singular integral operators on a half-line; Proceedings Nat. Acad. Sci. (1966) 1668-1673.

[CM_1] ______, and R. McOwen, The C^*-algebra of a singular elliptic problem on a noncompact Riemannian manifold; Math. Zeitschrift 153 (1977) 101-106.

[CM_2] ______________ ,Remarks on singular elliptic theory for complete Riemannian manifolds; Pacific J.Math. 70 (1977) 133-141 .

[CMe_1]______, and S.Melo, An algebra of singular integral operators with kernels of bounded oscillation and application to periodic band spectra; J. Diff. Equ. 75 (1988) 216-238.

[CH] R.Courant and D.Hilbert,Methods of Mathematical Physics, I: Interscience Pub. ltd. New York 1955; II: Interscience Pub. New York 1966.

[Dd_1] J.Dieudonne, Foundations of modern analysis,Acad.Press , New York 1964 .

[Dj_1] D.Z.Djokovic, An elementary proof of the Baker-Campbell-Hausdorff-Dynkin formula; Math.Z.143 (1975) 209-211.

[Do_1] H.Donnelly, Stability theorems for the continuous spectrum of a negatively curved manifold; Transactions AMS 264 (1981) 431-450.

[$DS_{1,2,3}$] N.Dunford and J.Schwarz, Linear operators, Vol.I,II,III, Wiley Interscience,New York 1958,1963,1971.

[Dx_1] Dixmier, Les C^*-algebres et leurs representations; Gauthier-Villars, Paris 1964.

[Dx_2] _______ Sur une inégalité de E.Heinz, Math.Ann.125 (1952) 75-78.

[Dv_1] A.Devinatz, Essential self-adjointness of Schroedinger operators; J.Functional Analysis 25 (1977) 51-69.

[Du_1] R.Duduchava, On integral equations of convolutions with discontinuous coefficients; Math. Nachr. 79 (1977) 75-98.

[Du_2] _______, An application of singular integral equations to some problems of elasticity; Integral Equ. and Operator theory 5 (1982) 475-489.

[Du_3] _______, On multi-dimensional singular integral operators,I: The half-space case; J.Operator theory 11 (1984) 41-76 ; II: The case of a compact manifold; J. Operator theory 11 (1984) 199-214.

[Du_4] _______, Integral equations with fixed singularities; Teubner Texte zur Math.,Teubner,Leizig,1979.

[Dn_1] J.Dunau ,Fonctions d'un operateur elliptique sur une variété compacte; J.Math.pures et appl. 56 (1977) 367-391

[Dy_1] A.Dynin, Multivariable Wiener-Hopf operators; I:Integ.eq.and operator theory 9 (1986), 937-556.

[Es_1] G.I.Eskin, Boundary-value problems for elliptic pseudodifferential operators; AMS translations vol.52, 1981.

[Eg_1]Yu.V.Egorov, The canonical transformations of pseudodifferential operators; Uspehi Mat Nauk 24:5 (1969)235-236.

[ES] ______, and M.A.Shubin, (ed's) Partial differential equations IV, microlocal analysis and hyperbolic equations;

Springer-Verlag, Berlin New York 1993.

[Er_1] A.Erkip, The elliptic boundary problem on the half-space; Comm. PDE 4(5) (1979) 537-554.

[ES] _____,and E.Schrohe, Normal solvability of elliptic boundary problems on asymptotically flat manifolds; J. Functional Analysis 109 (1992) 22-51.

[Es_1] G.Eskin, Boundary problems for elliptic pseudo-differential equations; AMS Transl. Math. Monogr. 52 Providence 1981 (Russian edition 1973).

[Fa_1] H.O.Fattorini, The Cauchy problem; Encyclopedia of Math. and its applications, Vol.18, Addison-Wesley 1983.

[Fr_1] K.O.Friedrichs, Symmetric positive linear differential equations; Commun.P.Appl.Math.11 (1958) 333-418.

[Fr_2] __________,Spectraltheorie halbbeschraenkter Operatoren Math.Ann.109 (1934) 465-487 ,685-713 ,and 110 (1935) 777-779 .

[Fr_3] ________, Pseudodifferential operators; lecture notes 1968, Courant Institute Math. Sci., NYU, New York.

[FL] ________, and P.D.Lax, On symmtrizable differential operators; Proc. Sympos. Pure Math. Chicago (1967) AMS 128-137.

[Fh_1] I.Fredholm, Sur une classe d'équations fonctionelles; Acta Math. 27 (1903) 365-390.

[Ga_1] L.Garding,Linear hyperbolic partial differential equations with constant coefficients;Acta Math.85 (1-62) 1950

[Ge_1] D. Geller, Analytic pseudodifferential operators for the Heisenberg group and local solvability; Math. Notes 37 Princeton Univ. Press, Princeton 1990

[GT] D.Gilbarg,and N.Trudinger, Elliptic partial differential equations of second order; 2-nd edition; Springer New York 1983.

[GS] I.Gelfand and G.E.Silov, Generalized functions; Vol.1,Acad. Press, New York 1964.

[Gl_1] I.M.Glazman, Direct methods of qualitative spectral analysis of singular differential operators; translated by Israel Prog. f. scientific transl. Davey, New York 1965.

[Gb] I.Gohberg, On the theory of multi-dimensional singular integral operators; Soviet Math. 1 (1960) 960-963.

[GK] I.Gohberg and N.Krupnik, Einfuehrung in die Theorie der eindimensionalen singulaeren Integraloperatoren;

Birkhaeuser, Basel 1979 (Russian ed. 1973).

[G_1] B.Gramsch, Relative inversion in der Stoerungstheorie von Operatoren und ψ-Algebren; Math. Ann. 269 (1984) 27-71.

[G_2] ________, Meromorphie in der Theorie der Fredholm Operatoren mit Anwendungen auf elliptische Differentialoperatoren; Math. Ann. 188 (1970) 97-112.

[Gr_1] G.Grubb, Problemes aux limites pseudo-differentiels dependent d'un parametre; C.R.Acad.Sci.Paris 292 (1981) 581-583.

[Gr_2] ________, Functional calculus of pseudidifferential boundary problems; Birkhaeuser 1986; Boston Basel Stuttgart

[GLS] A.Grossmann,G.Loupias,and E.Stein, An algebra of pseudodifferential operators and quantum mechanics in phase space; Ann.Inst. Fourier,Grenoble 18 (1968) 343-368.

[Hd_1] J.Hadamard, Lectures on Cauchy's problem, Dover pbl. inc. New York 19

[Hg_1] S.Helgason, Differential geometry, Lie groups, and symmetric spaces; Acad. Press, New York 1978.

[Hb_1] D.Hilbert, Integralgleichungen; Chelsea, New York 1953

[HP] E.Hille,and R.S.Phillips, Functional analysis and semi-groups Amer.Math.Soc.Coll.Publ. Providence, 1957.

[Hr_1] L.Hoermander,Linear partial differential operators;Springer, Berlin, Heidelberg, NewYork ,1963.

[Hr_2] L.Hoermander, Pseudo-differential operators and hypoelliptic equations; Proceedings Symposia in pure and appl. Math. Vol 10 (1966) 138-183.

[Hr_3] ________, The analysis of linear partial differential operators; Springer New York I: distribution theory and Fourier analysis, 1983; II: diff. operators with constant coefficients, 1983; III: pseudo-differential operators, 1985; IV, Fourier inte-operators, 1985.

[IZ] C.Itzykson, J.B.Zuber, Quantum field theory; McGraw-Hill, New York 1980.

[Ka_1] T.Kato, Perturbation theory for linear operators; Springer New York 1966.

[Ka_2] ________, Perturbation theory for nullity, deficiency and other quantities of linear operators; Journal d'Analyse Math. 6 (1958) 261-322.

[Ka_3] ________, Notes on some inequalities for linear operators;

Math.Ann.125 (1952) 208-212.

[Ka_4]________, Boundedness of some pseudo-differential operators; Osaka J. Math. 13 (1976) 1-9.

[Ka_5]________, Linear evolution equations of hyperbolic type; J. Fac. Sci., U. of Tokyo, Vol. 17 (1970) 241-258; part II; J. Math. Soc. Japan 25 (1973) 648-666.

[Ke_1] J.L.Kelley, General topology; Van Nostrand, Princeton 1955.

[KG_1] W.Klingenberg, D.Grommol, and W.Meyer, Riemannsche Geometrie im Grossen; Springer Lecture Notes Math. Vol. 55 New York 1968.

[KN] J.Kohn and L.Nirenberg, An algebra of pseudo-differential operators; Comm. Pure Appl.Math. 18 (1965) 269-305

[Kr_1] M.G.Krein,On hermitian operators with direction functionals; Sbornik Praz. Inst. Mat. Akad. Nauk Ukr. SSR 10 (1948) 83-105.

[Kg_1] H.Kumano-go, Pseudodifferential operators; MIT-Press 1982.

[LMg_1] J.Lions and E.Magenes, Non-homogeneous boundary value problems and applications; I, II, Springer, New York 1972.

[Lv_1] E.E.Levi, I problemi dei valori al contorno per le equazioni lineare totalemente ellitiche alle derivate parziali; Rome 1909.

[Lw_1] K.Loewner, Ueber monotone Matrixfunktionen; Math.Zeitschr. 38 (1934) 177-216.

[MO] W.Magnus, F.Oberhettinger, Formeln und Saetze fuer die speziellen Funktionen der Mathematischen Physik; 2-nd edit. Springer-Verlag, Berlin, Heidelberg, 1948

[MOS] W.Magnus, F.Oberhettinger and R.P.Soni, Formulas and theorems for the special functions of Mathematical Physics; Springer-Verlag, New York, 1966.

[MNP] W.G.Mazja, S.A.Nasarow, B.A.Plamenewski, Asymptotissche Theorie elliptischer Randwertaufgaben in singulaer gestoerten Gebieten. Akademie Verlag, Berlin 1991.

[Mr] J.Marschall, Pseudodifferential operators with coefficients in Sobolev spaces; Transactions AMS 307 (1988) 335-361.

[M_0] R.McOwen, Fredholm theory of partial differential equations on complete Riemannian manifolds; Ph.D. Thesis, Berkeley 1978.

[M_1] R.McOwen Fredholm theory of partial differential equations on complete Riemannian manifolds;Pacific J.Math.

87 (1980) 169-185 .

[M_2] R.McOwen On elliptic operators in $\mathbb{R}^n$;Comm. Partial Diff. equations, 5(9) (1980) 913-933.

[Me_1] R.Melrose, Transformation of boundary problems, Acta Math. 147 (149-236).

[MM_1] R.Melrose and G.Mendoza,Elliptic boundary problems on spaces with conic points; Journees des equations differentielles, St. Jean-de-Monts, 1981.

[MM_2] ________, Elliptic operators of totally characteristic type; MSRI-preprint, Berkeley, 1982.

[Mm_1] T. Muramatu, estimate for the norm of pseudo-differential operators by means of Besov spaces; Conf. on pseudodifferential operators; Springer L.N., Vol 1256 330-349; Springer-Verlag, New York

[Ng_1] M. Nagase, On sufficient conditions for Pseudodifferential operators to be L^p-bounded; Conf. on Pseudodifferential operators, Oberwolfach; Springer L.N.,vol 1256 (1987)350-359; Springer-Verlag; New York.

[Pa_1] K.Payne, Smooth tame Frechet algebras and Lie groups of pseudodifferential operators; Comm. Pure Appl. Math. 44 (1991) 309-337.

[Pl] B.A.Plamenewski, Algebras of pseudodifferential operators; Kluwer Acad. Pub. Dordrecht, Boston, London, 1989.

[Pr] S. Proessdorf, Einige Klassen singulaerer Gleichngen; Birkhaeuser Verlag, Basel, Stuttgart 1974

[Rb_1] V.S.Rabinovic, On the algebra generated by pseudodifferential operators on $\mathbb{R}^n$, operators of multiplication by almost periodic functions, and shift operators, Soviet Math. Dokl. 25(1982) 498-501.

[RT] J.Rauch and M.Taylor, Decay of solutions to non-dissipative hyperbolic systems on compact manifolds; Comm. Pure Appl. Math. 28 (1975) 501-523.

[R_1] C.E.Rickart, General theory of Banach algebras; v.Nostrand Princeton 1960.

[Rf_1] M.A.Rieffel, Deformation quantization for actions of R^d ; Memoirs AMS 106 , No 506 1993 .

[RN] F.Riesz and B.Sz.-Nagy, Functional analysis; Frederic Ungar, New York 1955.

[Ro_1] D.Robert, Proprietes spectrales d'operateurs pseudo-differentiels; Comm. PDE 3(9) (1978) 755-826.

[$Schr_1$] E.Schrohe, The symbol of an algebra of pseudo-differential operators; Pacific J. Math. 125 (1985) 211-224.

[$Schr_2$] ______, Potenzen elliptischer Pseudodifferentialoperatoren; Thesis, Mainz 1986.

[$Schr_3$] ______, Spaces of weighted symbols and weighted Sobolev spaces on manifolds; Conf. on pseudodifferential operators, Springer LNM 1256, 360-377, New York.

[$Schr_4$] ______, Complex powers on noncompact manifolds with singularities; Math. Annalen 281 (1988) 393-409.

[$Schu_2$] B.-W.Schulze, Mellin expansions of pseudo-differential operators and conormal asymptotics of solutions; Procedings, Oberwolfach conference on Topics in pseudodifferential operators 1986; Springer LNM vol.1256 (1987) 378-401.

[$Schu_2$] ______, Pseudodifferential operators on manifolds with singularities; North-Holland 1991, Amsterdam.

[$Schw_1$] L.Schwartz, Theorie des distributions; Hermann, Paris 1966

[Se_1] R.T.Seeley, Topics in pseudo-differential operators, CIME Conference at Stresa 1968.

[Se_2] ______, Integro-differential operators on vector bundles; Transactions AMS 117 (1965) 167-204 .

[Sh_1] M.A.Shubin, Pseudodifferential operators and spectral theory Springer-Verlag 1987, Berlin Heidelberg New York.

[S_1] H.Sohrab, The C^*-algebra of the n-dimensional harmonic oscillator; Manuscripta.Math.34 (1981),45-70 .

[S_2] ______, C^*-algebras of singular integral operators on the line related to singular Sturm-Liouville problem; Manuscripta Math. 41 (1983) 109-138.

[S_3] ______, Pseudodifferential C^*-algebras related to Schroedinger operators with radially symmetric potential; Quarterly J.Math.Oxford 37 (1986) 105-115.

[S_4] ______, A class of Pseudodifferential C^*-algebras; Quart.J. Math.Oxford (2), 40 (1989) 119-131.

[S_5] ______, Spatially nontemperate pseudodifferential operators sphere extensions, Fredholm theory; to appear in Rocky Mtn. J. Math.

[Sm_1] A.Sommerfeld, Atombau und Spektrallinien; Vieweg,Braunschweig 1957 .

[Sp_1] F.O.Speck, General Wiener-Hopf factorization methods; Pitman Adv. Publ. Prog. 119 (1985).

[Sv_1] L. Svensson, Necessary and sufficient conditions for the hyperbolicity of polynomials with hyperbolic principal part; Arkiv f. Mat. 8 (1969) 145-162.

[Tl_1] M.Taylor , Pseudo-differential operators; Springer Lecture

Notes in Math. Vol. 416, New York 1974 .

[Tl_2] ________, Pseudodifferential operators;Princeton Univ.Press Princeton 1981 .

[Tl_3] __________, Noncommutative harmonic analysis; Math. Surveys and Monographs 22 (1986) AMS, Providence.

[Tl_4] ________, Pseudodifferential operators and nonlinear PDE; Birkhaeuser, Boston, Basel, Berlin, 1991.

[Un_1] A.Unterberger, Quantification relativiste; Bull.Soc.Math. France 119 (1991) Memoire No 44-45.

[Tr_1] F.Treves, Introduction to pseudodifferential operators and Fourier integral operators; Plenum Press,New York and London 1980 .

[Vr] A.Voros, An algebra of pseudodifferential operators and the asymptotics of quantum mechanics; J. Functional Analysis 29 (1978) 104-132.

[Wm_1] J.Weidmann, Linear Operators in Hilbert space. Springer-Verlag, New York 1980

[W_1] A.Weinstein, A symbol class for some Schroedinger equations on $\mathbb{R}^n$; Amer.J.Math.107 (1985) 1-21.

[WZ] ______, and S.Zelditch, Singularities of solutions of some Schroedinger equations on $\mathbb{R}^n$, Bull.AMS 6(1982)449-452

[We_1] H.Weyl, Ueber gewoehnliche Differentialgleichungen mit Singularitaeten, und die zugehoerigen Entwicklungen willkuerlicher Funktionen; Math.Ann.68 (1910) 220-269.

[Yo_1] K.Yosida, Functional analysis; fourth edition; Springer-Verlag; New York 1974.

[Z_1] S.Zelditch, Reconstruction of singularities for solutions of Schroedinger equations;Ph.D.Thesis,Univ.of Calif. Berkeley (1981).

INDEX

Printed in the United States
By Bookmasters